Essentials of
CELL BIOLOGY

Aminul Islam

CELEBRATING 55 YEARS 1960-2015

Since 1960 BOOKS AND ALLIED

[Written according to new CBCS syllabus. Meant for the undergraduate students of Zoology, Botany, Microbiology, Genetics & other Bio-Sciences courses. Also useful for NET, SET, GATE and similar competitive examinations.]

Essentials of
CELL BIOLOGY

Aminul Islam, M.Sc., Ph.D., PDF(USA)

Ex-HOD, Department of Biological Sciences, Aliah University
Ex-Associate Professor, Department of Zoology and TIC,
K. C. College, Hetampur, Birbhum (W.B.)
Ex-visiting Scientist, Cancer Research Unit,
VA Medical Centre, Kansas, USA.

BOOKS AND ALLIED (P) LTD.

No.1-E(1) "Shubham Plaza" (1st Floor)
83/1, Beliaghata Main Road, Kolkata 700010 (India)
Tel: (033)23740062, 8274085530
e-mail: booksandallied1960@gmail.com, booksmktng@gmail.com

Essentials of
CELL BIOLOGY

First Published : July, 2018
Reprinted : 2019

Publisher : Ruba Sen
BOOKS AND ALLIED (P) LTD.
8/1 Chintamoni Das Lane, Kolkata 700009

Typesetter : BOOKS AND ALLIED (P) LTD.
8/1 Chintamoni Das Lane, Kolkata 700009

Printer : EUREKA
91/A, Baithakkhana Road, Kolkata 700009

ISBN **978-93-84294-13-7**

***Price* : ₹ 335.00**

Dedicated to
the departed soul of
MY PARENTS

Preface

Cell Biology has no boundary. It encompasses a wide field, covering all the aspects of the structure and functions of an organism from prokaryotes to mammals. As the basic principle of the structure and function of a cell is same in all the organisms, scientists from different corners of the globe are trying to unveil the mysteries of the structure and functions of different organelles from different view points. Cell Biology is not a single subject, rather it is an admixture of Cytology, Biochemistry, Biophysics, Biotechnology and so on. Therefore it is not an easy matter to cover all aspects in a single book. The present book is a sincere endeavour to cover relevant new inventions in the field of cytology and molecular cell biology in addition to the traditional matter. It sincerely tries to cover the traditional and modern aspects of the causomechanism of different molecular events of the cell. I really salute to the tireless research activities and feel proud to honour them by incorporating their inventions for the benefit of innumerable students. Moreover the readers must be acquainted the trends of the scientific achievements in their traditional field of thoughts and innovations.

This book is mainly aimed for the students who in addition to being acquainted with the modern views of the molecular cell biology can prepare themselves for traditional and competitive examinations. For their preparation different types of questions and glossary will no doubt be highly beneficial for them. Glossary portion will no doubt help them to answer the short questions. More lucid diagrams and flow charts will help them to visualize the intricate steps in a very simple way.

I gratefully acknowledge the assistance rendered by my colleagues for their help and inspiration during manuscript preparation. The students of the department of Biological Sciences, Aliah University, have constantly encouraged me during manuscript preparation. My wife Dolly has greatly sacrificed in the preparation of manuscript. I acknowledge the cooperation of Mr. Amit Ganguly who has been acting as a mediator between me and Mr. Arunabha Sen, Managing Director of Books & Allied (P) Ltd., in different spheres of publication of the book. I am really indebted to the staffs and technical personnels of this company.

At last, but not the least, I convey my heartfelt acknowledgement to Dr. S. Banerjee, Director, Cancer Research Unit, VA medical centre, Kansas, USA, who invited me to honour me as a Visiting Scientist in his laboratory for one year. During my tenure as a scientist in his laboratory I became interested to molecular biology and felt inspired to write a book in this subject.

Independence Day, 2018 **Aminul Islam**

About the Author

Aminul Islam obtained his M.Sc. degree from the University of Burdwan with Entomology as his special paper. He obtained his Ph.D. degree from the University of Burdwan in 1983. His main aspect of research was Biochemical estimation of the Haemolymph of some Insects under different experimental conditions. During his research period he attended some national symposia and published many papers in national and international journals. In 1983 he joined as a full time lecturer in Zoology in K. C. College, Hetampur, W.B. Shortly after joining in service he began to write text books on Zoology in Bengali and he wrote a number of books covering syllabi of most of the universities of West Bengal.

In 2003, he had an assignment to visit USA as a Visiting Scientist and carried out some post doctoral research works on molecular characterization of Esophageal Cancer in VA Medical Centre, Kansas, under Dr. S. Banerjee and Dr. A. P. Weston. During his tenure of PDF, he also served as ad hoc Reviewer of Research papers in the journal of Endocrinology. He also attended an International Symposium of AACR in Orlando, Florida (USA).

He came back in 2004 to join his original service in K. C. College, Hetampur, Birbhum (W.B.) In 2011, he had written a book on Cell Biology which had been revised in 2014. In his last 17 month's service tenure he adorned the chair of Teacher-in-charge of that college. On 31.03.2016 he had retired from service after completion of 32 years of teaching.

On 25th July, 2016, he again joined Aliah University as HOD of the department of Biological sciences. He continued in this position upto 28th Feb, 2018.

About the Book

In 2011, this book was first published as a *A Text book of Cell Biology*. The sphere of Cell Biology is so large that it was not possible to cover all aspects of Cell biology in one book. Due to its immense importance so many workers have plunged into the sea of this field. However keeping pace with the present trend the book was written covering different aspects of the need of the students. The book has been mainly designed to cover the undergraduate syllabi of all the Universities of India and other Asian countries. This edition is particularly aimed to incorporate the contents of the combined syllabus stipulated by UGC under the banner CBCS. Each chapter was designed to present the syllabi-wise answers in brief and scientific way accompanied with relevant diagrams. Most of the diagrams are drawn by the author so that it could be prim and scientific. Each chapter is followed by a set of short, broad and MCQ questions following the trend of most of the Universities and competitive examinations. The book also accompanies the glossary that would help the students to find answers of the short questions. Therefore it is a complete book for academic purpose, as well as, help book for the aspirant examinees of competitive examinations.

In comparison to other allied books in the market, it aims to incorporate the subject matters keeping pace with the recent research developments in the relevant fields. Therefore it will satisfy the urge of getting acquainted with the modern trend of the subjects as well as the traditional views.

Recent data on the respective field in one hand will satisfy the students and in other hand it will be good hand book for the research stations also. Therefore it aims to fulfil the lacuna created in the academic as well as research fields. Fulfilling the needs of the students the book was revised in 2014. I feel delighted when readers from other states send different queries through e-mails.

As time rolled on various suggestions from different corners of the country urged me to revise the book in a different way. Its essence would be judged by the students and teachers. Their comments and suggestions are expected to enrich this edition

Key points of this endeavour are as follows:

i) It encompasses the syllabi-wise discussion as per CBCS system stipulated by UGC.

ii) The contents are to the point, up to date and provided with adequate illustrations.

iii) Each chapter is followed by short, broad and MCQ questions.

iv) It aims to incorporate the questions suitable for competitive examinations.

At the end I would request readers from different corners of the country to communicate with me (09732032093) if they feel any discrepancy in the presentation of the subject matter. Any suggestion to improve the standard of the book would be highly appreciated. If satisfied tell others, if not, tell me.

About the Book

Brief Contents

Detailed Contents

Chapter 1

Generalized Features of Cells

Cell is the fundamental structural and functional unit of all living organisms, and are also called the "building blocks of life." Some organisms are unicellular, consisting of a single cell while others are multicellular. Since the discovery of cell many workers have advocated different definitions of a cell. However as a generalized definition, a cell is *"a protoplasmic mass, surrounded by a selectively permeable plasma membrane that depends upon its neighbouring cells for its existance and is capable of responding environmental stimuli and producing new cell in a livimg environment"*. However, on the basis of this definition, a protozoan which inspite of unicellular organization regulates all vital activities without the cooperation of neighbouring cell. Therefore it is not considered as a 'cell'. They are referred as 'acellular organisms' and their cell is referred as 'body'.

COMMON FEATURES OF CELLS:

Cytologists have recommended following features about a cell.

i) Cells are usually microscopic; average plant cell is about 50μ, while animal cell is about 10μ in diameter. Macroscopic cells are noted in plants (bast fibres of phloem, juice filled cells of citrus fruits) and animals (neurons, insect and avian egg cells). Size of bacterial cells ranges from 5μ to ultramicroscopic size.

ii) On the basis of the occurrence of true nucleus there are two types of cells - Prokaryotes which have no membrane bound nucleus. E.g. Bacteria and Archaeans while Eukaryotes which have a true membrane bound nucleus. e.g. Animals, plants, fungi and protists.

iii) Prokaryotic cells are the earliest and most primitive forms of life on earth.

iv) Cells contain the nucleic acid - DNA and RNA for directing cellular activities. In prokaryotes single DNA molecule remains coiled up in a region of cytoplasm called nucleoid. In eukaryotic cells, DNA molecules are located within the nucleus.

v) In eukaryotic cell functional diversification is accomplished by the separation of specific enzymes located within organelles. Prokaryotic cells however have no membrane bound organelles.

vi) Different types of cells reproduce through different methods - Prokaryotic cells reproduce by binary fission in which two identical cells are derived from a single cell. Eukaryotic organisms form two identical cells by mitosis and gametes are produced by meiosis cell division.

vii) Groups of similar cells form tissues – Ontogenetically and functionally similar cells held together by a sticky substance to form tissues. Different types of tissues can also be arranged together to form organs. Groups of organs can in turn forms systems.

viii) Cells within the multicellular body have different life spans based on their type and function. Certain cells of the digestive tract live for a few days, while some cells of immune system live up to 6 weeks while pancreatic cells live as long as a year.

ix) Surplus cells or damaged or infected cells, show self destruction by a process called apoptosis.

x) ATP is the source of all biochemical reactions.

xi) Protein is the prime constituent of all enzymes required for the regulation of biochemical reactions.

1.1. TYPES OF CELLS

On the basis of compartmentation of different organelles cell is divided into two main categories - prokaryotic cells and eukaryotic cells.

Archaea: Archaea are the oldest organism in this world. For much of the 20th century, prokaryotes and archaea were regarded as a single group of organisms but Archaea were first classified as a separate group of prokaryotes in 1977 by Carl Woese and George E. Fox based on the sequences of 16S rRNA genes.

Archaea were initially viewed as extremophiles living in harsh environments, such as hot springs and salt lakes, but now they are found in a broad range of habitats, including soils, oceans and marshlands. They are also found in the human colon, oral cavity and skin. Archaea are particularly numerous in the oceans.

Archaea and bacteria are generally similar in size and shape, (except flat and square-shaped cells of *Haloquadratum walsbyi*). They range from 0.1 - 15 °m in diameter and are commonly spherical, rod shaped, spiral, plate like, rectangular or needle-like filament or irregular shaped. Some species form aggregates or filaments of cells up to 200 °m long (e.g. *Pyrodictium*. Aggregates of *Thermococcus coalescens* cells fuse together in culture, forming single giant cells. They have 1-3 apical flagella. Their cell membranes composed of glycerol-ether lipids. In some archaea, the lipid bilayer is replaced by a monolayer. The cell membrane is bounded by a cell wall (except *Thermoplasma* and *Ferroplasma*) which is again bounded by capsule. They lack periplasmic space (except *Ignicoccus*, which possess a large periplasm containing membrane bound vesicles and is enclosed by an outer membrane). The cell wall shows protein molecules but no peptidoglycan. They lack interior membranes and organelles.

Archaea usually have a single circular DNA with varied base pairs (as many as 5,751,492 bp in *Methanosarcina acetivorans* while 490,885 bp in *Nanoarchaeum equitans*. Smaller independent *plasmids* are also found. Most archaeal genes lack introns, although there are many introns in their tRNA and rRNA genes and in a few protein-encoding genes. The chromosomes replicate from multiple starting-points (origins of replication) using DNA pol. that resemble the equivalent eukaryotic enzymes. Archaea reproduces asexually by binary or multiple fission, fragmentation, or budding; meiosis does not occur

Prokaryote cells: These are defined as *'protoplasmic mass encircled by plasma membrane and cell wall and are devoid of any membrane bound organelles and a definite nucleus enclosing chromosomes'*. They lack sub cellular membrane bound "organelles." They have no definite nucleus (genetic material is dispersed in the cytoplasm which is known as *nucleoid or genophore* but may have photosynthetic pigments, as is found in cyanobacteria ("blue bacteria"). Some have external flagella for locomotion or hair like pili or fimbriae for adhesion. They may be round (cocci), rod shaped (baccilli) or helical (spirilla or spirochetes).

Viruses are the tiny structures (100 nm) containing nucleic acids, surrounded by a proteinaceous coat but no other structures. They can only reproduce in contact of a living organism. They mostly cause different diseases in animals and plants. ***Due to their limitation in growth and reproduction without the help of an organism they are not considered as a living entity.***

Eukaryote cells: These exhibit definite membrane bound organelle and above all they show definite nucleus. In between pro- and eukaryotes some biologists consider Algae – *Gymnodinium, Peridinium etc.* and Protozoa – *Noctiluca* as ***mesokaryotes*** that show chromosomes with acidic non histone protein and amitotic cell division. However modern cell biologists consider them as eukaryotes though they possess some prokaryotic features. In multicellular organisms, all the cells are interconnected by some specialized proteins or polysaccharides which constitute *extracellular matrix (ECM).* Additionally cells are also glued together by some specialized *cell adhesion molecules (CAM).* In plants, cells are tied rigidly by cell wall. Animal cells are interconnected together by *gap junctions* allowing free exchange of ions and signaling molecules. Plant cells are interconnected together by plasmodesmata facilitating coordinated functioning of the cells in a tissue.

1.2. STRUCTURE OF PROKARYOTIC CELL

Prokaryotes are microscopic unicellular organisms of relatively simple construction, characterized by the lack of a definite nucleus and membrane bound intracellular organelles. Structurally, it has three main regions: *appendages* in the form of flagella and pili or fimbriae; a *cell envelope* consisting of a capsule, cell wall and plasma membrane and cytoplasm containing DNA, ribosomes and other inclusions (Fig. 1.1).

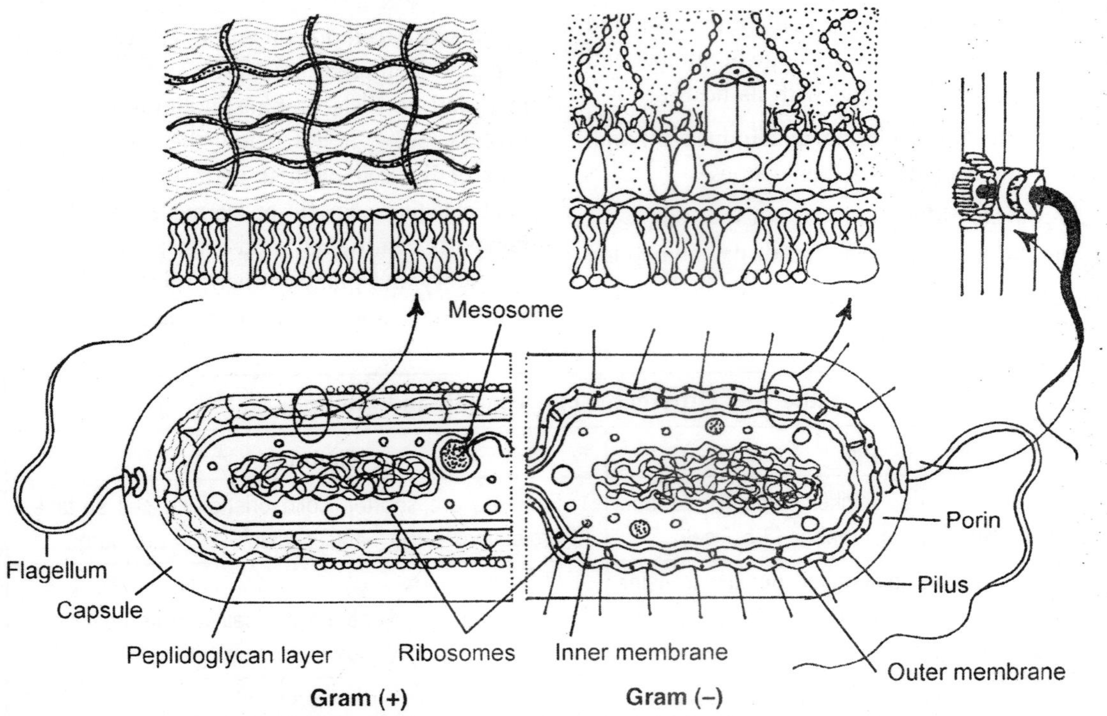

Fig. 1.1 : Schematic view of a Gram (+) & Gram (-) bacterium.

Most of the functions of mitochondria, chloroplasts and Golgi apparatus are carried out by plasma membrane. Their detailed structural organization is as follows:

Flagella: These are filamentous structures may be *polar or peritrichous*. Its diameter is about 20 nm. A motor appararatus consisting of several distinct *proteins*, a system of *rings* embedded in the cell envelope (the *basal body*) and a *hook-like structure* near the cell surface rotates the *flagellar filament*. The rings are divided into two main sets, the innermost rings (the M and S rings), located in the plasma membrane, comprise the motor apparatus while the outermost rings (the P and L rings) located in the periplasm and outer membrane support the rod where it is joined to the hook of the filament on the cell surface.

Fimbriae / Pili: *Fimbriae* and *pili* are short, hair like proteinous structures present on the cell surface of Gram (-) bacteria. Pili are less in number and are mainly used in sexduction and serves as a receptor for certain viruses. Fimbriae are also surface projection, 3-20 nm in diam. and 0.5 – 20 μ in length. They are large in number and are used to attach the cell with its host.

Cell Envelopes: Plasma membrane, cell wall and capsule constitute cell envelopes.

Capsules: Most prokaryotes contain a gelatinous layer of polysaccharide or polypeptide layer at the outside of the cell wall called *capsule*. Sometimes a thin layer of polysaccharide is observed on the surface of the growing cells, called ***glycocalyx***. It is considered as capsule if it is firmly attached to the cell wall or ***slime layer*** if loosely attached.

Capsule protects the cell from bacteriophages, phagocytes and lysozyme. It can be a source of nutrients and energy to microbes. It prevents cell from desiccation.

Cell Wall: Beneath the capsule there is cell wall that differs in two classes of bacteria. In one class, cell wall shows thick (15-80 nm) layer of peptidoglycan with teichoic acid. In another class, cell wall contains a thin layer (10nm) of peptidoglycan with no teichoic acid. It is followed by a periplasmic space outside which there is an outer membrane containing transport protein called porin. Outside the outer membrane there is a thick layer of lipopolysaccharide. If crystal violet stain is applied over the bacteria, and after removal of stain if the cell appears blue/purple or reddish then these are designated as Gram (+)/(-) cells respectively.

Plasma Membrane: It shows typical structure as per fluid mosaic model. It is about 5-10 nm thick and consists of phospholipids (40%) and proteins (60%). It lies below the peptidoglycan layer of the cell wall. It performs a variety of functions like osmotic or permeability barriers, transports nutrients and ions; carries out respiratory and photosynthetic roles, synthesizes membrane lipids, murein, secretes extra

Table 1.1 : Differences between Pili and Fimbriae

Pili	Fimbriae
1. Pili are longer, thicker tubular surface projections, less in number (3-5 / cell).	1. Fimbriae are shorter, solid bristle like cell surface projections, 300 -400 / cell. They are large in number.
2. Mainly used in conjugation (for the transfer of genes and may serve as receptor for certain viruses.	2. They are used to attach the cell with its host.
3. Found in Gram (+) bacteria.	3. Found in Gram (-) bacteria.
4. Its formation is goverened by plasmid gene.	4. Its formation is goverened by other gene.

Table 1.2 : Differences between Gram(+) and Gram(–) bacteria

Gram (+) bacteria	Gram (–) bacteria
1. Flagellum if present, contains only two rings.	1. Flagellum, If present, have four rings. The S-layer is directly attached to the outer membrane.
2. Cell wall contains a thick layer of peptidoglycan showing teichoic acids.	2. Outside the plasma membrane the peptidoglycan layer is very thin. It is followed by an outer membrane containing porin. Outside this layer there is layer of lipoprotein.
3. After staining with crystal violet followed by its removal, they look purple or bluish.	3. After staining with crystal violet followed by its removal, they look reddish.

cytoplasmic proteins, coordinates DNA replication and cell division; performs chemotaxis and sensing functions, regulates many enzymatic activities, helps in signal transduction, helps in cell wall synthesis and septum formation.

Mesosomes: The plasma membrane of Gram (+) bacteria forms vesicular, lamellar or tubular invaginations into the cytoplasm or form stacks or vesicles attached to the inner membrane surface. These are called *mesosomes* or *peripheral body* or *chondroid*. It contains light harvesting pigments and ATPase. They also help in DNA replication and segregation, cell wall synthesis. There are two types of mesosomes - *septal mesosome* (participate in the formation of cross-walls) and *lateral mesosome* (rich in respiratory enzymes).

Ribosomes: The granular appearance of the cytoplasm is due to the presence of free *ribosomes*. These are of 70S in size, being composed of 30S and 50S subunits. Ribosomes are involved in the process of translation (protein synthesis).

Nucleoid / Genophore: In the centre of the bacterial cell, there is a convoluted mass of supercoiled DNA with 10% RNA and proteins (mostly RNA polymerase). Around the whole mass there is a fine fibrous material but no membrane. It shows double stranded DNA forming a circular chromosome which is devoid of histone protein. But several basic proteins have also been detected along with the mass of chromosome. There are several small circular extranuclear bodies containing DNA called ***plasmids*** or ***sex factors*** or ***conjugons*** or ***extra chromosomal replicons*** or ***transfer factors.*** The DNA of plasmids replicate independently of the nucleoid. They contain about 2% of the total nucleic acids of the cell. Some plasmids remain closely associated with the nucleoid and are called ***episomes.***

Inclusion bodies: Cytoplasm contains two types of storage granules. Some are enclosed by membrane, includes *chlorosomes* (contain photosynthetic pigments), carboxysomes (involved in the regulation of carbon metabolism), **magnetosome** (containing iron in the form of magnetite) while some are devoid of membranes, includes glycogen granules, crystals and paracrystalline granules.

Endospores: These are dormant, tough, highly resistant, thick walled structure produced by a small number of bacteria produced during sporulation. They help the bacterium to withstand environmental stress. They occur in soil and water for long time. The position of the endospore may be terminal, subterminal or central. Lateral endospores are seen occasionally. A thin covering known as the *exosporium* overlies the endospore. Endospore usually occurs in *Bacillus, Clostridium, Desulfotomaculum, Sporolactobacillus etc.*

Cytoskeletons: Traditional eukaryote cytoskeletons have also been found in bacteria and other prokaryotes. FtsZ, MreB and CreS have been proved to be bacterial counterpart of eukaryotic tubulin, actin filaments and intermediate filaments, respectively (Wang *et al*, 2016). FtsZ assemble to a Z-ring at the cell division site and regulate bacterial cell division (Si *et al*, 2013). MreB are involved in maintaining cell shape, regulating chromosome segregation. CreS is crucial for cell morphology maintenance. There are also some prokaryotic unique cytoskeleton components playing crucial roles in cell division, chromosome segregation and cell morphology.

Gut flora and their role: Human body, consisting of about 10 trillion cells, carries about 30 trillion bacteria in the intestines. Microorganisms are located in the skin, nose, mouth and the gut. Bacteria make up most of the flora in the colon and 300 - 1000 different species of bacteria live in the gut. The relationship between gut flora and humans is of mutualistic type. Not all of the bacterial types found in the large intestine are beneficial. The gut bacteria play important roles in fermenting unused energy substrates, nutrient digestion, vitamin synthesis (e.g. Biotin and vitamin K), energy metabolism and immune responses. *E. coli* and *C. difficile* are opportunistic pathogens; under normal conditions, they live alongside other bacteria but once host's immune system is weakened or after a course of antibiotics they multiply rapidly, causing harm to the host. Some species are however pathogenic, causing disease by producing infection or increasing the risk for cancer.

Functions: **Carbohydrate fermentation and absorption:** Large intestinal lumen is mostly anaerobic. Here the bacteria turn polysaccharides into short chain fatty acids (SCFAs) by a form of fermentation called *saccharolytic fermentation* producing acetic acid, propionic acid and butyric acid providing the host cells a major source of useful energy as well as helping the body to absorb calcium, magnesium and iron. Gases and organic acids, like lactic acid are also produced by saccharolytic fermentation. Acetic acid is used by muscle, propionic acid helps the liver produce ATP and butyric acid provides energy to gut cells and may prevent cancer. Evidence also indicates that bacteria enhance the absorption and storage of lipids and facilitate the body to absorb vitamin K. Beneficial flora increase the gut's absorption of water, reduce counts of damaging bacteria, increase growth of human gut cells and stimulate growth of indigenous bacteria.

Trophic effects: They increase growth, proliferation and differentiation of intestinal epithelial cells. They also alter intestinal growth by changing the expression of Na^+/glucose transporters.

Repression of pathogenic microbial growth: They prevent colonization of harmful bacteria. Yeasts and *Clostridium* (the overgrowth of which causes pseudomembranous colitis) are unable to grow excessively due to competition from helpful gut flora adhering to the mucosal lining of the intestine. Symbiotic bacteria are more at home in this ecological niche and are thus more successful in the competition. Indigenous gut floras also produce bacteriocins - proteinaceous toxins that inhibit growth of similar bacterial strains and kill harmful microbes. Bacterial fermentation, producing lactic acid and fatty acids lower the pH in the colon, preventing proliferation of harmful but facilitating helpful bacteria. The pH also enhances the liberation of carcinogens.

Immunity: The bacteria stimulate the lymphoid tissue associated with the gut mucosa to produce antibodies to pathogens. The immune system recognizes and fights harmful bacteria, leaving the helpful species alone. Gut bacteria express toll-like receptors (TLRs) that help to repair damage due to injury. TLRs cause parts of the immune system to repair injury caused by radiation. TLRs help to discriminate between the pathogenic and commensal bacteria.

Metabolic function: The resident gut microflora mediate the syntheses of vitamins like biotin and folate, as well as absorption of Mg, Ca and Fe ions. The gut flora play a major role in metabolizing dietary carcinogens.

Allergy prevention: They are also implicated in preventing allergies, an overreaction of the immune system to non-harmful antigens. They also stimulate the immune system and "train" it to respond properly to antigens.

Preventing inflammatory bowel disease: Gut bacteria also help the immune system against Inflammatory Bowel Disease.

Cancer: *Bacteroides* and *Clostridium* are associated with an increase in tumour growth rate, while *Lactobacillus* and *Bifidobacteria* are known to prevent tumour formation.

Weight Loss: It initiates a shift in the gut flora. *Bacteroidetes* increase as weight loss progresses.

Colitis: It has been suggested that commensal bacteria (*C. difficile)* are responsible for the development of colitis.

1.3. DETAILED STRUCTURE OF EUKAYOTIC CELLS

Plasma membrane / Cell membrane: Every cell is surrounded by a fine, elastic, selectively permeable *membrane* that separates and protects its interior from its environment. It is found in both prokaryotes and eukaryotes. Its molecular architecture according to ***Fluid Mosaic model*** (Singer and Nicolson, 1972), shows that it consists of basically double layer of lipids and haphazardly distributed integral and peripheral proteins held in a jelly like matrix. The main lipid component is phospholipid followed by cholesterol, and galactolipids. Their polar hydrophilic ends are directed outwards while the inner hydrophobic ends are held together by van der Wall force. Both protein and lipid molecules can be exchanged between cytoplasm and plasma membrane.

Chemically plasma membrane shows lipid (27-79%), protein (20 – 70%), sugars (1 – 3%) and about 40 different types of enzymes. It performs

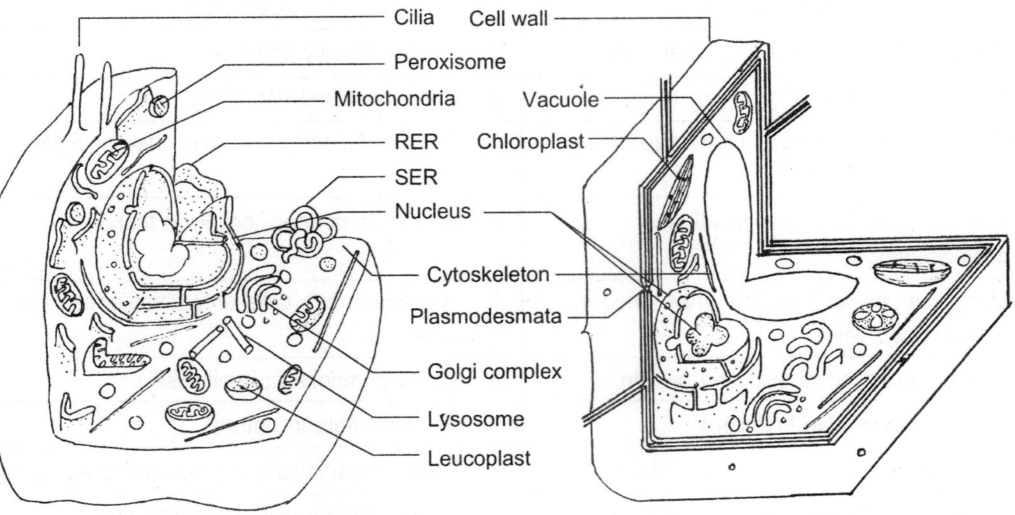

Fig. 1.2 : Typical animal (left) and plant cell (right).

transport of different substances by diffusion and active transport. It takes up solid or liquid substances by *phagocytosis* and *pinocytosis*, expels out secretory substances by *exocytosis*, performs intercellular communication by transducing various signals, receives hormones, transmits impulse. It exhibits different specializations like microvilli, tight junctions, gap junctions etc

Cell Wall: Most of the plant cells have a rigid wall surrounding the plasma membrane. Apart from protection of the cell, it maintains intercellular communication with the help of *plasmodesmata* providing intercellular bridges. In mature cells, it shows three layers – *middle lamella* the intercellular membranous colloidal, elastic structure; on each side of the middle lamella is the *primary cell wall* (1-3μ thick). On the inner side of the primary cell wall there is *secondary cell wall* (5-10μ thick). Middle lamella shows calcium and magnesium pectate, primary cell wall shows cellulose, pectin, hemicellulose, lignin etc. and secondary cell wall shows cellulose, hemicellulose, lignin, suberin, cutin, wax, tannin etc. organic and calcium carbonate, calcium oxalate, silica etc. inorganic molecules. Non-uniform deposition of secondary cell wall over the primary cell wall produces circular, spiral and reticular ornamentations and pit like structures.

At the end of telophase, electron dense particles accumulate in the interface of two cells forming a *cell plate*. *Phragmoplast*, released from the Golgi complex are aligned on each side of the cell plate forms primary cell wall on each side of which secretory substances accumulate to form secondary cell wall.

Protoplasm: All living cells possess a colloidal substance being encircled by the plasma membrane called protoplasm. It is made up of water, inorganic ions, simple carbohydrates (sugars), amino acids and lipids. Sugars, amino acids, fatty acids and glycerol are called *monomers* that are linked together in long chains called *polymers*.

The carbohydrates constitute glycogen, starch and cellulose. The amino acid polymers are called proteins. Carbohydrates provide energy for heat

Table 1.3 : Comparison of cell membrane and cell wall

Cell membrane	Cell wall
1. Present in both animal and plant cells.	1. Present in only plant cells.
2. Thin, flexible and living.	2. Thick, rigid and non-living.
3. Ornamentation over the membrane is absent.	3. Ornamentation occurs over the secondary cell wall.
4. May be modified into microvilli and desmosomes.	4. No functional modification is found.
5. Many intercellular junctions are formed in between two cells.	5. In between two cells only plasmodesmata are noted.
6. Mainly composed of lipid and protein.	6. Mainly composed of cellulose.
7. It is semipermeable in nature.	7. It is permeable in nature.
8. Participates in pino- and phagocytosis.	8. Does not participate in pino- and phagocytosis.
9. It iriginates by cleavage or furrowing of the existing cell membrane.	9. It originates by cell plate formation.

and movement. Fats store energy and are combined with proteins in cell membranes. Proteins are used for growth and repair of tissues and for the formation of enzymes to speed up and control chemical reactions in the cell. Proteins are also oxidized to provide energy.

Cytoskeleton: These are complex and dynamic cell component that maintains cell's shape; position, of the organelles, helps in endocytosis and movement of the cell. Three protein polymers - actin microtubules and intermediate filaments form a net-work of filaments that regulate many physiological functions.

Nucleus: It is the most conspicuous chromosome containing organelle found in eukaryotic cell. (except mammalian mature RBC and sieve tubes of the phloem in plants). Some cells are multinucleated In animals, these are called *syncitium* and in plants- *coenocytes.*

Nucleus is encircled by a double layered *nuclear envelope or karyotheca.* The outer membrane is thick with attached ribosomes. There is a 300A° thick membrane is attached to the inner side of the inner karyotheca, called *fibrous lamina.* Karyotheca shows numerous circular or octagonal pores. *Nucleoplasm* is additionally encircled by a fine membrane present inside the inner karyotheca. Nucleoplasm contains proteins, enzymes minerals, newly synthesized mRNAs, *perichromatin fibres, perichromatin granules* containing rRNAs and fine fibriller *chromatins* that later on condense to form *chromosomes* and one or more *nucleolus.* Nucleus performs the most crucial roles in the regulation of protein synthesis, cell division and integration of nucleo-cytoplasmic activities

Mitochondria: Mitochondria are self-replicating bilayered organelles that are primarily concerned with the generation of ATP. They are therefore called *'power house of the cell'.* They measure 3-7μ / 1-1.5μ. These are invested by two 60A° thick unit membranes that remain separated from each other by 60-80A° wide *perimitochondrial space.* It however remain united at about 100 *contact zones* that regulate the transport of materials between matrix and cytoplasm. From the inner membrane numerous finger like projections – *cristae* are given out. On the wall of the cristae there are numerous stalked granules - *oxysomes or F_1 particles.* Matrix contains 2-6 *circular DNA.* In human beings, mtDNA is 16,659 bp long and contains 37 genes. Matrix also contains many enzymes and ribosomes.

Chemically it shows 60-70% protein, 25-30% lipid, 0.5% RNA and minute amount of DNA, Vit-E, S, Fe and Cu. ATP generation is the principal function of mitochondria. Additionally they generate CO_2 and water.

Endoplasmic reticulum: It is the transport network for molecules targeted for specific destination. ER is of two types - the *rough ER* (RER), having ribosomes on its surface, and the *smooth ER (SER),* which lacks them. When RER and SER occur side by side in that case, in between two types, an intermediate form is noted, called *transitional element.* RER occur predominantly in protein synthesizing cells while SER in glycolipid or steroid synthesizing cells.SER shows many enzymes related with lipid metabolism, steroid hormone synthesis, glycogen metabolism and detoxifcation. SER are abundant in liver cells. In the striated muscle, a specialized type of ER called *sarcoplasmic reticulum (SR)* occurs. It stores Ca^{++} and plays a critical role in muscle contraction. ER shows parallel sac like cisternae, vesicles or tubules. After synthesis of proteins these are post translationally modified in ER. Additionally they play crucial role in the storage of protein, apoptosis, detoxification and maintenance of Ca^{++} balance in the cell.

Golgi apparatus: It is the central delivery system of the cell and is a a site mainly involved in the secretion, for protein processing, packaging and transport. These occur in between nucleus and plasma membrane. They appear as a heterogeneous assemblage of structural components, Around the whole organelle there is a clear zone, called *Zone of exclusion*. The whole organelle is roughly semicircular; its nuclear facing side is convex and is called *Forming or CIS face* while plasma membrane facing side is concave and is called *Maturing or Trans face*. The middle part is called *Medial region*. The whole organelle shows following 3 types of structural components –

- ***Cisternae*** – flat, double layered, parallel saccular structures showing 4-7 rows that are interconnected by electron dense structures.
- ***Vacuoles*** **- large sized globular structures present at the periphery of the organelle.**

Vesicles - minute globular structures present at the transface. Due to their occurrence this part looks fenestrated.

These are very dynamic structures and constant flow of vesicles is noted. In the trans region a network of tubular structures originate and form an anastomosing structure, called *Trans Golgi Network (TGN)*. They remain united with similar structures of neighboring Golgi. TGN remains associated with the ER and lysosomes to form

Golgi complex is associated with the formation and trafficking of secretory substances; formation of ECM proteins, lysosome, acrosome in sperms, cortical sacs in the ovum, cell wall synthesis in plant cells and apoptosis.

Ribosomes: These are small globular organelles found in both prokaryotes and eukaryotes. In eukaryotes, they are found either in free or attached to the wall of RER or nucleus. They are devoid of any membrane. On the basis of Sedimentation Coefficient values (S-values) three types of ribosomes are found –in prokaryotes these are of 70S; in eukaryotes 80S and in the mitochondrial and chloroplast matrix these are of 55S types. All ribosomes consists of two subunits – having different S values. 70S ribosomes exhibit 30S and 50S; 80S ribosomes 40S and 60S while 55S ribosomes exhibit 30S and 40S subunits. Both subunits unite with the help of Mg^{++} at the time of protein synthesis. During protein synthesis many ribosomes get attached with the mRNA to form a *polysome* or *polyribosome*. They may remain free or attached with the membrane of RER or nucleus. Ribosomes show rRNAs and proteins, and are generally referred as 'factory of protein systhesis'.

Plastids: In plant cells many double layered structures, involved in many metabolic activities particularly in the synthesis of glucose are called plastids. These are mainly of three types – chloroplasts, chromoplasts and leucoplasts. Chloroplasts are involved in the synthesis of sugar via photosysnthesis with the help of sun rays, CO_2 and water. Chromoplasts render the colour of different plant parts while leucoplasts store food materials for the future.

Chroloplasts are invested by two 50A° thick unit membranes. The space invested by the inner membrane is called *stroma* that contains protein, enzymes, iron-protein complexes, sugars, ribosomes, nucleic acids, Vit E, K and many inorganic ions (Mg^{++}, Fe^{++}, Cu^{++}, Mn^{++}, Zn^{++} etc). Stroma shows 60 – 80 disc shaped stacked inter connected structures called *grana*. Each granum contains 20-80 sac like *thylakoids* that contain many chlorophyll containing *quantasomes*.

Chromoplasts are circular and star shaped plastid with incompletely divided matrix. It shows carotene, lycopene and luteol pigments rendering the organ orange and xanthophill producing yellow colour of the organs. They are abundant in the petals and fruits.

Leucoplasts are circular or filamentous colourless plastids that mainly store starch, lipids or proteins. In the presence of light they can be converted into chloro- or chromoplasts.

Lysosomes : Lysosomes are often referred to as the garbage disposal system of a cell. Both organelles are somewhat spherical, bounded by a single membrane and rich in digestive enzymes, naturally occurring proteins that speed up biochemical processes. Lysosomes contain more than 36 enzymes for degrading proteins, nucleic acids, and polysaccharides. Most of the lyossomal enzymes are acid hydrolases which are active at a pH around 5.0.The pH of cytosol remains around 7.2. Thus even if accidentally the membrane of lysosome is ruptured it remains in an inactive state in normal pH range of the cytosol but if the acidity of the cytosol increases then it digests the whole cell which is called *autolysis.* For this reason it is referred to as *"Suicide bag of cell"*.

Lysosomes range between 0.4-0.8µ diam. Newly formed lysosomes that are loaded with mostly inactive enzymes are called *primary lysosomes.* Primary lysosomes are abundant in phagocytic cells like macrophages, neutrophils etc. One or more primary lysosomes fuse with a phagosome to form *secondary lysosomes.* Prior to digestion they aggregate around any dying or degraded cellular organelle or any foreign

Table 1.4: Comparative acoount of prokaroytic and eukaryotic cells

Parameters	***Prokaryote***	***Eukaryote***
Organization	Unicellular	Unicellular, colonial, higher multicellular organisms with specialized cells.
Capsule	A capsule externally protects the cell.	Capsule is absent.
Cell wall	It shows amino sugars, mucopeptides and muramic acid.	It occurs only in plant cells; composed mainly of cellulose
Flagella	Flagella may be present showing flagellin protein.	Flagella and cilia occur in many cases that are composed mainly of tubulin and dynein proteins.
Plasma membrane	In addition to proteins and lipids it shows respiratory enzymes and electron transport system.	In plant cells, plasma membrane occurs below the cell wall. In animal cells, protoplasm is in vested by plasma membrane. It is not provided with respiratory enzymes and ETS
Cytoplasmic organelles	No membrane bound organelles though thylakoids occur in photosynthetic bacteria.	Organelles may be membrane less, or provided with single/ double-layered membranes.
Cytoskeleton	Absent	Present
Ribosomes	70 (30S+50S) S type	80 (40+60) S type.
DNA	Circular double helical thread like, not provided with histone protein.	Linear, double helical, provided with acidic and basic proteins to form chromosomes.
Nucleus	No true nucleus DNA threads (nucleoid) are not encircled by membrane.	Chromosomes are encircled by membrane to form nucleus.
Plasmid	Present	Absent
Mesosome	Present	Absent
RNA-protein-synthesis	RNA and protein syntheses are coupled in cytoplasm, no intron.	Exons and introns are found, RNA synthesis occurs inside the nucleus; protein synthesis occurs in the cytoplasm.
Reproduction	Asexual by binary fission. No mitotic apparatus is formed.	Sexual and asexual. Mitosis and meiosis cell divisions are found. Microtubules form mito- tic apparatus by the MTOC.

microbes to digest them. Such structures are respectively known as *autophagosomes* or *heterophagosomes*. The residual part after digestion is called *residual bodies* that are thrown outside by exocytosis.

Peroxisomes are also small, circular (0.2-1.5μ in diam.) structure encircled by single membrane containing catalase and oxidase enzymes. They mainly protect the cells from oxygen toxicity, help in the conversion of lipid into sugars and help in the synthesis of glycine and serine amino acids. They also participate in gluconeogenesis by assisting in the β-oxidation of fatty acids.

Centrosome: Centrosome is a unique cytoplasmic domain of lower plants (in some fungi and algae) and animal interphase cells (except muscle cells, neurons and mature RBC). Mostly it shows one or two central granular *centrioles or diplosomes* surrounded by an electron dense matrix of pericentriolar materials (PCM) called *centrosphere.* Due to its unique role in the nucleation of microtubules (MT) during cell division it is called *microtubule-organizing centers* (MTOC). It is also crucial for cell motility (forming cilia or flagella), proliferation and differentiation. The wall of each centriole consists of 9 triplets of microtubules each exhibiting 3-sub tubules – A, B and C. The wall of the sub tubules contains filamentous subunits whose number is 13 in A while 10 in B and C sub tubules. The distal end of each centriole contains many distal and sub distal appendage like structures attached with the peripheral triplets by specific proteins. Centrioles are composed of α, β, δ and ε *tubulins, centrin* and Ca^{++} binding proteins are essential for centriole biogenesis and for spindle pole biogenesis.

1.4. VIRUSES

Structural Components of Viruses: Viruses in extracellular state are called *virions or virus particles.* They represent the complete assembly of the infectious virus and vary widely in size ranging from 0.02 to 0.3 nm. A virion consists of nucleic acid core surrounded by a protein coat (called capsid). The complete complex of nucleic acid and protein is packaged in nucleocapsid (nucleoprotein) that forms the virus particle. Such virions are called naked viruses. Few viruses are called enveloped viruses in which nucleocapsid is enclosed in a membrane.

Symmetry of Viruses: Viruses, so far studied, usually fall into three main symmetry groups on the basis of EM studies. These are icosahedral (or cubic), helical and complex.

(i) **Icosahedral (or Cubic) Symmetry:** Many of the viruses (e.g., herpesvirus, poliovirus, polyomavirus, adenovirus,) Influenza Virus, ϕX174bacteriophage, have icosahedral symmetry. They resemble small crystals and appear spherical in EMs and their capsomeres are arranged to form an icosahedron. An icosahedron is a regular polyhedron with 20 faces formed by equilateral triangles and 12 intersecting corners.

(ii) **Helical Symmetry:** Some viruses (e.g. TMV, bacteriophage M13, mumps virus, measles virus, rabies virus) resemble long rods and show helical symmetry. In these viruses the capsomeres are arranged in a helix around a single rotational axis. The capsomeres curve into a helix because they are thicker at one end than the other. The helical capsids may be naked (e.g., TMV) or surrounded by a loose membranous envelope (e.g., Influenza virus, Mumps virus, Plant viruses). with helical symmetry, usually appear rod-shaped in EM.

(iii) **Complex Symmetry:** This group of viruses shows complex or uncertain symmetries that result in complicated morphology. Viruses of this group can be divided into two subgroups as follows: (i) that illustrated by most of the bacteriophages that have a head of the shape of bipyramidal hexagonal prism attached to a tail consisting of a helical contractile sheath surrounding a central hollow core, and (ii) that illustrated by pox viruses that do not possess clearly identifiable capsids but have several coats around the nucleic acid.

Baltimore classification of Viruses: In this system of classification viruses are placed in one of seven groups depending on a combination of their nucleic acid (DNA or RNA), strandedness (single-stranded or double-stranded), Sense, and method of replication.

i) **dsDNA viruses** (e.g. *Adenoviruses, Herpesviruses, Poxviruses*).

ii) **ssDNA viruses** (+) sense DNA (e.g. *Parvoviruses*).

iii) **dsRNA viruses** (e.g. *Reoviruses*).

iv) **ssRNA viruses** (+) sense RNA (e.g. *Picornaviruses, Toga viruses*).

v) **ssRNA viruses** (") sense RNA (e.g. *Orthomyxoviruses, Rhabdoviruses*)

vi) **ssRNA-RT viruses** (+) sense RNA with DNA intermediate in life-cycle (e.g. *Retroviruses*).

vi) **dsDNA-RT viruses** (e.g. *Hepadnaviruses)*

For most viruses with RNA genomes and some with single-stranded DNA genomes, the single strands are said to be either positive-sense (called the '+' strand) or negative-sense (called the '-' strand), depending on whether or not they are complementary to the viral mRNA. Positive-sense viral RNA is the same as viral mRNA and thus at least a part of it can be immediately translated by the host cell. Negative-sense viral RNA is complementary to mRNA and thus must be converted to positive-sense RNA by an RNA-dependent RNA polymerase before translation. DNA nomenclature for viruses with single-sense genomic ssDNA is similar to RNA nomenclature, in that the *coding strand* for the viral mRNA is

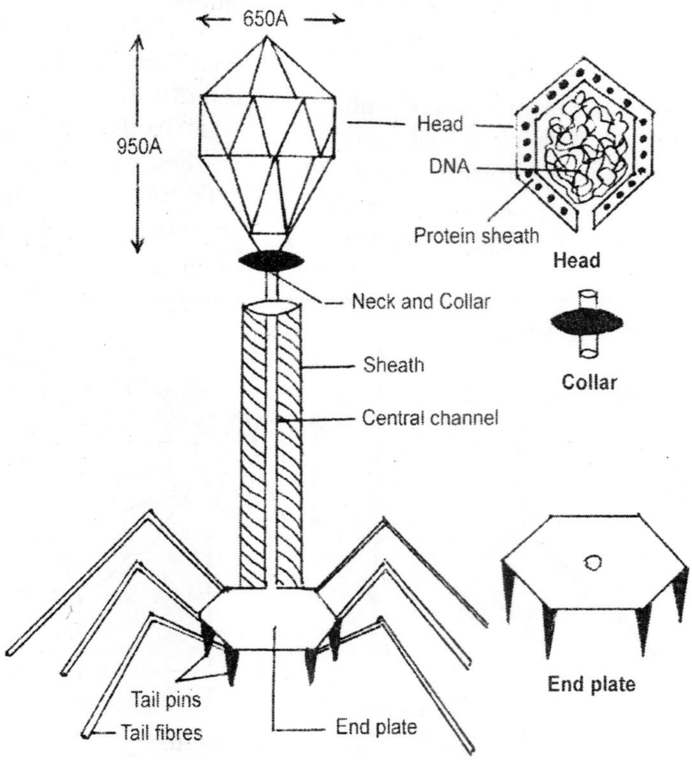

Fig. 1.3 : Structure of Bacteriophage

complementary to it ("), and the *non-coding strand* is a copy of it (+). However, several types of ssDNA and ssRNA viruses have genomes that are ambisense in that transcription can occur off both strands in a double-stranded replicative intermediate. Examples include Geminiviruses, which are ssDNA plant viruses and arenaviruses, which are ssRNA viruses of animals.

Typical examples of Viruses:

Bacteriophage: Those viruses that infect bacteria are called bacteriophage or Phage virus. Their shape and nature of nucleic acids vary. The phage that infects *E.coli* bacteria in the gut of man is called Coliphage or T_2 / T_4 bacteriophage. The structure of T_2 bacteriophage is as follows:

1. *Size* – Size of most phages range from 24-200/80-100 nm in length.
2. *Head or Capsid* - All phages contain a head which can vary in size and shape. Head has a weight of 194 MDa and is 1150Å long and 850Å wide. Head is an icosahedron having (20 sides) others are filamentous. Head or capsid shows manycopies of one or more different proteins which are arranged into two layers. It consists of 160 hexamers of gp23 (major capsid protein), 11 pentamers of gp24, and 1 gp20. Another structural element is a small protein, Soc. It forms an almost continuous mesh on the sur face of the gp23 hexamers. Inside the head is found the ds DNA. It is about 50m long showing 20,000 bp divided into 650 genes. Its MW is 130×10^6 dalton. It shows hydroxyl methyl cytosine instead of cytosine.
3. *Neck:* The constricted part below the head is called neck. It is encircled by a round collar pierced by a neck tube which communicates with head and tail.
4. *Tail:* Tail is 110 nm long and 18 nm in diam. It is composed of two concentric protein cylinders. The outer cylinder is contractile and the inner one forms the channel through which nucleic acid is transferred to the bacteria during infection. The inner tube has 90-Å outer and 40-Å inner diameter and is constructed of 144 copies of gp19. The outer part of the tail is called the *tail sheath.* It is built by 144 copies of gp18. The uncontracted tail is 1000Å long and 210Å in diameter. When contracted, it is only 360Å long and 270Å wide. At the lower end of the cylinders is the hexagonal base plate with a central hole and fibres. It is a multiprotein structure. The proteins form six wedges

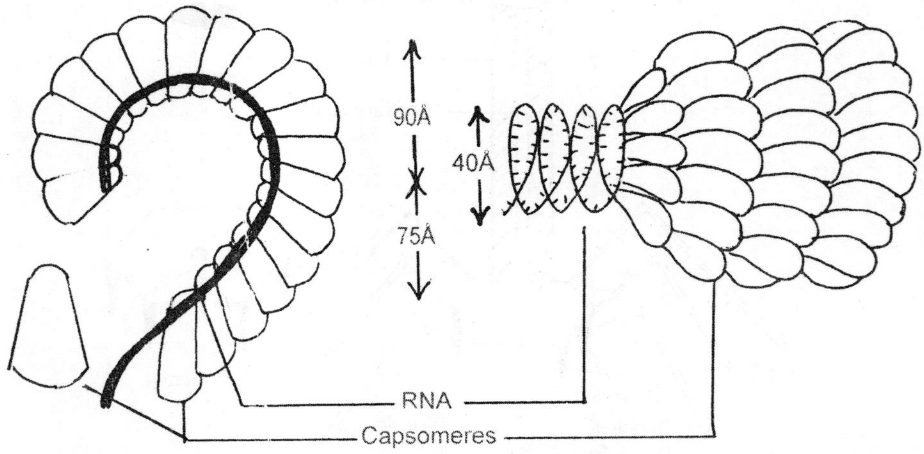

Fig. 1.4 : Structure of TMV.

surrounding the central hub with the help of two trimeric proteins, $(gp9)_3$ and $(gp12)_3$. Gp11, gp10, gp7, gp8, gp6, gp53, and gp25 combine sequentially to build up the wedges. Gp5, gp27, gp29, and probably gp26 and gp28 form the base plate's hub. Gp5 has a lysozyme domain which is necessary for the digestion of the bacterial peptidoglycan layer during the infection process. The base plate is provided with six fibres which are 150 nm long. They help the virus to adhere with the host pellicle. Each fibre shows two halves: the proximal half encoded by gene *34* and the distal encoded by genes *36* and *37*. These halves are connected by gp35.

Tobacco mosaic virus: Tobacco mosaic virus (TMV) is a positive-sense rod like ssRNA virus that infects.

Pants, especially tobacco and other members of the family Solanaceae. The infection causes characteristic patterns (mottling and discoloration) on the leaves (hence the name). TMV was the first virus to be discovered. The protein coat consists of 2130 identical polypeptide chains. Virions are ~300 nm in length and ~18 nm in diameter. Its capsid is made from 2130 molecules of coat protein or capsomeres each with a MW 17.500. The coat proteins (constituting of 95% of virus) self-assemble into the rod like helical structure (there are 130 turns in a complete virus capsid and 16.3 proteins per helix turn) around the RNA which forms a hairpin loop structure.

The protein monomer consists of 158 amino acids. The capsid is in the form of a tube with a cavity measuring about 2nm in diameter. The capsid encircles the genomic ss RNA 6400 bases long. The RNA has a molecular weight of $2,1 \times 10^6$. The RNA is located at a radius of ~6 nm and is protected from the action of cellular enzymes by the coat protein. The RNA strand forms 20 genes that encodes four proteins, which together orchestrate the life cycle of the virus. These include two proteins that replicate the viral RNA, a protein that transports the RNA from cell to cell, spreading the infection, and the capsid protein seen in the PDB structures.

Influenza Virus: These are RNA viruses that make up three genera- Influenza virus A, B and C

***Influenza virus A*:** This genus has one species, influenza A virus. Wild aquatic birds are the natural hosts for a large variety of influenza A. Occasionally, viruses are transmitted to other species and causes devastating outbreaks in poultry birds or man. These are the most virulent human pathogens and cause severe disease. The influenza A virus can be subdivided into serotypes - H1N1, H2N2, H3N2, H5N1, H7N7, H1N2, H9N2, H7N2, H7N3 and H10N7 based on the antibody response.

Influenza virus B: This genus has one species, influenza B virus that exclusively infects humans.

Influenza virus C: It has one species, influenza C virus, which infects humans, dogs and pigs.

Structure: Influenza viruses A, B and C are very similar in overall structure. The virus particle is 80–120 nm in diameter and is roughly spherical. These are made of a viral envelope containing two main types of glycoproteins, wrapped around a central core. The central core contains the viral RNA genome and other viral proteins that package and protect this RNA. RNA tends to be single stranded but in special cases it is double. Unusually for a virus, its genome is not a single piece of nucleic acid; instead, it contains 7 - 8 pieces of segmented (-) sense RNA, each piece of RNA containing either one or two genes, which code for a protein. Hemagglutinin (HA) and neuraminidase (NA) are the two large glycoproteins that occur on the outside of the viral particles. HA mediates binding of the virus to target cells and entry of the viral genome into the target cell, while NA is involved in the release of progeny virus from infected cells.

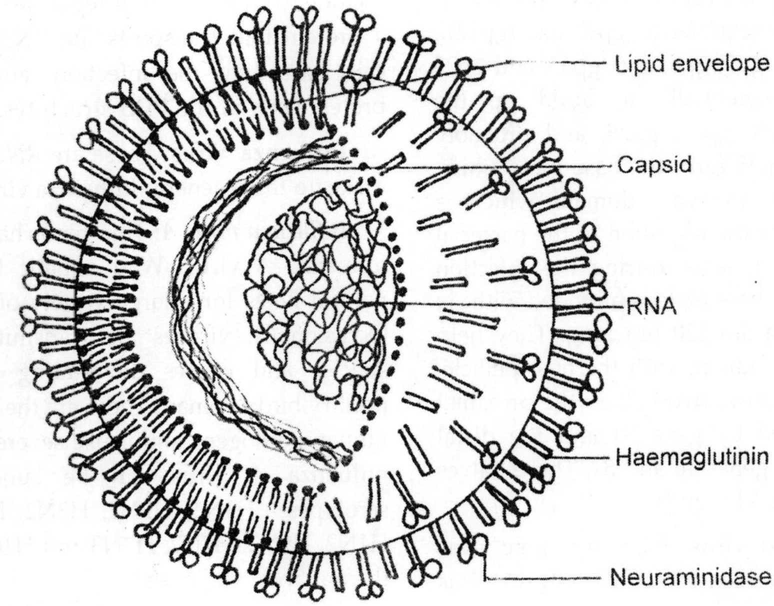

Fig. 1.5 : Structure of Influenza virus.

1.5. PRION

A prion is an infectious agent that, is composed entirely of proteins. As the infectious agent or the specific protein that the prion consists of is named PrP, an abbreviation for "prion protein". Its normal cellular role is as a copper dependent antioxidant. Prusiner was awarded the Nobel Prize in Physiology or Medicine in 1997 for his pioneering research in prions

Structure: PrP is found throughout the body, even in healthy people and animals. However, the prion protein found in infectious material has a different folding pattern and is resistant to proteases. The normal PrP is called PrP^C, while the infectious form is called PrP^{Sc} — the C refers to 'cellular' or 'common' PrP, while the *Sc* refers to 'scrapie', a prion disease occurring in sheep.

PrP^C: PrP^C is a normal prion protein found on the membranes of cells. It has 209 amino acids (in humans), one disulfide bond and a mainly alpha-helical structure. Its function has not been fully resolved. The significance of this is not clear, but it presumably relates to PrP structure or function. PrP^C is readily digested by proteinase K and can be liberated from the cell surface by the enzyme phosphoinositide phospholipase C

PrP^{Sc}: The infectious isoform of PrP^C, known as PrP^{Sc}. It is able to convert normal PrP^C proteins into the infectious isoform by changing their conformation. Although the exact 3D structure of PrP^{Sc} is not known, there is increased β-sheet content replacing normal α-helix. The end of a fiber acts as a template for the free protein molecules, causing the fiber to grow; only free protein molecules that are identical in amino acid sequence to the prion protein can be recruited into the growing fiber.

Prion disease: Prions cause neurodegenerative disease by aggregating extracellularly within the central nervous system to form plaques known as *amyloids,* that disrupt the normal tissue structure. This disruption is characterized by "holes" in the tissue with resultant spongy architecture due to the vacuole formation in the neurons. Neurodegenerative symptoms include convulsions, dementia, ataxia and behavioural or personality changes.

All known prion diseases, collectively called *transmissible spongiform encephalopathies* (TSEs), are untreatable and fatal. However, a vaccine has been developed in mice that may provide insight into providing a vaccine in humans to resist prion infections.

The following diseases are believed to be caused by prions.

In animals - Scrapie in sheep and goats. Bovine spongiform encephalopathy (BSE) in cattle (known as *mad cow disease*), Transmissible mink encephalopathy (TME) in mink, Chronic wasting disease (CWD) in elk and mule deer, Feline spongiform encephalopathy in cats, Exotic ungulate encephalopathy (EUE) in nyala, oryx and greater kudu, Spongiform encephalopathy of the ostrich.

In humans - Creutzfeldt-Jakob disease (CJD) and its varieties, Gerstmann-Str‰ussler-Scheinker syndrome (GSS), Fatal familial insomnia (sFI) and Kuru.

Transmission: It is often assumed that the diseased form directly interacts with the normal form. An unidentified cellular protein (Protein X) enables the conversion of PrP^C to PrP^{Sc} by bringing a molecule of each of the two together into a complex. The primary method of infection in animals occurs through ingestion. It is thought that prions may be deposited in the environment through the remains of dead animals and via urine, saliva, and other body fluids. They may then linger in the soil by binding to clay and other minerals.

1.6. MYCOPLASMA

Mycoplasma, means *Mykes* (= fungus) and *Plasma* (= formed). It was formerly called pleuro pneumonia-like organisms (PPLO) referring to organisms similar to the causative agent of Contagious Bovine Pleuropneumonia - CBPP and this name was proposed in the 1950's, replacing the term PPLO. It is a genus of bacteria that lack a cell wall. There are no free-living species of Mycoplasma, they parasitize a wide range of organism including humans, plants, animals and insects. Many species of Mycoplasma are commensal, living with their host as part of the natural flòra of the body having no detrimental effects. But when Mycoplasma acts as pathogens they cause chronic but mild infections that rarely, if ever, kill their host. They are usually surface parasites, although several species (M. penetrans) are intracellular pathogens. M. pneumoniae is an important cause of atypical pneumonia and other respiratory disorders in human; M. genitalium is believed to be involved in pelvic inflammatory diseases.

Mycoplasma measures <1m. They are considered as the perfect model of the minimal cell. They are believed to contain the absolute minimum structures necessary for survival and are considered as the model organisms for the essential functions of all living cells. Their cell shows plasma membrane, ribosomes, and a highly coiled circular chromosome. They appear flask shaped with polar extensions, the so-called 'tip-structure', protruding from their cell body. As most species of Mycoplasma are extracellular pathogens, they use their tip organelle to bind to their host cells, which has a high concentration of adhesins, to attach to the eukaryotic cell. The organelle is bounded by the cell membrane and contains a central rod-shaped structure. Mycoplasma that acts as intracellular pathogens, penetrate their host cell using their tip organelle. This structure also helps in the movement along solid surfaces (gliding motility) and in cell division. The lack of a rigid cell wall, may also facilitate contact between mycoplasma and their host cell by creating the possibility of fusion between the two membranes. This would enable the exchange of membrane and cytoplasmic components.

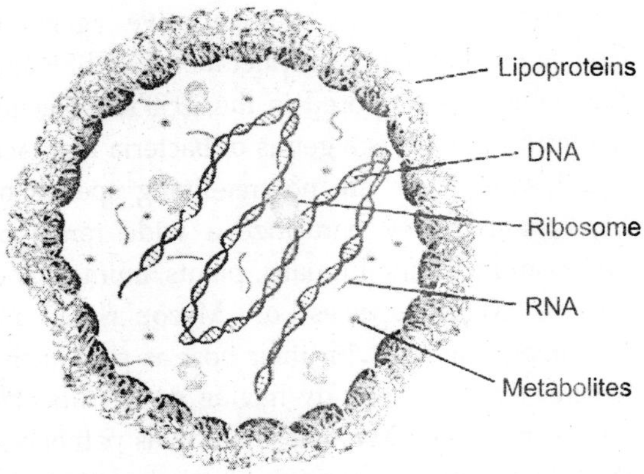

Fig. 1.6 : Structure of Mycoplasma.

Their genome size ranges from 5,80,070 bp – 9,96,422 bp long which results in drastically reduced biosynthetic capabilities and explains their dependence on a host. Additionally they use an alternate genetic code where the codon UGA is encoding for the amino acid tryptophan instead of the usual opal stop codon.

Pathophysiology: *Mycoplasma* species have been isolated from women with bacterial vaginosis. *M. genitalium* is found in women with pelvic inflammatory disease. In addition, infection is associated with increased risk of cervicitis, preterm birth, spontaneous abortion, and infertility. Mycoplasmae are associated with fetal respiratory distress syndrome, bronchopulmonary dysplasia, and intraventricular hemorrhage in preterm infants.

The correlation of *Mycoplasma* with cancer was first reported in 1960s. Since then, several studies tried to prove the connection between *Mycoplasma* and cancer. Several studies have shown that cells that are chronically infected with the bacteria go through a multistep transformation. The first visual sign of infection is the change of shape from normal to sickle shaped structure. They also become hyperchromatic due to an increase of DNA in the nucleus of the cells.

Cells infected with *Mycoplasma* show significant chromosomal abnormalities. These include the addition of chromosomes, the loss of entire chromosomes, partial loss of chromosomes, and chromosomal translocation. All of these may contribute to the process of malignant transformation. Chromosomal translocation and extra chromosomes help create abnormally high activity of certain proto-oncogenes. Another possible mechanism of carcinogenesis is RAC1 activation by a small GTPase-like protein fragment of Mycoplasma. Mycoplasma is associated with colon cancer, Gastric cancer, Lung cancer, Prostate cancer and Renal cancer.

SUMMARY

1. Cell is the building block of life. They may be of prokaryotic or eukaryotic type. The organisms may be unicellular or multicellular. However protozoans being the unicellular organisms are better referred as acellular organisms due to independent existence of their cells. Plant cells are slightly larger than animal cells. Usually cells are microscopic but very few plant and animal cells are macroscopic.
2. According to Fluid Mosaic model plasma membrane basically consists of bilayered lipid molecules and haphazardly oriented proteins. Proteins may be of integral and peripheral type. Both proteins and lipids show an exchange between cytoplasm and plasma membrane. Plasma membrane performs many vital roles like transport of molecules, intake of solid and liquid matters, expulsion of secretory substances, signal transduction, impulse transmission etc.
3. In most plant cells surrounding the plasma membrane, there is a multilayered rigid structure called cell wall. It shows middle lamella, primary and secondary cell wall layers and mainly composed of cellulose and related components. ER and Golgi complex play vital roles in the formation of cell wall at the end of cytokinesis.
4. All the cells possess a colloidal substance – protoplasm encircled by plasma membrane. It contains many simple chemical substances like sugars, amino acids, fats, inorganic ions and water.

These simple monomers unite to form polymers having specific functions. Sugars and lipids mainly provide energy, proteins regulate the biochemical activities by forming enzymes.

5. Cytoplasm is traversed by complex filamentous structures – actin, intermediate filaments and microtubules called cytoskeletons. They regulate a variety of physiological functions.
6. Usually all cells contain one or more nucleus having one or many nucleoli. In both plants and animals multinucleated cells are noted called coenocytes and syncitiums respectively. A porous double-layered membrane enclosing nucleoplasm containing perichromatin fibres, granules, nascent mRNAs, enzymes and chromatins invests nucleus.
7. Ribosomes are the main organelle for protein synthesis, usually remain free in the cytosol or remain attached to the wall of nucleus or ER. They are noted on prokayotes, eukaryotes and in the matrix of mitochondria.
8. Mitochondria are self replicating organelles mainly concerned with energy generation. They contain 2-6 circular DNA containing 37 genes.
9. Endoplasm shows reticular arrangement of sac like or vesicular or tubular structures. They play crucial role in the post translational modification of proteins. On the basis of the occurrence of ribosomes on their surface they are of two types – RER and SER. In muscle cells specific sarcoplasmic reticulum stores Ca^{++} that is crucially required for muscle contraction.
10. In between nucleus and plasma membrane there are many roughly semicircular structure that are mainly destined to carry out secretion of the cell are called Golgi complex. Its nuclear end is the forming and the opposite end is called the maturing face. The whole organ consists of cisternae, vacuoles and vesicles. In addition to secretion, they are concerned with the formation of lyso somes, acrosome, cortical sacs and cell wall. They also participate in apoptosis.
10. Choloroplasts are double-layered organelle found in plant cells. They are mainly concerned with the production of sugars by the process – photosynthesis. They are self replicating since they contain DNA and RNA.
11. Centrosome consisting of two centrioles, is the most diagonostic feature of the animal cell and plays a vital role in the nucleation of microtubules during cell division. It is therefore called microtubule organizing centre (MTOC). Its wall shows 9 triplets each of which contains 13/10 subtubules. It also regulates cell motility and differentiation.
12. Prokaryotes lack most of the cellular organelles except ribosomes. Their plasma membrane performs the roles of mitochondria, chloroplasts and Golgi complex. A prokaryotic cell has five essential structural components: a genome (DNA), ribosomes, cell membrane, cell wall, and some sort of surface layer which may or may not be an inherent part of the wall. Flagella lack eukaryotic 9+2 arrangement of microtubules but are provided with specific motor apparatus for their motility. Fimbriae are shorter and stiffer structures, common in Gram (-) bacteria. Pili are attachment organs. The cell wall of Gram (+) bacteria shows a thick layer of peptidoglycan. The Gram (-) wall appears thin and multilayered, consists of a thin peptidoglycan sheet between the plasma membrane and a phospholipid-lipopolysaccharide outer membrane. Most prokaryotes contain a varied polysaccharide or polypeptide layer outside the cell wall, called capsule.
13. Bacterial plasma membrane has the same basic structure as in eukaryotic plasma membarane but functionally it widely differs from that of eukaryotic plasma membrane. As they lack intra cellular organelle it performs the role of energy production, photosynthesis, signal transduction, cell wall synthesis and DNA replication
14. Mesosomes are coiled invaginated parts of the plasma membrane that contains the enzymes for energy production, CO_2 fixation, DNA replication, cell wall synthesis etc.
15. A large DNA represents chromosome and is located in the center of the cell called nucleoid. Additionally there is extra chromosomal DNA called plasmid. Cytoplasm also shows ribosomes of 70S size involved in the translation of protein.
16. Vegetative cells of bacteria in response to stress situations forms endospores tat germinate in favourable conditions by overcoming the environmental stresses.

QUESTIONS

Long type questions:

1. Describe the structure of a typical animal cell.
2. Describe the structure of a typical plant cell.
3. Compare animal and plant cell.
4. What are the main layers of cell wall? Compare their chemical constituents. Comment on the mode of formation of cell wall.
5. Describe the structure of a typical prokaryotic cell.
6. Describe the molecular structure of bacterial flagella. How it differs from eukaryotic flagella?
7. Describe the structure and functions of bacterial plasma membrane.

Short type questions:

1. Why protozoans are referred as acellular animals?
2. Discuss the main points of cell theory.
3. What are the main constituents of plasma membrane?
4. Name the amino acid that does not have D and L enantiomers.
5. What are macro and microelements?
6. What are the main constituents and functions of cytoskeletons?
7. What are endosomes?
8. What are main constituents of nucleoplasm?
9. What are synsitium and coenocytes?
10. Name a human non-nucleated and a syncitial cell.
11. What are cristae?
12. What is the length of mtDNA? How many genes are there in mt-genome?
13. What do you mean by Zone of Exclusion?
14. What do you mean by CGN and TGN?
15. What is the relationship of Golgi complex and lysosome?
16. What are thylakoids and quantasomes?
17. What do you mean by residual bodies?
18. What are major roles of pilli?
19. What are slime layers and biofilms? State their specific roles.
20. What are the vital roles of capsules?
21. State the major differences of the cell envelopes of Gram (+) and Gram (-) bacteria.
22. What are the major roles of bacterial plasma membrane?
23. Which bacterium lacks cell wall?
24. What do you mean by 'minimal cell model'?
25. What is genetic meaning of the codon in *Mycoplasma*?
26. What are the basic differences in the structure of chromosomes of pro-, meso- and eukaryotic cells?
27. What do you mean by totipotent cells?
28. What do you mean by stem cells?
29. What do you mean by microsomes?

MCQs:

1. **Middle lamella contains**
   a) cellulose, b) pectin c) calcium pectate.
2. **Primary cell wall contains**
   a) magnesium pectate, b) pectin, c) suberin.
3. **Wax is noted in**
   a) middle lamella,
   b) primary cell wall,
   c) secondary cell wall.
4. **Phragmoplasts are derived from**
   a) ER, b) Golgi complex, c) chloroplasts.
5. **Mitoribosomes belong to**
   a) 70S, b) 80S, c) 55S type.
6. **Mitochondrial DNA is of**
   a) B-DNA, b) Z-DNA , c) circular type.
7. **Acrosome is formed by**
   a) ER, b) lysosomes, c) Golgi complex.
8. **Cortical sacs in the ovum are formed by the**
   a) mitochondria, b) Golgi complex, c) ER.
9. **The main chemical substance responsible for photosynthesis is**
   a) grana, b) thylakoid, c) quantasomes.
10. **The organelle that protects the cell from oxygen toxicity is**
    a) Lysosome, b) Peroxisome, c) Quantasome.
11. **In *E. coli*, the sex pili mediate the role of**
    a) attachment,
    b) resistance against WBC,
    c) transfer of DNA during conjugation.
12. **The spores that help to overcome the environmental stresses are called**
    a) oospore, b) zygospore, c) endospore.
13. **Metabolically active cell shows**
    a) large nucleus, b) more mitochondria, c) small size.

14. **Inulin is found in**
a) root of wheat, b) root of daliah, c) seed of pea)
15. **Site of steroid genesis is**
a) RER, b) SER, c) lysosome.
16. **Totipotency of cells is lost due to**
a) differentiation, b) maturation, c) migration of cells.
17. **Lipid rich membrane in neuron occurs in.**
a) cyton, b) dendrite, c) axon
18. **The site of respiration in bacteria is**
a) mesosome, b) episome, c) plasmid.
19. **Pili in bacteria represent**
a) extra chromosomal genetic materal,
b) outgrowth of cell membrane,
c) ingrowth of cell membrane.
20. **Bacteria do not possess**
a) ribosome, b) mitochondria, c) plasma membrane.
21. **Bacteria are considered as plants due**
a) sedentary nature,
b) presence of cell wall,
c) can not multiply by fission.
22. **Nucleoid has**
a) ss-RNA, b) dsRNA, c) ds-DNA)
23. **The organelle covered by a single layered membrane is**
a) lysosome, b) peroxisome, c) both
24. **Detoxification site in the liver is**
a) RER, b) SER, c) hepatic sinusoid.
25. **Organelle involved in the transformation of the cell membrane is**
a) ER, b) lysosome, c) mitochondria)
26. **Microsome is**
a) cluster of ribosome,
b) organelle obtained after centrifugation,
c) disrupted fragments of ER.
27. **Fusion of lysosome and phagosome is called**
a) autophagosome,
b) secondary lysosome,
c) residual body.
28. **Hydrolytic enzyme loaded organelle is**
a) peroxisome, b) glyoxysome, c) lysosome.
29. **Plastidial ribosome is of**
a) 55S, b) 70S, c) 80S type.
30. **Vesicles are essential for Golgi complex because**
a) they contribute to the structural integrity,
b) they carry proteins and lipids to other organelle,
c) they provide energy.
31. **Autolysis is related with**
a) peroxisome, b) Golgi complex, c) lysosome.
32. **Mitochondria and chloroplasts are considered as endosymbionts because**
a) they possess their own nucleic acids,
b) they do not arise *de novo,*
c) both.
33. **ATP synthesis in mitochondria takes place in**
a) cristae,
b) intermembrane space,
c) the inner membrane.
34. **Two cellular organelle that transform energy are**
a) chloroplast & ER,
b) ER & Golgi complex,
c) chloroplast and mitochondria)
35. **Mesosome in bacteria is the**
a) intercellular connection,
b) folded cell membrane,
c) extra nuclear DNA
36. **Functions of ER are**
a) mechanical, b) protein modification, c) both.
37. **Hydrolytic enzymes occur in**
a) Peroxisome, b) Lysosome, c) Mitochondria)
38. **Which cell wall component is regarded as biological plastic?**
a) Hemicellulose, b) Cellulose, c) Lignin.
39. **Which is a membrane structure?**
a) Microsome, b) Oxysome, c) Liposome.
40. **The largest cell organelle in cell is**
a) Mitochondria, b) ER, c) Plastid.
41. **Widely distributed structure in the cell is**
a) DNA, b) RNA) c) Chloroplast.
42. **pH of nucleoplasm is**
a) Slightly basic, b) Slightly acidic, c) Neutral.
43. **Ca^{++} is not essential for**
a) Fungi, b) Pterydophytes, c) Gymnosperms.
44. **The marker enzyme present in lysosome is**
a) Lysozyme, b) Hyaluronidase, c) Acid phosphatase.
45. **On which side of the Golgi body the membranes are thin?**
a) Concave distal side,
b) Concave proximal side.
c) Convex proximal side.
46. **The lysosomal enzyme having antiseptic property is**
a) Lysozyme, b) Acid phosphatase, c) Lipofuscin.

47. **A multicellular organism contains**
a) Differentiated cells,
b) Undifferentiated cells,
c) Both.

48. **Bacteria do not possess**
a) DNA, b) RNA) c) Nucleus.

49. **Which one is Gram (+) bacteria?**
a) *E.coli,* b) *Pneumococcus,* c) *Pseudomonas.*

50. **Plasmid which is associated with nucleoid is**
a) Episome, b) Proplasmid, c) Cryptic plasmid.

51. **Which of the following is common to prokaryotic and eukaryotic cells?**
a) Cell organelles present.
b) Cell wall present,
c) Nucleus present.

52. **Which one of the following is not true about an eukaryote?**
a) Mitoribosome is of 80S type.
b) Mitochondria contain circular DNA)
c) Cytoplasmic ribosomes are of 80S type.

53. **Which dye is used to stain mitochondria?**
a) Fast green. b) Acetocarmine. c) Janus Green.

54. **Which dye is used to stain chromosome?**
a) Fast green. b) Acetocarmine. c) Safranin.

55. **The common feature of sieve tube and mammalian RBC is**
a) Absence of ribosome.
b) Absence of nucleus.
c) Absence of mitochondria)

56. **Which statement is false for cell membrane**
a) Bilayered lipid present.
b) Intrinsic and extrinsic proteins present.
c) Carbohydrates are absent.

57. **Feulgen stain is used for**
a) RNA, b) DNA, c) Proteins.

58. **Which is a cell within a cell?**
a) Golgi complex, b) Nucleus c) Mitochondria)

59. **Zone of Exclusion is associated with**
a) Chloroplast b) Peroxisome c) Golgi complex.

60. **Which is essential for binding the subunits of ribosomes?**
a) Mg^{++} b) Ca^{++} c) Cu^{+}

61. **Teichoic acid is present is**
a) Gram(+) bacteria,
b) Gram(-) bacteria)
c) Mycobacteria)

62. **Cells which loose their nucleus during differentiation is**
a) Nerve cell, b) Muscle cell, c) RBC)

63. **In dehydration saline drip is given because**
a) Cl^{-} is a component of plasma,
b) NaCl is a component of blood,
c) Na^{+} retains waster in the body and helps in transport through the cell membrane.

Answers to MCQs:

1. c; 2. b; 3. c; 4. b; 5.c; 6. c; 7.c; 8. b; 9. c; 10. b. 11. c; 12.c; 13.c, 14.b, 15.b, 16. a, 17. c, 18.a, 19.b, 20. b, 21. b, 22. c, 23. c, 24. b, 25. b, 26. c, 27.a, 28. c, 29. b, 30. b, 31.c, 32.c, 33. a, 34. c, 35. b, 36. c, 37.b, 38.b .39.c.40.c, 41. b. 42.a. 43.c .44.c .45.c .46.a. 47.c. 48.c .49.b.50.a.51. b; 52.a; 53.c; 54.b; 55.b .56.c; 57.b ; 58.c; 59.c; 60.a. 61.a. 62.c; 63.c.

Chapter 2

Plasma Membrane

The protoplasm of all cells (prokaryotic and eukaryotic) is encircled by a thin, elastic, semipermeable lipoproteinous membrane called **plasma membrane / cell membrane / plasmalemma**. In addition to maintaining the shape of the cell, it acts as a simple barrier between the extra cellular and intracellular media and plays a crucial role in the intercellular communication, in anchoring the cell with the ECM, intracellular signaling and transport of different molecules through it. The same membrane, surrounds many cellular organelles (ER, nucleus, mitochondria, Golgi bodies, lysosomes, chloroplast, peroxisome etc.) performing different physiological roles. The term plasma membrane was coined by Plowe (1931).

2.1. DIFFERENT MODELS EXPLAINING THE MOLECULAR STRUCTURE OF PLASMA MEMBRANE

A. *Unit membrane concept (Robertson, 1959):* According to this model :

i) Plasma membrane is a trilayered structure with outer and inner protein layers and intermediate lipid bilayer.

ii) Outer protein layer is 25Å thick and inner layer is 25Å - 35Å thick.

iii) Lipid layer is 25Å - 30Å thick. Both lipid layers show outer polar hydrophillic and inner non-polar hydrophobic ends. Polar ends remain bound with protein molecules by ionic linkages but the opposing ends are held together by van Der Wall force. Lipid layer mainly shows phospholipids and cholesterol.

This model is not only applicable to the plasma membrane but also to all the intra cellular membranes. For this reason it is called *Unit membrane* so that Protein-Lipid-Protein acts as structural unit of all the membranes. But the static nature of these molecules in this model cannot explain various physiological functions of plasma membrane for which this model has been rejected by the cytologists.

B. *Fluid Mosaic model (Singer and Nicolson, 1972)*: According to this model:

i) Plasma membrane is a bilayered structure consisting of mainly lipid molecules that are arranged on a jelly like matrix.

ii) The width of the lipid layer is not constant.

iii) The main lipid component is phospholipid,

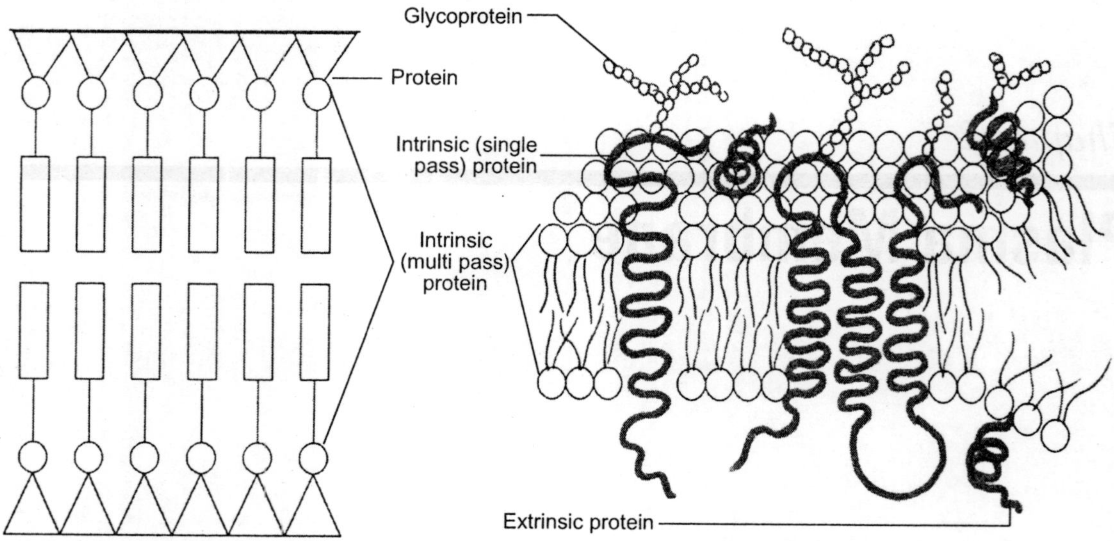

Fig. 2.1 : Structure of plasma membrane as per A- Unit membrane concept. B- Fluid mosaic model.

discontinuously arranged, being interrupted by protein molecules. Their polar hydrophilic ends are directed outwards while the inner hydrophobic ends are held together by van Der Wall force.

iv) Protein molecules are haphazardly oriented. Three types of proteins are noted, some are loosely attached or partially embedded in the lipid layer, and their outer ends protrude to either extracellular matrix or in the cytoplasm. Such proteins are called *peripheral proteins.* Other type of proteins are arranged through out the breadth of the membrane protruding on either side of the membrane, these are called *integral proteins.* Other type of proteins are cytoskeletal or associated signaling proteins at the inner membrane surface and certain glycoproteins or glycosaminoglycans at the outer membrane surface. These are thought to be involved in stabilizing cell membranes in maintaining membrane function and dynamics, like cell–cell adhesion, cell motility, endocytosis, exocytosis and many other important cellular activities.

v) Both protein and lipid molecules are dynamic in nature and exchanges of these components always occur between cytoplasm and plasma membrane.

vi) Lipid molecules can migrate from one monolayer to another. This is called *flip-flop movement.* Moreover individual lipid molecules can exchange places within a monolayer showing *lateral diffusion* or they can *rotate* in their longitudinal axis or their hydrocarbon chains can show *bending* or *flexion*. The phospholipids can also exhibit alternate contraction and relaxation of the tails. This is called *flex movement*.

vii) Each transverse portion of the membrane is called *'leaf-let'* which shows similar disposition of the molecules but biochemically they are asymmetrical.

C. ***Corral model (Leitner, Frank, Brown and Wilson, 2000)***: This model emphasizes the role of cytoskeletons in the localization and movement of membrane proteins. They proposed that there

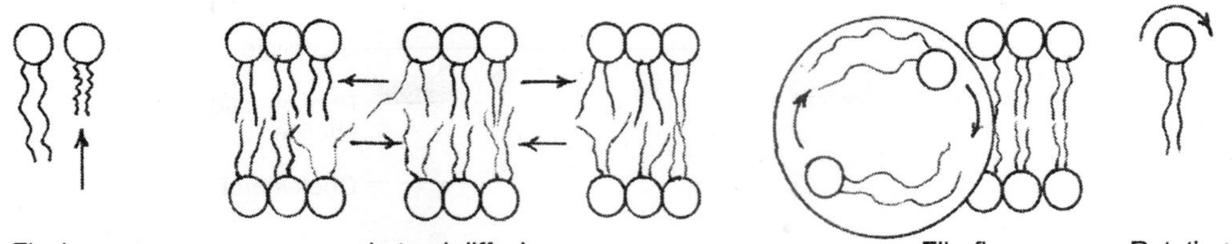

Fig. 2.2 : Different types of movement of lipid molecules.

Table 2.1 : Basic differences between Unit membrane concept and Fluid Mosaic model

Unit membrane concept	Fluid mosaic model
1. Plasma membrane is a trilayered structure.	1. Plasma membrane is a bilayered structure.
2. Both lipid and protein molecules are serially.	2. Protein and lipid molecules are haphazardly and regularly arranged. arranged.
3. There is no differentiation of protein molecules.	3. Protein molecules are differentiated into peripheral and integral proteins.
4. Arrangement carbohydrates is not specific.	4. Glycolipids or glycoproteins occur towards extracellular face.
5. The whole membrane exhibits a monolayer configuration.	5. The whole membrane exhibits biochemically assymetrical bilayer configuration.
6. The position of respective molecules is fixed.	6. The position of both protein and lipid molecules are dynamic and furnish exchange between cytoplasm and plasma membrane.

are some specific parts in the cytoskeleton beneath the membrane called *'corral'* which binds proteins and renders them immobile. Unbound proteins are also temporarily arrested due to interaction with cytoskeletal networks rendering their slow diffusion. Corrals act as gates which when open permit trapped proteins to escape to a neighboring corral in the cytoskeletal network.

2.2. MICRODOMAINS OF PLASMA MEMBRANE

Lipids often cluster to form microdomains rich in cholesterol, glycollipids, sphingolipids and specific proteins. These are referred to as *caveolae, glycosphingolipid enriched membrane (GEM) and polyphosphoinositol rich rafts.*

Caveolae are very stable flask shaped invaginated regions of mammalian membranes rich in free cholesterol, glycoshingolipid, arachidonic acid. The protein components of caveolae are the caveolins, cavins 1-4, integrins and flotillin. Caveolae regulate growth, cell division, adhesion, integrin signal transduction, cholesterol trafficking, capillary permeability and hormonal response (Hansen and Nichols, 2010; Briand *et al*, 2011). Caveolin acts as a scaffold that surrounds the caveolar bulb. Caveolin interacts, both directly and indirectly, with a large number of signaling molecules and regulate the endocytosis of molecular cargo by caveolae (Root *et al*, 2015).

Rafts are flat tightly packed structure rich in cholesterol and sphingolipid. The fatty-acid

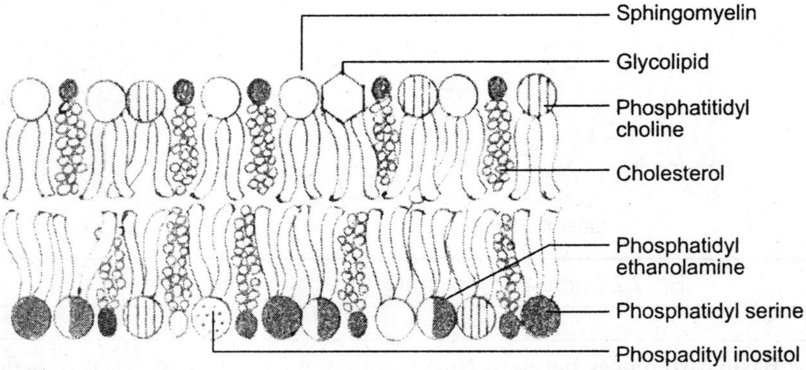

Fig. 2.3 : Structure of plasma membrane with raft and caveolus.

chains of lipids within the rafts extend and create tightly packed domains. It is therefore thought that rafts exist in a separate ordered phase that floats in a sea of poorly ordered lipids. Flotillin is one of the specifically located proteins in rafts that is anchored in the outer leaflet of the membrane through covalent attachment to a special glycolipid - glycosyl phosphatidyl inositol (GPI). It seems to be an important component of the plasma membrane of the parasites and is suggested to play a crucial role in the invasion of the parasite (Gallego *et al*, 2007). Rafts are very unstable being dynamically produced and degraded. Many actin-binding proteins (gelsolin) are known to link the actin cytoskeleton and rafts. Since these contain several signaling molecules, they play an important role in receptor-medicated signal transduction, endocytosis, cholesterol transport and pathogen entry (Hara-Chikuma and Verkman, 2008).

2.3. CHEMICAL COMPOSITION OF THE PLASMA MEMBRANE

Its chemical composition is as follows:

Lipid : It comprises 28 – 79% of all the constituents in the plasma membrane. A common characteristic of the membrane domains is their association with cholesterol. Lipid rafts and caveolae are examples of cholesterol enriched domains. There is an intrinsic asymmetry between the two monolayers of the bilayer. In addition, some lipids tend to be enriched in particular regions of the membrane. Membrane lipids have important roles in determining the physical properties of the membrane, in modulating the activity of membrane-bound proteins and in certain cases being specific secondary messengers that can interact with specific proteins. Certain domains recruit specific proteins involved in signal transduction. Of the different lipids phospholipid is the main component (up to 35%) followed by cholesterol (23%) (Epand, 2015). Notable lipid components are as follows:

Phosphoglycerides / Phospholipids: Of the phospholipids, phosphatidyl ethanolamine (7-35%), phosphatidyl choline 0-40%), sphingomyelin (5-19%) and phosphatidyl serine are the main components. A large part uniteswith the polysaccharides to form glycolipids or protein to form glycoproteins The head parts of phospholipids form the hydrophilic polar ends that unite with proteins or glycoproteins and the tails as hydrophobic non-polar ends. Phospholipids show a bit of side ways movement and alternately contract or relax their tails to prevent close packing and promote fluidity; this is called *flex movement*.

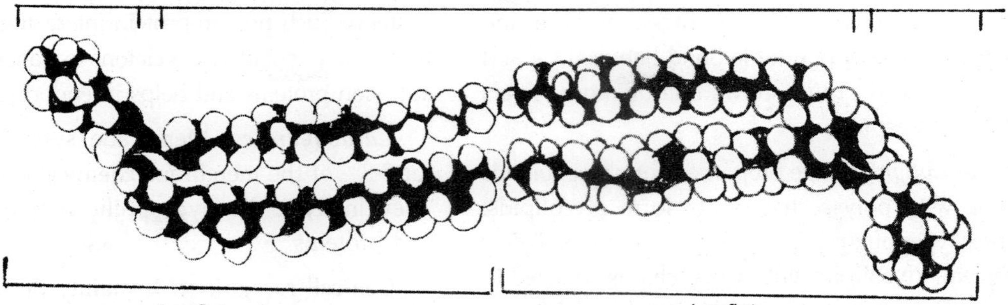

Fig. 2.4 : Position of lipid molecules in two leaflets.

Cholesterol: Sterols along with temperature, electrical charges, pH protein / phospholipids ratio, phospholipids / cholesterol ratio and degree of fatty acids unsaturation are responsible for maintaining the fluidity of the membrane (Gulu *et al*, 2011). Tight association of cholesterol, sphingolipid and flotilin is the major driving force for the formation of rafts that are important for receptor-mediated signal transduction, endocytosis and pathogen entry. Cholesterols inhibit the transformation of the fluid state of the plasma membrane to gel or crystalline state (*phase transition*) and also decrease the permeability of the lipid bilayer to small water-soluble molecules. Krause *et al* (2014) and Wang *et al* (2015) reported that the overall driving forces correspond to a strong pull (attraction) between the high-melting lipids and cholesterol, a push (repulsion) between the low-melting and high-melting lipids and a significant push between the low-melting lipids and cholesterol. In a broader context, these results provide strong support for the notion that repulsive forces play a major role in the formation of lipid rafts. Cholesterol is thought to combine with high-melting lipids to form "lipid rafts" that float in a "sea" of low-melting lipids. Cholesterol also regulate the fluidity of the membrane by regulating the phospholipid tail length by its alternate contraction and relaxation. This is called *flex movement* (longer the tail, less fluid is the membrane). Phospholipids also show rapid lateral movement called *flip flop movement*.

Table 2.2 : Different roles of membrane proteins

Types of proteins	Functions
Structural proteins	They provide fibrous frame work. Collagen and elastin in connective tissue, keratin of hair, horns, feathers and integumentary appendages provide the structural support. Some peripheral proteins are anchored to components of ECM, some are exposed at the interior face of the plasma membrane and are closely associated with the actin microfilaments. Some intercellular proteins act as adhesion molecules.
Storage proteins	Plasma membrane of the seed cells in wheat and beans stores amino acids.
Transport proteins	Transmembrane proteins facilitate passage of different molecules into and outside the cell.
Act as receptors	Several transmembrane proteins act as receptors playing active roles in signal transduction. Integrins in addition to acting as receptors, mediate intercellular interactions and intra or extracellular signal transduction.

Sphingolipids: These are the main sugar containing lipids with a nitrogenous base and fatty acid showing double bond between C4 and C5. The head group shows one or more sugars but no phosphate.

Glycolipids: Large proportions of phospholipids unite with polysaccharides to form glycolipids. Three types of glycolipids are noted – sphingolipids, glycerol glycolipids with sugar chains attached to C_3 and glycosyl phosphatidyl inositol (GPI).

Proteins: Membrane proteins fundamentally rule all the different biochemical processes that take place via lipid bilayers. These are of two types –integral and peripheral proteins.

Integral proteins / Transmembrane protein: These are firmly united with the membrane, span the membrane once (e.g. glycophorin) or several times (e.g. Band 3 protein of human RBC). Some transmembrane proteins show multiple subunits (e.g. Mitochondrial porins have 3 subunits). Most cells show several integral proteins that connect the cell with the ECM. Integral proteins mainly participate in the transportation of molecules and ions. Two integral proteins – Band 3 and glycophorin anchor a network of actin binding proteins.

Peripheral proteins: These remain loosely attached to either side of the membrane. Many proteins bind cytosolic domains of many integral proteins. Such protein-protein interaction provides firm attachment of cytoskeletons to transmembrane adhesion proteins and helps in endocytosis.

Carbohydrates: Membrane sugars accounts for 3-8% of the membrane chemical constituents. They increase the hydrophilic nature of lipids and proteins.

Cell-cell recognition is mainly performed by the glycoproteins. Tissue recognition is regulated by glycolipids.

2.4. STRUCTURE AND FUNCTIONS OF DIFFERENT PUMPS AND TRANSPORTERS

1. *Na^+-K^+ Pump:* It shows 8 or 10 transmembrane domains. The cytosol of animal cells contains 20 times higher concentration of K^+ than in the extracellular fluid which contains 10 times higher concentration of Na^+ than cytosol. causes a potential difference across the cell membrane which is called membrane potential. The same transporter - the Na^+/K^+ pump transports 3 Na^+ out of the cell for each $2K^+$ pumped into the cell.

2. *H^+-K^+ ATPase:* The parietal cells of stomach transport H^+ from a concen tration of

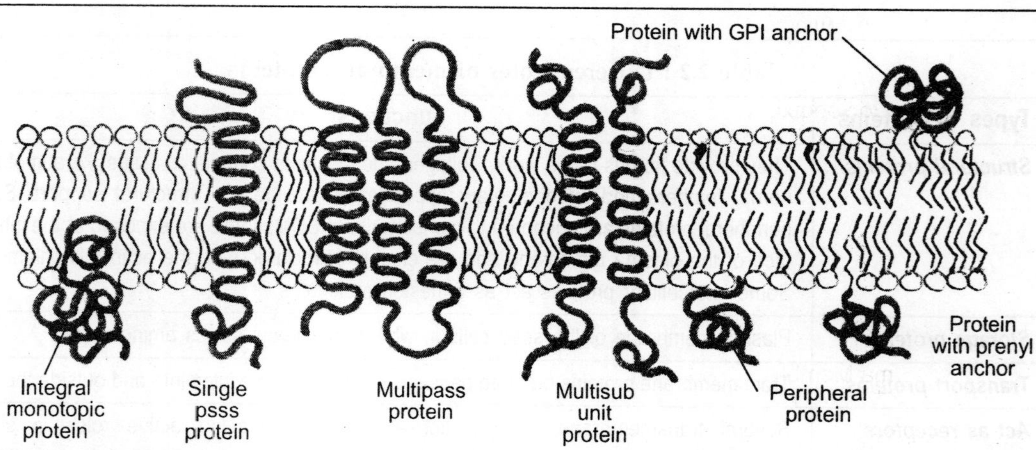

Fig. 2.5 : Different membrane proteins in plasma membrane.

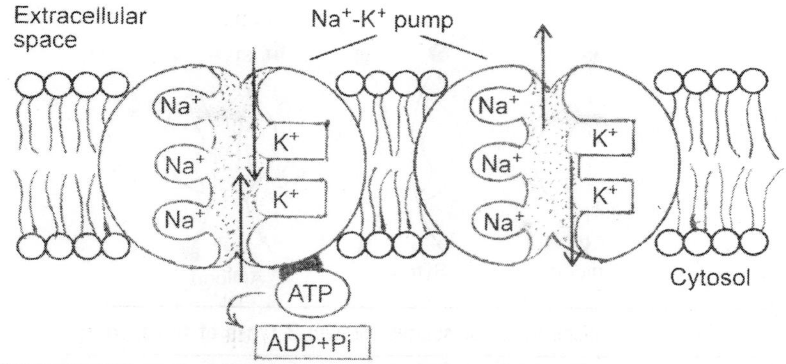

Fig. 2.6 : Transport of Na^+ and K^+ through Na^+-K^+ pump.

about 4 x 10^{-8} M into the cell to a concentration of about 0.15M in the gastric juice (giving it a pH close to 1).

3. *Ca^{++} ATPase:* This transport protein transports Ca^{++} from the cytosol into the ER lumen to keep low cytosolic Ca^{++} level. In resting skeletal muscle, there is a much higher level of Ca^{++} in the sarcoplasmic reticulum (SR) than in the cytosol. Activation of the muscle fiber allows some of this Ca^{++} to pass into the cytosol where it triggers contraction. After contraction, Ca^{++} is pumped back into the SR by Ca^{++}-ATPase.

Membrane Transporters: Transportation of varied substances is performed by a number of proteins called transporters. These are of two types - ***carriers*** and ***channels.*** Carrier proteins show a single polypeptide whose N and C termini project in to the cytoplasm. (e.g. GLUT1 is a glucose carrier). Channels are large sized transporters wi a central canal. On the basis of the directions of transporters are of three types:

Uniporters: These mediate transport of solute moving down its concentration gradient in one direction only. E.g GLUT1(Valinomycin a carrier of K^+).

Symporter (Cotransporter): These bind two dissimilar solutes and transport them together in the same direction across a membrane. It is sometimes referred to as *secondary active transport*. The Na^+ / glucose transporter allows Na^+ and glucose to enter the cell together. The Na^+ flows down their concentration gradient while the glucose molecules are pumped into the cytosol.

Antiporter: These carriers transport two solutes in opposite directions. A sodium-driven antiport pump transports Na+ into the cell and Ca^{++} out of cells. Antiport pumps in the vacuole of some plants allow the outward facilitated diffusion of protons in lieu of active inward transport of Na^+. This Na+/proton antiport pump enables the plant to sequester Na^+ in its vacuole.

2.5. FUNCTIONS OF PLASMA MEMBRANE

Plasma membrane transports a variety of molecules, metabolites and ions into and out of the cell or organelles.

Communication of cells with their environment and participates in intracellular as well as transcellular signaling. Various functions are as follows:

1. Transport of different substances through the membrane:

Plasma membrane contains many pores (7Å in diam.), through which substances smaller than the pores can pass easily. On the basis of

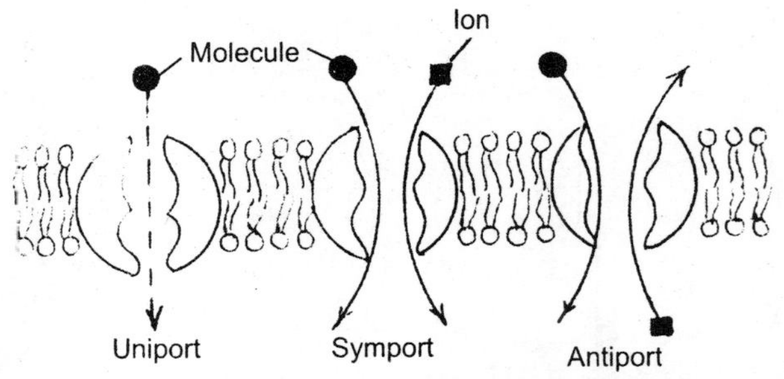

Fig. 2.7 : Transportation of solutes through 3 types of transporters.

permeability plasma membrane may be *impermeable* (e.g. plasma membrane of the unfertilized egg of fishes), *semipermeable* - allowing the passage of only water or solutes (plasma membrane of the fertilized eggs or swim bladder of fishes), *selectively permeable* – allowing the passage of selective ions or molecules (plasma membrane of most of the animal cells) and *dialyzing* – allowing the passage of different molecules in response to high hydrostatic pressure (basal membrane of the endothelial cells). Plasma membrane acts as a selectively permeable barrier that maintains a constant internal environment within the cell or organelles exhibiting a definite concentration of different ions and molecules those are necessary for normal physiological processes.

Types of transport: It is broadly divided into passive and active transport.

Passive transport: It involves transport of substances through the membrane from higher to lower concen tration without the help of energy.

Diffusion: Transport of molecules or ions from higher to lower concentration without involvement of energy is called diffusion. It occurs in gaseous and liquid media. The pressure exerted by the diffusible particles is called *diffusion pressure.* It is inversely proportional with the concentration. As the concentration increases the diffusion pressure falls. This is called *diffusion pressure deficit* (DPD). The rate of diffusion is

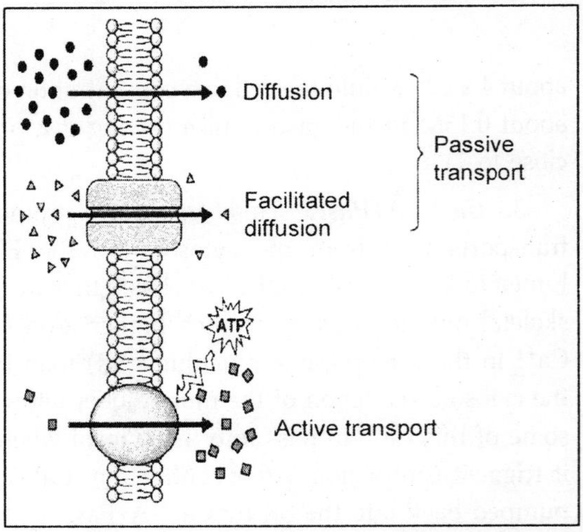

Fig. 2.8 : Representation of passive and active transport.

influenced by the concen tration gradient (in case of liquid), pressure gradient (in case of gas) and electrical gradient (in case of ions). Moreover, molecular size (bigger the molecular size, slower is the rate of diffusion), temperature (rate is directly proportional to the temperature up to a certain limit), medium of diffusion and density of the diffusing material regulate the rate of diffusion.

According to ***Graham's law of diffusion*** the rate of diffusion is inversely proportional to the square root of the density of the diffusing materials. As per ***Fick's law*** substances soluble in lipids can be transported depending on the concentration of the substance on both sides of the membrane, rate of solubility of the substance in the lipid and thickness of the membrane.

Osmosis: It involves transport of water molecules across the selectively permeable membrane driven by a difference in solute concentrations on the two sides of the membrane till the attainment of equilibrium. Water may enter into the cell or into the extracellular medium from the cell that is referred as *endosmosis* and *exosmosis* respectively. In plant cell due to excessive exosmosis the cytoplasm shrinks so that the cell squeezes. This condition is called *plasmolysis.*

Such cells if placed in hypotonic solution then water enters into the cytoplasm and the cell again attains the normal form, this condition is called *deplasmolysis.* A different concentration of solute molecules leads to different concentrations of free water molecules on either side of the membrane. Thus water flows from the solution with the lower solute concentration into the solution with higher solute concentration till equilibrium is reached. To demonstrate the status of osmosis, some standard terminologies are used to describe these differences:

Isotonic: The solutions being compared have equal concentration of solutes.

Hypertonic: The solution with the higher concentration of solutes.

Hypotonic: The solution with the lower concentration of solutes.

The term *osmolarity* is used to describe the number of solute particles in a volume of fluid. Movement of water across a membrane generates a pressure called *osmotic pressure.* As a result of entry of water, in plant cells, a pressure is developed in the cytosol and the cell membrane is pressed onto the cell wall. This pressure is called *turgor pressure.* To counter this pressure an opposite pressure is applied on the membrane by the cell wall, called *wall pressure.* If the pressure in the compartment into which water is flowing, is raised to the equivalent of the osmotic pressure, movement of water will stop. This counter pressure is often called *hydrostatic ('Water-stopping') pressure.*

Facilitated diffusion: Transmembrane proteins create a water-filled pore through which ions and some small hydrophilic molecules can pass by diffusion.

Facilitated diffusion of Ions: Facilitated diffusion of ions takes place through proteins or assembles of proteins, embedded in the plasma membrane. These transmembrane proteins form a water-filled channel (that can be opened or closed) through which the ion can pass down its concentration gradient. They are said to be "gated" which may be ligand gated, mechanically gated and voltage gated.

Table 2.3 : Basic differences of diffusion, osmosis and facilitated diffusion

Processes	Passage through membrane	Working principle
Diffusion	Through lipid bilayer	Random movement of molecules from areas of high concentration to low concentration; e.g. movement of oxygen into the cell.
Osmosis	Through lipid bilayer	Movement of water from areas of low concentration to high concentration across a selectively permeable membrane; e.g. movement of water into plant leaf cell.
Facilitated diffusion	Through carrier protein	Molecules bind to carrier protein and is transported from areas of high to low concentration e.g. movement of Ca^{++} in to the cell.

Table 2.4 : Basic differences between active and passive transport

Active transport	Passive transport
1. It is operated against the concentration gradient.	1. It is operated down the concentration gradient.
2. It is operated with the help of transporter and energy.	2. It is operated without the help of transporter and energy.
3. It occurs in only one directions in selective way.	3. It occurs in both directions in non selective way.

Active Transport: Active transport is the pumping of molecules or ions through the membrane against their concentration gradient with the help of transporter and energy (ATP).

2. Endocytosis and exocytosis:

Endocytosis involves engulfment of solid or liquid substances in response to some specific inductions. Plasma membrane forms vesicles called *endosome* cytosol. If the substance is around the substances that is pinched off into the solid with size range 1-2 μ in diam. then the process is called *phagocytosis* or *cell eating* but if the substance is liquid with the size range 0.1 – 0.2 μ in diam. then it is called *pinocytosis* or *cell drinking* (Fig.2.9).

Exocytosis involves expulsion of secretory liquid substances (encircled by specific vesicles) to the outside produced in the cell. The vesicles bind with the membrane with the help of annexin VII and Ca^{++} forming exocytic pore allowing expulsion of substance. Trafficking of protein and lipid cargo through the secretory pathway is mediated by membrane-bound vesicles. Fusion of secretory vesicle with the plasma membrane is mediated by a multi-subunit protein complex called *exocyst*.

Emiocytosis is the forceful expulsion of substances which is literally called *'vomiting of the cell'*.

3. Signal transduction:

Membrane lipids and the cytoskeleton, play vital roles in signal transduction. It is initiated by protein-protein interactions between ligands, receptors and kinases. Lipid rafts take part in this

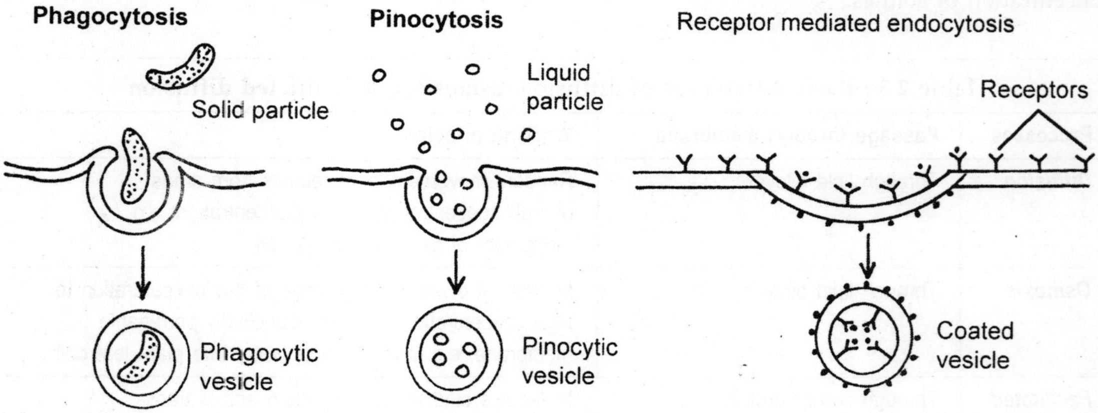

Fig. 2.9 : Diagrammatic presentation of different types of endocytosis.

Table 2.5 : Comparison between pinocytosis and phagocytosis

Pinocytosis	Phagocytosis
1. Liquid substances of the size range–0.1-0.2μ (diam) are taken inside the cell by plasmalemma.	1. Solid substance of the sige range – 1-2μ (diam) are taken inside the cell by plasmalemma.
2. Plasmalemma forms a tubular pinosome during the process.	2. Plasmalemma forms a sac like phagosome during the process.
3. Lysosome may or may not be involved in post drinking assimilation.	3. Lysosome is involved in post eating assimilation.

process. 20 different families of transmembrane protein receptors interact with different ligands to receive extracellular stimuli. Proteins involved in signal detection and transduction are organized in specific domains at the plasma membrane. and play vital roles in signaling pathways.

4. Hormone reception:

Definite receptors in the plasma membrane recognize the hormones and activate different signal transduction pathways. It leads to a conformational change in the receptor followed by relay and amplification of the signal causing definite biochemical reactions within the cell.

5. Antigen specific recognition:

Definite antigenic portions of the plasma membrane proteins recognize specifc antibody of another plasma membrane. This antigen-antibody recognition helps to stitch the operated organs.

6. Impulse transmission:

Nerve or electrical impulses are transmitted through the plasma membrane.

7. Chemoreception:

Definite receptors on the sensory cells are activated for specific chemicals.

SUMMARY

1. The protoplasm of all cells (prokaryotic and eukaryotic) is encircled by a thin, elastic, semipermeable lipoproteinous membrane called plasma membrane The same membrane performing different physiological roles, surrounds many cellular organelles (ER, nucleus, mitochondria, Golgi bodies, lysosomes, chloroplast peroxisome etc.). It acts as a selectively permeable barrier that maintains a constant internal environment within the cell or organelles exhibiting a definite concentration of different ions and molecules necessary for normal physiological processes.
2. According to Unit membrane concept plasma membrane is a trilayered structure with outer and inner 20Å thick protein layers and intermediate 35Å thick lipid layer. Both lipid layers show outer polar hydrophobic (that remain joined with protein molecules by ionic linkages) and inner non-polar hydrophilic ends (held together by Van der Wall force). Lipid layer mainly shows phospholipids and cholesterol. This model is also applicable to all intracellular membranes.
3. According to Fluid Mosaic model plasma membrane is a bilayered structure consisting of mainly lipid (phospholipid, cholesterol etc.). Proteins are haphazardly oriented as transmembrane and peripheral proteins. Exchanges of protein and lipid molecules occur between cytoplasm and plasma membrane. They also show lateral movement or rotation.
4. According to Polymeric network model plasma membrane acts like a network and diffusion of membrane components depends on their size, dimension, molecular weight and interaction of protein and lipopolysaccharides.
5. According to Membrane skeleton fence model the cytoskeletal networks beneath the membrane regulate the movement of the integral proteins.
6. According to Jacobson's model some membrane

proteins diffuse to a very limited distance and act as adhesion molecules, some are held by cytoskeletal materials, some show random diffusion while some show limited directed motion.

7. According to Corral model the movement of the membrane proteins are solely regulated by the cytoskeleton and the specific parts – corral beneath the membrane.

8. Lipids often cluster to form microdomains - caveolae, glycosphingolipid enriched membrane (GEM) and rafts. Caveolae are stable flask shaped invaginated regions rich in free cholesterol and caveolin1, 2. They regulate growth, cell division, adhesion etc. GEMs are merely a caveolae without caveolin-1. Their main function is signaling. Rafts are flat tightly packed cholesterol and sphingolipid rich unstable structure linked with the protein flotilin. They participate in receptor-medicated signal transduction, endocytosis, cholesterol transport, and pathogen entry.

9. Lipids comprise 28 – 79%, proteins 18-76%, and sugars 3-8% of the chemical constituents of the membrane. Of the different lipid categories phospholipid is the major component (up to 35%) followed by cholesterol (23%). Phospholipids show a hydrophilic head and a hydrophobic tail showing two fatty acids. Both leaflets of the plasma membrane are chemically asymmetrical. Cholesterol is mainly responsible for maintaining the fluidity of the membrane.

10. There are two main types of proteins integral (trans membrane) and peripheral proteins. Integral proteins may pass once or several times through the membrane. Integral proteins mainly participate in the transportation of molecules and ions. Peripheral proteins help in signal transduction, adhesion, and interaction with ECM.

11. Plasma membrane contains many pores of 7Å diam. through which smaller substances can pass easily. On the basis of permeability it may be impermeable allowing the transport of only gasses, semi permeable - allowing the passage of only solutes, selectively permeable – allowing the passage of selective ions or molecules and dialyzing – allowing the passage of different molecules in response to high hydrostatic pressure. Passive transport involves transport of substances through the membrane from higher to lower concentration without the help of any transport system or energy.

12. Osmosis is the transport of water molecules across the selectively permeable membrane till the attainment of equilibrium. Water may enter into the cell or into the extra cellular medium from the cell that is referred as endosmosis and exosmosis respectively. In plant cell due to excessive exosmosis the cytoplasm shrinks, it is called plasmolysis. Such cells if placed in hypotonic solution then water reenters into the cytoplasm and the cell again attains the normal form, this condition is called deplasmolysis.

13. In osmosis, to denote the solute concentrations between two solutions some terminologies are used: Isotonic - Both solutions have equal solute concentration; Hypertonic: The solution with higher concentration of solutes and Hypotonic: The solution with the lower solute concentration.

14. During osmosis, translocation of water generates osmotic pressure and counter pressure exerted by the compartment is called hydrostatic pressure.

15. Facilitated diffusion involves transport of ions and small hydrophilic molecules through proteins down its concentration gradient. The membrane proteins form water-filled channels that can be opened or closed in response to ligand binding, or mechanical stress or changes in voltage.

16. Active transport is the pumping of molecules or ions through the membrane against their concentration gradient with the help of a protein and energy derived from the hydrolysis of ATP.

17. Symport is the driving of ion (Na^+) and the molecule through the membrane protein in the same direction in contrary to the transport in opposite direction in antiport pump.

18. Endocytosis involves engulfment of solid or liquid substances through the plasma membrane with active involvement of cytoskeleton. Exocytosis involves expulsion of secretory substances to the outside.

19. Plasma membrane proteins play important role in signal transduction, hormone reception and transmission of its impulse inside the cell, antigen-antibody recognition, impulse transmission and chemoreception.

QUESTIONS

Long type questions

1. Describe different models explaining the molecular structures of plasma membrane.
2. Describe Unit membrane concept and Fluid mosaic model explaining the molecular structure of plasma membrane. What are the basic differences between these two models?
3. Describe the types, structures and functions of rafts.
4. Describe the main lipid fractions and their roles in different physiological processes.
5. Describe the main protein fractions and their roles in various physiological processes.
6. Describe the role of plasma membrane in the transport of different molecules and ions.
7. Describe the main functions of Plasma membrane.
8. Prove that membrane leaflets are biochemically asymmetrical.
9. Describe the structures associated with intercellular communication.

Short type questions

1. What the basic differences between Unit membrane concept and Fluid mosaic model.
2. What are the main constituents of the lipid bilayer?
3. What do you mean by peripheral and integral proteins? Give examples.
4. What do you mean by flip-flop and flex movements?
5. What is a 'leaflet'?
6. What do you mean by asymmetry of the leaflets? Explain with examples.
7. What are the main constituents of rafts?
8. What do you mean by 'membrane skeleton'?
9. What do you mean by semi permeable and selectively permeable plasma membrane?
10. What are the basic differences between passive and active transport?
11. What is Fick's law?
12. What do you know about exosmosis and endosmosis?
13. What happens when RBC is immersed in hypo and hypertonic solutions?
14. What are osmotic and hydrostatic pressures?
15. What is facilitated diffusion? Which molecules cross the membrane by this process?
16. What do you mean by ligand gated, mechanically gated and voltage gated ion channels?
17. What are direct and indirect active transports?
18. How Na^+/K^+ ATPase regulates the balance of Na^+ and K^+ in the cell?
19. What are the main physiological roles of H^+/K^+ ATPase?
20. What do you mean by uniport, symport and antiport?
21. What are basic differences between phagocytosis and pinocytosis?
22. What do you mean by exocytosis and endocytosis?
23. What do you mean by tight junctions and desmosomes?
24. What are micovilli and glycocalyx?
25. What do you mean by lateral diffusion and flexion of the lipid molecules?

Fill up the blanks

1. Unit membrane concept was advocated by Robertson in ———.
2. Fluid mosaic model was proposed by Singer and Nicolson in ———.
3. The outer ends of lipid molecules are ——— in nature.
4. The hydrophilic ends of the lipid molecules are held by ——— force.
5. A membrane less organelle is ———.
6. Each half of the lipid bilayer is called ———.
7. The most common single pass protein is ———.
8. Band-3 protein in the membrane of human RBC is an example of ———.
9. Porin is a characteristic transmembrane protein in the membrane of ———.
10. Annexin is a common ——— protein.
11. The characteristic protein of membrane skeleton is ———.
12. impermeable plasma membrane is noted in ———.
13. Excess exosmosis from plant cell leads to ———.
14. Hemidesmosomes attach the epithelial cells to the

15. Rate of diffusion is ——— proportional to the density of the medium.

MCQs

1. **Unit membrane concept was published in**
(a) 1957 (b) 1958 (c) 1959.
2. **Fluid mosaic model was proposed in**
(a) 1971, (b) 1972, (c) 1973.
3. **For producing red cell ghost RBC are treated with**
(a) isotonic (b) hypotonic (c) hypertonic solution.
4. **Thickness of unit membrane is**
(a) 75Å, (b) 70Å and (c) 80Å·
5. **Exchange of lipid molecules between the leaflets is called**
(a) flexion, (b) flip-flop movement, (c) flexion.
6. **Lipid molecules are joined with protein molecules by**
(a) ionic linkage, (b) hydrogen bonds, (c) van der Wall force.
7. **The main function of GEM is**
(a) adhesion, (b) Cholesterol trafficking, (c) signaling.
8. **Flotilin is the characteristic protein of**
(a) caveolae, (b) raft, (c) Gem.
9. **Rafts are rich in**
(a) cholesterol, (b) phospholipid (c) protein.
10. **The bondage between the integral proteins and lipid bilayer is done by**
(a) van der Wall force,
(b) hydrogen bonds,
(c) ionic linkage.
11. **Osmosis is basically referred to the transport of**
(a) ions, (b) solutes, (c) molecules.
12. **The pressure generated due to the passage of solutes during osmosis is called**
(a) hydrostatic (b) osmotic, (c) diffusion pressure.
13. **The balance of Na^+ and K^+ in the cell is maintained by**
(a) facilitated diffusion,
(b) active transport,
(c) osmosis.
14. **Exocytosis involves the incorporation of**
(a) Ca^{++}, (b) Na^+, (c) Cl^-
15. **Unit membrane is absent in**
(a) lysosome, (b) bacterial cell, (c) ribosome.
16. **Vomiting of cell is called**
(a) exocytosis, (b) emeiocytosis, (c) exosmosis.
17. **Plasmalemma are fused at**
(a) tight junctions,
(b) desmosomes,
(c) middle lamella.
18. (a) desmosome,
(b) middle lamella,
(c) plasmodesmata **is not a cell junction.**
19. **Desmosomes areconcerned with**
(a) cell adherens, (b) cell division, excretion.
20. **Glycocalyx serves**
(a) insulation, (b) protection, (c) both.
21. **Fluidity of the membrane decreases at**
(a) 0°, (b) –1°, (c) 50°(C)
22. **Unit membrane concept and Fluid mosaic model differs in types of**
(a) lipids, (b) proteins, (c) arrangement of proteins.
23. **An enzyme carrier protein associated with membrane is**
(a) catalase, (b) permease, (c) lipase.
24. (a) Sarcolemma, (b) Myelin sheath, (c) middle lamella **is composed of lipiprotein.**
25. **Lipids and protein contents in all membranes are in the ratio of**
(a) 40:58, (b) 60:40, 50:48.
26. **The main amino acids in the cell membrane are**
(a) arginine and lysine,
(b) arginine and leucine,
(c) arginine and glycine.
27. **In hypertonic solution**
(a) water,
(b) solute,
(c) both water and solute exit from the cell.
28. **If a hypertonic (A) and a hypotonic (B) solutions reaches equilibrium,**
(a) water becomes higher in B,
(b) water becomes higher in A,
(c) water content in both A and B remain same.
29. **Glycoproteins and glycolipids in cell membrane serves**
(a) cell adhesion, (b) cell-cell recognition, (c) fusion.
30. **All membrane processes are carried out by**
(a) lipids, carbohydrates, (c) proteins.
31. **A cell increases its volume if placed in**
(a) isotonic, (b) hypertonic, (c) hypotonic medium.
32. **Membrane proteins are responsible for**
(a) masaicism, (b) fluidity, (c) cell-cell recognition.

33. **In Na+-K+ pump**
(a) 3 Na^+ are pumped out and $2K^+$ are pumped in,
(b) $3Na^+$ are pumped in and $2K^+$ are pumped out,
(c) $3Na^+$ and $2K^+$ both are pumped out.

34. **Cell membrane shows maximum permeability to**
(a) K^+, (b) Na^+, (c) Ca^+.

35. **A plant cell placed in hypotonic medium becomes**
(a) plasmolysed, (b) flacid, (c) turgid.

36. **Ions diffuse across the membranes down their**
(a) electrochemical,
(b) electrogenic,
(c) concentration gradient.

37. **The proportion of intrinsic and extrinsic proteins in the membrane is**
(a) 30:70, (b) 70:30, (c) 50:50.

38. **Membranes function as a locus of biochemical reactions as**
(a) lipids catalyze various reactions,
(b) they hold enzymes of reactions in close proximity,
(c) separate reactant from the products of enzymatic reactions.

39. **In facilitated diffusion, transportation occurs**
(a) against their concentration gradient,
(b) through a channel,
(c) by a membrane protein that undergoes conformational changes.

40. **Plasmolysis occurs when a cell is kept in**
(a) hypotonic, (b) hypertonic, (c) isotonic solution.

41. **The main difference of active transport (AT) and facilitated diffusion (FD) is**
(a) AT needs energy,
(b) AT involved tranpost proteins,
(c) FD needs energy.

42. **The Na^+/K^+ pump transports**
(a) $3Na^+$ into and $2K^+$ out of the cell,
(b) $2Na^+$ into and $3K^+$ out of the cell,
(c) $3Na^+$ out and $2K^+$ into the cell.

43. **The structural lipid of cell membrane is**
(a) Steroids, (b) Sphyngolipids, (c) Phospholipids.

44. **Plasma membrane shows**
(a) Protein-phospholipid.
b Phospholipid-extrinsic protein-Intrinsic proteins.
(c) Lipid- hecelluloses.

45. **Lipid molecules in the cell membrane are arranged in**
(a) Alternately, (b) Scattered way, (c) Parallel way.

46. **Cell membrane is not permeable to**
(a) Glucose, (b) Glucose-6-Ph.(c) Glutamic acid.

47. **Cell membrane is selectively maximum permeable to**
(a) Na^{++}, (b) Ca^{++}, (c) K^+

Answers of MCQs:

1. c, 2. b, 3. b, 4. a, 5. b, 6. a, 7. c, 8. b, 9. a, 10. a, 11. b 12. b 13. b, 14. a, 15. c, 16. b, 17. a, 18. c, 19. a, 20. c, 21. a, 22. c, 23. b, 24. b, 25. a, 26. a, 27. a, 28. b, 29. b, 30. c, 31. c, 32. a, 33. a., 34. b, 35. c, 36. a, 37. b, 38. b, 39. c, 40. b, 41. a, 42. c, 43. c. 44. b. 45. c, 46. b, 47.c.

Chapter 3

Cell Wall

Plant cells encase themselves within a complex polysaccharide cell wall, which constitutes the raw materials of textile, paper and other industries. The plant cell wall is also the primary source of cellulose. The cell wall not only strengthens the plant body, but also has key roles in plant growth, cell differentiation, intercellular communication, water movement and defense mechanisms. Except some lower plants and gametes cell wall is of universal occurrence in plants.

3.1. STRUCTURE OF CELL WALL

The cell wall basically consists of three layers – middle lamella, primary cell wall and secondary cell wall (Fig. 3.1).

Middle lamella: It is laid out during the early phase of telophase in the former equatorial plane of the mitotic apparatus. It is a colloidal plate like structure; highly elastic consists of mainly protein and calcium and magnesium pectates (salts of picric acid). Electron microscopic studies show that it develops by the fusion of numerous vesicles derived from the Golgi complex, containing electron dense material called ***phragmoplast*** on both sides of the equatorial plate.

Primary cell wall: On each side of the middle lamella, secretions of the protoplasm

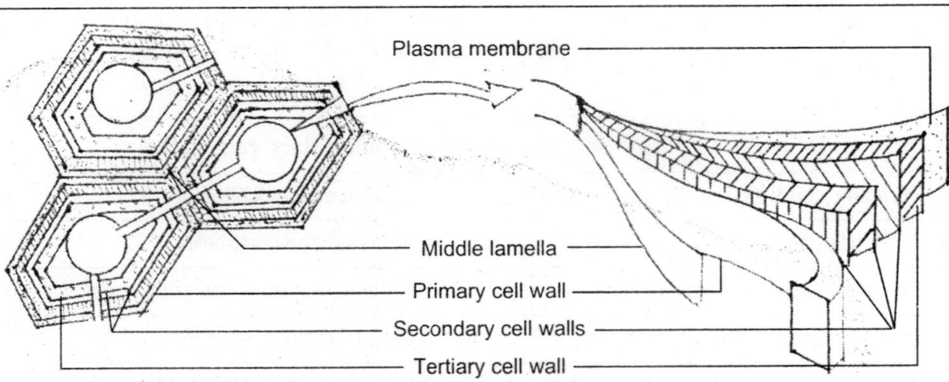

Fig. 3.1 : Structure of cell wall.

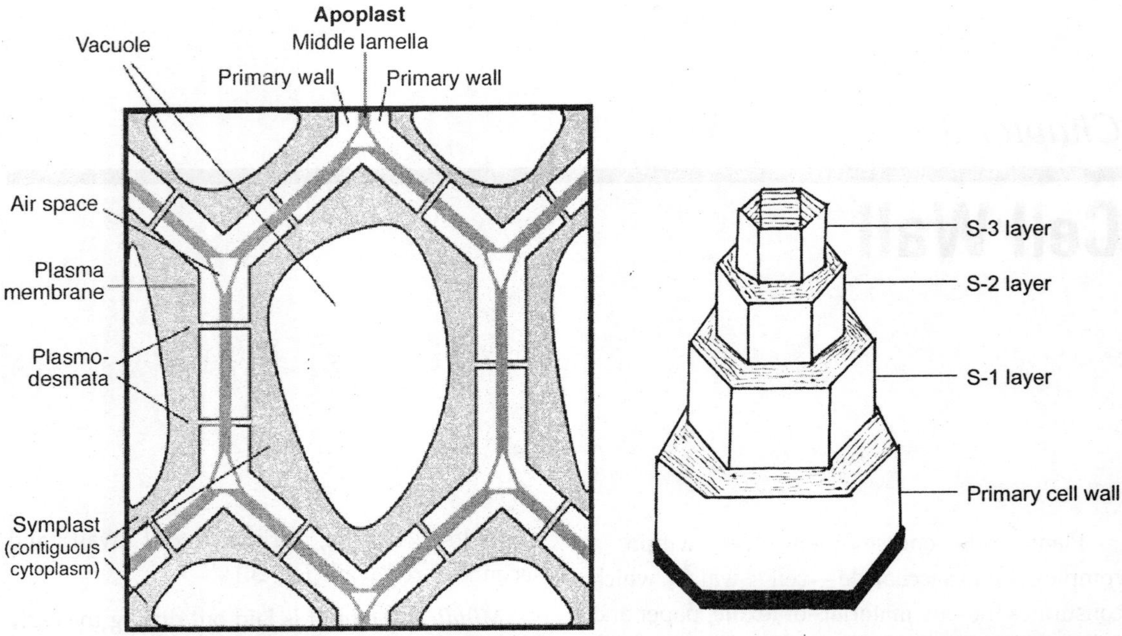

Fig. 3.2 : Different aspects of cell wall (A), position of secondary cell wall sublayers (B).

accumulate at all angles to form primary cell wall. It is 1-3μ thick and composed of cellulose microfibrils that are held together by hydrogen bonds to provide a high tensile strength. Insoluble cellulose fibers are meshed in to a matrix of pectin, lignin and hemicelluloses. They give the plant strength and support. In meristematic cells and cambium it is the only cell wall layer. No cell increases in volume after the formation of this layer.

Secondary cell wall: On the inner side of the primary cell wall cellulose, hemicellulose, lignin, suberin, cutin, tanin, wax, and various salts of calcium and silica are deposited to form 5-10μ thick layer called secondary cell wall. Unlike the primary wall, the microfibrils are aligned mostly in the same direction and with each additional layer the orientation changes slightly. This layer shows external (S1) internal (S3) and middle (S2) sublayers (Fig. 3.2). This layer is usually noted in the wall of vascular bundles and dead cells.

Table 3.1 : Comparison of primary and secondary cell wall

Primary cell wall	Secondary cell wall
1. It occurs in all plant cells.	1. It occurs in matured plant cells.
2. It lies at the inner part of the middle lamella.	2. It lies at the inner part of the primary cell wall.
3. It is elastic, single layered, 1-3μ in thickness.	3. It is rigid, 3 or more layered, 5-10μ thick.
4. Consists of cellulose (5%), Hemicellulose ((50%), Protein(5%) and Lipid (5-10%).	4. Consists of cellulose (50-90%), Hemicellulose (25%), Protein and Lipid as trace amount.
5. Cellulose molecules are loosely arranged.	5. Cellulose molecules are firmly arranged.

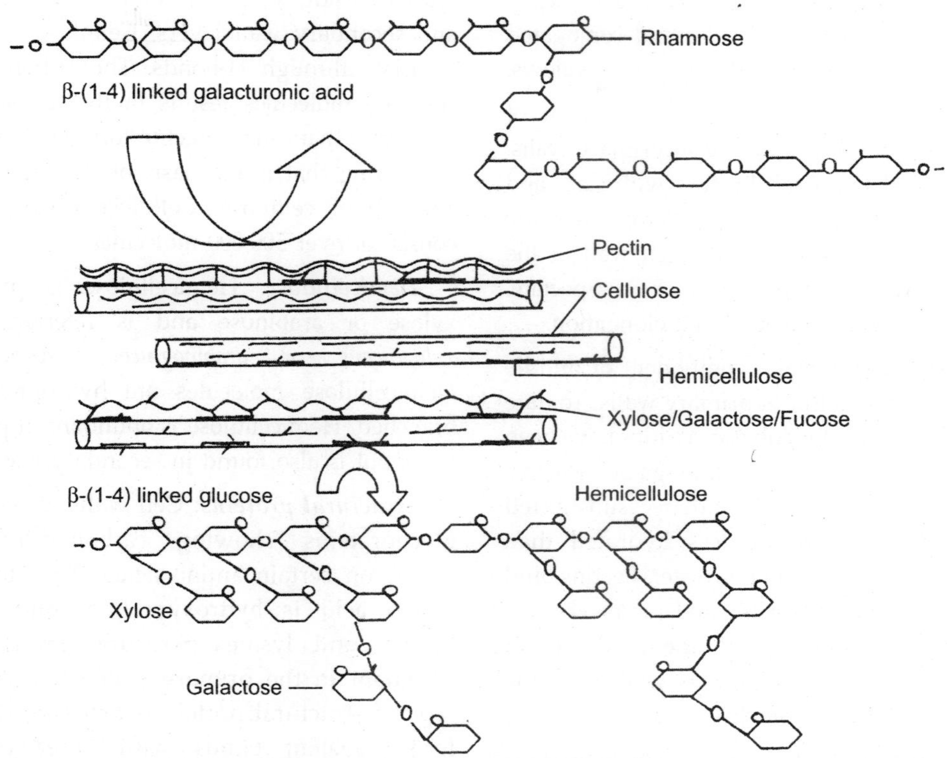

Fig. 3.3 : Ultra structure of cell wall.

Tertiary cell wall: On the innerside of the secondary cell wall xylan is deposited to form tertiary cell wall.

Ultra structure of cell wall:Primary and secondary cell walls mainly consist of microfibrils of *cellulose* (diam. 250Å), aligned in parallel or vertical rows having regular or irregular arrangements. In between the microfibrils, pectin, lignin, hemicellulose, cutin, suberin, callose, chitin, tannin, mucilage etc. are deposited. Each cellulose microfibril consists of 20 *micelle or elementary fibrils* (diam. 100Å). Each micelle again shows 100-170 *sub fibrils* that are composed of glucose molecules (Fig. 3.3). The secondary cell wall materials are deposited over the primary wall asymmetrically as circular rings, spiral thread like structure, network like structure, in staircase like fashion and as pits. Primary walls also contain lesser amounts of structural glycoproteins (hydroxyproline-rich extensins), phenolic esters (ferulic and coumaric acids), ionically and covalently bound minerals (e.g. Ca and B) and enzymes. In addition, primary wall contains hydroxyproline rich lycoproteins called extensins, the arabinogalactan proteins, the glycine-rich proteins, and the proline-rich proteins. Except glycine-rich proteins, all the previously mentioned proteins are glycosylated and contain hydroxyproline. Wall proteins are believed to have a role in regulating wall expansion. Lignin, a macromolecule composed of highly cross-linked phenolic molecules, is a major component of secondary walls. The relative proportion of carbohydrates and proteins vary between plants, cell type and age.

Chemical structure of cell wall: Chemically cell wall shows polysaccharides and conjugated proteins. Its different constituents are as follows:

Pectic acid: It is the major component of middle lamella but also found in primary walls. They form salts and salt bridges with Ca^{++} and Mg^{++} that are insoluble gels. Peroxidases are reported to be localized in the cell wall of the seedlings having a great affinity for Ca^{++}-pectate. They seem to be involved in cell elongation.

Pectin: It is the major component of middle lamella but also found in primary walls. To date three classes of pectic polysaccharides have been characterized. Homogalacturonans, rhamnogalacturonans and substituted galacturonans. These are less hydrated than pectic acid but soluble in hot water. Pectins bind to cellulose microfibrils and may be of considerable significance in the modeling of primary cell walls of plants as well as in the process of cell wall assembly.

Cellulose: It is the polymer of glucose – typically consisting of 1,000 to 10,000 1-4-b-D-glucose residues - major component of primary and secondary wall layers. Cellulose polymers associate through H-bonds. The H-bonding of cellulose molecules results in the formation of *microfibers* that can interact to form *fibers*. It is the H-bonding that is the basis of the high tensile strength of cellulose. Cellulose fibers usually consist of over 500,000 molecules.

Hemicellulose: Hemicellulose is primarily xylose or arabinose and is referred to as *xyloglucans* or *arabinoglucans,* respectively. Hemicellulose molecules are hydrophilic and branched. Hemicellulose is abundant in primary walls but is also found in secondary walls.

Structural proteins: Cell walls also contain glycoproteins showing carbohydrate side chains on certain amino acids. The abundant amino acid is hydroxyproline followed by proline and lysine.. Strucural proteins are abundant in the primary wall layer. Another type of structural protein is ***extensin,*** that can form covalent bonds with other extensin proteins through tyrosine.

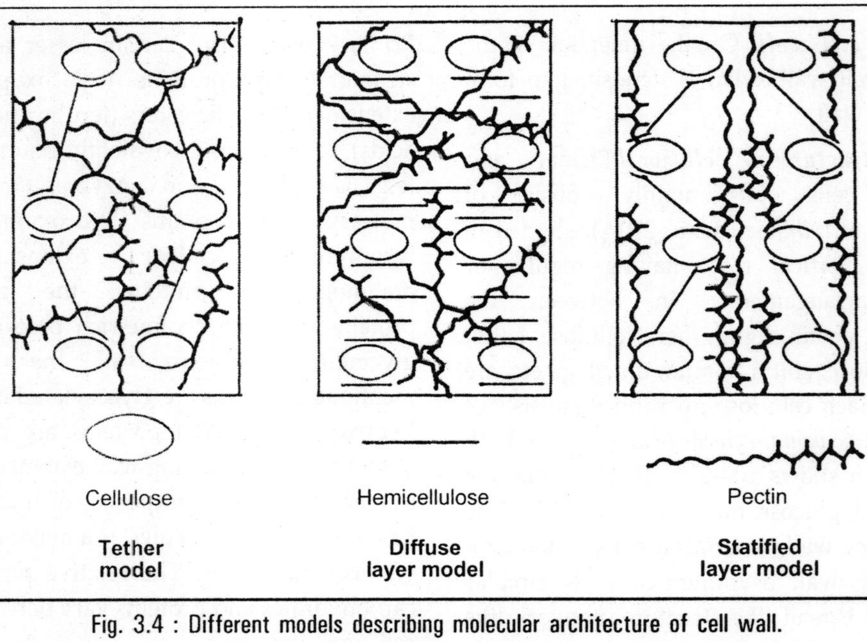

Fig. 3.4 : Different models describing molecular architecture of cell wall.

3.2. MODELS EXPLAINING THE ULTRASTRUCTURE OF CELL WALL

Regarding the organization and interactions of wall components several models have been proposed to account for the mechanical properties of the wall.

Tether model: Xyloglucan (hemicellulose) molecules are held by hydrogen bonds and cross-link with cellulose microfibrils. The cellulose-xyloglucan network is emeshed in a non-covalently cross-linked pectic net work.

Diffuse layer model: Xyloglucan molecules are hydrogen bonded to the surface of cellulose microfibrils but do not directly cross-link them. A layer of polysaccharides surrounds the tightly bound xyloglucan. The cellulose and xyloglucan are embedded in a pectic matrix.

Stratified layer model: Xyloglucan molecules are hydrogen bonded to and cross-link cellulose microfibrils. Strata of pectic polysaccharides separate the cellulose-xyloglucan lamellae. Much research is still required to provide a complete description of the molecular structure of primary wall. Moreover, primary walls are dynamic structures whose composition and architecture change during plant growth and development.

Sticky network model: The cell wall of growing plant tissues show inextensible cellulose microfibrils 'tethered' by hemicellulose polymers attached to the microfibrils by hydrogen bonds. Cell wall growth occurs when tethers are broken or 'peeled' off the microfibril surface by expansion.

Cell expansion involves loosening of existing wall materials and production of new material. Cell wall loosening can occur by at least 3 mechanisms:

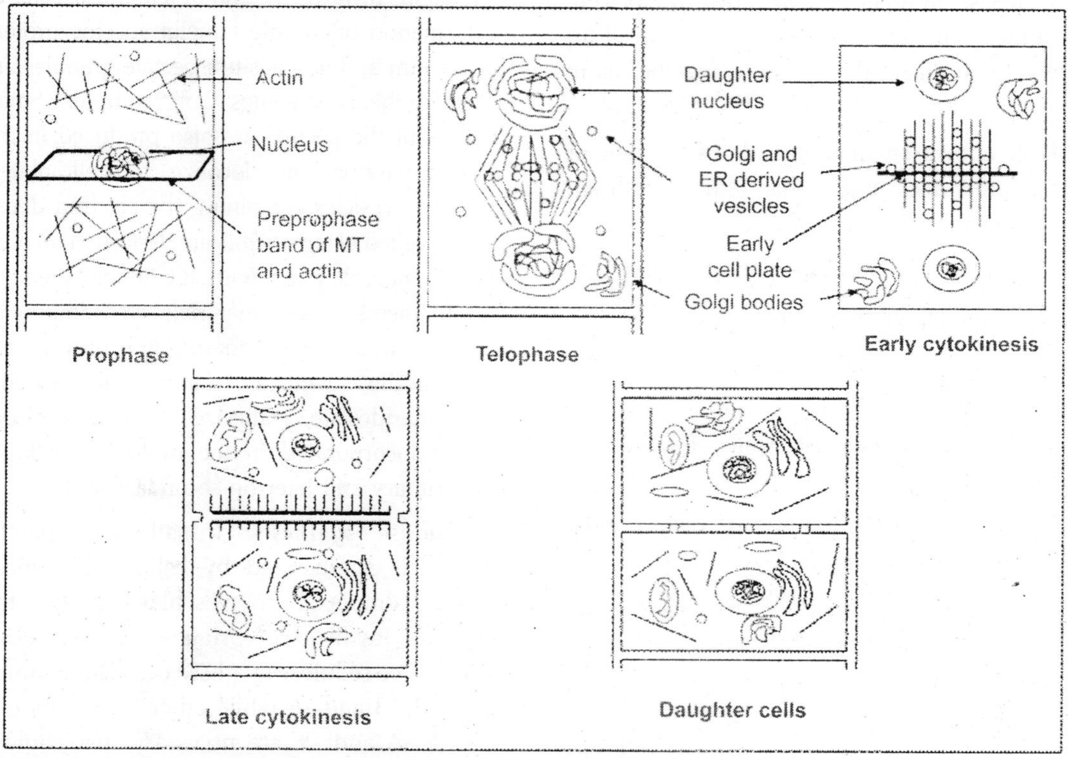

Fig. 3.5 : Secretion and deposition of wall forming substances.

1) *Wall acidification* - H^+ATPase in plasma membrane 'pumps" H^+ from cytoplasm into cell wall. The pH of the wall drops and carboxylic acids become protonated and 'salt bridges" are broken.
2) In addition, the enzyme "expansin" is activated and causes cellulose microfibrils to slip (mechanism of expansin action is unknown) resulting in cell wall "creep".
3) Hydrolytic enzymes like cellulase and pectinase, "degrade" cell walls by breaking polymers into smaller subunits or by breaking cross-links.

3.3. FUNCTIONS OF THE CELL WALL

1. Provide structural and mechanical support and a chemically buffered environment.

2. Maintain and determine cell shape.

3. Resist internal turgor pressure of cell. When the vacuoles within the cell wall swell up with water, it makes the cell walls together more stiff, improving the overall rigidity of the cell.

4. It allows circulation and distribution of water, minerals, and other small nutrient molecules into and out of the cell.

5. Control rate and direction of growth.

6. Maintains plant architecture and form.

7. Regulate diffusion of material through the apoplast.

8. Carbohydrate storage walls of seeds may be metabolized.

9. Protect against pathogens, dehydration, and other environmental factors.

10. Source of biologically active signaling molecules and cell-cell interactions.

11. Primary walls are the major textural component of plant-derived foods.

12. The ripening of fruits and vegetables is associated with changes in wall structure and composition.

13. Plant-derived beverages often contain significant amounts of wall polysaccharides.

14. Some wall polysaccharides bind heavy metals, stimulate the immune system or regulate serum cholesterol.

15. Wall polysaccharides are used commercially as gums gels, and stabilizers.

16. Thus, cell wall structure and organization is of interest to the plant scientist, the food processing industry and the nutritionist.

17. Proteins of plant cell walls serve as structural macromolecules and play important roles in morphogenesis; development and can play an active role in defense as a source of peptide signals for systemic wound signaling.

3.4. FORMATION OF CELL WALL

It begins towards the end of telophase. Prior to the formation of middle lamella specific protein and calcium and magnesium pectate granules are formed by the dictyosomes (Golgi complex). Small amounts of the pectins are also produced in the Golgi apparatus and localize to cell plates. Numerous vesicles containing the electron dense material called *phragmoplasts* are deposited on both sides of the cell plate with the involvement of actin filaments. These materials cross-link with each other in a colloidal matrix and form a plate like middle lamella. Later on specific polysaccharides are synthesized in the protoplasm and are incorporated in the middle lamella to form primary and later on secondary wall.

Cellulose synthesis: In all land plants, cellulose is synthesized by cellulose synthase (CESA) complex. In vascular plants the complexes involved in primary wall synthesis contain three cellulose synthase catalytic subunits – CESA-1, -3 and –6 while other three subunits *CESA 4, -7,* and *-8,* are necessary for cellulose deposition.

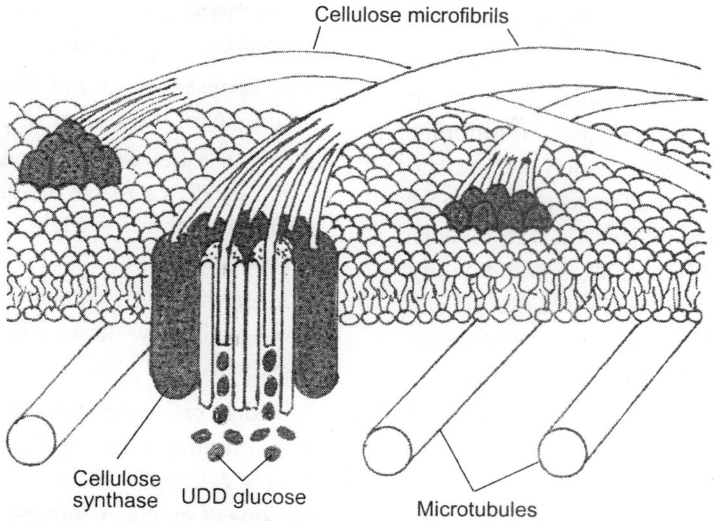

Fig. 3.6 : Disposition of cellulÅse microfibrils and their synthesis from cellulose synthase.

During secondary cell wall formation. The wall fragments are synthesized and assembled within the cisternae of the Golgi complexes where they appear inactive but as they move to the plasma membrane along the MTs, they become activated and are translocated in different directions that coincide with the orientation of microtubules. Non-lignified fibre cells develop a very thick S2 layer within the secondary cell wall, which is characterized by

(i) cellulose microfibrils largely parallel to the longitudinal axis of the cell, and

(ii) a high proportion of galactose-containing polymers among the non-cellulosic polysaccharides. Galactans accumulate in specialized Golgi-derived vesicles during fibre cell-wall thickening. The secretion of galactans plays a role in the orientation of cellulose microfibrils, and secondly the assembly of S2 layer. Plants must have evolved regulatory mechanisms controlling biosynthesis, targeted secretion and assembly of wall components to achieve the heterogeneity in cell walls. A number of factors, including hormones, the cytoskeleton, glycosylphosphatidyl inositol-anchored proteins, phosphoinositides, and sugar nucleotide supply have been implicated in the regulation of cell wall biosynthesis or deposition.

3.5. OTHER CELL WALLS

1. **Fungal cell wall:** Most true fungi have a cell wall consisting largely of chitin and other polysaccharides. True fungi do not have cellulose in their cell walls. (Webster and Weber, 2007). The fungal cell wall is a matrix of three main components: chitin / chitosan, glucans and proteins.

2. **Algal cell walls:** Like plants, algae have cell walls. Algal cell walls contain either polysaccharides (cellulose or a variety of glycoproteins or both. In addition, mamman, xylans, alginic acid, sulfonated polysaccharides (agarose, carrageenan, porphyran, furcelleran and funoran) occur in many algae. Other compounds that may accumulate in algal cell walls include sporopollenin and calcium ions. The group of algae - the diatoms synthesize their cell walls from silicic acid. In brown algae,

phlorotannins may be a constituent of the cell walls.

3. **Cell walls of Water molds:** The group Oomycetes (= water molds) are saprotrophic plant pathogens. They were widely believed to be fungi, but structural and molecular evidence has led to their reclassification as heterokonts, related to autotrophic brown algae and diatoms. They typically possess cell walls of cellulose and glucans rather than chitin, although some genera E.g. *Achlya* and *Saprolegnia*) do have chitin in their walls. Oomycetes cell walls also contain the amino acid hydroxyproline, which is not found in fungal cell walls.

4. **Cell walls of Slime molds:** The dictyostelids are slime molds formerly classified among the fungi. They show a reproductive stalk and sporangium under certain conditions. The reproductive stalk, as well as the spores possess a cellulose wall. The spore wall has been shown to possess three layers, the middle of which is composed of cellulose, and the innermost is sensitive to cellulase and pronase.

5. **Bacterial cell walls:** Outside the cell membrane of the bacteria is the cell wall. These are made up of peptidoglycan (also called murein), which is made from polysaccharide chains cross-linked with peptides containing D-amino acids. Bacterial cell walls are different from the cell walls of plants, fungi and archaea which are made up of cellulose, chitin and lack of peptidoglycan respectively. There are two different types of cell wall in bacteria, - Gram (+) and Gram (-). Gram (+) bacteria possess a thick cell wall containing many layers of peptidoglycan and teichoic acids while Gram (-) bacteria have a relatively thin cell wall consisting of a few layers of peptidoglycan surrounded by a second lipid membrane containing lipopolysaccharides and lipoproteins.

6. **Archaeal cell walls:** There are four types of cell wall currently known among the Archaea.

a) In some methanogens (*Methanobacterium* and *Methanothermus)* cell wall consists of pseudopeptidoglycan.

b) In *Methanosarcina* and *Halococcus* cell wall consists of thick layer of polysaccharides.

c) In *Halobacterium*, and some methanogens cell wall consists of glycoprotein, the proteins show a high content of acidic amino acids.

d) In *Methanomicrobium* and *Desulfurococcus*, cell wall may show only of surface-layer proteins

7. **Other cell coverings:** Many protists and bacteria produce other cell surface structures apart from cell walls. Many algae have a sheath or envelope of mucilage outside the cell made up of exopolysaccharides. Diatoms build a frustule from silica extracted from the surrounding water. Many green algae (e.g. *Halimeda* and the Dasycladales) and some red algae, the Corallinales, encase their cells in a secreted skeleton of $CaCO_3$. In each case, the wall is rigid and essentially inorganic.

SUMMARY

1. The polysaccharide walls secreted on the inner sides of plasma membrane surrounding the protoplasm is called cell wall. It consists of three main layers – middle lamella, primary cell wall and secondary cell wall. The main chemical constituents of the primary and secondary walls are cellu lose, hemicellulose etc. In addition, cell wall contains proteins rich in proline, hydroxypro line and lysine. The cells that have hard walls contains a specific protein called extensin.
2. Cell wall provide structural and mechanical support, determine cell shape, resist internal turgor pressure, regulate growth, intercellular transport, signal transduction etc.
3. Cell wall components, after synthesis, are stored in the ER and later on are biochemically altered in the Golgi complex and as electron dense phragmoplasts accumulate in the interface of the cell to form cell plate or middle lamella. The secretions gradually accumulate on each side of the middle lamella to give rise primary and secondary cell wall.

4. Some algae are devoid of cell wall. Cell wall in algae may contain glycoproteins, an extensin like protein, silica, calcium etc.
5. Cell wall in fungi shows cellulose, glucosamine, and chitin providing rigidity and strength of the cell preventing osmotic lysis. It also limits the entry of toxic molecules.
6. The composition of cell wall differs between Gram (+) and Gram (-) bacteria. In Gram (+) bacteria a thick peptidoglycan layer follows the plasma membrane while in Gram (-) bacteria; there is thin peptidoglycan layer in between two lipo-protein membranes. Additionally another layer of lipo-polysaccharide occurs outside the outer membrane.

QUESTIONS

Long type questions:

1. Describe the structure and function of cell wall.
2. Describe different models regarding the molecular architecture of cell.
3. Give an account of the structure of cell wall in algae.
4. Describe the structure of bacterial cell wall.

Short type questions:

1. What are basic constituents of middle lamella?
2. What do you mean by phragmoplast?
3. In which cells primary cell wall is the only cell wall layer?
4. What are the constituents of secondary cell wall?
5. Give an account of the structure of micelle lamella.
6. Name the proteins present in primary and secondary cell wall layers.
7. Write down the functions of primary cell wall.

Fill up the blanks:

1. Phragmoplast is derived from ———.
2. No cell increase in volume after the formation of ——— cell wall.
3. Secondary cell wall layers are mostly found in ———.
4. The ultimate component of cellulose is ———.
5. ——— is the major component of secondary cell wall.
6. ——— are the only bacteria that lacks cell wall.
7. ——— is not present in the cell wall of Archaebacteria.
8. Peptidoglycan is the major component of Gram ——— bacteria.
9. What is the main role of peptidoglycan in the cell wall of bacteria.

MCQs:

1. **Middle lamella is formed by**
   (a) Cellulose,
   (b) Calcium salts,
   (c) Calcium and magnesium pectates.
2. **Wax is noted in**
   (a) Primary, (b) Secondary, (c) Tertiary cell wall layer.
3. **Xylan is deposited in**
   (a) Primary, (b) Secondary, (c) Tertiary cell wall layer.
4. **Cell wall formation starts towards the end of**
   (a) Metaphase (b) Anaphase, (c) Telophase.
5. **Silicon is the component of the cell wall of**
   (a) Vascular bundle, (b) Algae, (c) Fungi.
6. **Calcium is the main component of the cell wall of**
   (a) Fresh water, (b) Estuarine, (c) marine algae.
7. **Endosperm cells have**
   (a) Only primary cell wall,
   (b) Both primary and secondary cell walls,
   (c) No cell wall.
8. **Cell wall is**
   (a) Permeable, (b) Impermeable, (c) Semi permeable.
9. **During cell wall formation the outermost layer is**
   (a) primary cell wall,
   (b) secondary cell wall, (c) middle lamella.

Answers to MCQs:

1. c, 2. b, 3. c, 4. c, 5. b, 6. c, 7. c, 8. a, 9. a.

Chapter 4

Intercellular Junctions

Multicellular animals are separated from the external environment by a layer of epithelial cells whose integrity is maintained by intercellular junctional complexes. Cells within tissues and organs must be attached to one another and anchored to the components of the extra cellular matrix (ECM). Plasma membrane have developed several types of junctional complexes to serve these functions, and in each case, anchoring proteins extend through the plasma membrane to link cytoskeletal proteins of one cell to cytoskeletal proteins in neighboring cells as well as to proteins in the ECM. These physical connections impart mechanical stability to the tissues. Such interconnecting cells are called *polarized cells*. Moreover the plasma membranes of polarized cells show functional specializations e.g. the pancreatic acinar cells that remain in close contact with the ductules, show two specialized regions – the membrane that lies in close proximity with the ductutes is called *apical membrane* to which bind the secretory vesicle and the membrane of the opposite portion is called *basolateral membrane* which is laden with hormone receptors that trigger the secretion of digestive enzymes. Each region contains unique lipids and proteins that enable to perform specialized functions Four types of intercellular junctions are noted – *tight junctions* (it completely seals the intercellular space); *gap junctions* (through which intercellular communication is restored); *zonula adherens* and *desmosomes* (through which intercellular adhesion is maintained); *hemidesmosomes* and *focal contacts* (maintain adhesion with extra cellular matrix). These junctions show specificities of transmembrane proteins and their association with ECM or cytoskeletal elements.

4.1. OCCURRENCE OF INTERCELLULAR JUNCTIONS

Each tissue adopts definite intercellular junctions in relation with its physiological activities. Columnar epithelial tissues adopt all four types of junctions (Fig. 4.1). Epithelial cells form a barrier between the luminal surface and the underlying tissue with the help of all types of junctional elements e.g. belt desmosomes, tight junctions and adherens junctions encircle the apex of the cells; desmosomes and gap junctions form patch like lateral connections between the cells and hemidesmosomes connect the cells with basal lamina. They not only regulate paracellular

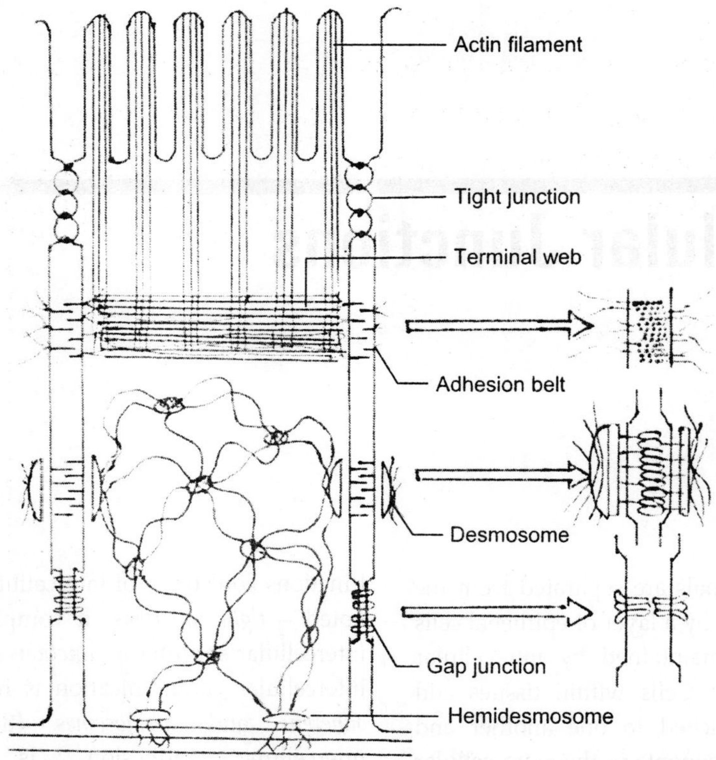

Fig. 4.1 : Occurrence of different Intercellular junctions in two columnar cells.

solute and water trans portation but also integrate gene transcription, tumor suppression, cell proliferation, and cell polarity. In stratified epithelial tissue, cells adhere with each other with the help of desmosomes. Muscle cells are surrounded by basal lamina and are linked together by desmosomes and adherens junctions. Hemidesmosomes and adherens junctions link muscle cells to tendons. Gap junctions join smooth and cardiac muscle cells. They have profound importance in development. The expression of cytokeratin-19 plays a vital role in its differentiation. During initial phases, keeping pace with the permeable nature of the epithelium and meager presence of desmosomes, wide intercellular spaces are formed. Junctional epithelium attaches to the cement surface by hemidesmosomes and basal lamina.

4.2. TIGHT JUNCTIONS (ZONULA OCCLUDENS)

Tight junctions (TJs) ("kissing points") are specialized plasma membrane microdomains (Fig. 4.2) that encircle the apical pole of each epithelial and endothelial cell, separating apical from basolateral side. They form an intercellular diffusion barrier (or gate) regulating the passage of ions, water, and various macromolecules through the paracellular spaces, and a fence restricting the apical/basolateral diffusion of membrane proteins and lipids.

4.2.1. General structure and occurrence

Tight junctions (TJs), are the characteristic structures of one-layered epithelia and of endothelia. These are considered as intramembranous "fences" and as hydrophobic "barriers" between lumina represented by liquid- or gas-filled spaces on one side and the mesenchymal space on the other side. This prevention facilitates the passage of molecules and ions only through the apical membrane by diffusion or active process. It directly distinguishes the physiological features between apical and basolateral portions of the plasma membrane. They have long been thought to be absent from stratified epithelia but presence of TJ proteins (claudins 1 and 4, occludin, cingulin, symplekin) in stratified squamous epithelia, mainly in the uppermost stratum granulosum is reported. Calcium plays a crucial role in the formation and maintenance of mammary TJs.

TJ structures have been detected in the stratum granulosum, interspersed between desmosomes. A tight junction occurs in several points exhibiting alternate grooves and elevations. The grooves represent thin bands showing two rows of protein particles each belonging to adjacent cells. In EM, TJs show a series of contacts exhibiting continuous strands of intramembranous particles forming a branching network. The architecture of these stands determines the tightness of the domain. The strands represent the integral membrane protein called *occludin*. In additions to it a family of eight proteins called *claudin* constitutes the structural proteins of tight junctions. The C-terminal cytoplasmic segment of occludin binds a peripheral protein – ZO-1 that associates with ZO-2 and spectrin. ZO-1 occurs in association with cadherins in the adheren junctions in fibroblasts and cardiac muscle cells. In addition to ZO-1 and 2, cingulin and GTP binding proteins also associate with tight junctions. Claudin-1, occludin and ZO-1 colocalize in the epidermis. Claudin-1 occurs in all cell layers, ZO-1 in the

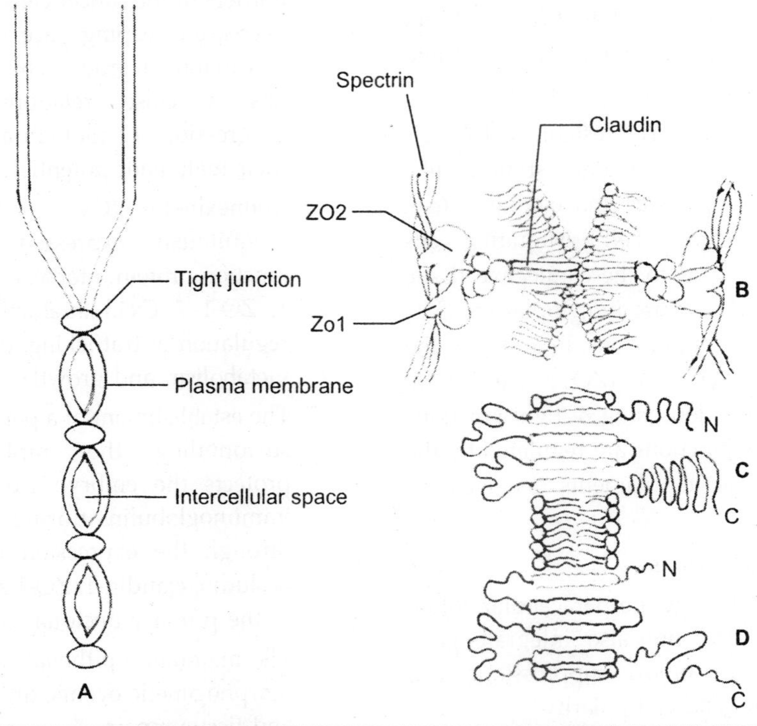

Fig. 4.2 : Diagramatic representation of tight junctions(A) and its molecular structure (B).

uppermost layers, and occludin in the stratum granulosum. Threonine-203 of claudin-1 is required to promote the barrier function of TJ. Na-K-ATPase is reported to be involved in the formation of tight junctions. Apoptosis in epithelial cells is associated with cleavage of TJ proteins – occludin, ZO-1 and ZO-2.

In invertebrates, these are called *septate junctions;* they differ in morphology from TJs and associate with proteins sequence homology to both ZO-1 and ZO-2. Septate junctions have a characteristic ladder-like appearance whereby adjacent cells are connected by septa. These also mediate barrier to paracellular permeability in sea urchin embryos. Low Ca^{++} levels and important signaling molecules involved in the formation and the functions of TJs are protein kinase C, G-proteins, MAP kinases and other kinases. The epidermal growth factor (EGF) is over expressed in a wide spectrum of epithelial cancers. It regulates TJ function through selective and differential regulation of claudins since EGF activation significantly inhibits Claudin-2 but induces increased expression of claudin-1, -3, and -4 that correlates with metastasis in cancer.

The physiological barrier, established by TJs is regulated by extracellular stimuli like hormones (e.g. vasopressin) and cytokines (e.g. tumour necrosis factor – TNF). Migrating cells moving from the blood to the connective tissue induces a localized increase in cytoplasmic Ca^{++} that creates leaky tight junctions. In addition, the bacterial toxin – ZO toxin (ZOT) released by *Vibrio cholerae* causes loosening of tight junctions in diarrhea. Leaky junctions are regulated by the tensions in the actin filaments as well as depletion of ATP.

Role of tight junctions:

1. They not only regulate paracellular solute and water flux but also integrate gene transcription, tumor suppression, cell proliferation, and cell polarity.
2. Several TJ proteins are involved in signal transduction. e.g. , the interaction of TJ protein ZO-1 with the transcription factor ZONAB (ZO-1 associated nucleic acid protein) is important in the regulation of expression of the proto-oncogene Erb-2 as well as epithelial proliferation/ differentiation. The signaling proteins are also reported to cause dissociation of the TJs proteins that are transported to the nucleus and regulate gene expression. These in turn induce epithelial proliferation and differentiation.
3. Different proteins (the small G-proteins of the Rab family) involved in membrane trafficking are associated with TJs. They regulate specific membrane transport events in both endocytic and exocytic pathways.
4. TJs emerge as a platform used to coordinate and regulate cell polarity, and proliferation/ differentiation.
5. Occludin was also found at the apical borders of the cancer cells, their expression decreased keeping pace with the increase in carcinoma grade. Thus the loss of TJs has a close relationship with the progression of endometrial carcinomas and their malignant potential.
6. Connexin-43 (Cx - 43), the most ubiquitously expressed vertebrate gap junction protein interacts directly with ZO-1. ZO-1 / Cx43 interaction plays in the regulation of trafficking, development, cell metabolism and growth.
7. The establishment of a permeability barrier surrounding the implanting embryo protects the embryo from the maternal immunoglobulins during early pregnancy through the expression of TJs proteins, occludin, claudin-1, ZO-1 and ZO-2 in cells of the primary decidual zone.
8. TJs maintain epithelial integrity during morphogenetic events, organ regeneration, and tissue repair.

9. Mutations in the claudin14 and16 genes result in hereditary deafness and hypomagnesaemia, respectively. Some pathogenic bacteria and viruses target and affect the tight-junction function, leading to diseases. *Helicobacter pylori* infection increases paracellular permeability by disrupting the association of claudin-4 and 5 of TJs and may lead to gastroduodenal ulcers and gastric cancer. Disruption of TJs also leads to intestinal hyper permeability (the so-called "leaky gut") causing systemic inflammatory response syndrome (SIRS), inflammatory bowel disease, type 1 diabetes, allergies, asthma, and autism.
9. Different types of signaling proteins and transduction pathways are associated with these junctions. They receive and convert signals to regulate junction assembly and function, and transmit signals to the cell interior to modulate gene expression and cell behaviour.
10. The acquisition of a cancerous phenotype by epithelial cells seems to involve the disruption of tight junctions (TJs) (Gonzalez-Mariscal, *et al* 2007).

4.3. GAP JUNCTIONS

Gap junctions (GJs) are microdomains constructed of assembles of channel proteins called connexins in vertebrates and innexins or pannexins in invertebrates providing direct intercellular communication pathways (Fig. 4.4). The pore-forming subunits bind to auxiliary channel subunits and associate with scaffolding proteins that play essential roles in channel localization and activity allowing rapid exchange of ions and metabolites up to approximately 1 kD in size. Except nerve and skeletal muscle cells, all metazoan cells exhibit GJs. These are the avenues through which intercellular cytoplasmic continuity is maintained. In vertebrates these are commonly expressed in epithelial cells of the skin, various glands, gastro intestinal tracts, urinary tract, muscles, bones, neurons etc.

In Gjs scaffolding proteins link signaling enzymes, substrates, and potential effectors (such as channels) into multiprotein signaling complexes that may be anchored to the cytoskeleton. In fact, GJs are actively involved in virtually all aspects of the cellular life cycle, ranging from cell growth to cell death, transcriptional and cytoskeletal regulation and even in carcinogenesis (Herve *et al*, 2007).

4.3.1. Occurrence and General structure of Gap junctions

Gap junction plaques contain a few to thousands of channels. Half channels in each membrane are called *connexons*, showing 6 protein subunits called connexins that collectively form an aqueous channel across the lipid bilayer. Connexins are named by molecular wt. like connexin-43. In invertebrates, the protein subunits of the connexons are called *innexins*. The connexions of adjacent cells unite firmly with each other. The cylindrical transmembrane channel is 10 nm long with 1.2 nm in diam. Twisting of the connexin subunits along with trans junctional voltage, cytoplasmic H^+ and Ca^{++} concentration and protein kinase regulate the opening and closing of the connexons. Gap junctional communication is conditional depending on the number of channels and the percentage of open channels. Phosphorylation of C-terminal tail of some connexins by cAMP-protein kinase (PKA) also regulates their opening and closing. The channels allow the transport of ions and metabolites up to 1 kD in size.

Usually identical connexons of two adjacent cells unite to form homotypic gap junctional channels through which molecules can pass in both directions but unusually non identical gap junctions can also unite to form heterotypic channels that pass ions in one direction only (e.g. neuronal gap junctions that pass action potential only in one direction). As the metabolic states of

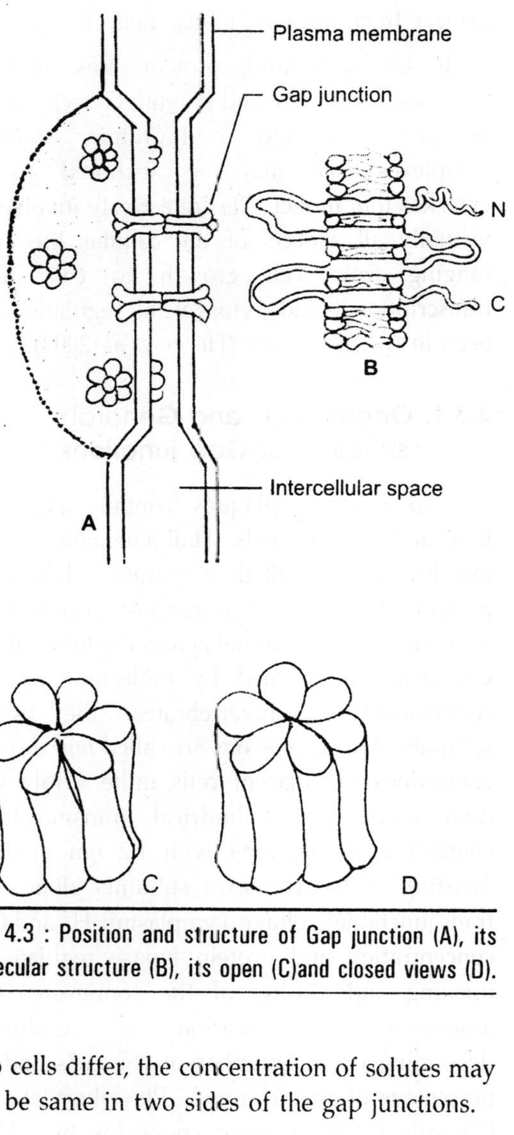

Fig. 4.3 : Position and structure of Gap junction (A), its molecular structure (B), its open (C)and closed views (D).

two cells differ, the concentration of solutes may not be same in two sides of the gap junctions.

4.4. PLASMODESMATA

Plants lack gap junctions but most of them cells maintain cytoplasmic continuity through membrane-lined channels across the cell wall, called plasmodesmata (Fig. 4.5). These are actually minute pores lined by specialized proteins and modified ER. The structure of plasmodesma may be simple or branched. Conversion of simple to branched plasmodesmata is noted in relation to the

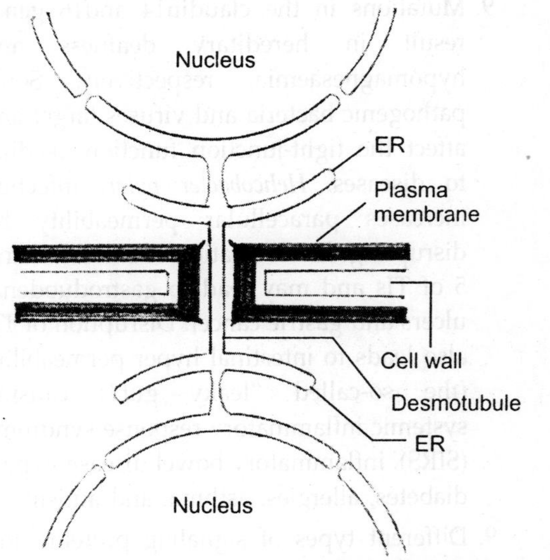

Fig. 4.4 : Relationship of ER, cell wall, plasmodesmata and nucleus.

capacity of macromolecular trafficking in leaf tissues. Dividing cells usually exhibit simple but non-dividing cells exhibit more branched plasmodesmata (Fig. 4.6). These are located in narrow areas of cell walls called primary pit fields and are so dense in these areas (up to one million / sq. mm) that they make up 1% of the entire area of the cell wall.

4.4.1. Structure of plasmodesmata

The outer leaflet of the plasma membrane is continuous with the cell wall while the inner leaflet is continuous with the plasmodesmal pore (Overall and Blackman, 1996). Within the plasmodesmal pore, a tightly wound cylinder of membrane termed the ***desmotubule*** runs the length of the plasmodesma. Thus, desmotubule is essentially a tube within a larger tube bordered by the plasma membrane.

Tilney *et al* (1991) suggested that the desmotubule provides rigid stability to plasmodesmata and confers a fixed diameter and pore size to the plasmodesmal canal, much like a

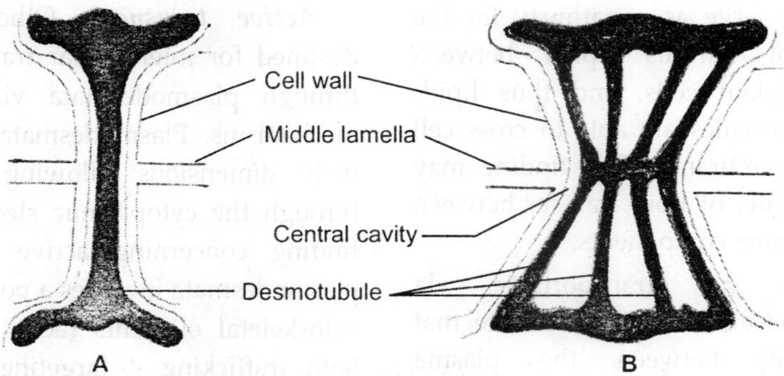

Fig. 4.5 : Diagrammatic view of simple and branched plasmodesmata.

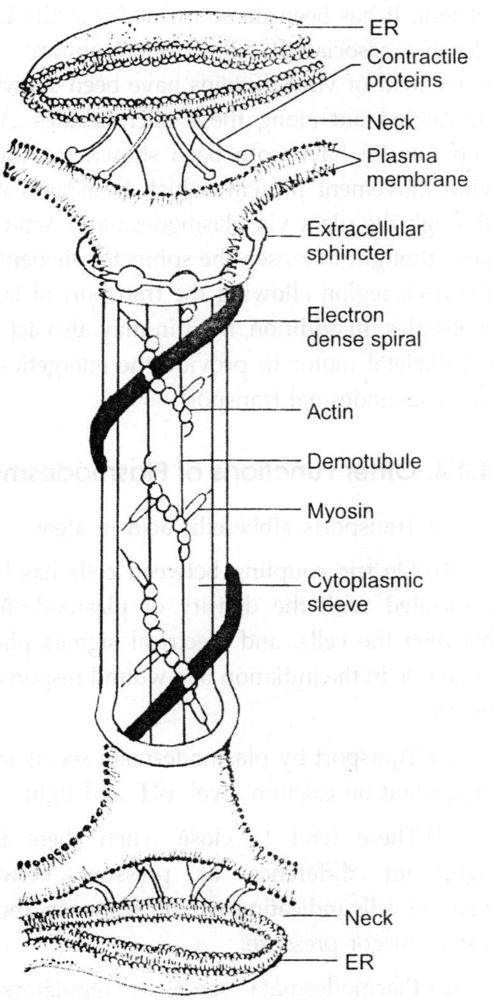

Fig. 4.6 : Structure of plasmodesmata (as per Overall and Blackman model).

cytoskeletal structure. Desmotubule is linked to the ER in each of the adjacent cells, forming a *dynamic endomembrane continuum.* The space between the plasmalemma and the desmotubule is the ***cytoplasmic sleeve*** or ***cytoplasmic annulus*** and transport through plasmodesmata has been proposed to occur either through the lipid portions of the desmotubule or this cytoplasmic sleeve (or both).

In some plasmodesmata, there is a constricted region at each end of the plasmodesmal channel called the ***neck region,*** where the plasma membrane component of the plasmodesmata closely associates with the central desmotubule. The neck region is proposed to contain several, spoke-like protein subunits that are located both extra cellularly and between the desmotubule and the plasmalemma, and these proteins can then act as a sphincter to regulate the passage of materials through the plasmodesmata.

There is an actin-myosin network associated with plasmodesmata. Overall and Blackman (1996) suggest that actin and myosin wrap around the central desmotubule and help to stabilize the desmotubule and consequently the entire plasmodesma (Fig. 4.7).

Grabski *et al* (1993) provided a model of lipid diffusion and transport between contiguous soybean root cells via the ER. This investigation offered two important ideas about a dynamic membrane continuation between plant cells –

i) The ER can serve as a pathway for the transport of various lipids between contiguous plant cells, and thus lipid-signaling molecules are able to cross cell walls and membranes. This finding may lead to a model of lipid transfer between endomem brane components.

ii) Lipids are not transportable via plasmalemma and there is a chance that the linkages between the plasma membrane and plasmodesmata are protein-protein associations not unlike those found in tight junctions in animal cells. This notion leads to the idea that these interactions are inhibiting the leakage of cytoplasmic components around the plasmodesmata.

4.4.2. Signaling and Transport Mechanisms of Plasmodesmata

The average size of mole cules passing through the plasmodesmal annulus is approximately 3 nm. Molecules smaller that 1 kD can diffuse through the plasmodesmata. Exchange of metabolites and transcription factors enable the cells to interact metabolically that is essential during root and shoot organogenesis. Protein trafficking through plasmodesmata is correlated with the physiological state of the cell. Plasmodesmata shows definite receptors and undergoes dynamic changes during development and may participate in long-range communication through the transmission of RNA signals. These are considered as major gatekeepers of signaling molecules that regulate developmental programmes maintaining physiological status, and respond to pathogens.

Passive transport: Specific molecules can pass through the neck region and can be transported passively. Highly sensitive nanotubular intercellular structures facilitate the transfer of membrane vesicles and organelles but impede the flow of small molecules.

Active transport: Other macromolecules destined for intercellular transport are shuttled through plasmodesmata via active transport mechanisms. Plasmodesmata proteins can alter their dimensions allowing larger molecules through the cytoplasmic sleeve. Another major finding concerning active transport through plasmodesmata involves a possible association of cytoskeletal elements (actin and tubulin) with both trafficking / targeting molecules to the plasmodesmata. Actin filaments regulate the pore size and cytoskeletons actually form intracellular networks that play a vital role in this trafficking system. It has been demonstrated that the ER is closely associated with microtubules, and movement of viral proteins have been shown to be carried out along these microtubules. Actin and tubulin have both been shown to bind to viral movement protein, which transports itself through the plant via plasmodesmata. Actin has been thought to loosen the sphincter elements of the neck region allowing the transport of larger molecules. In addition, myosin may also act as a cytoskeletal motor to provide the energetics for this plasmodesmal transport.

4.4.4. Other Functions of Plasmodesmata

(a) Transports gibberellic acid in algae.

(b) Electric coupling between cells has been associated with the density of plasmodesmata between the cells, and electrical signals play a large role in the initiation of a wound response in plants.

(c) Transport by plasmodesmata seems to be dependent on calcium level, pH, and light.

(d) These tend to close when there is a significant difference of pressure between adjacent cells indicating wound response upon a loss of turgor pressure.

(e) Plasmodesmata serve as regulators of plant growth and development and may help to determine a program of cell differentiation, such

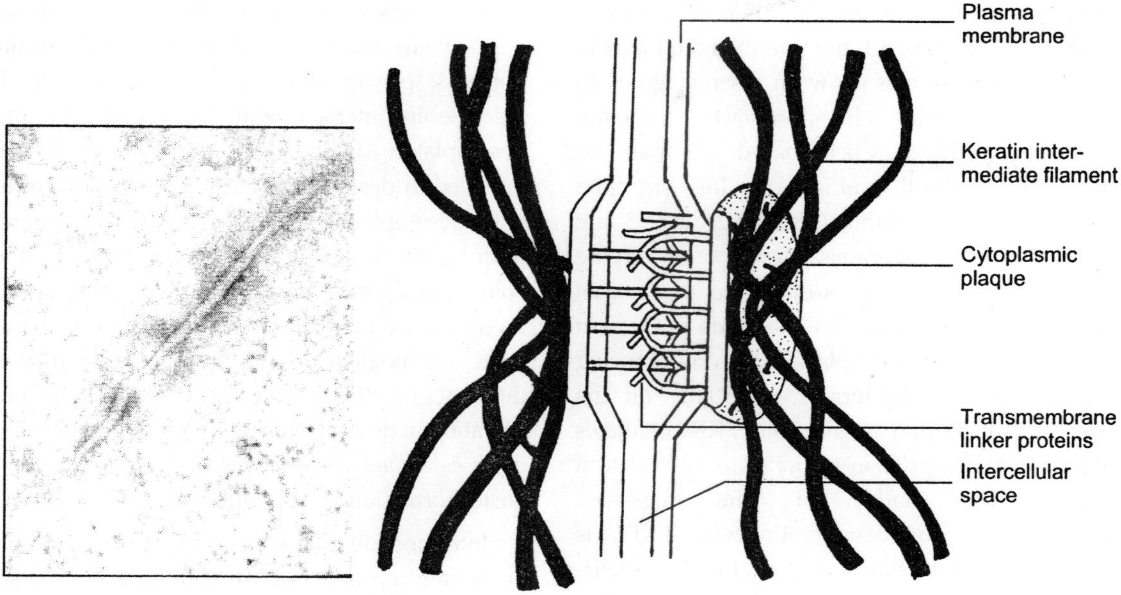

Fig. 4.7 : Diagrammatic structure of Desmosome.

as sealing off root and stem epidermal cells from the rest of the plant. Cell-to-cell communication via plasmodesmata is a very common feature in *Arabidopsis* embryogenesis between embryonic and maternal tissues but it ceases as development proceeds. A novel family of plasmodesmata located proteins (PDLP1) regulates cell-to-cell communication (Thomas *et al*, 2008).

(f) Spreading of RNA silencing in plants occurs through plasmodesmata which is caused from a cell-to-cell movement of a short-range silencing signal, most probably 21-nt siRNAs (Kalantidis *et al*, 2008).

4.4.5. Origin of plasmodesmata

During cytokinesis of plant cells, the ER associates with the dictyosomal vesicles of the forming phragmoplast. Sometimes, portions of the ER are trapped within these vesicles as they fuse, these tubular ER elements manage to traverse the cell plate, and form plasmodesmata.

4.5. DESMOSOMES

Desmosomes are morphologically and biochemically defined intercellular junctions that are required for maintaining the mechanical integrity of epithelial tissues and particularly in those that frequently undergo mechanical stress. They are common in epithelia, bladder,cardiac muscle, uterus, vagina, skin and the heart in adult mammals. Their presence renders the tissues more elastic without any damage to the cell.

4.5.1. General structure and occurrence of desmosomes

Desmosomes constitute critical adhesion complexes between cadherins of adjacent opposing cells. α-catenin, α-cadherin-associated protein, links cadherin / β-catenin and cadherin /plakoglobin complexes to the actin cytoskeleton. Interaction of α-catenin and actin also facilitates the assembly of desmosomes in epithelial cells. Classical cadherins interact via α- and β-catenin with the actin cytoskeleton, while desmosomal cadherins interact with the intermediate filament system. Desmosomes show an assemblage of an array of proteins. In fibroblasts, vimentin interacts with the actin-based complex to promote cell-cell adhesion. In stratified epithelium plakophilin (PKP) 1,2 and 3 are expressed in the suprabasal layers of stratifying epithelia. PKP-1 is expressed in the supra basal layers of the epidermis, its loss leads to the skin fragility-ectodermal dysplasia syndrome showing trauma induced blisterring, especially on the weight bearing parts of the sole. PKP-1regulates desmosome integrity in response to low calcium concentrations and alters the migratory properties of wounded keratinocytes. PKP-3 is expressed in both simple and stratified epithelia. It binds the desmosomal cadherines desmogleins and desmocollins. Periplakin, a desmosomal component of the epidermal cornified envelope hetero dimerises with envoplakin and intermediate filaments (Fig. 4.8; 4.9).

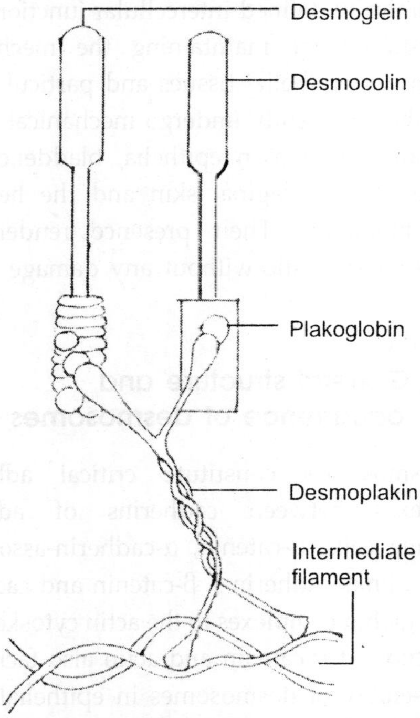

Fig. 4.8 : Some of the molecular components of desmosome.

Two types of desmosomal cadherins – desmogleins and desmocollins show cytoplasmic domains placoglobin and desmoplakin II, I. Placoglobin interacts with both cadherin and desmoplakin. Desmoglein 1 and 3 both are involved in desmosome formation and colocalize with desmoplakin but at high-level expression, desmoglein 1 disrupts desmosomes. In the epidermis, desmoglein-1 is expressed in the upper layers but desmoglein-3 is expressed in the basal layer. Desmoglein 1 and desmoglein 3 play a different functional role in cell-cell adhesion. Mutation in desmoglein leads to two different skin diseases, *pemphigus foliaceus* and SSSS (showing blister formation in the superficial epidermis).

Intermediate filaments composed of keratin or desmin are attached to membrane-associated attachment plaque on the cytoplasmic face of the membrane to which cadherin molecules are anchored and bind strongly to cadherins of the adjacent cell. They maintain close adhesion between adjacent keratinocytes in skin, maintaining an intact epidermis. They however migrate and differentiate to maintain the stability of the epidermis. Desmosomes thus show a dynamic capacity to assemble and disassemble. Hereditary skin disorders resulting from desmosome gene pathology may involve the palms and soles. Even in normal skin, the basis of the differences in desmosome number, size or structural organization in palm and sole is not clear. Relative differences in the structure and the expression profiles of the desmosomal proteins between these sites possibly reflect the functional differences of palms and soles in relation to mechanical stress.

By tethering intermediate filaments (IFs) to sites of intercellular adhesion, desmosomes facilitate formation of a super cellular scaffold that imparts mechanical strength to a tissue. IF-membrane attachments regulate adhesive strength and suggest that actin-IF-based junctions act synergistically to strengthen adhesion. Desmosomes are of three types:

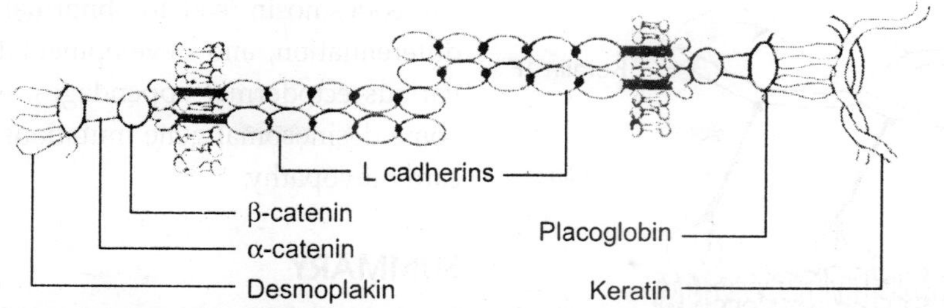

Fig. 4.9 : Structural components of Zonula adherens.

4.5.2. Belt desmosome (Zonula adherens)

These are band like structure made up of actin filaments; occur at the peripheral region of the cell below the tight junctions (Fig. 4.10). At belt desmosomal parts, the plasma membranes of adjacent cells appear thicker and slightly (15-20 nm) apart from each other. This continuous junction maintains the physical integrity of the epithelium. Homophilic interactions between densely clustered E-cadherins bind adjacent cells at these points. β-catenin and placoglobin bind the cytoplasmic domains of the E-cadherin with actin filaments via α-catenin in these points. Adherens junctions and desmosomes are responsible for mechanically coupling myocytes in the heart and are found closely apposed to gap junction plaques at the intercalated discs of cardiomyocytes.

4.5.3. Spot desmosome (Macula adherens)

These are like rivets that tightly hold the epithelial cells. They appear as circular button like structures (0.5m in diam.) called plaque. Many intermediate filaments – tonofilaments made up of keratin converge towards the plaque. The intercellular portion shows protein and mucopolysaccharides (Fig. 4.10).

4.5.4. Hemidesmosomes

These are similar to spot desmosomes but attach epithelial cells to the basal lamina instead of intercellular adhesion. These are integrin based adhesive junctions linking cytoplasmic intermediate filaments to the basal lamina. They also show a dense surface that anchors loops of intermediate filaments. Two proteins $\alpha_6\beta_4$ integrin and type XVII collagen are essential for the stability of hemidesmosomes. Extracellularly the integrin binds laminin-5 in the basal lamina. Intracellularly plectin links the integrin to keratin intermediate filaments (Fig. 4.11).

4.5.5. General role of desmosomes

1. Desmosomes play a critical role in the maintenance of normal tissue architecture.

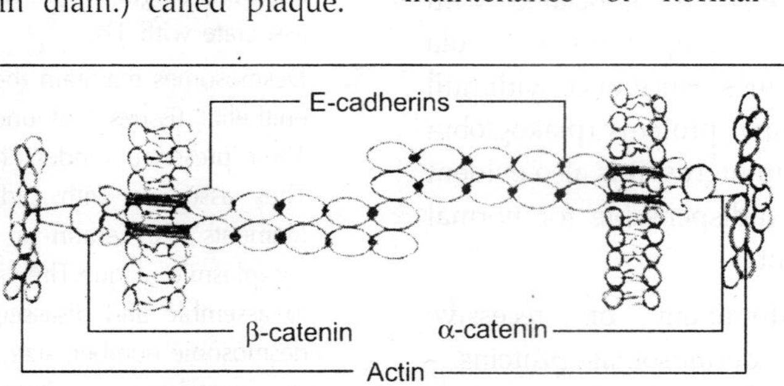

Fig. 4.10 : Structural components of spot desmosome.

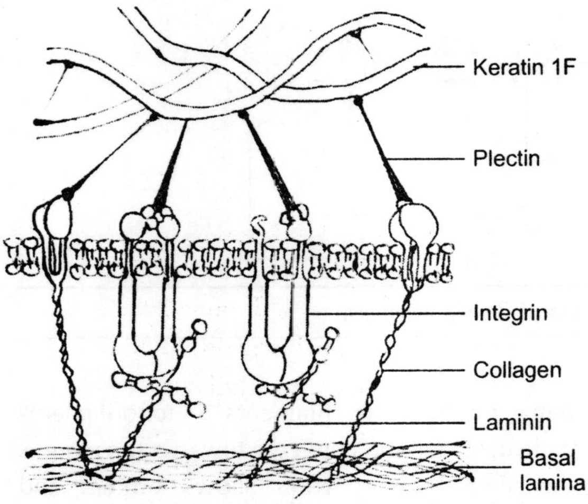

Fig. 4.11 : Structural components of hemidesmosome.

Inherited mutations in genes encoding desmosomal constituents can adversely affect the skin, causing pemphigus and heart abnormalities.

2. Desmosomes may have a tumour suppressor function: expression of desmosomal components is reduced in some human cancers, and desmosomal cadherins have the capacity to suppress the invasiveness of cells in culture.

3. Desmosomal adhesion is reported to play crucial role in tissue morphogenesis, autoimmunity, infectious and genetic diseases. Desmocollin and desmoglein regulate epidermal diffentiation. Desmosomal adhesion is regulated by protein kinase C.

4. The morphological changes observed in apoptotic epithelial cells are associated with specific fragmentation of components of zonula adherens and desmosomes. Since mice with null mutations in desmosomal proteins (plakoglobin and desmoplakin) die in uterus, it is also evident that desmosomes are indispensable for normal embryonic development.

5. Autosomal dominant or recessive mutations of several desmosomal proteins - plakophilin 1, plakophilin 2, desmoplakin, plakoglobin, desmoglein 1, desmoglein 4 and corneodesmosin lead to abnormal skin fragility, differentiation, and developmental anomalies of various ectodermal appendages, especially hair. Some desmosomal gene mutations may result in cardiomyopathy.

SUMMARY

1. Intercellular junctions maintain tight barrier as well as communicating channels that directly regulates intercellular transportations as well as other physiological functions for the maintenance of stability in the tissues. The junctional complexes comprise tight junctions, adherens junctions, and desmosomes, whereas gap junctions provide the avenue for intercellular communication. They not only provide intercellular junctions but anchor the cells with extra cellular matrix also. In plants intercellular communication is maintained by plasmodesmata
2. These junctions show specificities of transmembrane proteins and their association with ECM or cytoskeletal elements.
3. Tight junctions (TJs) are specialized plasma membrane micro domains that encircle the apical pole of the cell. They form an intercellular diffusion barrier regulating the passage of ions, water, and various macromolecules through the para cellular spaces. Leaky TJs are created by tensions in the actin filaments, increased cytosolic Ca++ and ZO toxin released by Vibrio cholerae.
4. In EM, TJs show a series of contacts representing the integral proteins- occludin and claudin. Occludin binds a protein –Zonula Occludens-1 (ZO-1) that in turn associates with ZO-2 and spectrin. ZO-1 associates with cadherins. In addition, cingulin and GTP binding proteins also associate with TJs.
5. Desmosomes maintain the mechanical integrity of epithelial tissues that undergo mechanical stress. Their presence renders the tissues more elastic. They associate with cadherins and intermediate filaments like keratin or desmin, attached to the cytoplasmic plaque.They show a dynamic capacity to assemble and disassemble. Differences in the desmosome number, size, structure and expression of proteins - desmogleins, desmocollins, desmoplakin, plakoglobin and plakophilin 1 are

noted in palm and sole. Placophilin-1 (PKP-1) is absent in patients with ectodermal-dysplasia skin-fragility syndrome.

6. Belt desmosomes are band like structure made up of actin filaments present at the apical part of the cell below the TJs. E-cadherins along with β-catenin, α-catenin and placoglobin bind the cells at these points.
7. Spot desmosomes are rivets like adhesion junctions showing tonofilaments (made up of keratin) in association with cadherins (desmoglein and desmocollin). Dysfunction of desmoglein results the skin disease – pemphigus foliaceus.
8. Hemidesmosomes attach epithelial cells to the basal lamina instead of intercellular adhesion. These are integrin based adhesive junctions linking cytoplasmic intermediate filaments to the basal lamina. Extra cellularly the integrin binds laminin-5 and intra cellularly plectin links integrin to keratin intermediate filaments. Desmosomes play a critical role in the maintenance of normal tissue architecture in skin; they suppress tumours, help in tissue morphogenesis, epidermal differentiation, apoptosis, and even are indispensable in embryonic development in mice.

QUESTIONS

Long type questions:

1. What are junctional complexes? Describe the structure, occurrence and physiological role of tight junctions.
2. Describe the structure, occurrence and physiological role of gap junctions.
3. Describe the structure, occurrence and physiological role of desmosome.

Short type questions:

1. What do you mean by polarized cells?
2. What are basic differences between apical and basolateral membranes of the polarized cells?
3. What are the basic difference between hemidesmosomes and spot desmosomes.
4. What do you mean by focal contact?
5. What are the intercellular kissing points?
6. What are the main proteins of tight junctions?
7. ZO-1 is a common TJ protein. What does ZO signify?
8. Which enzyme is involved in the formation of TJs?
9. What are the major signaling mole cules associated with TJs?
10. Which TJ proteins are associated with diseases?
11. What are the common proteins of Gap Junctions?
12. What are the differences among connexon, connexin and innexin?
13. Which connexons are associated with deafness?
14. Which connexin is associated with normal development of mice?
15. Which intracellular organelle is related with plasmodesmata?
16. What are the main proteins associated with desmosome?
17. Which proteins are expressed in hemidesmosomes?
18. Name the extra cellular and intracellular proteins linking integrins in hemidesmosomes?
19. What are the main physiological roles of desmosomes?

Fill up the blanks:

1. Plasmodesmata are equivalent to ——— of animal cells.
2. ——— link muscle cells to tendons.
3. In invertebrates TJs are known as ———.
4. Connexons contain — connexin subunits.
5. Mutation in ——— is associated with Charcot-Marie tooth disease.
6. Desmosomes maintain the ——— of the epithelial cells.
7. ——— is the main CAM that associates with desmosomes.
8. In stratified cornified epithelium, ——— and——— are the main proteins.
9. ——— desmosomal protein is absent in skin fragility syndrome.
10. Belt desmosomes are mainly composed of ——— filaments.
11. Mutation of ——— cadherin protein results pemphigus foliaceus skin disease.

MCQs:

1. Tight junction
   (a) gap junction
   (b) desmosomes
   (c) **maintain intercellular communication.**
2. Columnar epithelial
   (a) squamous epithelial
   (b) muscle cells
   (c) express **all four types of junctional tissues.**
3. **Cingulin is a common**
   (a) tight junction,
   (b) gap junction,
   (c) desmosomal protein.
4. **The sizes of the molecules that pass through the gap junctions are**
   (a) 2 kD, (b) 1 kD, (c) 3 kD.
5. **Mutation in** (a) cx26, (b) cx30, (c) cx50 **is associated with hereditary cataract.**
6. **Belt desmosomes show**
   (a) actin-catenin,
   (b) myosin-catenin,
   (c) spectrin-catenin associations.
7. **Tonofilaments are associated with**
   (a) belt desmosomes
   (b) hemidesmosomes
   (c) spot desmosomes.
8. **Desmogleins and desmocollins are**
   (a) integrin, (b) cadherin, (c) selectin.
9. **Hemidesmosomes are**
   (a) cadherin,
   (b) integrin,
   (c) IgCAM based junctions.
10. **Hemidesmosomes attach the**
    (a) lateral,
    (b) apical,
    (c) basal portions of the cells with basal lamina.

Answers to MCQs:

1. b, 2. a, 3. a 4. b, 5. c. 6. a. 7. c. 8. b, 9. b. 10. c.

Chapter 5

Cellular Adhesion Molecules

In multicellular organisms, cells are not an isolated entity, rather they tend to stick to each other or with non-cellular components of their environment. During ontogeny, intercellular adhesion is highly essential for regulated programme of the developmental profiles of tissues and organs. Some adhesive interactions are stable as in skin and muscle cells but in some cases it is unstable as in binding of W.B.C. to endothelial cells at the site of inflammation. Cell adhesion is mediated by protein molecules; so the study of cell adhesion is the study of proteins and the molecules that mediate interaction with other cells and with extracellular matrix (E.C.M.). Cellular adhesion is a dynamic process for the recruitment, binding, and reorganization of signaling proteins in the plasma membrane. Cell adhesion molecules (CAMs) can influence various developmental events, including cell migration, proliferation, and differentiation. Recent advances have provided evidence that in addition to their adhesive properties, CAMs can affect intracellular signaling and synaptic function (Nikonenko and Skybo, 2007). Adhesion of a cell to another cell or the extracellular matrix involves complex couplings between cell biochemistry, structural mechanics and surface bonding. The interactions are dynamic and act through association and dissociation of bonds that change considerably under stress. Cell signaling processes strengthen adhesion bonds and forces applied to cell-surface bonds act on intracellular sites to catalyze chemical processes or switch molecular interactions on and off (Evans and Calderwood, 2007).

5.1. CELL ADHESION PROTEINS/CELL ADHESION MOLECULES-CAMS

Cell adhesion proteins are transmembrane receptors that extend across the plasma membrane and have domains that extend into both the extracellular and intracellular space. The extracellular domain can bind to other soluble molecules like hormones, growth factors etc., molecule that might be either on the surface of an adjacent cell (cell-to-cell adhesion) or part of the E.C.M. (cell-to-ECM adhesion). The molecule that binds to it, is called its *ligands* (Fig. 5.1). Cell adhesion proteins are characterized on the basis of the structures of the adhesion proteins and their ligands. Adhesion between two copies of the same adhesion protein is called *"homophilic binding"* while between different adhesion proteins is called *"heterophilic binding"*.

The binding of the intracellular domain of the adhesion protein to a cytoskeletal protein leads to tight adhesion between the cytoskeleton and the ligand. This connection may be direct or indirect e.g. cadherin (a cell adhesion protein) is coupled to the cytoskeleton by special linking proteins called "catenins".

5.2. PRINCIPLES OF CELLULAR ADHESION

1. Cells express definite plasma membrane receptors with ligand binding capacity. Sometimes expression of receptors depend upon some extra-cellular stimuli (e.g. inflammatory hormones or endotoxins activate to express the receptor - E-selectin on the endothelial cells.

2. Adhesive protein and ligand relationship do not follow any definite rule. Some proteins bind only with one ligand, some receptors bind with many ligands while sometimes many ligands bind to a single type of receptors.

3. Sometimes cells express receptors after being activated by proper stimuli. e.g. the endothelial cells store P-selectin in the membranes of cytoplasmic vesicles but when activated by inflammatory cytokines the vesicles fuse with the cell membrane exposing the receptors on the plasma membrane.

4. Sometimes receptor-ligand relationship appears very unstable.

5. Many receptors interact with cytoskeleton or intermediate filaments. These interactions provide mechanical strength to the cell allowing them to transmit forces and resisting disruption of the cells.

6. Receptor-ligand binding may activate intracellular signal transduction pathways resulting gene activation, cellular differentiation, secretion, motility, cell division etc.

5.3. TYPES OF ADHESION MOLECULES (Fig. 5.2)

Cell surface glycoproteins actually play active roles in adhesion. Each protein is restricted to specific parts of the plasma membrane at a specific time during embryonic development. These can be divided into four main groups: ***the integrin family, the immunoglobulin superfamily, selectins and cadherins.*** Adhesion proteins are usually named CD ("clusters of differentiation") followed by a number. CD classifies cell surface

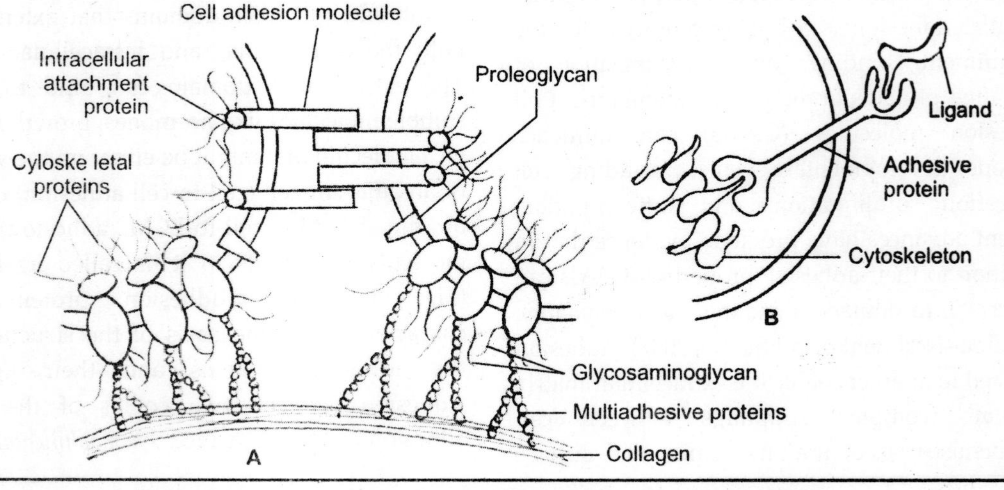

Fig. 5.1 : General relationship of CAM, cytoskeleton, ECM and ligand.

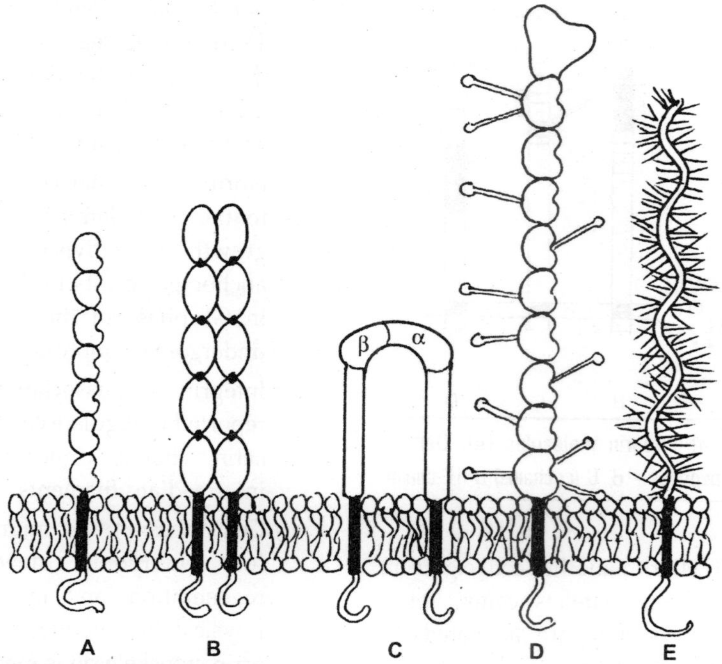

Fig. 5.2 : Different cell adhesion molecules. A -Immunoglobulin family, B-Cadherin, C-Integrin, D- Selectin, E- Mucin.

antigens recognized by monoclonal antibodies. Hence all adhesion proteins have a CD number. The binding sites are different for each of these adhesion molecules.

5.4. INTERCELLULAR BINDING THROUGH INTEGRINS

Integrins are transmembrane glycoproteins; and except RBC they occur as vital adhesion receptor in almost all cells linking them to other cells as well as to ECM. They perform adhesion function in calcium and magnesium dependent manner to a specific ligand on the target cell. Integrins show an extracellular ligand in the ECM and intracellularly they bind with the actin cytoskeleton. Lymphocytes can interact via integrins with the major components of ECM - proteoglycans and glycosaminoglycans to perform a multitude of physiological roles. Some integrins bind to cytosolic intermediate filaments like keratin.

Structure of Integrins: (Fig. 5.3) Integrins are heterodimers of two polypeptides - α and β chains both of which contribute in the ligand binding. Both α and β chains are attached with the membrane by definite stalks. All β and some α chains have special inserted domain (I domain). All a chains have N terminal β propeller domain that interact with the I-domains of β chain to form the stable dimer. Integrins generally have a low affinity for extracellular ligands. Its cytoplasmic domains interact with a variety of signaling and structural proteins. The adapter proteins talin and vinculin link the cytoplasmic domains of β chains to the actin filaments. Paxillin links integrins to signaling proteins. Several types of integrins may associate laterally in the plane of the bilayer with other transmembrane proteins like CD47 and Ig-CAM.

Physiological roles of Integrins:

i) Some integrins may require activation in order to bind their ligand and anchor the

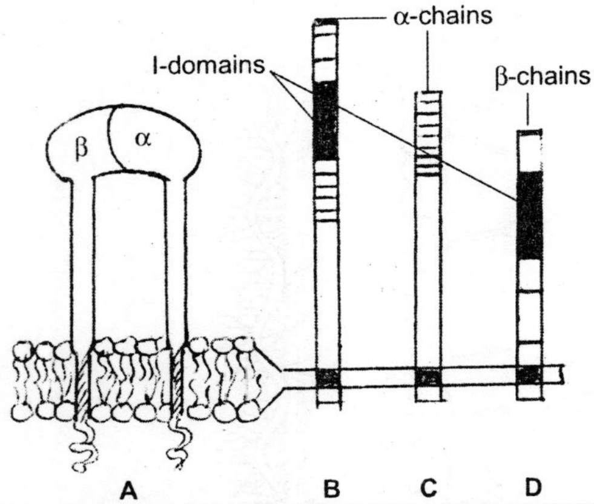

Fig. 5.3 : Structure of integrin molecules (A). Domain models of integrin polypeptides – B, C (α-chains) D (β -chain).

cell to the ECM or to another cell. e.g. α-m, β-3 integrin from platelets cannot bind fibrinogen unless they are activated by collagen or thrombin.

ii) Play multiple roles in differentiation and cell communication.

iii) They also play definite roles in signal transduction relaying information to control cell growth and structure.

iv) They help fibroblasts and WBCs to adhere to fibronectin and collagen as they move through the ECM.

v) They bind epithelial and muscle cells to laminin in the basal lamina, providing physical relationship necessary to transmit internal forces to the ECM and to resist external forces.

vi) Integrins help platelets to adhere to the basement membrane collagen and to each other via plasma fibrinogen during repair of small vessels.

vii) Integrins participate in the binding of mouse sperm with the egg during fertilization.

viii) Integrins facilitate the adhesion of WBC to endothelial cells at the sites of inflammation.

ix) Cell multiplication in the absence of integrin-derived adhesive signals is the phenotypic hallmark of malignancy. Therefore, up-regulation of some integrins (β-4 integrin) has been noted in human carcinomas. This suggests the use of β- (4) as a target for anticancer therapy.

x) Normal epithelial cells require anchorage to the basal lamina by β-4 integrins for growth and division. Loss of integrin anchorage results in the arrest of the cells in G1 phase of the cell cycle eventually undergoing apoptosis.

xi) Integrin in association with Ca^{++} is reported to regulate cell shape mediated by actin fibers, microtubules and by intermediate filaments.

xii) Cell adhesion and migration are essential for embryonic development, tissue regeneration, and immune defence. The physical link between the ECM and the actin cytoskeleton is mediated by receptors of the integrin family and many adapter proteins. During cell migration this physical link is dynamically modified, allowing the cell to sense and adapt to the microenvironment. This includes the formation of integrin clusters at the front and their subsequent disassembly at the rear of the cell. Genetic defects of integrins and their adapters lead to cellular dysfunction and generate pathological situations.

5.5. INTERCELLULAR BINDING THROUGH IMMUNOGLOBULIN (IG-CAM) SUPER FAMILY (Fig. 5.5)

These are single pass transmembrane proteins having 1-7 extracellular domains anchored to the plasma membrane by a single hydrophobic transmembrane sequence. Some Ig-CAM consists of single polypeptide but others are multimeric having 2 or 4 subunits. In some cases the intracellular domains bind to the cytoskeletal system while extracellular domains bind to other similar or dissimilar molecules.

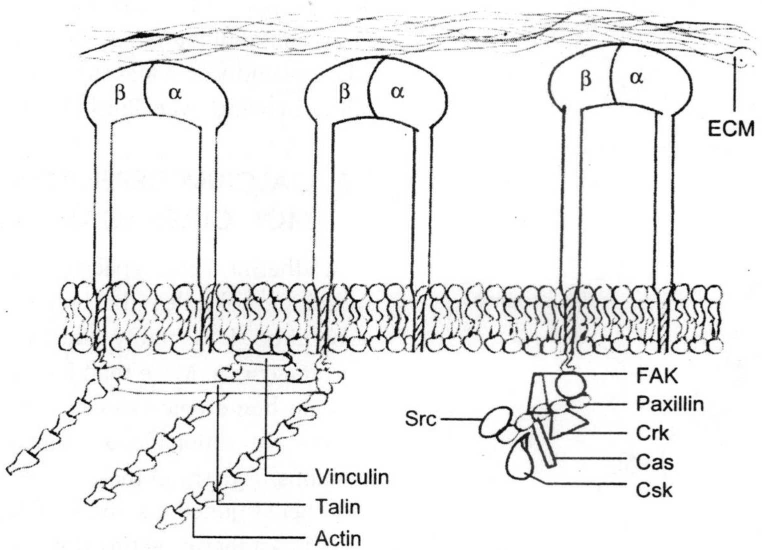

Fig. 5.4 : Interaction of integrins with ECM, cytosolic adaptor proteins and cytoskeleton.

These are responsible for Ca^{++} independent intercellular adhesion. One of the best-characterized members of this family is neural cell adhesion molecule (N-CAM), expressed by a variety of cell types. N-CAM is actually expressed by a variety of cell types and binds by homophilic interactions. Intercellular adhesion molecules (ICAM) and L1 are also important examples of immunoglobulin super family. ICAM is expressed on activated endothelial cells use heterophilic mechanism of cell adhesion; L1 is a neural cell-to-cell adhesion molecule important in developing nerves. One of the members of the Ig superfamily is CD2, a glycoprotein receptor of T-cells. It has two extracellular I-domains and is also required for T-cell activation. Like other

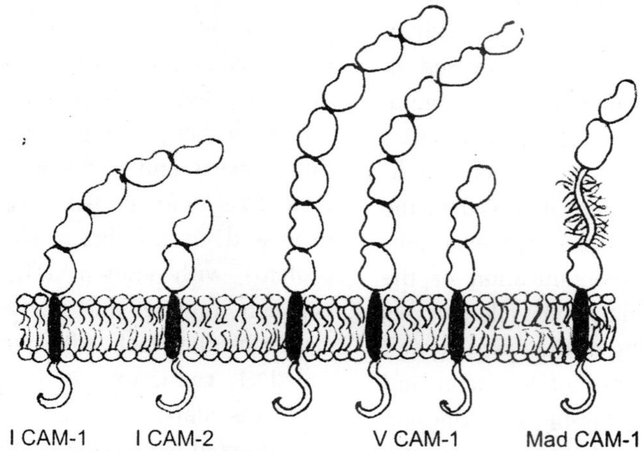

Fig. 5.5 : Molecular structure of some Ig-CAM.

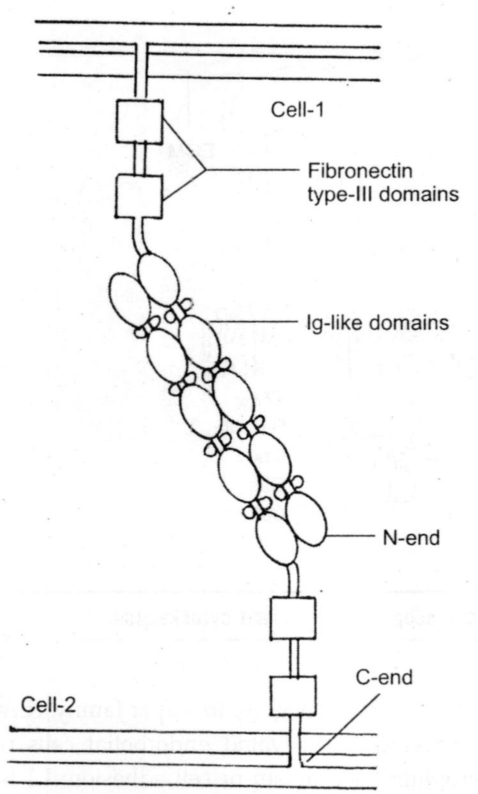

Fig. 5.6 : Homophilic interaction of N-CAM mediated inter cellular adhesion.

adhesion proteins Ig-CAMs participate in signaling process. TSLC-1, 2 are new member of the IgG superfamily and are actually tumor suppressor protein. TSLC-1 is associated with an actin-binding protein. TSLC-2 is expressed in prostate cancer, kidney, bladder and brain. Nectins are Ca^{++}-independent immunoglobulin-like cell-cell adhesion molecules, interact and form cell-cell adhesion and are involved in the formation of adherens junctions in epithelial cells and fibroblasts through reorganization of the actin cytoskeleton. The migration of leukocytes across the endothelial lining of the vascular wall requires a complicated series of adhesion and signaling events. Endothelial Ig-like cell adhesion molecules (Ig CAMs) such as intercellular adhesion molecule-1 play an important role, not only as ligands for leukocyte integrins, but also as signaling initiators. IgCAMs activates Rho-like GTPases, induces cytoskeletal changes to mediate cell-cell contact (van Buul *et al*, 2007).

5.6. CALCIUM DEPENDENT ADHESION MOLECULES (CADHERINS)

Cadherins are specific Ca^{++} dependent adhesion molecules that regulate intercellular recognition. It is highly effective during animal morphogenesis. More than 80 types of cadherins mediate homotypic cell-cell adhesion. They are present in intercellular junctions, adherens junctions, desmosomes, cardiac junctions, endothelial junctions, and synaptic junctions. Specific cadherin expression is noted during ontogeny that stabilizes the architecture of the respective organs. Almost all vertebrate and some invertebrate cells express one or more cadherins.

Types of Cadherins and their features (Fig. 5.7): Specialized cells express characteristic cadherins e.g. osteoblasts express OB-cadherin, kidney cells express K-cadherin and muscle cells express M-cadherin, placental tissues express P cadherins and embryos over express E-cadherins. E and P cadherins bind epithelial cells and occur in the adhesion belts. They show intracellular binding proteins like catenins, α-actinin and actin filaments. Cells in the nervous system express N-cadherin and protocadherins having a unique extra cellular domain consisting of 7 CAD domains. They occur as adhesion belts exhibiting catenin, α-actinin and vinculin in the interce llular space and actin in the cytosolic side. *Protocadherins* are non-classical cadherins that are expressed in the brain alone. These include proteins with known adhesive functions. In between muscle and nerve cells there are cadherins that lack adhesive function and are called T cadherins. It is also expressed in cardiovascular tissues. During embryonic development expression of different cadherins is noted e.g. the ectoderm at the end of gastrulation

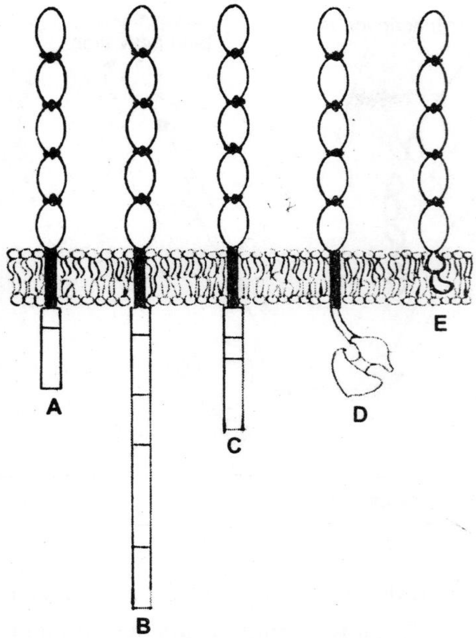

Fig. 5.7 : Diagrammatic representation of cadherin domains. A- E-cadherin, B- Desmoglein-1,C- Desmocolin 1a, D-RET, E-T-cadherin.

exhibits E-cadherin while the ectoderm in the neural tube express. *Desmosomal cadherin* connects epidermal and placental tissues, occur as desmosomes; show intercellular binding proteins like desmoplakins I & II and placoglobin. They exhibit keratin and desmin in the cytosolic side. N-cadherin. However classical and non-cassical cadherins collectively form a cadherin superfamily.

Structure of cadherins: These are single-pass transmembrane glycoprotein composed of about 700-750 residues. Their extra-cellular domains contain about 100 amino acids and calcium binding sites. The characteristic feature of the cadherins is the presence of CAD domain of about 115 residues. The classic cadherins have 5 extra-cellular CAD domains and the single hydrophobic sequence α-helix that spans the membrane (Fig. 5.8). Cytoplasmic domains vary in size, sequence and binding sites for associated proteins. E-cadherins lack cytoplasmic domain. N terminal parts of CAD project outwards while C terminal parts project inwards and participate in interdomain association. Five CAD domains stack laterally and end-to-end to form the classic rod shaped extracellular domain. The Ca^{++} are positioned between each pair of repeats and their stiffness depends upon the concentration of Ca^{++}.

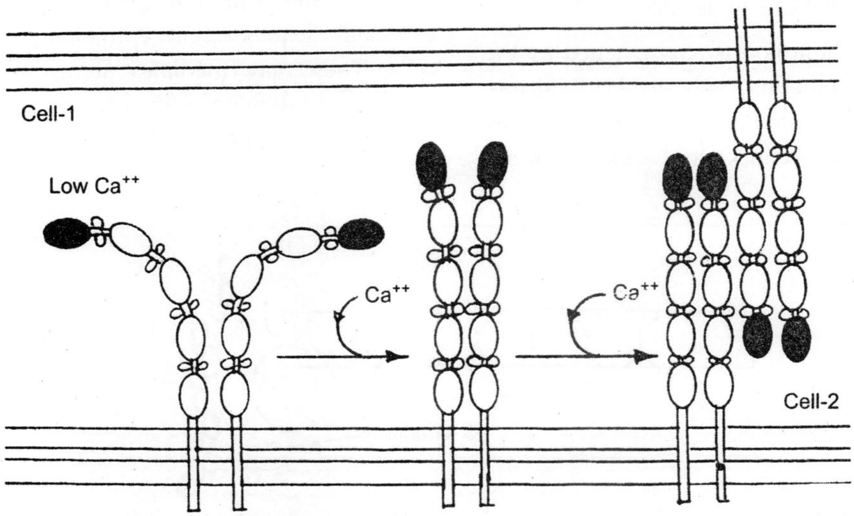

Fig. 5.8 : Increase in the stiffness of the CAD domains on the basis of Ca^{++}.

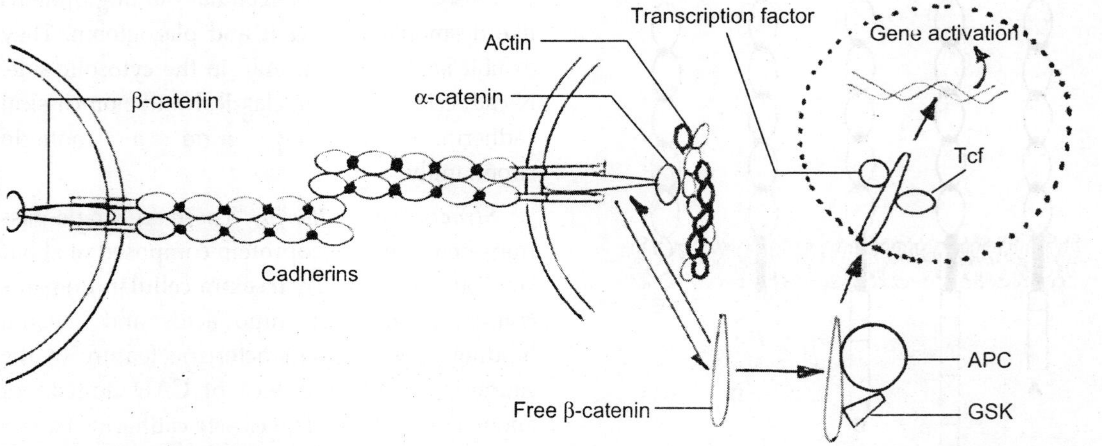

Fig. 5.9 : A model of intercellular links by cadherin.

Mechanism of cellular adhesion by cadherin (Fig. 5.9): There are three processes that regulate cell adhesion by cadherins.

(a) *Homophilic binding*: It involves binding of cadherin domains between two neighbouring cells.

(b) *Heterophilic binding*: It involves binding of cadherin of one cell with other molecules of neighbouring cell.

(c) *Linker dependent binding*: Cadherins of neighbouring cells are attached together by some linker ECM molecules.

Intercellular adhesion is strengthened by the interaction of cadherins with the cytoskeleton. Adapter proteins - catenins links the cytoplasmic tails of cadherins to actin. Cytoplasmic tail at first binds with the β-catenin that in turn is attached with actin via α-catenin. A related protein – vinculin helps to link cadherins to actin (Fig. 5.10). Most of the membrane β-catenins is bound to cadherins but additionally a cytoplasmic pool of β-catenin remains bound to APC (**a**denomatous **p**olyposis **c**oli), GSK (**g**lycogen **s**ynthase **k**inase) and nuclear β-catenins recruit transcription factor to regulate gene expression. Loss of APC results in excess cytosolic β-catenin that ultimately activates the nuclear receptors and activates different genes. Cadherins are especially expressed in epithelial and muscle cells. They may mediate interaction of their cytoplasmic domain with actin or intermediate filaments strengthening the physical integrity of

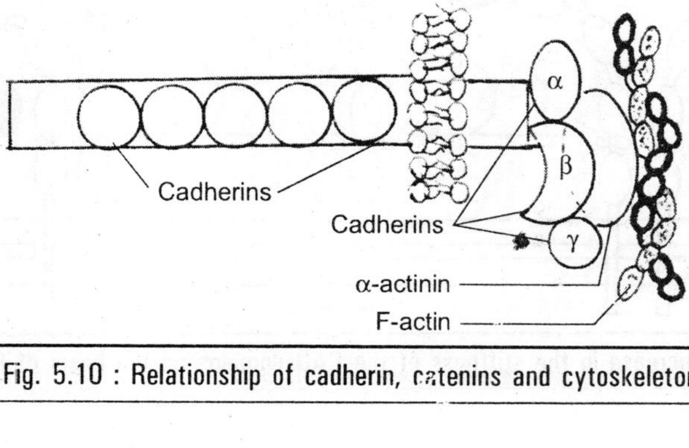

Fig. 5.10 : Relationship of cadherin, catenins and cytoskeleton.

the cell but they do not interact with ECM and bind in a homophilic pattern, i.e. one cadherin binds to the same type of cadherin on adjacent cell.

Role of cadherin in intercellular association and intracellular signal transduction:

i) In intercellular binding, CAD1 domains of adjacent cells link with each other creating a molecular zipper to bind cells together. It is reported that a strand exchange process from the second amino acid at the N-terminus of the cadherin molecule, docks into a hydrophobic pocket domain of the opposing cadherin.

ii) Cadherin and catenin participate in the transduction of extra-cellular signals. They regulate cellular behavior and the integration of information received from these signaling determines cellular phenotype and behavior.

iii) Cadherin function is critical in normal development and any alterations in cadherin function leads to tumorigenesis. During early phases of mammalian development E-cadherins paly a crucial role in the compaction of the blastomeres. During neural tube separation, when it becomes a tubular struture the E cadherins in the overlying epidermis continues to main the sructure while the underlying neural tube lose connection with the E cadherin and establish new connection with other cadherins. In the latter part of development when the cells aggregate to form ganglion they again express N cadherin.

iv) It indirectly inhibits the spread of cancer cells. Moreover expression of E-cadherin checks the transformation of benign tumour to malignant tumour. Reduced E-cadherin/β-catenin results in the loss of cell adhesion that lead to the metastasis of cancer cells. The expression of membranous β-catenin decreases in relation to the progression of carcinoma.

v) Desmosal cadherins (called desmocolins and desmogleins) have complicated cytoplasmic domains that interact with the proteins – desmoplakin and placoglobin. Desmoplakin links cadherins to keratin (intermediate filament).

5.7. CELL-TO-CELL SURFACE CARBOHYDRATE BINDING PROTEINS (SELECTINS) (Fig. 5.11)

Selectins are trans membrane mosaic proteins recognizing carbohydrates. They consist of small cytosolic domain, single trans membrane helix, EGF-domain and N-terminal calcium dependent lectin domain. Lectins are structurally and functionally are of two types- *C-type CRDs* (*CRD* = Carbohydrate Recognition Domain) are Ca^{++} dependent. *I-type lectins* that contain two or more I-domains they recognize terminal sialic acid residues. Selectin, expressed on WBCs is L-selectin.

Relatively low affinity-binding sites for monosaccharides are formed at shallow grooves on lectin surface. Metal ions may play a role too. Higher selectivity is achieved by extending binding sites through additional contacts between oligosaccharides and protein surface. Lectin binds with O-linked sulfated oligosaccharides containing sialic acid and fucose. They are present on the surface of WBCs, platelets and also the cells lining the blood vessels (endothelial cells). Natural ligands for selectins are mucin like glycoproteins expressed on endothelial cells and WBCs. Selectins interact with the mucin by its CAD. L-selectin plays an important role in binding WBCs to the endothelial cells lining the blood vessels. This interaction enables the WBCs to migrate out of the blood vessel into the tissues.

The two other selectins (P and E selectins) are found on endothelial cells and due to their presence they can attract leukocytes. When a tissue is infected, it secretes cytokines (IL-1, TNF)

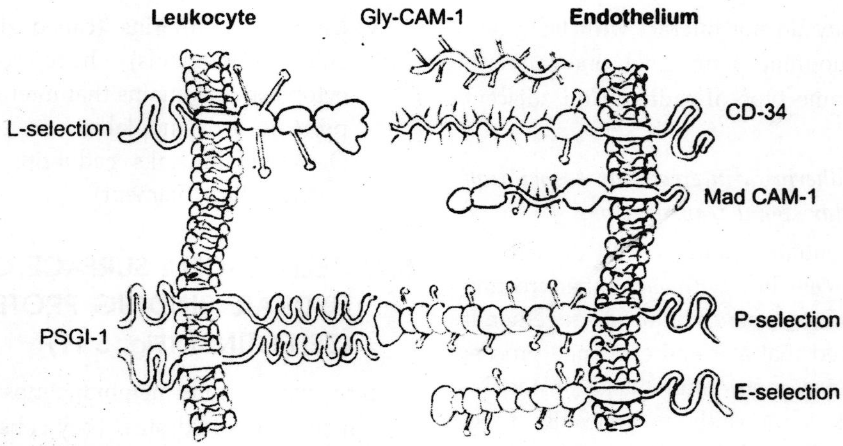

Fig. 5.11 : Structure of selectins and their mucin ligands.

that stimulate endothelial cells in the venules to express P and E selectins. They appear on endothelial cells at different times and recruit different types of WBCs. P-selectins are expressed earlier (a few minutes after the infection begins) because endothelial cells have stockpile of them. The WBCs adhere to these selectins because carbohydrate coats contain complementary structures. Thus leukocytes can leave the vessels and migrate into the infected tissue. They draw leuko cytes that act during the early phases of the immunologic defense. Soluble P-selectins are elevated in the serum of patients suffering from coronary artery disease. E-selectins are synthesized only when they are required. Thus it takes longer to appear (about four hours after the beginning of infection; then the level of E-selectins gradually fades away). They are involved in recruiting neutrophils and macrophages to the inflammation sites. Selectins act in association with integrins, selectins bind to specific oligosaccharides on glycoproteins and glycolipids but integrins bind to specific proteins. But the interaction of WBCs to the tissues is relatively weak since binding of the lectin domain of the selectin to the carbohydrate ligand is of low affinity. This regulates weak adhesion of WBCs with the endothelium as well as with the tissues. Possible role of E-selectin in urological malignancy is also known. Selectins and integrins are reported to play crucial roles in the development of atherosclerosis. Little is known about the role of the selectins in signal transduction.

5.8. OTHER ADHESION RECEPTORS

Advances in research activities have identified following additional adhesion receptors.

Mucins: Selectins bind to cell surface glycoproteins called *mucins*. Their extracellular domains are rich in serine and threonine. They are expressed on endothelial cells or WBCs. Endothelial mucin interacts with L-selectin on WBC while endothelial P-selectin interacts with WBC PSGL-1 mucin.

Galactosyl transferase: This enzyme is required for protein glycosylation in Golgi complex. It has two isoforms, the longer enzyme moves to the cell surface and acts as receptors of those ligands that terminate in N-acetyl glucosamine. On cytosolic end it interacts with actin. During fertilization it mediates the initial contact of mouse sperm with the matrix surrounding the egg. This binding induces secretion of acrosomal content that destroys the transferase-binding site so that the sperm can proceed through the zona pellucida to fuse with the egg. Intercellular attractions during the early part of embryogenesis also requires this enzyme receptor.

Cartilage link proteins: Hyaluronan is a high MW glycosaminoglycan polysaccharide is present in the tissue matrix and body fluids of all vertebrates. It plays a fundamental role in the regulation of cell migration and differentiation. The majority of hyaluronan binding proteins belong to the cartilage link-proteins super family. These can be divided into two groups, those that are true cellular adhesion molecules such as CD44 and those that are components of ECM like cartilage link protein itself. They are characterized by the presence of a link domain that serves as a binding site for the glycosaminoglycan, hyaluronan. CD44 has a number of glycosylation sites those determine its capacity to bind to fibroblast growth factor and human chorionic growth factor and its role in cell migration and metastasis in tumours.

Adhesion receptors with leucine rich repeats (GP IB-IX): It is expressed in platelets and mainly participates in the reception of von Willebrand factor (vWF). It shows leucine rich repeats in the ECM that mainly participate in the binding of the receptor-vWF. It acts as a bridge between a specific glycoprotein on the surface of platelets (GPIB-IX) and collagen fibrils. Binding of vWF to the receptor generates a signal that activates the platelets.

Dystroglycan/Sarcoglycan complex (Fig. 5.12): In the striated muscle plasma membrane there is a large protein – dystroglycan having an extra cellular α and a transmembrane β-subunit. α subunit unites with α2 laminin and agrin that makes intimate contact with ECM. The β-subunit binds cytosolic element dystrophin that in turn joins with syntrophin complex. The β-subunit also bind with sarcoglycan and sarcospan. These

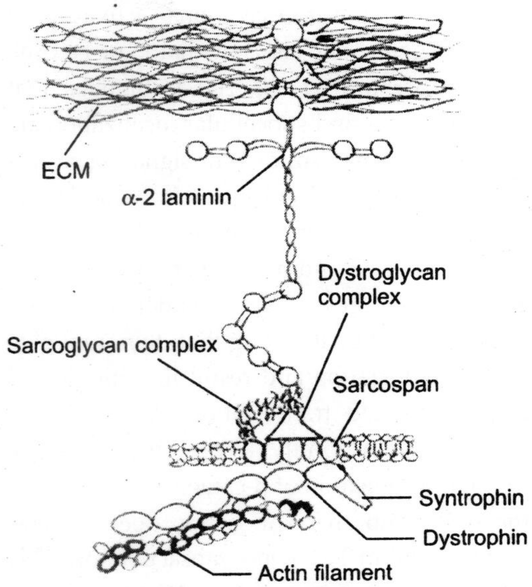

Fig. 5.12 : Dystroglycan/ sarcoglycan complex in the sarcolemma.

protein associations stabilize the sarcolemma from inside and outside. Mutation in these proteins forming genes cause muscular dystrophy.

Trophinin: Trophinin mediates cell adhesion between trophoblastic cells and endometrial epithe lial cells, which are the initial attachment, step in human embryo implantation. Trophinin localizes to the cytoplasm particularly at the nuclear envelope (NE).

Focal adhesion: The conversion of physical signals into chemical signaling events is a fundamental cellular process that occurs at cell-ECM contacts, known as focal adhesions. At these sites, transmembrane integrin receptors are associated via their cytoplasmic domains with the actin cytoskeleton. A sub membrane focal adhesion plaque consisting of numerous cytoskeletal and signaling molecules, mediates this interaction with actin.

β-catenin as a mediator of cell cell adhesion: During embryonic development, α-catenin is central both for the transcriptional activation of Wnt target genes and as a mediator of cell-cell adhesion (Bernard *et al*, 2007), Signals that regulate its levels and subcellular localization are critical. One mechanism of Wnt signaling results in stabilization of β-catenin protein, which leads to its translocation into the nucleus where it interacts with TCF and activates transcription of target genes. Less well understood are mechanisms of Wnt signaling that do not involve β-catenin stabilization and result in inhibition of β-catenin-mediated transcription. A member of the Wnt protein family, Wnt4, regulates the subcellular localization of β-catenin, redirecting it to the cell membrane. This action does prohibit its involvement in TCF gene transactivation. This novel mechanism suggests that Wnt4 acts as a switch between the two modes of β-catenin function, transcriptional activation and cell-cell adhesion.

5.9. GENERAL ROLE OF CELL ADHESION PROTEINS

1. Cell adhesion proteins are important for the normal functioning of living organisms.

2. They are important for the function of migratory cells like WBCs.

3. Regulation of cell adhesion proteins is important during embryonic development for the process of morphogenesis.

4. Blistering diseases result from inherited molecular defects in genes for adhesion proteins.

5. Mutations in genes for adhesion proteins that result in abnormal cell-to-cell interactions and tumor growth.

6. Cell adhesion proteins are important for interactions that allow viruses and bacteria to cause damage to humans.

7. Cell adhesion proteins hold synapses together and the regulation of synaptic adhesion is involved in learning and memory. In Alzheimer's disease there is abnormal regulation of synaptic cell adhesion. Synapses are specialized intercellular junctions whose specificity and plasticity provide the structural and functional basis for the formation and maintenance of the complex neural network in the body. The existence of cellular and molecular properties that determine the connectivity of each neuron in the nervous system. Recent evidence has elucidated that several families of CAMs including neuroligins, SynCAM, N-CAM, L1-CAM, cadherins, protocadherins, and integrins regulate these key features of the synapse. Neurexins of the presynaptic membrane interacts with postsynaptic neuroligins to promote adhesion between dendrites and axons bridging the synaptic cleft (Craig and Kang, 2007).

SUMMARY

(1) Attachment of cell by by protein molecules with each other or with non-cellular components of their environment is called cell adhesion. Cellular adhesion is a dynamic process for the recruitment, binding, and reorganization of signaling molecules in the plasma membrane.

(2) Cell adhesion proteins are transmembrane receptors that extend across the plasma membrane and have both the extra cellular and intra cellular domains. The extracellular domain can bind to other molecules causing intercellular adhesion or with the extra cellular matrix (cell-to-ECM adhesion). The binding molecule is called ligand. Adhesion between two copies of the same adhesion protein is called "homophilic" binding while between dissimilar molecule is "heterophilic" binding.

(3) The binding of the intracellular domain of the adhesion protein to a cytoskeletal protein leads to tight adhesion between the cytoskeleton and the ligand. It may be indirect i.e. coupled by special linking proteins called "catenins" or may be direct.

(4) Expression of receptors occurs naturally or may depend upon extra cellular stimuli. Adhesin protein and ligand relationship does not follow any definite rule. Some binds only one ligand while some bind many ligands while many ligands may bind a single type of receptors. Receptor-ligand

relationship may be unstable or stable. Many receptors interact with cytoskeleton providing mechanical strength to the cell. Receptor-ligand binding may activate intracellular signal transduction pathways.

(5) Four main membrane glycoproteins the integrin family, the immunoglobulin superfamily, selectins and cadherins play active roles in adhesion. These are usually named CD followed by a number. The binding sites are different for each of these adhesion molecules.

(6) Integrins are main cellular adhesion receptors linking the cells to other cells as well as to ECM in Ca^{++} and Mg^{++} manner. They bind with ECM protein in the basal lamina while intracellularly with the actin cytoskeleton. Some bind to intermediate filaments like keratin in the cytoplasm.

(7) Integrins are heterodimers of two polypeptides - a and β chains both of which contribute in the ligand binding at least in two sites. Both α and β chains are attached with the membrane by definite stalks. Integrins generally have a low affinity for extra cellular ligands. The cytoplasmic domains of integrins interact with a remarkable variety of signaling and structural proteins like talin, vinculin, paxillin and actin filaments.

(8) Some integrins may require activation in order to bind their ligands. In general they play multiple roles in differentiation, intercellular communication, signal transduction, transmission of internal forces to the matrix. They also play role in the binding of mouse sperm with the egg during fertilization. They facilitate the adhesion of WBC to endothelial cells at the sites of inflammation, cell proliferation during carcinoma, in the apoptosis of the cell, and regulate cell shape. They seem to be essential for embryonic development, tissue regeneration, cellular migration and immune defense.

(9) Immunoglobulin type of cell adhesion molecules is single pass transmembrane protein having 1-7 extra cellular domains. Some Ig-CAM consists of single polypeptide but others show 2 or 4 subunits. The intra cellular domains may bind to the cytoskeletal system while their extra cellular domains may bind to other similar or dissimilar molecule. Ig-CAMs may participate in signaling process, tumor-suppression. Nectins are a family of Ca(2+)-independent Ig-CAM involved in the formation of adherens junctions in epithelial cells and fibroblasts through reorganization of the actin cytoskeleton.

(10) Cadherins are Ca^{2+} dependent adhesion molecules that regulate intercellular recognition and are fundamental to the process of animal morphogenesis. Cadherins are generally present in intercellular junctions, adherens junctions and desmosomes. Almost all vertebrate and some invertebrate cells express one or more cadherins. Specialized cells express characteristic cadherins. During embryonic development expression of different cadherins at different period of development is noted. Cadherin-related neuronal receptor (CNR) proteins are a diverse set of synaptic protocadherins that plays role in synaptogenesis

(11) Cadherines are of four types (a) E and p-Cadherin – bind epithelial cells via catenins, α-actinin and actin filaments and occur as adhesion belts. (b) Desmosomal cadherin - connects epidermal and placental tissues and occur as desmosomes. (c) N-cadherin - occur as adhesion belts, they exhibit catenin, α-actinin and vinculin in the intercellular space and actin in the cytosolic side.

(12) These are single-pass transmembrane glycoprotein composed of about 700-750 residues. Their extra cellular domains contain about 100 amino acids, 5 extra cellular CAD domains and calcium binding sites. In intercellular binding, CAD1 domains of adjacent cells link with each other creating a molecular zipper to bind cells together.

(13) Intercellular adhesion is strengthened by the interaction of cadherins with the actin cytoskeleton through the adapter proteins α and β-catenins and vinculin. In addition to membranous catenins, cytoplasmic pool of β-catenin remains bound to APC and GSK. Nuclear α-catenins recruit transcription factor to regulate gene expression.

(14) They connect cells at specialized junctions like adherens junctions and desmosomes. They participate in the transduction of extra cellular signals.

(15) They indirectly inhibit the spread of cancer cells, checks the transformation of benign tumour to malignant tumour and metastasis of cancer cells.

(16) Desmosal cadherins have comparatively complicated cytoplasmic domains that interact with the proteins – desmoplakin and placoglobin linking cadherins to keratin (intermediate filament).

(17) Selectins are membrane-attached mosaic proteins recognizing carbohydrates. They consist of small cytosolic domain, single trans membrane helix, EGF-domain and N-terminal calcium dependent lectin domain that are involved in recognizing of different carbohydrates and sorting of glycoproteins. Metal ions may play a role too. They determine the endothelial cells to which a lymphocyte will adhere. They are present on the surface of white blood cells, platelets and also the cells lining the blood vessels. Selectin, expressed on WBCs is L-selectin. The two other selectins (P and E selectins) are found on endothelial cells and due to their presence they can attract leukocytes. P and E selectins appear on endothelial cells at different times and recruit different types of WBCs. P-selectins are expressed earlier while E-selectins are synthesized only when they are required.

(18) Selectins bind to cell surface glycoproteins called mucins; their extra cellular domains are rich in serine and threonine.

(19) Membrane bound galactosyltransferase enzyme acts as receptors of those ligands that terminate in N-acetyl glucosamine. On cytosolic end it interacts with actin.

(20) Hyaluronan plays a fundamental role in the regulation of cell migration and differentiation.

(21) GPIB-IX is expressed in platelets and mainly participates in the reception of von Willebrand factor that acts as a bridge between a specific glycoprotein on the surface of platelets and collagen fibrils.

QUESTIONS

Long type questions

1. What do you mean by cell adhesion molecules? Describe the general principle and role of adhesion molecules.
2. Describe the structure and physiological role of integrin.
3. Describe the structure, cellular adhesion and physiological role of Ig-CAM.
4. Describe the structure, cellular adhesion and physiological role of cadherins.
5. Describe the structure, cellular adhesion and physiological role of selectin.

Short type questions:

1. What do you mean by stable and unstable adhesive interactions?
2. What is ligand?
3. What do you mean by homophilic and heterophilic binding?
4. What do you mean by CD?
5. Name the adapter proteins that link the integrins with the cytoskele ton?
6. Which adhesion protein helps the fibroblasts and WBCs to adhere to fibronectin and collagen?
7. Which adhesion protein helps in fertilization?
8. Which adhesion protein participates in apoptosis?
9. To which cell adhesion protein family does fibronectin belong?
10. Which cadherin subtypes are expressed in nervous system?
11. How many CAD domains occur in usual cadherins?
12. Which cadherin occurs in cardio vascular system?
13. Which cadherin occurs in placenta?
14. Which cadherin lacks cytoplasmic domain?
15. Which adhesion protein recognizes carbohydrates?
16. What is the full form of CRD?
17. Which adhesion molecules are expressed in atherosclerosis?
18. Which glycoprotein binds to selectin?
19. Which amino acids are common in the extra cellular domain of mucin?
20. Which adhesion protein mediates embryo implantation in the uterus?
21. What do you mean by focal adhesion?

Fill up the blanks:

1. Cell adhesion is mediated by ——— molecules.
2. Cell adhesion mediates interaction with other cell and with ———.
3. Stable adhesive interactions are noted in ——— and ——— cells.
4. Binding of leucocytes to endothelial cells at the site of inflamation is an example of ——— adhesive interaction.
5. Indirect linking of adhesive protein and actin is done by ———.

6. Integrins link the cells to other cells as well as to ———
7. Up regulation of ——— occurs in malignancy.
8. ——— is a neural cell-to-cell adhesion molecule in developing nerves.
9. ——— is the Ig-CAM required for T cell activation.
10. Osteoblasts express ——— cadherin.
11. ——— cadherin is expressed in muscles.
12. Placoglobin is the principal adhesion protein in ———cadherin.
13. Cytoplasmic pool of β-catenin is regulated by ——— and ———.
14. Hischsprug's disease is caused due to mutation in the CAD domain of ———.
15. ——— type selectins are expressed in WBC.
16. Cell adhesion by selectin depends upon ———-ion.
17. vWF is received by ——— adhesion protein.
18. Dystroglycan is expressed in———.

MCQs:

1. **The adapter protein that links actin and cadherin is**
   (a) Catenin, (b) Flotilin, (c) Caveolin.
2. **Integrins perform adhesion function in**
   (a) Ca-Cl, (b) Na-Ca, (c) Ca-Mg dependent manner.
3. **Cadherins are** (a) Ca^{2+}, (b) Na, (c) Mg **dependent adhesion molecules.**
4. **K-cadherin is expressed in**
   (a) Colon, (b) Kidney, (c) Cornea
5. **E-cadherin is expressed in**
   (a) Eye, (b) Ear, (c) Embryo.
6. **Cadherins**
   (a) Activates,
   (b) Inhibits,
   (c) Plays no role in the spread of cancers.
7. **During metastasis**
   (a) Reduced,
   (b) Enhanced,
   (c) normal levels of β-catenin is noted.
8. **Endothelial cells express**
   (a) E-selectin, (b) P-selectin, (c) P and E- selectins.

Answers to MCQs:

1. a, 2. c, 3. a, 4. b, 5. c. 6. b, 7. a, 8. c.

Chapter 6

Extracellular Matrix (ECM)

One of the basic features of all multicellular organisms is the interdependence of the cells. Each cell must have to depend upon the neighboring cells for mutual coexistence and maintenance of physiological activities of the cell. Thus intercellular adhesion is one of the basic features of the cells. Most cells, even unicellular organisms exhibit molecular mechanism for intercellular adhesion. Many protozoans even adhere during mating, they normally bind microorganisms on their surface and engulf them by phagocytosis. In multicellular organisms each cell secretes a certain varieties of macromolecules that form a particular environment around the cell which govern all types of physiological activities of the cells. This type of extracellular medium comprises many fibrous proteins, linker proteins, complex polysaccharides and an array of adhesion molecules that make specific interaction with the complimentary proteins on the surface and varieties of cells that float in a semisolid environment, which is collectively called ***extra cellular matrix (ECM)***. The nature and composition of ECM varies considerably in different tissues; it is simplest in basal lamina present beneath the epithelial cells and surrounding muscle cells, providing a sort of rug, abundant, sparse and shows diverse types of cellular and fibrous structures in connective tissues. The nature of ECM varies widely in connective tissues, it is liquid in blood and lymph, hard in teeth and bone, fibrous in tendons and ligaments while intermediate in haemopoietic tissues; it is opaque in most cases while transparent in cornea. The abundance, organization and proportion of macromolecular components determine the mechanical properties of ECM. It supports relatively defense less epithelial cells, muscle cells and neuronal cells. In tendons, it provides tensile strength due to the presence of collagen, in bone it is incompressible and rigid owing to the presence of calcified collagen matrix, while it is resilient in skin and blood vessels because of numerous elastic fibres. ECM also provides an arena for cellular defense system housing professional phagocytic cells and immune system cells those roam in the connective tissues to invade the foreign cells. In multicellular organisms each cell interacts with the ECM macromolecules. In fact the connective tissue cells, epithelial cells and neurons secrete collagen fibrils, elastic fibres, adhesion proteins and complex polysaccharides that constitute the ECM.

ECM of vascular network is a series of linked blood vessels composed of the endothelium, a monolayer of cells lining the vessel lumen and surrounding layer(s) of mesenchymal cells (vascular smooth muscle, pericytes and fibroblasts). The ECM is a major constituent of blood vessels and provides a framework in which various cell types are attached. The composition and organization of vascular ECM is controlled by the mesenchymal cells, and is also responsible for the mechanical properties of the vessel wall, forming complex networks of structural proteins. ECM also plays a central role in cellular adhesion, differentiation and proliferation.

6.1. TYPES OF ECM

6.1.1. Connective tissue cells:

The immediate environment surrounding a cell is composed of connective tissue showing a variety of cells that secrete ECM, defend against infection and maintain energy stores in the form of lipid. Some of the cells are indigenous from the very period of their formation while some cells are formed elsewhere and are carried to the connective tissue via blood and lymph. These are called *immigrant cells*.

Indigenous connective tissue cells: ***Mesenchymal cells:*** These are undifferentiated pleuripotent stem cells. During development they differentiate to give rise to all other indigenous cells of the connective tissue. A small number of them may persist into adulthood. These are smaller than fibrocytes and difficult to detect in histological sections. They may regenerate blood vessels or smooth muscle, which have been lost as a conesquence of tissue damage.

Fibrocytes/Fibroblasts: *Fibrocytes* are the most common cell type in connective tissues. They are the "true" connective tissue cells. These are moderate sized spindle shaped cells with many irregular projections Usually the cytoplasm of a resting (i.e. inactive) fibrocyte does not contain many organelles and shows oval or flattened nucleus. But once they are stimulated in response to damage to the surrounding tissue they are transformed into a *fibroblast*, which contains large amounts of RER and large Golgi apparatus that are necessary for the synthesis and secretion of proteins needed to repair the tissue damage. Fibrocytes do not usually leave the connective tissue. They are, however, able to perform amoeboid movement. In response to tissue damage they migrate and proliferate to synthesize new matrix to restore the integrity of the tissue. The terms fibrocyte and fibroblast refer here to the inactive and active cells respectively - at times you will see the two terms used as synonyms without regard for the state of activity of the cell.

Mast cells: These are secretory cells and mediate immediate hypersensitivity reactions by secreting specific chemical. Numerous large vesicles fill the cytoplasm of mast cells. They discharge the contents of these vesicles in response to contact with specific antigens present on the surface of an invading bacterium, as the in allergic reactions, in response to antigens found on the surface of pollen grains, mechanical trauma, radiant energy etc.

The most prominent substances contained in the vesicles are heparin (30%), proteases (30%) and histamine (10%). By facilitating access to the area, mast cells facilitate an immune response to the antigen that triggered the release histamine and heparin. They increase blood flow to the injury site and cause the blood vessels to leak plasma constituents and other white blood cells. It also leads to contraction of the smooth muscles.

Adipocytes: Fat cells or adipocytes are fixed cells in loose connective tissue. There are two types of adipose tissue, which derive their names from the colour of the tissue (white or brown) and/or the number of lipid droplets found in the adipocytes.

White adipocytes contain one large lipid droplet.

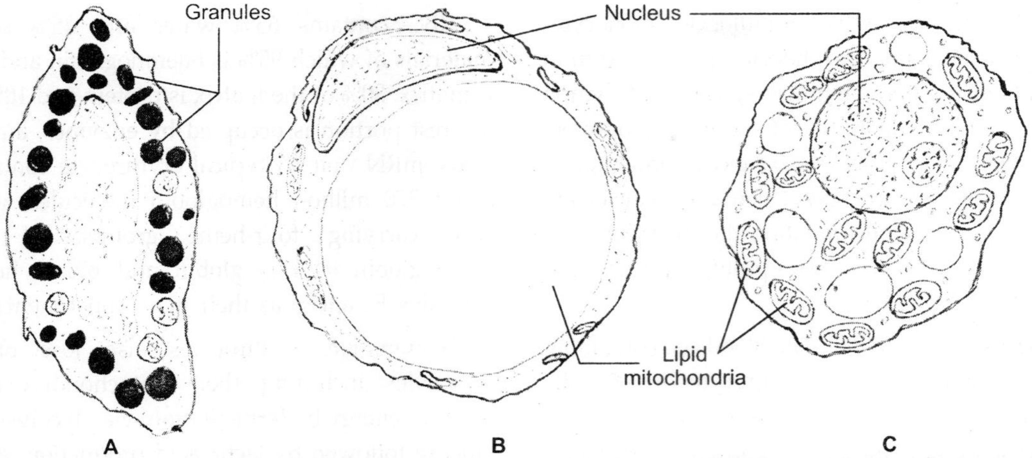

Fig. 6.1 : Some indigenous connective tissue cells A- Mast cell, B- While adipose cell, C- Brown fat cells.

Brown adipocytes contain many comparatively small lipid droplets. White adipose tissue does not only function in the storage of lipids but additionally they perform other roles. For example, in the palms of the hands, on the sole of the feet and in the buttocks it has a structural, cushioning function. In these regions, strong connective tissue fibres surround accumulations of adipocytes. Also, the distribution of white adipose tissue is different in males and females and is part of the secondary sexual characteristics. The storage/mobilisation of lipids does require quite some metabolic activity of the tissue. Consequently, adipose tissue has a rich supply of capillaries. Their main role is the storage of lipids. If "well fed" the cytoplasm only forms a very narrow rim around a large central lipid droplet. The flattened nucleus may be found in a slightly thickened part of this cytoplasmic rim. A "starving" adipocyte contains multiple small lipid droplets. Adipocytes are very long-lived cells. Their number is determined by the number of preadipocytes (*lipoblast*) generated during foetal and early postnatal development. Lipid storage/mobilisation is under nervous (sympathetic) and hormonal (insulin) control. Adipocytes also have an endocrine function -they secrete the protein - leptin that induces the brain regulating fatty food intake appetite. Leptin deficiency results in obesity. Brown adipose tissue occurs mainly during development and may account for 2-5% of the body weight in a newborn. In adult individuals, most of the brown fats differentiate into white fat. Adipocytes in brown fat contain plenty of mitochondria. A very rich capillary supply and the cytochromes found in the mitochondria give the tissue its characteristic colour. A protein (UCP-1 or thermogenin) found in these mitochondria decouples the oxidation of fatty acids from the generation of ATP. Instead, these cells generate heat. The location of the brown fat reflects its heat-generating function. It is located in the axilla (arm pits), between the shoulder blades, in the region of the neck and along large blood vessels. The heat generated by the brown fat warms the blood which supplies nearby organs or which re-enters the trunk from the limbs.

6.1.2. Immigrant connective tissue cells:

These cells include blood cells and macrophages.

Blood cells: Vertebrate blood contains a variety of cells that perform specialized functions. These are derived from common peuripotent stem cells –

haemocytoblast rise RBC, granulocytes, monocytes and platelets while lymphocytes are derived from lymphoid tissues (thymus, spleen and lymph nodes). Minute quantities of specific glycoprotein growth factors regulate the balance of proliferation of these cells. Destruction of stem cells may lead to *aplastic anaemia* while mutation altering the growth control of a stem cell causes proliferative disorder – *leukemia*.

Erythrocytes (RBC): The blood cell containing haemoglobin is called erythrocyte or RBC. Its different features are as follows:

Number: In embryonic condition it is 78 lacs / 100 cu mm blood, in infants it varies from 60 – 70 lacs / 100 cu mm blood and in adult males it is about 50 and in females it is about 45 lacs /100 cu mm blood. If this number declines the condition is called *anaemia* or *oligocythemia* or if it increases substantially it is called *polycythemia*.

Mature erythrocytes in mammals are *anucleate* while the erythrocytes of nearly all other vertebrates have nuclei; the only known exception is *Batrachoseps* (Salamanders). Erythrocytes are circular, except in the family Camelidae, where they are oval. The diameter of a typical human RBC is 6-8μ (av. 7.2 μ).

Adult humans have roughly 2–3 × 10^{13} red blood cells at any given time (women have about 4 to 5 million erythrocytes per microliter (cubic millimeter) of blood and men about 5 to 6 million; people living at high altitudes with low oxygen tension will have more). Red blood cells are thus much more common than the other blood particles: There are about 4,000–11,000 (leucocytes) white blood cells and about 150,000–400,000 platelets in each microliter of human blood. RBC store collectively about 3.5 grams of iron. Mammalian erythrocytes are biconcave, flattened and depressed in the center, with a dumbbell-shaped cross section. This shape (as well as the loss of organelles and nucleus) optimizes the cell for the exchange of oxygen with its surroundings.

RBC contains 65% water and 35% solid materials of which 95% is haemnoglobin and 5% is matrix. 80% of the matrix is protein and 10% is fat; rest portion is occupied by enzymes, amino acids, mRNA etc. A typical erythrocyte contains about 270 million hemoglobin molecules, with each carrying four-heme groups. Of the haemoglobin 94% is globin and 6% is heme (contains Fe and S as their main constituent).

Mammalian erythrocytes also lose other organelles including their mitochondria and produce energy by fermentation, via glycolysis of glucose followed by lactic acid production. As a result of the lack of nucleus and organelles, the cells cannot produce new structural or repair proteins or enzymes and their lifespan is limited (usually 120 days). In large blood vessels, red blood cells sometimes occur as a stack, flat side next to flat side. This is known as *rouleaux formation*, and it occurs more often if the levels of certain serum proteins are elevated, as for instance during inflammation.

6.1.3. Macrophages:

Macrophages arise from monocytes. These originate in the bone marrow from where they are released into the blood stream. They are actively mobile and leave the blood stream to enter connective tissues, where they differentiate into macrophages. Macrophages change their appearance depending on the demand for phagocytotic activity.

Origin of macrophages: Monocytes grow in the bone marrow and circulating monocytes respond to chemical mediators of inflammation. Upon activation by these mediators, monocytes squeeze through the endothelium and gradually transform into macrophages.

Roles of macrophages: Macrophages are the scavenger in the connective tissues.† They phagocytose pathogens, dead cells and cellular debris. Macrophages also present fragments of antigens on their cell membranes. After ingestion,

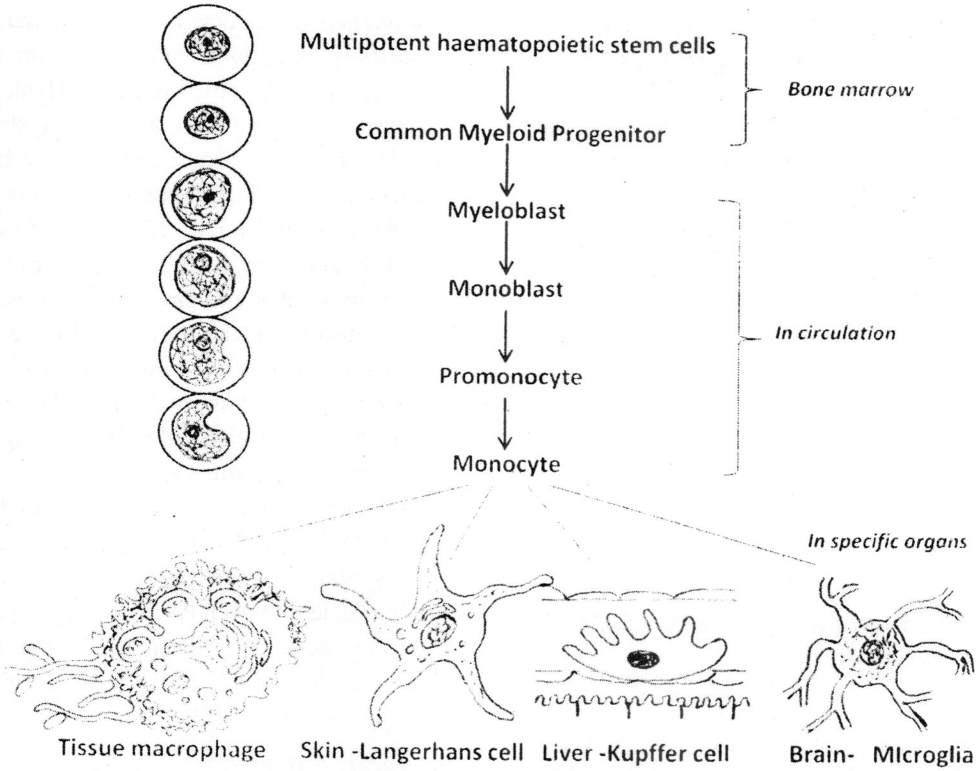

Fig. 6.2 : Different types of macrophages.

the pathogen becomes trapped in a food vacuole, which then fuses with a lysosome and in turn are digested. They are also part of the innate immune system.

Types of macrophages:

A. ***Dendritic cells (DC)*** - are immune cells and form part of the immune system. They occur at a low frequency in the skin (where they are often called *Langerhans cells*) and the lining of nose, lungs, stomach and intestines. In immature state, they can also be found in blood. Once activated, they migrate to the lymphoid tissues where they interact with T and B lymphocytes to initiate the immune response.

In vertebrates, following different types of DCs are known:

a) *Myeloid dendritic cells (MDC)* - are similar to monocytes. MDC are divided into two subsets, MDC-1 and the rare MDC-2, which may have a function in fighting wound infection.

b) *Plasmacytoid dendritic cells* **(*PDC*)** - they look like plasma cells, but have certain characteris tics similar to myeloid dendritic cells. They are also known as *interferon producing cells (IPC)* as they produce high amounts of interferon a.

c) *Follicular dendritic cells* **(*FDC*)** are probably not of hematopoietic origin, but simply look similar to 'true' dendritic cells.

Altered function of dendritic cells is also known to play a major or even key role in allergy and autoimmune diseases like lupus erythematosus.

B. ***Glial cells:*** Glial cells provide support and protection for neurons. They are thus known as the "glue" of the nervous system. Threse are of three types - *astrocytes, oligodendrocytes,* and *microglia.* They mainly surround neurons and hold them in place, supply nutrients and oxygen to neurons, insulate one neuron from another, and destroy and remove dead neurons.

C. ***Microglia:*** These are a kind of glial cell that act as the immune cells of the central nervous system (CNS). They are the smallest of the glial cells and act as phagocytes, cleaning up CNS debris. Microglia are thought to play a role in neurodegenerative disorders such as Alzheimer's disease, dementia, multiple sclerosis and Amyotrophic lateral sclerosis.

6.1.4. Glycosaminoglycans (GAGs):

As the name implies these components show long disaccharides (usually hexuronic acid and hexosamine) and polysaccharides, linked with amino group. These were formerly called mucopolysaccharides. Almost all GAGs are covalently linked to proteins to form proteoglycans. These constitute 10% of the total fibrous proteins of the ECM. They form gel like consistency and attract cations, which are osmotically active and thus hold water in the ECM. They can withstand tremendous pressure.Their deficiency, of synthesis may lead to dwarfness, joint, bone and muscle pain, which collectively lead to premature aging. Hyaluronan (formerly called hyaluronic acid) is the simplest GAG. It can be produced very quickly by the enzymes on the cell surface and therefore plays vital role in wound repair. Networks of densely charged hyaluronan restrict water flow, limit diffusion of solutes and impede the passage of microorganisms. Hyaluronans also act as lubricants in joint cavities and as an optically transparent space filling medium in the vitreous body of the eye. It also helps in cell migration and differentiation during organogenesis and eventually excess hyaluronan becomes degraded by hyaluronidase. Synthesis of GAG chains involve synthesis of core protein in the RER. The transfer vesicles between ER and Golgi apparatus translocate these core proteins. In the Golgi, oligosaccharides are added to serine or asparagines residues of the core protein.

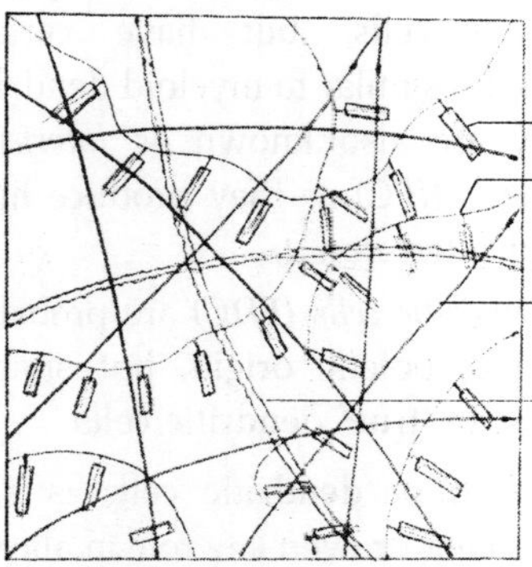

Fig. 6.3 : Disposition of proteoglycan hyaluronan, type II collagen and elastic fibres.

6.1.5. Proteoglycans (Fig. 6.4):

These are specific types of glycoproteins and consist of linear chains of repeating disaccharide units whose one monomer has an amino group. All these constituents eventually form a web like structure. Cells secrete proteoglycans in the ECM but they may retain some types on the plasma membrane. The core protein varies in size from 100 to 4000 amino acids. The core protein shows intimate association with GAGs. The number of GAGs attached to the core protein varies from one (decorin) to more than 200 (aggrecan). The nature of attached GAGs depends upon the core protein e.g. fibroglycan, glypican and versican have same but aggrecan, serglycin and syndecan have different GAGs. Proteoglycan and collagen ratio determines the consistency of the ECM; cartilage with high proteoglycan content is soft, tendon with very low proteoglycan but high collagen is tough while basal lamina with predominance of proteoglycan provide structural support having

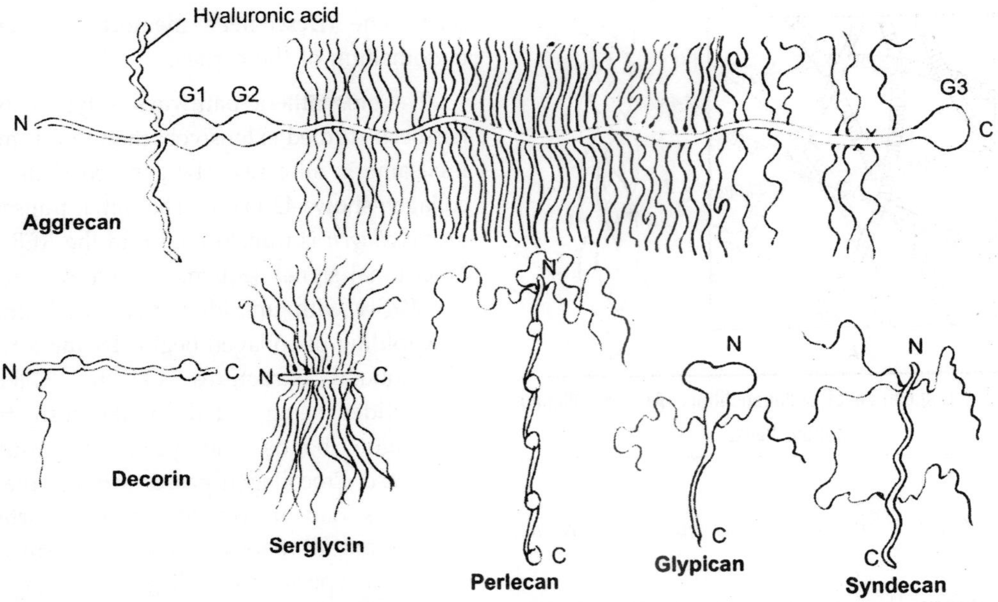

Fig. 6.4 : Different types of proteoglycans.

sieving role. Some of the common proteoglycans are aggrecan, present in cartilages, providing mechanical support; β-glycan, occurring in the cell surface; decorin present in the connective tissues and help in binding TGF-β; periecan present in the basal lamina; serglycin present in secretory vesicles that help in packaging and storage of secretory molecules; syndecan-1, present in fibroblast and mainly help in the cell adhesion. Some proteoglycans form a coating surrounding the cell and help to maintain the identity of the cells, some mediate intercellular connections and some promote growth of neighbouring cells while some mediate the binding of the growth factors to their corresponding receptors. Some proteoglycans including syndecan and CD44 remain concentrated in the plasma membrane having exposed GAGs on the cell surface. Syndecan can also link cells to fibronectin and connective tissue collagens. Lymphocyte plasma membrane express syndecan early in their maturation when they adhere in the bone marrow matrix and later when mature cells adhere in the matrix of the lymph nodes. Retention of water by aggrecan-keratan sulfate/chondroitin sulfate is essential for cartilage. Along with hyaluronans proteogly cans also act as lubricants in joints and act as transparent space filling medium in the vitreous body of the eye. The tissue and age specific distribution of proteoglycans suggests their specific functions. They can promote and inhibit cell motility or adhesion. Decorin and fibromodulin regulate the assembly of collagen fibrils. Expression of specific proteoglycans at specific times during wound healing and embryonic development suggests their specific role in the signal transduction in the organization of connective tissues. Many polypeptide growth factors (FGF, bFGF, PDGF, TGF-β) bind to proteoglycans in the ECM and release them at specific time that is important in many aspects of embryonic development.

6.1.6. Fibrous ECM proteins:

One of the major constituents of ECM is fibre, which provides tensile strength to the ECM.

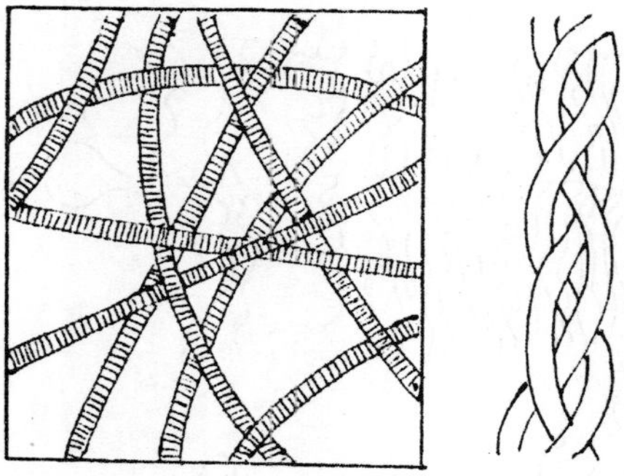

Fig. 6.5 : Disposition of collagen fibrils (A) and collagen tripple helix.

Some of the common fibrous elements are as follows:

Collagen (Fig. 6.5.): Collagens are diverse family of ECM proteins and form wide range ECM proteins with remarkable mechanical properties. 25% of the fibrous ECM proteins are collagen. Humans show 100 genes while nematodes show 160 genes for transcribing collagen. In different animals 15 different types of collagen molecules forming 25 different categories are known which are also glycoproteins, rich in proline and glycine amino acids. These show triple helix of polypeptide a chains showing repeating domains of glycine-proline-hydroxyproline that wound around one another to form a super helix. The a chains often form homotrimers but mostly they form heterotrimers with different types of collagens. The collagen molecules primarily assemble to form 300 nm long banded collagen fibrils (10-300 nm diam), which further aggregate to form collagen fibres (0.5-3 nm diam.) (Fig. 6.5). The size and shape of collagens vary according to functions and they are referred numerically like type I, II III, IV on the basis of their discovery. Type IV collagen form a net like polymer that polymerizes into sheets which surround organs, epithelia and even the whole animals. Hexagonal nets of type VIII collagen form special basement membrane (Descemet's membrane) under the endothelium of the cornea.

The biosynthetic pathways of type I collagen has been studied extensively though it is more or less same in all cases. The gene encoding type I collagen shows 42 exons. The initial transcript or *preprocollagen* is translocated into the RER where N-terminal signal sequence is removed to yield *procollagen* with unfolded α-chains. Within the RER folding of collagen begins by the formation of globular structures of the C-terminal propeptides being aided by the formation of disulphide bonds, catalyzed by disulphide isomerase. Procollagen passes through the Golgi apparatus and moves to the cell surface in vesicles and ultimately is secreted outside. Some cells have specialized collagen assembly sites. Outside the cell procollagen proteases cleave the propeptides forming mature collagen molecule that gradually self-assembles into fibrils. Weak noncovalent bonds between collagen molecules specify the self assembly of fibrils but it provides little tensile stength therefore covalent cross linking is required for reinforcement which is catalyzed by the enzyme lysyl oxidase.

Fibrillar and sheet forming collagens are linked to other structures by connecting and anchoring collagens. Type VII collagens form homotrimers, they self associate to form antiparallel dimers, several dimmers associate laterally to form so called anchoring fibrils that link type IV collagen of the basal lamina to the underlying connective tissue. Type IX collagen links glycosaminoglycans to type II collagen fibrils. Type IX collagens do not polymerize but associate laterally with type II collagen fibrils.

The fibrils provide tensile strength to tendons, ligaments, bones and dense connective tissues. They also form fabric for cartilages and vitreous body in the eye. These are organized in different ways in different tissues. Beneath the skin, in loose connective tissues an open network of individual fibrils or small bundles of fibrils are interwoven in different directions as mats to

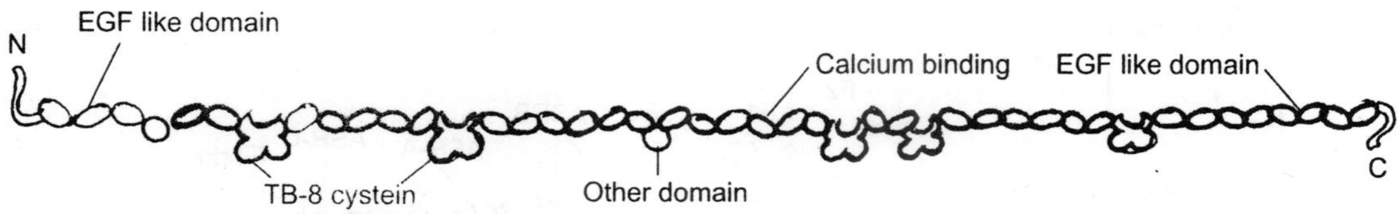

Fig. 6.6 : Domain organization of fibrillin.

resist tensile strength. In tendons, these are aligned as parallel bundles to resist the tension in longitudinal direction. In bone, type I collagen fibrils, reinforced by calcium phosphate crystals are aligned as parallel bundles that are arranged in right angles to each other. In cartilage and vitreous body of the eye type II collagen fibrils along with glycosaminoglycans and proteoglycans retain enough water in the matrix to resist the compression. In the cornea layers of orthogonal collagen fibers make the organ transparent to provide an optically clear path for light.

Elastin: In contrast to the non-extensible collagen, ECM of most of the pulsatile organs like lungs, heart, wall of arteries and in the skin there occurs a network of elastic fibres called elastin. It is also proline and glycine rich protein showing alternate arrangement of hydrophobic and alanine/lysine rich segments. Basically it consists of network of fibrillin microfibrils, embedded in an amorphous core of cross-linked elastin. Fibroblasts produce both fibrillin and elastin. In the animal kingdom only vertebrates produce elastin, Molluscs produce elastic fibres composed of abductin while an elastic fibre of insects consists of resilin proteins. Fibrillin microfibrils (Fig. 6.6) measures 10 nm in length that actually shows two proteins Fibrillin-I and II (encoded by two genes) along with several glycoproteins. Intermolecular disulphide bonds contribute to the structural integrity and insolubility. Elastin core shows tropoelastin proteins that assemble on the surface of elastic fibres. After assemblage lysyl oxidase oxidizes lysines of tropoelastin, which gradually condense into desmosine, ring that cross-links tropoelastin molecules. The hydrophobic segments between the cross links form extensible random coils that account for the elastic properties of the fibrils. Only embryonic and juvenile fibroblasts synthesize elastic fibres. Therefore many tissues (skin) gradually loose elasticity keeping pace with age, which is the main cause of wrinkle ness of the skin in old age.

Along with another protein, fibrillin, elastin provides integrity to the elastic fibres. Sometimes collagen fibres are physically interwoven with elastin to increase the tensile strength bearing capacity. Elastin is predominant in the wall of aorta, and arterial elastic fibres withstand more than 2 billion cycles of stretching and recoil in the whole life. Mutation in fibrillin-1 gene leads to Marfan's syndrome that result in fragility of the wall of aorta. In such patients elastic fibres are poorly formed which along with other symptoms show enlargement (called aneurysm) and eventually rupture of the blood vessels. In some cases defective mitral valve in the heart causes regurgitation of blood from left ventricle back into the left atrium. Weak suspensory ligaments in the eye result in the dislocation of the lens and impaired vision. Mutation in fibrillin-II causes contractural arachnodactyly showing joint stiffness. However most of these patients are heterozygous. A dominant mutation in the elastin gene causes a disease called *cutis laxa*.

Fibronectin: These are very large (460 kD) fibrous adhesive glycoprotein showing two dimers joined by two disulphide bonds. Each monomer shows a number of domains and each domain consists of serially repeated smaller modules. It has several binding sites whose

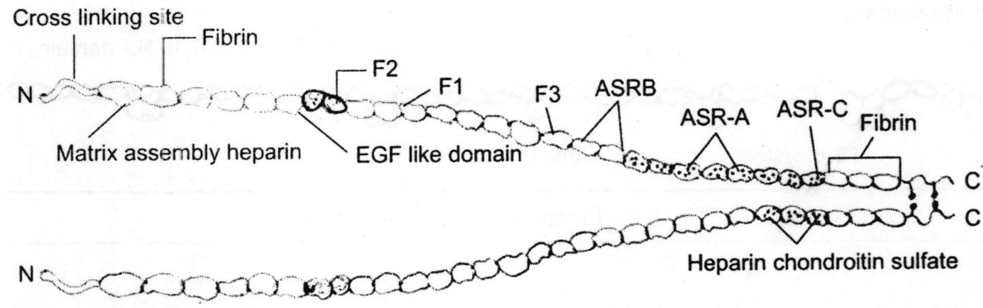

Fig. 6.7 : Domain organization of fibronectin.

interaction with appropriate molecules lead to proper alignment of cells with ECM (Fig. 6.9). Fibronectin helps in cell migration during gastrulation and development of heart.

Laminin: (Fig. 6. 8) It is a major constituent of basal lamina and and provides binding of cells with the basement membrane. It is a large multichain glycoprotein, showing three polypep tide chains – A, B_1 and B_2 (Fig.6.8). These three chains are linked through disulphide bonds to form asymmetric cross structures, however they exhibit great diversity in different organisms. All the three chains of to type IV collagen, heparan sulphate, and entactin and laminin. Laminin show a number of functional domains that binds receptors. Entactin in fact acts as a bridge between type IV collagen and laminin in basal lamina. It is the first ECM proteins that appear during development and plays vital roles in cell attachment, movement, differentiation and growth. Laminin also act as a substrate adhesive molecule for that help in adhesion of epithelial cells to the basal lamina.

Tenascin / Cytotactin: It resembles fibronectin and is found in several ECMs during embryonic development. However it does not always promote cell adhesion (Fig. 6.9).

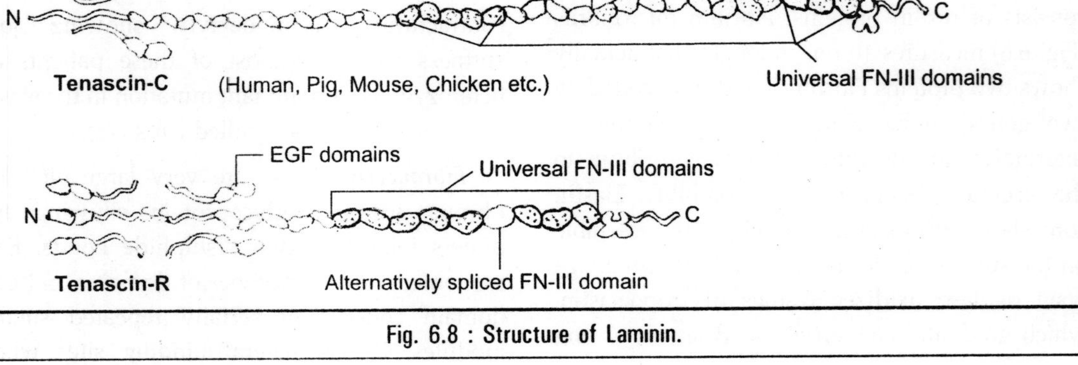

Fig. 6.8 : Structure of Laminin.

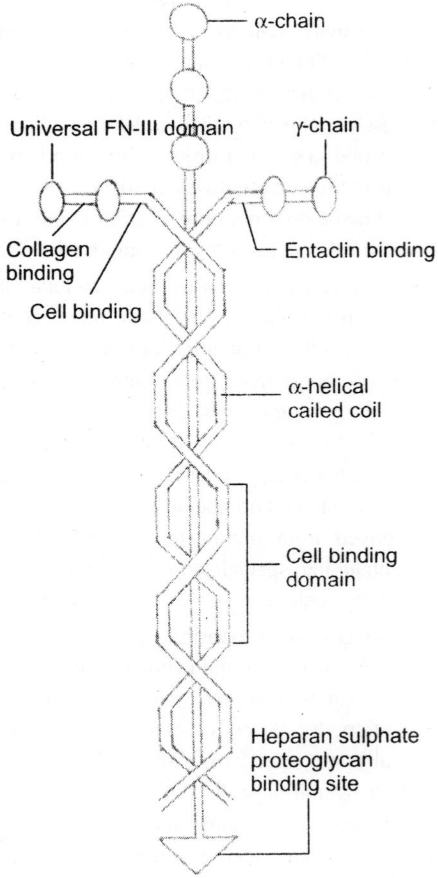

Fig. 6.9 : Domain organization of three isoforms of Tenascin.

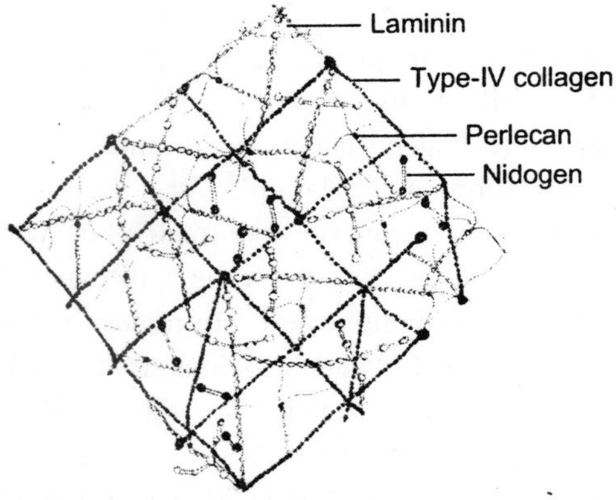

Fig. 6.10 : Struture of basal lamina.

6.1.6. Basal lamina

In between epithelial cell sheets and connective tissues, the matrix is transformed in to tough, thin membranous structure celled basal lamina. In muscle cells, fat cells and Schwan cell basal lamina surrounds individual cells, in nephron basal lamina act as selective filter. Basal lamina differs from other ECM in biogenesis to provide its flexibility as sheet like multi-layered networks. Mostly it shows two distinct layers – *lamina lucida or rara,* present towards the plasma membrane of epithelial cells and *lamina densa,* lying below the lamina lucida. Sometimes a third layer – *lamina fibroreticularis* connects the basal lamina with underlying connective tissue. These three layers are collectively called basement membrane. The composition of basal laminae differs in different tissues though the most important constituent is the type IV collagen, a heparan-perlecan, laminin and entactin (Fig. 6.10).

Important functions of basal lamina include *a) Filtration of macromolecules-* The GAG chains of heparan sulpate filters most of the physiologically important macromolecules from the blood during the formation of urine. *b) Act as barrier-* Basal lamina act as barrier against different ECM cells and keep the epithelial cells quite distinct from other ECM constituents. *c) Wound repair-* basal lamina directly participates in wound repair. *d) Cell migration-* Basal lamina acts as a medium through which cells migrate during embryonic development.

6.2. FORMATION OF ECM

ECM represents the common medium of cell, fibres and matrix. In fact the cytoskeletons of the ECM cells actually determine the orientation of the macromolecules produced by the cells. Mostly these are secreted by the fibroblasts or cells belonging to fibroblast family members like chondroblasts in cartilage or osteoblasts in bones.ECM macromolecules are divided into two main categories – polysaccharides and proteins. Polysaccharides mainly belong to the class glycosaminoglycans (GAGs), which are linked with the proteins in the form of proteoglycans. Proteins fall into two main categories – structural proteins (collagen, elastin etc.) and adhesive proteins (fibronectin, laminin etc.). GAGs and proteoglycans form gel like matrix in which fibrous proteins causes rapid diffusion of nutrients, metabolites and hormones between blood and cells. Collagen provides strength while elastin provides elasticity to the ECM. An adhesive protein promotes attachment of respective cells with the ECM (e.g. fibronectin helps in the adhesion of fibroblasts with ECM and laminin provides the basis of attachment of epithelial cells with the basal lamina.

SUMMARY

1. Most cells, even unicellular organisms exhibit molecular mechanism for intercellular adhesion. In multicellular organisms each cell secretes a certain varieties of macromolecules (protein and polysaccharides) that form a particular environment around the cell which govern all types physiological activities of the cells. It is called Extracellular matrix. The nature and composition of ECM varies considerably in different tissues; it is simplest in basal lamina present beneath the epithelial cells and surrounding muscle cells, providing a sort of rug. abundant, sparse and shows diverse types of cellular and fibrous structures in connective tissues. ECM also provides an arena for cellular defense system and professional phagocytic cells and immune system cells roam connective tissues to invade the foreign cells.

2. The immediate environment surrounding a cell is composed of connective tissue showing a variety of cells that secrete ECM, defend against infection and maintain energy stores in the form of lipid. Some of the cells are indigenous from the very period of their formation while some cells are formed elsewhere and are carried to the connective tissue via blood and lymph.

   Mesenchymal cells are undifferentiated pleuripotent stem cells. During development they differentiate to give rise to all other indigenous cells of the connective tissue. Fibrocytes do not usually leave the connective tissue. Mast cells are secretory cells and mediate immediate hypersensitivity reactions by secreting specific chemical. Numerous large vesicles fill the cytoplasm of mast cells. Adipocytes are fixed cells in loose connective tissue. White adipocytes contain one large lipid droplet. Brown adipocytes contain many comparatively lipid droplets. White adipose tissue does not only function in the storage of lipids. Adipocytes are very long-lived cells.

   Immigrant connective tissue cells include blood cells and macrophages. Vertebrate blood contains a variety of cells that perform specialized functions. These cells are derived from common peuripotent stem cells. Macrophages arise from monocytes. Kupffer cells or Browicz-Kupffer cells are spe cialized macrophages located in the liver that form part of the reticuloendothelial system.Macrophages are derived from monocytes. (Some monocytes differentiate into specialized cells such as dendritic cells or microglia).

   Macrophages are responsible for phagocytosis of pathogens, dead cells and cellular debris. Helper T cells recognize this and release a lymphokine notification to B cells. Macrophages become especially attracted to cells with antibodies attached. Dendritic cells (DC) are immune cells and form part of the mammal immune system. The name veiled cells refers to immature or maturing dendritic cells; these cells do not yet possess the dendrites, but rather larger cytoplasmic 'veils'.

   Myeloid dendritic cells are similar to monocytes. Plasmacytoid dendritic cells look like plasma cells, but have certain characteristics similar to myeloid dendritic cells. Follicular dendritic cells that are probably not of hematopoietic origin, but simply look similar to 'true' dendritic cells.

   Glial cells include three types of CNS supporting cells are Astrocytes, Oligodendrocytes, and microglia. Microglia are a type of glial cell that act as the immune cells of the Central nervous system (CNS). Microglia are close cousins of other phagocytic cells including macrophages and dendritic cells. Microglia are derived from myeloid progenitor cells (as are macrophages and dendritic cells) which come from the bone marrow.

   ECM represents the common medium of cell, fibres and matrix. ECM macromolecules are divi ed into two main categories – polysaccharides and proteins. Glycosaminoglycans (GAGs) show -ing di or polysaccharides and proteoglycans form gel like matrix in which fibrous proteins cause rapid diffusion of nutrients, metabolites and hormones between blood and cells. Proteins fall into two main categories – structural proteins (collagen, elastin etc.) and adhesive proteins (fibronectin, laminin etc.). Collagen provides strength while elastin provides elasticity to the ECM. Adhe sive proteins promote attachment of respective cells with the ECM (e.g. fibronectin helps in the adhesion of fibroblasts with ECM and laminin provides the basis of attachment of epithelial cells with the basal lamina.

3. Almost all GAGs are covalently linked to proteins to form proteoglycans. These constitute 10% of the total fibrous proteins of the ECM. Cells secrete proteoglycans in the ECM but they may retain some types on the plasma membrane. The core protein shows intimate association with GAGs. Syndecan can also link cells to fibronectin and connective tissue collagens. Decorin and fibromodulin regulate the assembly of collagen fibrils. 25% of the fibrous ECM proteins are collagen. Collagens are diverse family of ECM proteins and form wide range ECM proteins with remarkable mechanical properties. Humans show 100 genes while nematodes show 160 genes transcribing collagen. The a chains often form homotrimers but mostly they form hetero trimers with different types of collagens. The collagen molecules primarily assemble to form 300 nm long banded collagen fibrils (10-300 nm diam), which further aggregate to form collagen fibres (0.5-3 nm diam.). Hexagonal nets of type VIII collagen form special basement membrane (Descemet's membrane) under the endothelium of the cornea.

4. The gene encoding type I collagen shows 42 exons. Some cells have specialized collagen assembly sites. Outside the cell procollagen proteases cleave the propeptides forming mature collagen molecule that gradually self-assembles into fibrils. Type VII collagens form homotrimers, they self associate to form antiparallel dimers, several dimmers associate laterally to form so called anchoring fibrils that link type IV collagen of the basal lamina to the underlying connec tive tissue. Type IX collagen links glycosaminoglycans to type II collagen fibrils. Type IX colla gens do not polymerize but associate laterally with type II collagen fibrils.

The fibrils provide tensile strength to tendons, ligaments, bones and dense connective tissues. Elastin core shows tropoelastin proteins that assemble on the surface of elastic fibres. Only embryonic and juvenile fibroblasts synthesize elastic fibres. Along with another protein, fibrillin, elastin provides integrity to the elastic fibres. Sometimes collagen fibres are physically interwoven with elastin to increase the tensile strength bearing capacity. Entactin in fact acts as a bridge between type IV collagen and laminin in basal lamina. However it does not always promote cell adhesion.

In between epithelial cell sheets and connective tissues, the matrix is transformed in to tough, thin membranous structure celled basal lamina. In muscle cells, fat cells and Schwan cell basal lamina surrounds individual cells, in nephron basal lamina act as selective filter. Mostly it shows two distinct layers – lamina lucida or rara, present towards the plasma membrane of epithelial cells and lamina densa, lying below the lamina lucida. Basal lamina act as barrier against different ECM cells and keep the epithelial cells quite distinct from other ECM constituents. Basal lamina directly participates in wound repair. It also acts as a medium through which cells migrate during embryonic development.

QUESTIONS

Long questions:

1. Describe different constituents of connective tissue.
2. Describe the role of GAGs and proteoglycans in maintaining the structure of the tissue.
3. Describe the structure and functions of fibrous ECM?

Short questions:

1. Name some indigenous and immigrant connective tissue cells.
2. Differentiate fibrocyte and fibroblast.
3. Which substances are released from the mast cells?
4. Differentiate white and brown adipocytes.

5. What do you mean by polycythemia?
6. Which dendritic cells are not derived from hematopoietic organ?
7. Which cells are called 'glue' of the nervous system?
8. Which cells are the scavengers of the nervous system?
9. What are the major roles of Glycosaminoglycans?
10. Which amino acids predominate in the collagen?
11. Describe the disposition of collagen in different organs maintaining their structure.
12. What is the cause of wrinkle ness in the skin in old age?
13. How Murfan's syndrome affects the arterial wall?
14. Which ECM component plays vital role in the cell migration during gastrulation?
15. In which cells basal lamina surrounds individual cells.
16. Describe the structure and functions of Basal lamina.

Fill up the blanks:

1. Fibroblasts show large amounts of —— and ——.
2. Adipocytes secrete —— that regulates fatty food intake.
3. Brown adipocytes occur mainly in the —— babies.
4. Heat generation is mainly related with —— adipocytes.
5. The macrophages of liver are ——.
6. Macrophages are part of the —— immune system.
7. Macrophages are the —— in the connective tissue.
8. Dendritic cells are also known as ——.
9. Proteoglycans are proteins combined with the polysaccharide ——.
10. Fibronectin is an —— protein.
11. Elastin is a —— protein.
12. GAGs were formerly known as ——.
13. The simplest GAG is ——.
14. The GAG present in the vitreous body of the eye is ——.
15. Cartilage and tendon show —— and —— proteoglycan content respectively.
16. A mutation in the elastin gene causes the disease called ——.
17. Mutation in fibrillin-1 gene leads to —— syndrome.
18. —— is the major constituent of basal lamina binding cells with the basement membrane.
19. —— acts as a bridge between type IV collagen and laminin in basal lamina.

MCQs:

1. **Mesenchymal cells are**
(a) Unipotent, (b) Multipotent, (c) pleuripotent cells.
2. **Histamine is released from**
(a) adipocytes, (b) fibrocytes, (c) mast cells.
3. **Macrophages are derived from**
(a) basophills, (b) monocytes, (c) lymphocytes.
4. **Decorin, proteoglycan shows**
(a) 50, b, 100, (c) 1 attached GAG.
5. **The ratio of proteoglycan and**
(a) collagen,
(b) elastin,
(c) decorin determines the consistency of the ECM.
6. **The common proteoglycan in the basal lamina is**
(a) aggrecan, (b) serglycin, (c) periecan.
7. **The ECM of heart and lung shows**
(a) collagen, (b) elastin (c) GAG.
8. **Glycine occurs at every third amino acid position in collagen. Substitution of any other amino acid results in unstable collagen molecules because of**
(a) a decrease in the number covalent crosslinks of interchain,
(b) destabilization of intrachain hydrogen bonding,
(c) weakened ionic interactions between oppositely charged amino acids.
9. **Kupffer cell phagocytize**
(a) nonfunctional RBC, (b) bacteria, (c) both.

Answers to MCQs:

1. c, 2. c, 3. b, 4. c, 5. a, 6. b, 7. b, 8. b, 9. c.

Chapter 7

Cytoskeleton

With advent of modern cytological tools it is clear that morphological and physiological parameters of a cell is regulated by an intricate anastomizing network of filaments that not only regulate cell shape and resist cellular deformities during various morphological changes but also maintain cellular motility, intracellular transportation of various proteins as well as movement of chromosomes during cell division. They support the fragile plasma membrane and provides mechanical strength that helps the cell to withstand the stresses and strains. It enables sperms to swin while fibroblasts and WBCs to crawl across the blood vells and participate in wound healing. Cytoskeleton consists of three major components – actin filaments, microtubules and intermediate filaments. The prokaryotic origins of all the components have now been established.

Actin filaments are flexible, so they must be cross linked into bundles to bear compression forces. High tensile strength allows them to bear forces. *Microtubules* (MTs) are rigid, hollow rod like structures that can sustain both compression as well as tension. *Intermediate filaments* are flexible cables having considerable tensile strength but little capacity to resist compression. They check excessive stretching of cells by external forces. The dynamic nature of cytoskeletal polymers allows cells to change shape within fraction of seconds. The ability of actin and microtubules to resist mechanical deformations also helps cells to maintain cell shape. In fact together with extracellular matrix, the shape of individual cell defines the shape of tissues and organs.

Irrespective of the qualities of the cytoskeletal filaments, their usefulness actually depends upon a large number of *accessory proteins* that link the filaments to other cell components. In addition, some *motor proteins* (a special class of enzyme) help to move the organelles or other components or even the filaments to different locations within the cell. They use energy from ATP to slide along the MTs.

7.1. ACTIN FILAMENTS OR MICROFILAMENTS

Actin is a globular protein that polymerizes helically forming ***actin filaments*** (about 7-8 nm in diameter). These are being the thinnest of the cytoskeletal filaments, are also called ***micro filaments***. By mass these are dominant structural component in many eukaryotic cells accounting for as much as 15% of the total cellular protein. It is a highly conserved protein. The amino acid sequence of actin from *Acanthamoeba* a soil amoeba is 95% identical with vertebrate isoforms. The actin microfilament network is important in maintaining cell shape and function in eukaryotic cells. It has a multitude of roles in cellular processes like cell adhesion, motility, cellular signalling, intracellular trafficking and cytokinesis. Even in rod shaped bacteria, proteins are structurally very similar to actin.

Structure of actin and the microfilament (Fig. 7.1): The actin monomer is approximately pear shaped showing four domains but folded into two asymmetric domains with a large cleft called ATPase fold that almost bisect the molecule. Towards the bottom of the cleft there are binding sites for Mg^{++} and ATP or ADP. Actin has higher affinity for ATP than ADP. ATP is hydrolyzed immediately after the molecule is incorporated into an actin filament. The ADP is trapped in the actin filament until it depolymerizes. Both N and C termini are near the bottom of the small domain. Post translational modification of actin includes acetylation of the N-terminus and methylation of histidine-68 or phosphorylation on tyrosine-211. Most organisms have genes for several actin isoforms. Humans have 6; *Dictyostelium* has more than 10 while budding yeast have1actin coding genes. Plant actins differ from animal actins and have diverged among themselves to a greater degree. Actin isoforms are biochemically highly similar, humans express β and γ isoforms in non-muscle cells and four α and β isoforms in various muscle cells. Many non muscle cells express both β and γ isoforms while RBCs show only β actin.

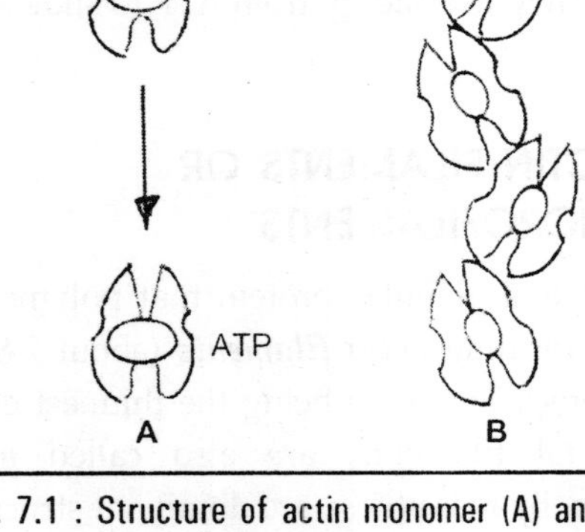

Fig. 7.1 : Structure of actin monomer (A) and actin filament (B).

The exact atomic model of the microfilament is yet to be defined but EM studies reveal that the filament appears as double stranded helices and due the asymmetrical structure of the monomers (G-actin) the ends of the filament (F-actin) are referred to as 'pointed' or (-) end (the end at which ATP binding cleaft is exposed) and 'barbed' or (+) end (at which the cleft contacts the neighbouring actin subunit). In the filament, each monomer is rotated 166° with respect to the neighboring monomer. It means that after every 13 subunits the filament rotates. In physiological conditions, actin molecules spontaneously self associate to form microfilament at both ends of the filament. The overall rates of monomer assembly at both ends differ; the barbed end is ten times preferred for monomer addition over the pointed end where disassembly is more favoured. This phenomenon is known as *treadmill effect*. If a solution containing myosin is added to actin filaments, they will "decorate the filaments" and they will be pointing in the direction of the plus end.

More than 60 families of actin binding proteins regulate the assembly and dynamics of actin based structures in the cytoplasm. Cells are endowed with a rich variety of *actin related proteins (Arps)* (Arp1, 2, 3 .) that participate in a variety of molecular interaction with actin.

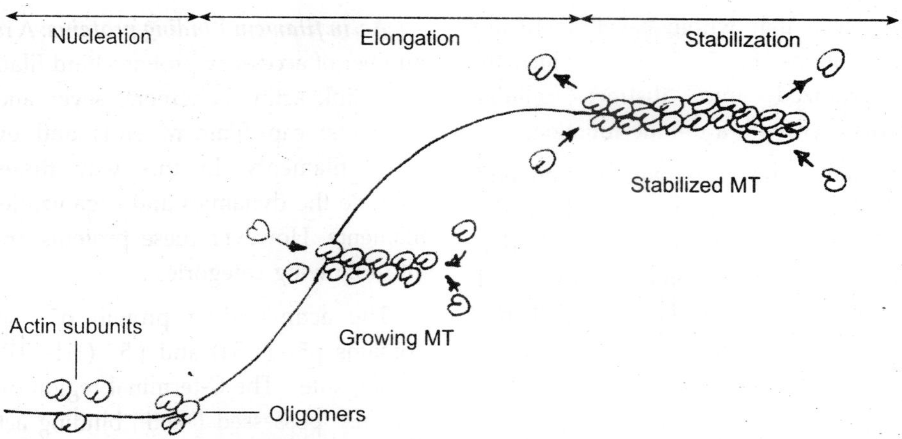

Fig. 7.2 : Different phases of the polymerization of G-actin into F-actin.

Microfilament assembly: The process of *actin polymerization or nucleation,* (Fig. 8.2) starts with the weak binding of two actin monomers but addition of a third stabilizes the complex forming a trimer. This trimer then adds additional molecule and forms a *"nucleation site"*. This is the *slow or lag phase* of the polymerization process. It is followed by *elongation* phase. ATP-actin then binds the plus (+) end and the ATP is subsequently hydrolyzed, which reduces the binding strength between neighboring units and generally destabilizes the filament. This is called *steady state phase.* ADP-actin dissociates from the minus end and the increase in ADP-actin stimulates the exchange of bound ADP for ATP, leading to more ATP-actin units. The '+' end elongates 5-10 times fast than the '-' end.

Nucleating proteins: The protein *cofilin* binds to ADP-actin units and promotes their dissociation from the minus end and prevents their reassembly. The protein *profilin* reverses this effect by stimulating the exchange of bound ADP for ATP. Profilin and ATP hydrolysis both are required for nucleation mediated by the *formin.* In addition, ATP-actin units bound to profilin will dissociate from cofilin and are then free to polymerize. Another important component in filament production is the *Arp2/3 proteins,* which serve as sites for nucleation, stimulating the formation of G-actin trimers. All of these three proteins are regulated by cell signaling mechanism. After a period of growth, an equilibrium phase is reached in which depolymerization controls the length as new monomers are added. Formins nucleate actin filaments with variable efficiencies, as high as one actin filament /formin complex. Formin coordinates actin filaments and microtubules.

Actin filaments are assembled in two general types of structures - *bundles* and *networks.* Actin-binding proteins dictate the formation of either structure since they cross-link actin filaments. Actin filaments have the appearance of a double-stranded helix. In non-muscle cells actin bundles are held together by actin-bundling proteins. Bundles play a role in many cellular processes such as cell division (cytokinesis) and cell movement. Polymerization of actin filaments in to bundles might be mediated by the proteins like *villin, fimbrin, fascin, or α-actinin.* These could fill thin cellular processes. Villin is almost entirely responsible for causing bundle formations in the microvilli of intestinal cells. These bundles can form networks regulated by filamin that clips to connect the filaments at the cross-over points. All actin filaments are not the same and there is growing evidence that within a cell there are different populations of actin filaments which are spatially organized into distinct cellular compartments each with a unique function.

Some proteins prevent polymerization and depolymerization. Cytochalasin D binds to the '+' end of F-actin and prevents further addition of G - actin. Latrunculin binds G-actin and inhibits it from adding to a filament. Phalloidin [from the poisonous mushroom -*Amanita* (angel of death)] prevents actin filaments' depolymerization.

Actin monomer binding proteins: These proteins bind actin monomers and work together to control the pool of unpolymerized actin in the cytoplasm.

β-*Thymosins:* These are the most abundant actin binding protein in some animal cells. They have higher affinity for ATP-actin monomers than ADP-monomer. Thymosin β-4 binds G-actin and checks its addition to the actin filament. They, in fact inhibit actin polymerization.

ADF/Cofilin: These are ubiquitous eukaryotic protein family that bind ADP-actin with higher affinity and destabilize the filament. They bind to the side of the filament with a relatively rare conformation called 'defect' and causes change in the helical twist of the filament leading to the destabilization and severing of the filament producing two uncapped ends. For this role it is also called *actin depolymerizing factor*. Usually the rate of actin filament treadmilling is slow but cofilin binding causes an increase in the treadmilling rate enhancing the turnover of actin filament.

Profilins: These are also universal actin binding protein that has highest affinity for nucleotide free actin monomers. It inhibits actin filament elongation at the pointed end; stimulates assembly of actin filaments. It can bind with G-actin and attract more monomers to the (+) end. Thus, it speeds up the nucleation process.

Actin filament binding proteins: A remarkable number of accessory proteins bind filament ends, cross link actin monomers, sever and stabilize filaments; cap filament ends and even move along filaments. In this way these proteins regulate the dynamics and organization of actin filaments. However these proteins are grouped into following categories:

The actin-binding protein p57 shows two domains p57 (1-34) and p57 (111-204) as actin-binding sites. The C-terminal region of p57 (297-461) also possessed F-actin binding activity. The leucine zipper domain at the C-terminus of p57 (297-461) was essential for this actin-binding activity. Annexins, that bind in a calcium-dependent manner to membrane phospholipids were proposed to regulate dynamic changes in membrane architecture and to organise the interface between secretory vesicles and the membrane. Certain proteins from this family were also identified as actin binding proteins making them ideal mediators in cell membrane and cytoskeleton interactions (Konopka-Postupolska, 2007).

Actin filament capping proteins: These proteins bind either to pointed or barbed end and check the addition or dissociation of subunits. Capping is regulated by various intracellular signals. It is noted that '+' end capping is mostly regulated by the level of PIP_2 since an increase in

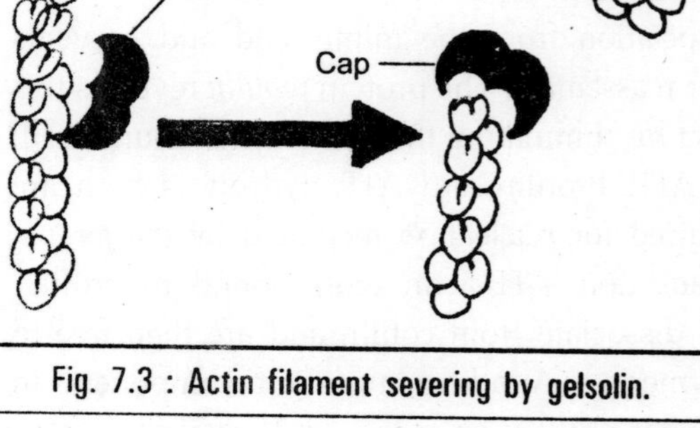

Fig. 7.3 : Actin filament severing by gelsolin.

PIP_2 level in the cytosolic leaflet uncaps the '+' end. This uncapping makes the '+' end available for elongatin. Capping proteins are of following types:

Gelsolin: They basically have six domains. Domains 2-6 bind tightly to the side and barbed ends of the actin filament increasing the possibility that domain 1 can be inserted between actin subunits causing disruption of the filament. Their severing activity is induced by high levels of Ca^{++}. In resting platelets, short actin filaments capped by CapZ remain surrounded by a large pool of profilin bound actin monomers. During the activation of platelets massive influx of Ca^{++} activates gelsolin which cleaves capped filaments into many short filaments each capped by gelsolin (Fig.7.3).

Arp2/3 complex: Arp2 and Arp3 proteins jointly cap the pointed end of the actin filament. In addition to capping it participates in the formation of branched filament. Arp2 and Arp3 may be oriented by their accessory proteins in such a way that resembles '+' end of an actin filament. Like an original filament actin subunits can then assemble to this filament. Actin filament nucleation continues at the side of the existing filament that orients at 70° angle relative to the original filament. Repeated round of branching results in web like structure formed by actin filament.

Tropomodulin: They cap the '-' end of the stable actin filaments in muscle, RBC and other euka ryotic cells. Its capping activity is enhanced by the presence of tropomyosin which suggests that the two proteins act collectively to stabilize the filament.

CapZ: They bind the '+' end of the actin filament in muscles independently of Ca^{++}. Its activity is controlled by some signaling pathways that control cofilin and profilin.

Carmil: It is a capping protein may play an important role in the physiological regulation of actin assembly.

Actin filament severing proteins: In some cases an existing actin filament can be severed into a number of smaller filaments. Thus many '+' and '-' ends can be generated which nucleate new filaments. In addition it can change the physical and mechanical property of the cytosol. It thus results the alteration of stiff gel into fluid sol. These include three classes of proteins – *Gelsolin, ADF/Cofilin* and *Fragmin/Severin*. Gelsolin and ADF/cofilin are already described. *Fragmin / Severin* are reported from molds and show Ca++ dependent severing activity.

Actin filament cross linking proteins: Actin filament cross linking proteins are divided into two categories *bundling proteins* and *gel forming proteins.*

Actins, along with many actin-binding proteins (such as the Arp2/3 complex and filamin) form a complex cortical network.

α-actinin, ankyryn, spectrin, dystrophin proteins forming cortical network help to bind actin to the plasma membrane. *Dystrophin* (the most abundant protein in the muscle) and *utrophin* bind actin filaments to the sarcolemma with sub-micromolar affinities though the molecular epitopes involved.

Fimbrin and α-actinin are widely used as crosslinker protein. Fimbrin is used to associate actin filaments in the filopodia. Tight association of actin and fimbrin excludes myosin (which is responsible for contractility) for which filopodia are not contractile in nature. Looser association of actin and α-actinin allows myosin to be incorporated. Therefore the stress fibre showing *α-actinin* is contractile in nature. α-actinin holds the ends of the stress fibre in the *focal contacts* at the plasma membrane.

Spectrin is long flexible protein made of 2α and 2β polypeptide chains. In the RBC it occurs just beneath the plasma membrane where it forms a two dimensional web held by short actin filaments. Spectrin links this web to the peripheral membrane proteins of the plasma membrane.

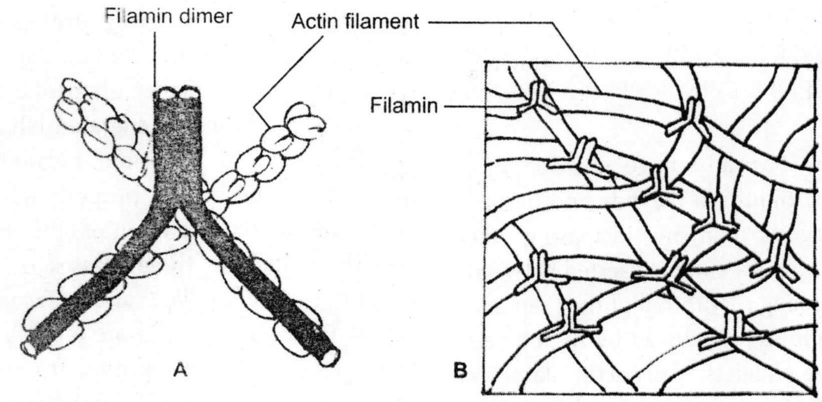

Fig. 7.4 : Filamin cross linking actin filaments.

Filamin (Fig. 7.4) cross links filaments in the cortex of many cells and anchors them to the integrins. It has two actin binding domains joined together by a long bent linkage. It results the formation of a loose and highly viscous gel by binding actin filaments roughly at right angles. These are abundant in lamellipodia. These often tend to form bundles some of which are large and may contain myosin. These bundled arrangements are called *stress fibres* those mainly aggregate towards the cortical regions of the cell. The cortical networks interacts with a number of different proteins associated with the membrane as well as proteins in signaling pathways. It may also be traversed by myosin. It is also suggested that cortical actin network serve as barriers for molecular diffusion. In resting cells about half of the actins get polymerized and the pool of unpolymerized actins allows cells to respond rapidly to growth factors and chemoattractants.

Proteins, regulating stabilization of actin filaments and interaction with myosin: The structure and function of the actin cytoskeleton is primarily regulated by the following associated actin binding proteins.

Tropomyosin (Tm) is an intrinsic component of most actin filaments and over 40 isoforms have been identified in non-muscle cells. They determine the stability of the actin filaments in the cell. It increases the tensile strength of actin filament. In skeletal muscle it is an important component of the Ca^{++} sensitive regulatory

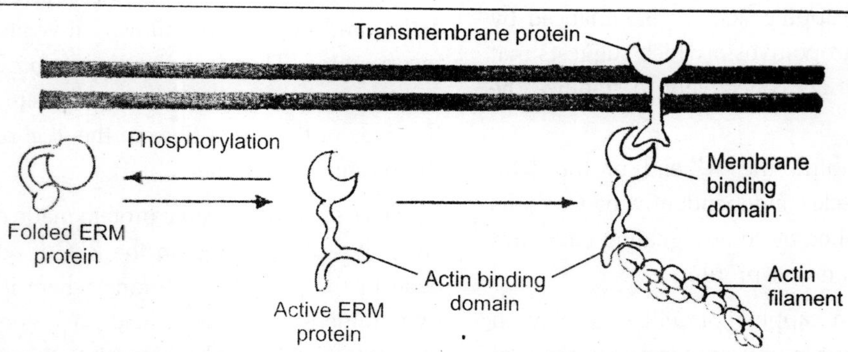

Fig. 7.5 : ERM protein attachment with actin filament and membrane protein.

machinery controlling the interaction of actin with myosin.

Nebulin often regulates the length of actin filament.

Caldesmon, together with tropomyosin regulate actin-myosin interaction mediated by Ca^{++}-calmodulin in smooth muscles and non muscle cells. It also reorganizes actin filaments during mitosis.

Adapter proteins: Some multidomain proteins act as adapters between actin and other proteins (GTP binding proteins of the Rho family and other signaling proteins).

Wiskott - Aldrich syndrome causing protein – WASp activate Arp2/3 complex to nucleate new filament on the side of the existing filament. Another protein – *VASP* interact with membrane protein and facilitate actin filament assembly by inhibiting barbed end capping.

Organization of actin within cells: In non muscle cells actin is found in both G-form and F-form. Due to the presence of a variety of actin binding proteins the individual monomers are pooled in the cytoplasm and participate in the formation of F-actin whenever required. Actin filaments are organized ito two types of structures – bundles and gel like networks. Actin along with other crosslinking proteins form a cortical network in the form of plasmagel while in the endoplasm it mainly forms bundles to maintain the normal activities of the cell (Fig. 7.5).

Many intracellular proteins like ezrin, radixin and moesin (these proteins belong to the ERM family) mediate the attachment of actin filaments with with the plasma membrane. Their C terminal domains binds directly to the sides of the actin filament and N terminal domains bind with the cytoplasmic face of the transmembrane glycoproteins. ERM proteins nomally remain in an inactive form but in response to an extracellular signal, phosphorylation or binding of PIP_2 exposes their binding sites to mediate binding between actin and membrane glycoprotein. Mutation to the ERM proteins lead to the genetic disease – neurofibromatosis showing multiple benign tumours in many parts of the nervous system. It is also reported that Rho GTPases activate WASP and formins which mediate actin polymerization and its interaction with the membrane.

In skeletal muscles all the actins occurs as a thin filament in alternate fashion with myosin that occurs as a thick filament. Actins occur as two spirally twisted helices showing 13-14 monomers in each helix. However actins always occur in association with actin binding proteins. Tropomyosin runs along the length of the filament extending up to 7 monomers. At the apex of the tropomyosin, three troponin molecules occur in association with Ca^{++}. Nebulin extends through out the length of the actin filaments and regulates their lengths in striated muscles. CapZ binds to the barbed end while tropomodulin bind to the pointed end of the actin filaments. These proteins inhibit severing of the actin filament.

Muscle fibres contract in response to an electrical signal generated at the neuromuscular junction and each muscle fibre is innervated by a motor nerve. In response to the signal, action potential is generated that causes Ca^{++} release from the sarcoplasmic reticulum. Ca^{++} binds with the Ca binding sites of the troponins that lead to the contact of myosin heads with actin. Troponin directly binds to an actin filament in skeletal muscle even in the absence of tropomyosin. It can transform an actin monomer within a filament into an off-state without blocking the myosin-binding sites with tropomyosin molecules. Due to activation of the head, ATP is hydrolyzed into ADP and Pi which causes gradual inward movement of myosin heads leading to sliding of the actin fibres. Repeated effects of such deflection of the myosin heads lead to the contraction of muscles.

Actin extends longitudinally in bundles in the microvilli. Villin and fimbrin bind 20-30 bundled actin filaments while actin filaments at the core of the microvilli are attached to the plasma membrane with the help of myosin I. At the base of the microvilli actin filaments that run perpendicularly bind with the horizontal filaments to form a web like structure called *"terminal web"* that are connected by spectrin molecules. Each terminal web ends at a specialized junction called an *"adherent junction"*.

Actin filaments regulate peripheral nucleus (chromosome) migration, cortical spindle anchorage, homologous chromosome separation, cortex development /maintenance, peripheral cortical granule migration, polarity establishment, and first polar body emission during oocyte maturation. However movement of mitochondria is mediated by microtubules. During fertilization, microfilaments are involved in sperm incorporation, spindle rotation, cortical granule exocytosis, second polar body emission and cleavage ring formation. Many of the events are driven by the dynamic interactions between myosin and actin filaments whose polymerization is regulated by RhoA, Cdc42, Arp2/3 and other signaling molecules. Studies have also shown that oocyte cortex organization and polarity formation mediated by actin filaments are regulated by mitogen-activated protein kinase, myosin light-chain kinase, protein kinase C as well as PAR proteins. The completion of several dynamic events, including homologous chromosome separation, spindle anchorage, spindle rotation, vesicle organelle transport and pronuclear apposition requires interactions between microfilaments and microtubules. In addition, actin polymerization aided profilins by sequestering actin monomers in association with other actin-related proteins (Arps) both within the cytoplasm and inside the nuclei of developing mammalian embryos play a vital role in the flaw less development..

General major roles of actin filaments:

i) Form a band just beneath the plasma membrane that provides mechanical strength to the cell, links transmembrane proteins (e.g., cell surface receptors) to cytoplasmic proteins, anchors the centrosomes at opposite poles of the cell during mitosis and pinches dividing animal cells apart during cytokinesis Balanced collaboration of Adf1, profilin, myosin-II and capping proteins are required in assembling the contractile ring at the division site by depolymerizing actin.

ii) Generate cytoplasmic streaming in some cells.

iii) Regulates locomotion in cells such as WBC and the amoeba.

iv) Interacts with myosin ("thick") filaments in skeletal muscle fibers to provide the force of muscular contraction.

v) Plays crucial role in various cell junctions

vi) Actin filaments are also known to play vital roles in endocytosis.

vi) The cortical cytoskeleton provides a structural framework for rapid signal transduction from the cell periphery into the nucleus.

vii) In plant cells, the secretory vesicles are delivered to their target membranes by interaction with actin microfilaments.

viii) Existence of a functional Arp2/3 complex in higher plants is reported. They play a pivotal role in determining the shape of plant cells. They also regulate interlinked processes of cortical-actin organization, growth-site selection, organelle motility and actin-microtubule interactions during plant cell morphogenesis.

7.2. MICROTUBULES

A microtubule (MT) is a hollow cylinder about 24 nm in diameter. Along the microtubule axis, tubulin heterodimers are joined end-to-end

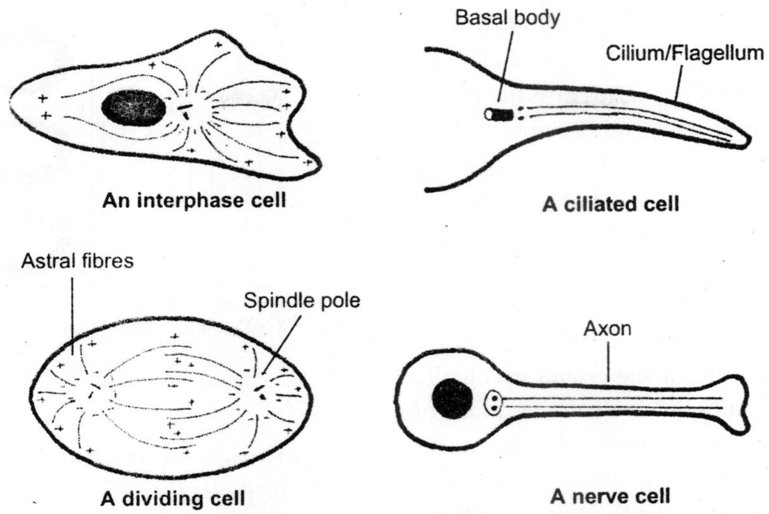

Fig. 7.6 : Occurrence of microtubules in different locations in different cells.

to form *protofilaments*, with alternating α and β subunits. Staggered assembly of 13 protofilaments yields a helical arrangement of tubulin heterodimers resulting a *microtubule*. The basic structural unit of microtubules is α, β-tubulins. These are homologous but not identical. Each has a nucleotide-binding site. α-tubulin has a bound molecule of GTP that does not hydrolyze. β-tubulin may have bound GTP or GDP but it hydrolyzes its bound GTP to GDP and Pi.

Microtubule Formation: Regulation of MT nucleation sites is an essential step in MT organization. Cortical MT arrays in green plant cells at interphase are organized in a distinct manner—the array is formed in the absence of pre viously recognized organelle for MT nucleation - the centrosome and spindle pole body. MTs in the cortical array were recently found to be nucleated as branches on pre-existing MTs via recruitment of cytosolic γ-tubulin (Murata and Hasebe, 2007). The first stage of MT formation is called *"nucleation"*. The process requires tubulin, Mg^{++} and GTP. This stage is relatively slow until the microtubule is initially formed. It is followed by the second phase, called *"elongation"* that proceeds much more rapidly. During "nucleation", an α and a β-tubulin molecule joins to form a heterodimer. Then these attach to other dimers to form oligomers, which elongate to form *protofilaments*. Each dimer carries two GTP molecules. However the GTP that appears to function binds to the b-tubulin

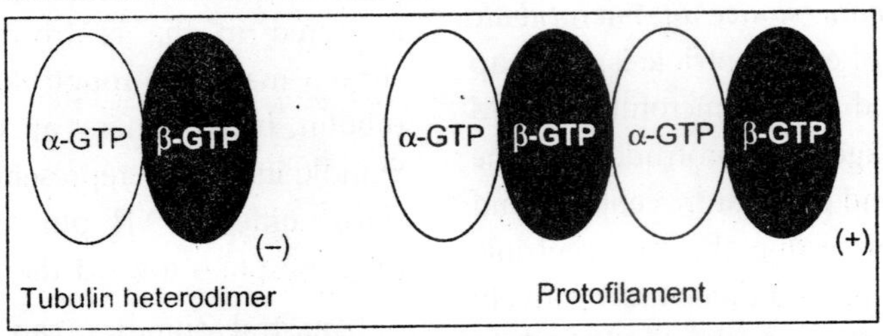

Fig. 7.7 : Tubulin monomers and protofilaments.

molecules. When a tubulin molecule adds to the microtubule, the GTP is hydrolyzed to GDP + Pi. Eventually the oligomers join to form the ringed microtubule 25 nm in diameter. In cross section, each ring consists of 13 beads. The rows of beads are called *protofilaments*. EM study of microtubules indicates. a "3 start helix" in which each turn of the helix spans 3 tubulin monomers (a, b, a). Thus, instead of aa and bb contacts, a subunits laterally join with b subunits. These associate laterally to form sheets and eventually (MTs). *In vitro* addition and dissociation of heterodimers can occur at either end of the microtubule, but there is greater tendency for subunits addition at the plus (+) end. Subunit addition brings β-tubulin into contact with α-tubulin This promotes hydrolysis of GTP bound to the β-tubulin into GDP and Pi Pi dissociates, but β-tubulin within a microtubule cannot exchange its bound GDP for GTP. The GTP on α-tubulin does not hydrolyze. The minus (-) end of α-tubulin may contribute an essential residue to the catalytic site of β-tubulin. Thus the (-) end of a α subunit may serve as GAP (GTPase *a*ctivating protein) for β-tubulin of the adjacent dimer in a protofilament. MTs appear as singlet, doublet, and triplet MTs. Doublet MTs occur in the axoneme of cilia and flagella, while triplet MTs occur in the basal bodies and centrosomes. In doublet and triplet microtubules the wall of one microtubule partly consists of the wall of an attached microtubule. The 'A' tubule of a doublet or triplet microtubule is a complete cylinder, made of 13 protofilaments while 'B' or 'C' tubules are made of usually 10 protofilaments (Fig. 7.8).

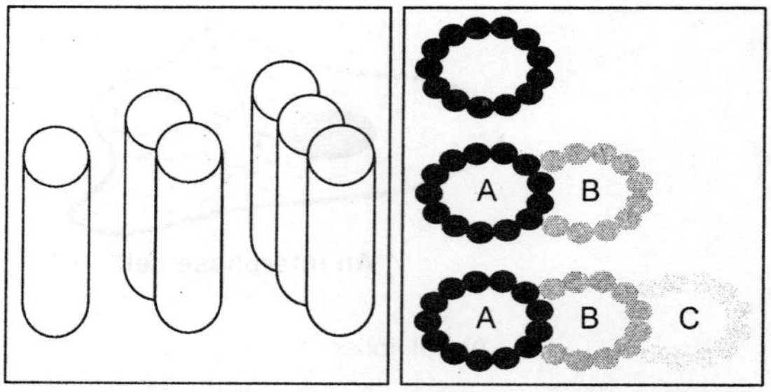

Fig. 7.8 : Singlet, doublet and triplet microtubules in side view and T.S.

Centrioles, the main source of microtubule nucleation: The wall of each centriole is made up of nine interconnected triplet microtubules. EM study shows appendages that protrude from the outer surface at one end of a mature centriole and fibrous structures connecting the two centriole cylinders. Centriolar microtubules are relatively stable. The mass of protein which include centrin, pericentrin, ninein, cenexin, CEP110, CEP 250 (C-Nap1) and others called the *microtubule organizing center (MTOC) or pericentriolar material*, surrounds centrioles in animal cells. Such proteins and additional tubulins, γ, δ, ε, ζ and η present in centrioles aggregate along lateral margins of the centriolar cylinder forming a tube. MT nucleation requires γ-tubulin, which exists as two main complexes: the *γ-tubulin small complex* and the *γ-tubulin ring complex*. During mitosis, these complexes accumulate at the centrosome to support spindle formation. γ-tubulin complexes are also present at non-centrosomal MT nucleation sites both in and in interphase mitosis. γ-tubulin complexes may be involved not only in MT nucleation, but also in regulating MT dynamics (Raynaud- Messina and Merdes, 2007). The *γ-tubulin*, nucleates microtubule assembly within the centrosome. Several (12-14) copies of γ-tubulin associate in a complex with other proteins called *"grips" (γ ring proteins)*.

During cell division, the duplicated centrosome organizes the mitotic spindle.MTs grow out from the MTOC, with minus (-) ends anchored in the centrosome. MTs grow and shrink mainly through addition and loss of tubulin heterodimers at their plus (+) ends. Spindle itself may represent a source of new MT since numerous MT plus ends grow from acen trosomal poles toward the metaphase plate.

Anastral spindles can be assembled in a centrosome independent manner in cells that contain centrosomes. It is assumed that the MTs that build these acentrosomal spindles originate over the chromatin.

The onset of non centrosomal MT nucleation coincides with the end of chromosome condensation and NE breakdown. Thus in centrosome containing cells both centrosomal and non centrosomal MTs contribute to the formation of cell division spindle. In interphase, non-centrosomal nucleation enables the formation of MT bundles or networks of branched MTs. γ-tubulin complexes may be involved not only in MT nucleation, but also in regulating MT dynamics.

Microtubules exhibit dynamic instability: They may grow steadily and then shrink rapidly by disassembly at the plus end. The rapid disassembly is referred to as *catastrophe*. The tendency to grow or shrink is attributed to the tubulin concentration. Instability of microtubules may be regulated by interaction with other proteins at the (+) end. For example, during prophase of mitosis, if the (+) end of a MT makes contact with a chromosome, it becomes stabilized. Otherwise rapid disassembly at the (+) end occurs and the tubulin dimers become available for growth of other microtubule.

Microtubule-associated proteins – MAPs: These are diverse classes of cell and tissue specific proteins that bind along the sides of microtubules. MAPs can stabilize MTs against disassembly. Some MAPs mediate the interaction of MTs with other cellular components. Such MAPs form the core of the axons and dendrites. These MAPs have a MT binding domain and another projecting domain. Binding of MAPs may be regulated by phosphorylation / dephosphorylation. Each MAP shows two domains, one binds with tubulin and other with the vesicles, granules or other IF/MTs.

MAPs are of two types – Type I and Type II. Type I includes MAP- 1A and 1B. They bind with tubulin, intermediate filaments. They mainly control the spacing of MTs.

Type-II includes MAP2, MAP4 and Tau. Their structure resembles Type I types. MAP2 has a long but Tau shows a short projecting domain. MAP2 and Tau are confined to the selected cell types which have a dominant role in MT dynamics. MAP4 is essential for cell division. Tau organizes MT into bundles.

Motor microtubule associated proteins: The MAPs include kinesins and dynein which "walk" along the moving microtubules in opposite directions. Kinesin is a double-headed motor protein that moves along MTs. The energy for their movement is provided by the hydrolysis of ATP. Different vesicles are seen attached to MAPs and the microtubule conveyer belt. Kinesins move the vesicle towards the plus end and dynein walks towards the minus end. These proteins show head regions that bind to both MTs and ATP. The head domains are thus ATPase motors and tail domain binds to the organelle to be moved. ATP is needed for both binding and movement. Kinesin can bind directly to cargo at its light chain end; however dynein requires a complex of proteins. Kinesins are involved in a variety of intracellular transport processes through the modulations of α and β tubulin structure.

Two models have been proposed to explain kinesin movement: *hand-over-hand* and *inchworm models*. Yildiz *et al* (2004) strongly support hand-over-hand mechanism for their movement. Some MAPs stabilize MTs, inhibiting depolymerization. XMAP215, a highly conserved MAP, stabilizes the (+) ends of MTs, preventing catastrophic shrinkage. They bind along the sides of the MTs stabilizing the free ends of the MTs and inhibit their shrinking from the growing state. They cross link with IF or PM or with other MTs. They stabilize MT during growth and division. Phosphorylation of XMAP-215 at the onset of mitosis lead to increased MT instability causing liberation of tubulin dimers forming the mitotic spindle.

Some destabilizing proteins - catastrophe-promoting proteins - *catastrophins* have been identified which sever MTs, generating new ends that lack a GTP cap. *Stathmin* is such a MT destabilizing protein noted in some cancer cells. MTs are even reported to be transported down the axon as short pieces by molecular motor proteins - dynein. Anterograde MT transport involves actin filaments. *Katanin* is another MT severing protein..They break thirteen bonds, one for each protofilament. Katanin shows one small and one large subunits, the smaller one hydrolyzes ATP to separate the protofilaments while the larger one releases the MTs from the MTOC. At the end of mitosis, katanin performs the role of depolymerization of MTs in this way.

Role of microtubules in the formation of Cilia/Flagella: Cilia and flagella are directly formed from the centrioles. Before generation of cilia, the centriole divides and forms basal bodies. They are transformed to some extent from the typical centriolar structure. They and are designated as A and B subfibres. All the protofilaments. They show peripheral doublets not triplets; moreover in the centre there is a doublet. All these structures actually represent protofilaments having 13 beads. The doublets show 13 and 10 beads are however connected by a MAP – nexin and other MAP dynein extending from the A subfibre towards the B subfibre for functional coordination.

Epithelial tight junction (TJ) and adherens junction (AJ) form the apical junctional complex (AJC) that regulates cell-cell adhesion, paracellular permeability and cell polarity. The AJC is anchored on cytoskeletal structures including actin microfilaments and microtubules. Such cytoskeletal interactions are thought to be important for the assembly and remodeling of apical junctions. Calcium depletion resulted in disruption and internalization of epithelial TJs and AJs. MT polymerization directly induces junctional complex assembly).

Microtubule-organelle interaction: Several biochemical processes in animal cells are confined to distinct membrane-bound compartments and proper balance in the functioning of the cell depends on the inter compartment transfer of materials which is mediated by carrier vesicles that are targeted to their correct destinations. Microtubules are involved both in the intracellular transport processes and in the organization of cytoplasmic organelles. The positioning of Golgi complex and lysosomes depends on microtubule network; moreover Golgi elements, tubules of the ER, exocytic or secretory vesicles and lysosomes move along microtubules.

7.3. INTERMEDIATE FILAMENTS (IF)

Intermediate filaments are found in all vertebrates, nematodes and molluscs but their presence in fungi, plants and other lower unicellular eukaryotes is controversial. Their presence is all cell types is also debatable since in oligodendrocytes (that make myelin in the vertebrate CNS lacks Ifs. In epidermal cells and the axons of neurons, IFs occur abundantly. Organization of IF and their association with nuclear and plasma membranes suggest that their principal role is structural to reinforce cells and to organize cells into tissues. In epithelium IFs provide mechanical support for the plasma mem brane where it comes in in contact with other cells or with ECM.

The dead remnants of epidermal cells in claws and hair are composed largely of IFs. Unlike MFs and MTs, IFs do not participate in cell motility. They may stabilize organelles, like the nucleus, or they may be involved in specialized junctions. IF are stable, durable, range in diameter from 8-10 nm (intermediate in size compared with thin microfilaments and microtubules). They are prominent in cells that withstand mechanical stress and are the most insoluble part of the cell. IFs can easily be distinguished from MFs and MTs. IFs are extremely stable, their diameter -10-nm is smaller than MTs (24 nm) but larger than MFs (7 nm). In contrast to the beaded filaments and hollow tubules formed by the globular actin and tubulin subunits, respectively, IF subunits are α-helical rods that assemble into ropelike filaments. Finally, IF subunits do not bind ATP or GTP as noted during polymerization of G-actin and αβ-tubulin.

Types of Intermediate filaments: In higher vertebrates, the subunits composing IF are divided into five major classes whose expression is characteristic of a certain cell/tissue.

Types I and II proteins: Acidic and basic *keratins* constitute type II and I IF proteins. They have subtypes that are unique to different epithelial cells (bladder, skin, etc) or even different subsets of one cell type (like basal epidermal cells). These are typically expressed in epithelial cells. Each type of epithelium always expresses a characteristic combination of type I and type II keratins. Keratins are the most diverse classes of IF proteins, with a large number of isoforms. About 10 keratin isoforms are specific for "hard" epithelial tissues like nails, hair, and wool while about 20 (called *cytokeratins)* are found in the epithelial lining of internal body cavities.

Type III proteins: Four proteins vimentin, desmin, glial fibrillary acidic protein (GFAP), and peripherin are classified as type III IF proteins. The most widely distributed of these IF proteins is *vimentin*, which is typically expressed in leukocytes, blood vessel endothelial cells, some epithelial cells, and mesenchymal cells such as fibroblasts. Vimentin filaments support cellular membranes, help to keep nucleus and other organelles in a defined places within the cell. It is frequently associated with MTs. Strong vimentin expression was found in neuroepithelial (ependymal) cells and in the radial glia of the spinal cord throughout all stages of development.

Desmin is found in muscle cells, connecting Z disks and may connect the center of contractile units and are responsible for stabilizing sarcomeres. It is also found connecting to desmosomes in cardiac muscle. *Glial fibrillary acidic protein* forms filaments in the glial cells that surround neurons and astrocytes. *Peripherin* is found in neurons of the peripheral nervous system, but little is known about it.

Type IV proteins: These include neurofilament L, M, or H (named for low, medium or high molecular weight. These are linked by plectin cross bridges to each other and to MTs. It helps to strengthen the neurofilaments (NFs). NF proteins also influence the functions of axons. In contrast to microtubules, which direct the elongation of an axon, NFs are responsible for the radial growth of an axon and thus determine axonal diameter. Another type IV IF is "*internexin*" and some non-standard IV's are found in lens fibers of the eye (*filensin and phakinin*).

Type V proteins: It includes lamins which is exclusively found in the nucleus. Lamins are vital to the re-formation of the nuclear envelope after cell division. From evolutionary view point this was probably the first IFs made. They have a very long rod domain and carry a nuclear transport signal because they occur just under the inner nuclear envelope. They form a lattice-like array providing support to the inner nuclear membrane except at the nuclear pore complex. They are difficult to distinguish from the nearby dense heterochromatin. Lamins are phosphorylated at the end of prophase and cause to disassemble the nuclear envelope also breaks down. After prophase lamin filaments reassemble around each set of chromosomes, under the inner nuclear membrane to reform nuclear envelope.

IF polymerization: The monomer and polymerization: All IF subunit proteins have a common central α-helical rod domain structure separated by three "spacer" elements and terminal N- and C- domains. The α-helical

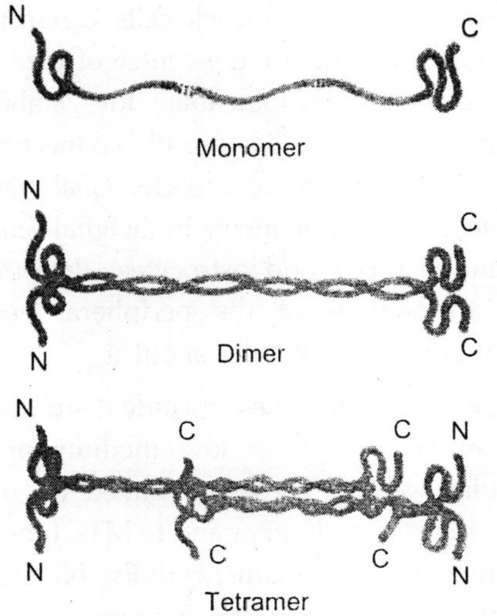

Fig. 7.9 : IF monomer, dimer, tetramer and protofibrils.

segments pair to form a coiled-coil dimer which appears as a rod like molecule with globular domains at the ends. A pair of dimers associates laterally into a tetramer called *protofilament*. The dimers in a tetramer have an antiparallel orientation. Actually the tetramer is the subunit for assembly of IFs. Tetramers bind end to end, forming protofilaments 2-3 nm thick, which pair together into *protofibrils*. Finally, four protofibrils form a single rope like IF - 10 nm in diameter.

The IF, unlike actin filament or a microtubule has no polarity.Initially it was speculated that the N- and C-terminal domains are not involved in IF assembly but experimentally it has been proved that the N-terminal domain plays an important role in assembly of most IFs. However the C-terminal domain is dispensable for IF assembly, it seems to affect the stability of the filament. Thus these domains may control lateral interactions within an IF, as well as interactions between IFs and other cellular components.

IF proteins can form homopolymeric or heteropolymeric filaments. *Vimentin, desmin, glial fibrillary acidic protein,* and *NF-L* assemble into homopolymers or heteropolymers while acidic and basic keratins, which form only obligate heteropolymeric filaments. However, keratins do not form heteropolymers with other classes of IF proteins. Vimentin and NF proteins can assemble into a heteropolymer while NF-L self-associates to form a homopolymer. In a neurofilament however NF-H, NF-M and NF-L co-assemble. NF-H and NF-M can extend outwards from the surface of the assembled IFs and bind to a neighbouring filament. Thus neurofilaments form can be held together by multiple lateral contacts to form robust filaments.

Dynamic properties of IFs: The exact pathway for the assembly of IF is yet to be unveiled. IFs exhibit dynamic property in the cell. Several experiments reveal that IF proteins exchange with the IF cytoskeleton. If type I keratin is injected into an epithelial cell; within 2 hours, it becomes incorporated into the already existing keratin cytoskeleton. IF subunits in a soluble pool are able to incorporate themselves to preexisting filaments. Moreover subunits are also able to dissociate from intact filaments. Although IFs exhibit dynamic properties within cells, they clearly are more stable than MTs and microfilaments. Under the control of Cdc2 kinase, filaments of vimentin, desmin, and lamins disassemble in early mitosis and reassemble after telophase.

IF associated proteins: *Intermediate filament associated proteins (IFAPs)* cross-link IF with one another, forming a bundle (*tonofilament*) or a network and with other cell structures, including the plasma membrane. Only a few IFAPs have been identified to date, but unlike actin-binding proteins or MAPs, none of the known IFAPs sever or cap IFs, sequester IF proteins in a soluble pool, or act as a motor protein. The organization of IFs and their supportive function in various cells types depends largely on their linkage to other cell structures via IFAPs. A physical linkage between IFs and MTs has been proved.

Plectin is a common IFAP that interacts with other cytoskeletal proteins, including spectrin,

MAPs, and lamin B. It also crosslinks MTs and vimentin. A family of proteins related to plectin is implicated in forming cross-links between MFs and IFs or IFs and myosin II as well as IF bundles with the adhesive structures at the plasma membrane. Plectin gene mutation results *epidermolysis bullosa* showing disruption of desmin filaments and neurodegeneration.

Filaggrin bundles keratin filaments in differentiating cells of the epidermis to provide toughness to the outermost layer of the skin.

Relationship of IFs and intercellular junctions: The epithelial cells are held together by specialized cell junctions called *desmosomes* and *hemidesmosomes*. Desmosomes mediate cell-cell adhesion, and hemides mosomes are responsible for attaching cells to the underlying base ment membrane. Both junctions appear as darkly staining plaques, bound to the cytosolic face of the plasma membrane and attached to bundles of IFs. The IFs in one cell thus are directly connected to the IFs in a neighboring cell via desmosomes or to the extra cellular matrix via hemi desmosomes. Connections between IFs and these cell junctions provide strength and rigidity to the entire epithelium. Without the supporting network of intermediate filaments, an epithelium remains intact, but the cells are easily damaged by abrasive forces.

As actin filaments are attached to the specialized regions of the plasma membrane, intermediate filaments form a flexible but resilient framework that gives structural support to an epithelium.

A network of IFs is often found adjacent to cellular membranes, where it provides mechanical support. The best example is the nuclear lamina along the inner surface of the nuclear membrane. This supporting network shows lamin A and lamin C filaments cross-linked into an orthogonal lattice, which is attached via lamin B to the inner nuclear membrane. In addition to forming the nuclear

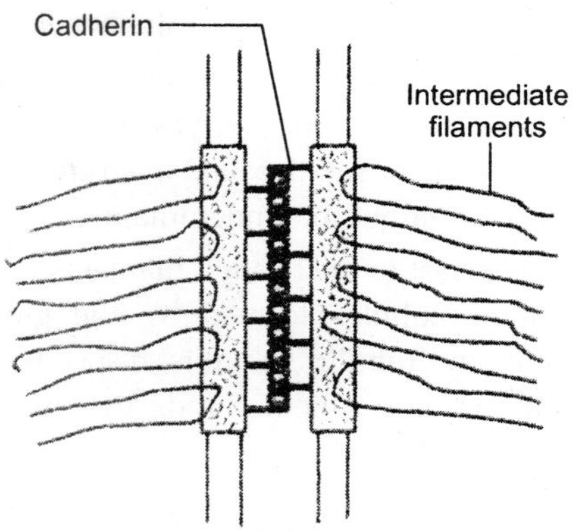

Fig. 7.10 : Position of IF in desmosomes.

lamina, IFs are typically organized in the cytosol as an extended system that stretches from the nuclear envelope to the plasma membrane. Some IFs run parallel to the cell surface, while others traverse the cytosol; together they form an internal framework that helps support the shape of the cell. At the plasma membrane, vimentin filaments bind two proteins: ankyrin and plectin; vimentin filaments are also attached to the nuclear membrane, thereby providing a flexible structural support.

General role of IFs: The organization of intermediate filaments and their association with plasma membrane suggests that their principal function is structural to reinforce cells and to organize cells into tissues. Indeed, their supportive role is carried to an extreme in claws and hair, the dead remnants of epidermal cells, which are composed largely of IF proteins. Perhaps the most important function of intermediate filaments is to provide mechanical support for the plasma membrane where it comes into contact with other cells or with the extra cellular matrix. Several features distinguish IF from microfilaments and microtubules.

i) These are extremely stable. Even after extraction with solutions containing detergents and high concentrations of salts, most IFs in a cell remain intact, whereas microfilaments and microtubules depolymerize into their soluble subunits.

ii) IFs differ in size from the other two fibers. Indeed, their name derives from their 10-nm diameter smaller than microtubules (24 nm) but larger than microfilaments (7 nm).

iii) In contrast to the beaded filaments and the hollow tubules formed by the globular actin and tubulin.subunits, respectively, IF subunits are α-helical rods that assemble into ropelike filaments.

iv) IF subunits do not bind nucleotides, and their assembly into intermediate filaments does not require nucleotide (GTP or ATP) hydrolysis, as in the polymerization of G-actin and αβ-tubulin.

SUMMARY

1. Cytoskeletons constitute proteinous structure that maintain the structural integrity of the cell. It consists of three major components – actin filaments, microtubules and intermediate filaments.
2. Actin is a globular protein that polymerizes helically forming actin filaments. Most organisms have genes for several actin isoforms. Actin isoforms are biochemically highly similar, humans express β and γ isoforms in non muscle cells and four α and β isoforms in muscle cells. Many non muscle cells express both β and γ isoforms while RBCs show only β actin. More than 60 families of actin binding proteins regulate the assembly and dynamics of actin based structures in the cytoplasm. Cells exhibit a rich variety of actin related proteins (Arps) (Arp1, 2, 3..) that participate in a variety of molecular interaction with actin. In physiological conditions, actin molecules spontaneously self associate to form microfilament at both ends of the filament.
3. The process of actin polymerization or nucleation, starts with the weak binding of two actin monomers but addition of a third stabilizes the complex forming a trimer. ADP-actin dissociates from the minus end and the increase in ADP-actin stimulates the exchange of bound ADP for ATP, leading to more ATP-actin units.Formin nucleates actin filaments with variable efficiencies. Formin coordinates actin filaments and microtubules. Actin filaments are assembled in two general types of structures - bundles and networks. Actin-binding proteins dictate the formation of either structure since they cross-link actin filaments. Actin filaments have the appearance of a double-stranded helix. In non-muscle cells actin bundles are held together by actin-bundling proteins. Bundles play a role in many cellular processes such as cell division (cytokinesis) and cell movement. The proteins like villin, fimbrin, fascin, or α-actinin might mediate polymerization of actin filaments in to bundles. Some proteins prevent polymerization and depolymerization. Many proteins bind actin monomers and work together to control the pool of unpolymerized actin in the cytoplasm.
4. β-Thymosins are the most abundant actin binding protein in some animal cells. ADF/Cofilin are ubiquitous eukaryotic protein family that bind ADP-actin with higher affinity and destabilize the filament. A remarkable number of accessory proteins bind filament ends, cross-link actin monomers, sever and stabilize filaments; cap filament ends and even moves along filaments. In this way these proteins regulate the dynamics and organization of actin filaments. Arp2/3 complex shows Arp2 and Arp3 – two proteins that jointly caps the pointed end of the actin filament. There are several actin filament-severing proteins. Several proteins actually cross-link randomly forming net works of actin filaments and stabilize their higher order structures. Fimbrin and Villin- stabilize the actin filament bundles in microvilli. Spectrins and Dystrophin link actin filaments of the membrane skeleton with integral membrane proteins.

   Three proteins regulate stabilization of actin filaments and interact with myosin. Tropomyosin increases the tensile strength of actin filament. Nebulin regulates the length of actin filament. Caldesmon, together with tropomyosin regulate actin-myosin interaction mediated by Ca^{++}-calmodulin in smooth muscles and non muscle cells.
5. Some proteins act as adapters between actin and other proteins. In non-muscle cells actin is found in both G-form and F-form. F-actins exist as fine meshwork of microfilaments forming gels by a variety of actin binding proteins. Actins, along with many actin-binding proteins form a complex

cortical network. α-actinin, ankyryn, spectrin and dystrophin proteins help to bind actin to the plasma membrane. Dystrophin (abundant protein in the muscle) and utrophin both bind acting filaments. CapZ binds to the barbed end while tropomodulin bind to the pointed end of the actin filaments. These proteins inhibit severing of the actin filament. Actin extends longitudinally in bundles in the microvilli. Actin filament dynamics are critical in cell motility. Endoplasm contains a pool of unpolymerized actin bound to the protein profilin. Activation of cortical proteins. induces actin polymerization in the endoplasm that generates the force for protrusion. Growing actin filaments push the plasma membrane forward until capping proteins terminate elongation of the actin filaments. Actin filament cross-linking protein stabilizes the advancing pseudopodium. Actin filaments mediate oocyte cortex organization and polarity formation.

6. Actin forms a band just beneath the plasma membrane that provides mechanical strength to the cell; links transmembrane receptors to cytoplasmic proteins; anchors centrosomes at opposite poles of the cell during mitosis. Balanced collaboration of Adf1, profilin, myosin-II and capping proteins are required in assembling the contractile ring at the division site by depolymerizing actin. Plays crucial role in various cell junctions and endocytosis.
7. Microtubules are hollow cylinder consists of α-tubulin and β-tubulin. The g-tubulin, nucleates microtubule assembly within the centrosome. There are diverse classes of cell and tissue specific proteins that bind to microtubules. These are called microtubule-associated proteins – MAPs. Some MAPs stabilize MTs, inhibiting depolymerization. Some MAPs cross-link adjacent MTs or link MTs to membranes or to IFs.
8. Intermediate filaments exhibit intermediate diameters between actin and microtubules. In higher vertebrates, IFs are divided into five major classes. Acidic and basic keratins constitute type I and II IF proteins. Four proteins vimentin, desmin, glial fibrillary acidic protein (GFAP), and peripherin are classified as type III IF proteins. The most widely distributed of all IF proteins is vimentin that is typically expressed in leukocytes, blood vessel endothelial cells, some epithelial cells, and mesenchymal cells such as fibroblasts. Vimentin filaments help support cellular mem branes. Glial fibrillary acidic protein forms filaments in the glial cells that surround neurons and astrocytes. IF proteins can form homopolymeric or heteropolymeric filaments. IFs exhibit dynamic property in the cell. Several experiments reveal that IF proteins exchange with the IF cytoskeleton. Intermediate filament associated proteins (IFAPs) cross-link IF with one another, forming a bundle (tonofilament) or a network and with other cell structures, including the plasma mem brane. A family of proteins related to plectin is implicated in forming cross-links between microfilaments and intermediate filaments.
9. The organization of intermediate filaments and their association with plasma membrane suggests that their principal function is structural to reinforce cells and to organize cells into tissues.

QUESTIONS

Long questions:

1. Describe the actin monomer. How is the polymer formed? What ingredients do you need to make an actin filament?
2. Actin filaments, like microtubules, have polarity. How is polarity defined in actin? Which end of actin filaments polymerizes more rapidly?
3. Describe the process of nucleation and elongation of actin filaments. How different actin cross linking proteins forms either bundles or networks of actin filaments.
4. Describe the role of actin fibres in the contraction of skeletal mus cles.
5. Describe the role of actin in the movement of unicellular animals.
6. Describe the role of actin in the maturation of oocytes and fertilization.
7. Describe the structure of protofilaments formation and the steps of microtubule formation.
8. What do you mean by MAPs? How they are assembled and point out their roles.
9. What are basic characteristics of Inter mediate Filaments and IF monomers?:
10. Name the five major types of Intermediate filaments and give examples of tissues that contain each type.

11. Describe the formation of an Intermediate filament protofilament. How do these organize to form the fibrils?
12. How could you regulate the assembly (or disassembly) of intermediate filaments?

Short questions:

1. Different proteins bind actin to the plasma membrane. Give some examples.
2. What are effects of cytochalasin D, latrunculin, or phalloidin on actin filaments?
3. What is the role of thymosin b-4 with respect to actin filament assembly?
4. What do you mean by 'treadmill effect'?
5. What are 'lag' and 'elongation phase' of the microfilament assembly?
6. 'Cofilin and profilin has reverse action in the microfilament assembly'– discuss.
7. Which actin binding proteins regulate actin bundle formation?
8. Which protein forms actin network?
9. What are capping proteins? Describe the roles of different capping proteins.
10. Name some proteins those bind actin filaments with plasma membrane?
11. What do you mean by 'stress fibres'?
12. How is actin organized in microvilli?
13. How do myosins interact with actin? What is required for binding?
14. Discuss the roles of a-actinin, titin, nebulin, troponin and tropomyosin in muscle contraction.
15. Compare the "molecular events of the front and rear of a moving cell".
16. Point out the genera roles of actin filaments.
17. How myosin I and myosin II are involved in cell movement?
18. Describe the process of MT nucleation from MTOC.
19. Name one MT catastrophic protein.
20. Name some drugs which depolymerize MTs.
21. Explain how MTs are involved in intercellular junctions?
22. If you add protofilaments of an intermediate filament to the cytoplasm, could they be added to the existing cytoskeleton? How could you prove this?
23. Do Intermediate filaments associate with other proteins? What are some functions associated with this?
24. Where are lamins found? How are they unique?
25. A number of different forms of keratins are found? How is this useful for tumor diagnoses?
26. What types of junctions are formed with keratin or desmin as the intermediate filaments?
27. Describe the components and structure of each junction.
28. Types III and IV Intermediate filaments are found in what types of tissue? What extra proteins are found that add strength to the neuro filaments?

Fill up the blanks:

1. Actin filaments are cross linked into ——— to bear ——— forces.
2. Microtubules are hollow structures that sustain both ——— as well as ———.
3. Intermediate filaments are flexible with ——— strength but little ——— capacity.
4. Actin filaments are the ——— filaments of all the cytoskeletons.
5. The pointed and barbed ends of actin filament are also known as ——— and ——— ends.
6. The protein ——— regulate actin network formation.
7. The tensile strength of microfilament is regulated by ——— protein.
8. ——— ion plays a vital role in muscle contraction.
9. An actin filament is ———, a MT is ——— and an IF is ——— in diameter.
10. GTP hydrolysis is a common feature in ——— tubulin.
11. The diameter of a protofilament is ——— nm.
12. Each protofilament contains ——— beads.
13. During nucleation in MTs subunits addition occurs in ——— end.
14. Mutation in keratin forming gene causes the disease ———.
15. Intermediate filaments are prominent in cells that withstand ——— stress.

• MCQs:

1. **Cytoskeleton, capable of bearing compression forces –**
   (a) actin, (b) microtubule, (c) intermediate filament.
2. **Cytoskeleton that bears compression as well as tension –**
   (a) actin, (b) microtubule, (c) intermedia te filament.
3. **Cytoskeleton that mainly bear tensile strength –**
   (a) actin, (b) microtubule, (c) intermediate filament.
4. **The protein responsible for the formation of actin network is**
   (a) fascin, (b) profilin, (c) villin.
5. **Which of these proteins acts as capping as well as severing protein?**
   (a) tropomodulin, (b) fimbrin, (c) gelsolin.
6. **α-actinin is a**
   (a) severing, (b) capping, (c) cross linking protein.
7. **The IF present in wool is**
   (a) tubulin, (b) actin and (c) keratin.
8. **The cytosketon present in WBC is**
   (a) lamin, (b) keratin, (c) vimentin.
9. **Desmin is found in**
   (a) epithelium, (b) muscle cells and (c) neurons.

Answers to MCQs:

1. a, 2. b, 3. c, 4. c, 5. c, 6. c, 7. c, 8. c, 9. b.

Chapter 8

Molecular Motor Proteins

Cytosol of a eukaryotic cell is not a static entity; rather it depicts a dynamic nature showing transportation of different macromolecules from one organelle to another and even to the outside of the cell. Additionally different macromolecules are internalized by endocytosis and participate in diverse physiological activities. This complex itinerary involves movement of macromolecules or even organelles along a definite cytoskeletal tracks by some complex motor proteins. These are a class of molecular motors able to move along the surface of a suitable substrate powered by the hydrolysis of ATP and convert chemical energy into mechanical work.

8.1. GENERAL STRUCTURE OF MOTOR PROTEINS

Most eukaryotic motor proteins consist of two distinct domains - a motor head portion with ATPase function and a tail part that can either form fibers (muscle myosin) or attach to a cargo (e.g. chromosomes during anaphase of mitosis, performed by kinesin or vesicles during endocytosis carried out by dynein). The head domain carries out the movement by binding to a specific site on the cytoskeleton and changing conformation depending on ATP hydrolysis. The tail domain normally binds adaptor proteins that allow for stable interactions with the cargo to be moved along the substrate. These motor proteins typically form a complex of longer "heavy chains" with motor head domains and shorter "light chains" for stabilization.

Motor proteins, utilizing the cytoskeleton for movement, fall into two categories based on their substrates – Actin motors that allow movement of substrate (e.g. myosin) through interaction with actin. Microtubule motors that allow movement of substrate (e.g. dynein and kinesin) along microtubules through interaction with tubulin. Microtubule motors are of two types - plus-end motors and minus-end motors, depending on the direction in which they "walk" along the microtubules within the cell.

8.2.1. Actin motor

8.2.1.1. Myosin:

Myosin is the protein responsible for generating muscle contraction. Initially it was thought that myosin occurs only in muscle cells but recent workers confirm that not only muscle cells, virtually all-eukaryotic cells, express a

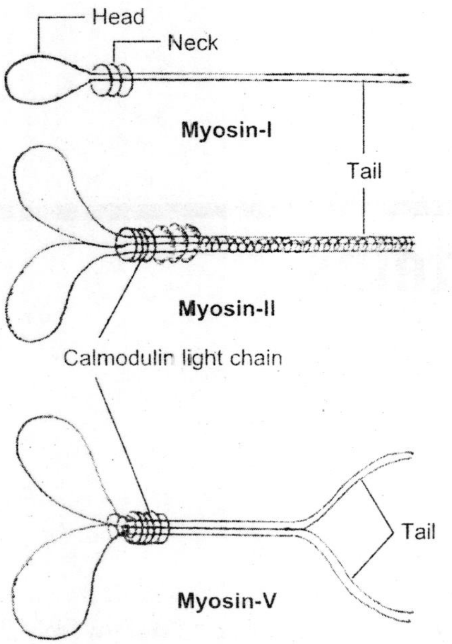

Fig. 8.1 : Structure of some myosin molecules.

multitude of different myosin molecules (Fig. 8.1). In general, myosins are divided in to 10 sub types. Myosin I occurs in *Acanthamoeba*, *Drosophila*, mouse, rat and Yeast; myosin II in *Plasmodium*, all vertebrates, Yeast and *Arabidopsis*(Plant); myosin III, in *Drosophila;* myosin V in *C. elegans*, *Drosophila*, mouse and yeast; myosin VI in *C. elegans*, *Drosophila*, mouse and rats; myosin VII in *C . elegans*, *Drosophila*, man and mouse; myosin IX in humans and mouse; myosin XI and XIII in *Arabidopsis* and maize; while myosin XIV in *Plasmodium* and *Toxoplasma*. The nematode C. elegans possesses 7 while human genome shows about 40 genes encoding myosins.

Irrespective of their minor structural differences, they show two heavy and four light chains. Each heavy chain (showing about 2000 amino acids) shows a conserved motor domain having separate head and neck regions at their N terminus. Two different light chains are associated with each head domain. The neck region consists of an -helix of variable length that is followed backwards as coiled coil tail. Tail domains contain amino acid sequences that are also found in other proteins. The head region contains the ATP- and actin-binding sites and exhibits ATPase activity. Neck region is proposed to serve as a lever arm for force production. The tail domain serves to regulate signal transduction by the small Ras-related G protein Rho that regulates the organization of the actin cytoskeleton and various other cellular processes. The N and C terminal ends of the myosins vary widely. Many myosins form dimers with two motor domains per molecule while myosin I, IX and XIV function as monomers with one motor domain. All the myosins move towards the '+' end (only myosin VI moves towards the '-' end) of the actin filament. The exact role of myosins is yet to be fully determined; myosin II generates force

8.2.2. Microtubule motors:

8.2.2.1. Kinesin:

Kinesins are a group of related motor proteins with diverse structures (Fig. 8.2) and use microtubule track to "walk." They are vital for the movement of chromosomes during mitosis and are also responsible for shuttling mitochondria, Golgi bodies and vesicles within eukaryotic cells. Kinesins typically contain two heavy chains with motor heads which move along MTs towards the plus-end or the minus-end, depending on the type of kinesin. Kinesins have a conserved MT-binding site in their motor domain but two homologous kinesins - *KatB* and *KatC* in *Arabidopsis thaliana*, show a second MT-binding site in their tail domains (Jiang *et al*, 2007).

Fourteen distinct kinesin families are known, with some additional kinesin-like proteins that cannot be classified into these families. Grossly these are categorized into three types – Conventional kinesins or Kinesin-I, Kinesin – II and Monomeric kinesins or Unc104/KIF1.

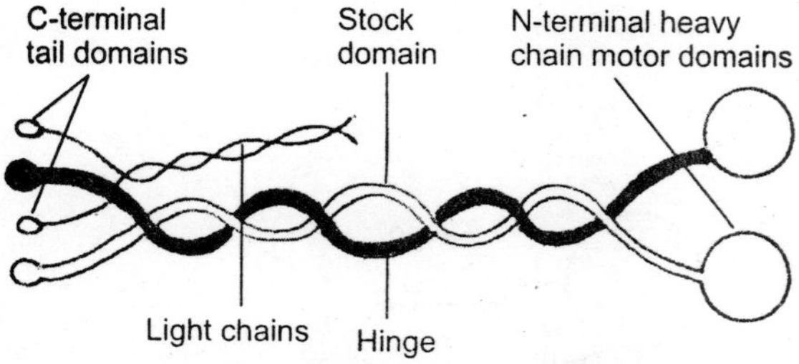

Fig. 8.2 : Diagramatic structure of Kinesin-I.

Conventional Kinesin or Kinesin I: It shows an N-terminal *motor domain,* a long coiled coil *stalk domain* having a central *hinge* and a globular *tail* domain. Stalk domains of the heavy chains interact in a α-helical coiled coil that extends from the heavy chain neck to the tail. It shows two copies each of a heavy chain and a light chain. Each heavy chain includes a globular ATP-binding motor domain at the N-terminus.

C terminal domains of heavy chains also participate in binding kinesins to cargo. In the absence of cargo, the stalk domain of the heavy chains fold at hinge region, bringing heavy chain tail domains into contact with the motor domains. In filamentous fungi, they transport membrane vesicles to the tip of the growing hyphae. In higher metazoans, they are reported to be involved in the transport of different organelles, mRNA and even intermediate filaments. Metazoan kinesin I structure reveals that a light chain binds to the tail domain of the heavy chain. Light and heavy chains are known to transport distinct cargoes.

Kinesin II: It contains two distinct polypeptide chains that formed a heterodimer and a subunit called KAP, found to be tightly bound to its tail domain. Due to the presence of these three subunits this motor is referred as heterotrimeric kinesin II. This motor is important for intra flagellar transport (IFT) of tubulin, radial spoke proteins etc. along the axonal MTs just beneath the axolemma. IFT is crucially important for the genesis of flagella or maintaining the length of flagella in different physiological conditions.

Unc 104/KIF1. Unlike other kinesins. Unc104/KIF1 kinesins are predominantly monomeric. The cargo-transporting role of these motors is also encounterd in the metazoans. Unlike other kinesins it lacks the coiled coil domain. These motors transport mitochondria and synaptic vesicles to the presynaptic terminals of motor neurons. A few kinesins have their motor domain located in the interior of the heavy chain sequence or at the C-terminus, instead of at the N-terminus. Those with the motor domain at the C-terminus, e.g., *Ncd,* move along MTs toward the minus end; *BimC,* a kinesin related protein (KRP) involved in mitosis, has a tail domain that allows it to assemble into antiparallel dimers that can mediate sliding of MTs relative to one another.

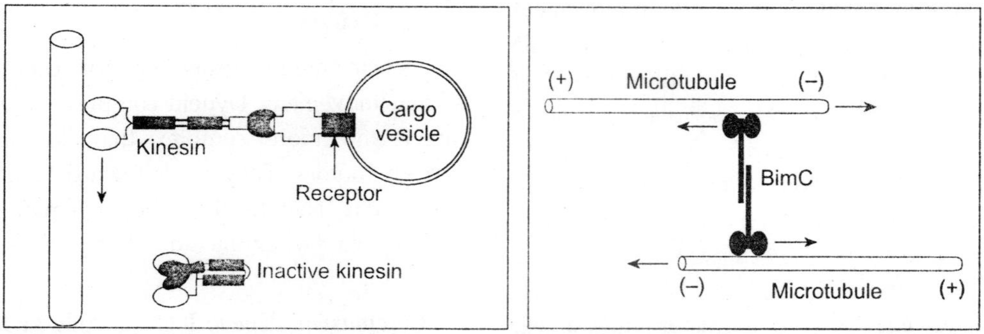

Fig. 8.3 : Transport of cargo by kinesin (left) and sliding of MTs by BimC during anaphase (right).

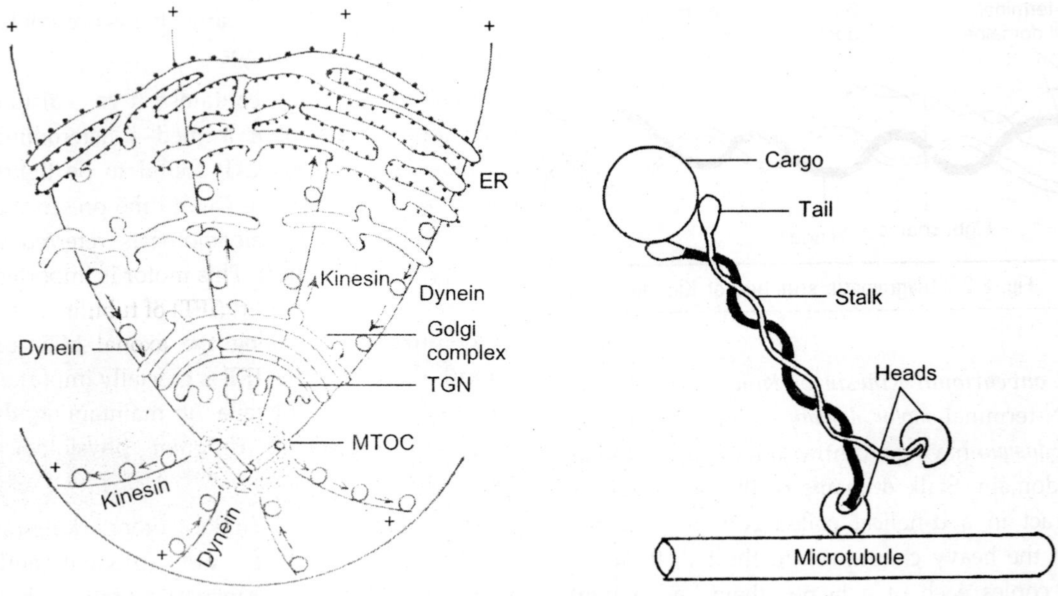

Fig. 8.4 : Role of kinesin and dynein during vesicular transport.

Kinesin's motor domain exhibits structural similarity to that of myosin. Kinesin and myosin both have nucleotide (ATP) binding domains. However kinesin has short connecting loops while the myosin head has longer stretches of amino acids. Both may cooperate in vesicle transport. Cargo particles do not roll along the MT. Instead kinesin walks along the MT. Each kinesin motor domain binds tightly to a MT when it has bound ATP. Movement of the kinesin involves alternate association of two heads along the MT without dissociation.

Conventional kinesin has a role in movement of vesicles and lysosomes, from the vicinity of the Golgi apparatus toward the plus ends of microtubules in the cell periphery. Kinesin transports vesicles along MTs within the axon to the plus ends. Various kinesins function in mitosis (cell division). Some kinesins promote shortening of MTs. The catastrophe-promoting kinesin MCAK is located in the kinetochore, where plus ends of spindle microtubules attach to chromosomes.

The disassembly-promoting kinesin KLP10A is found at minus ends of spindle MTs. During metaphase, there is treadmilling in MTs connecting kinetochores to the poles. During anaphase A, chromosomes move to the spindle poles, as MTs linking kinetochores to the poles shorten due to dissociation of tubulin dimers at the kinetochore. During anaphase B the mitotic spindle poles move apart due to the sliding of antiparallel spindle MTs mediated by the movement of BimC motor domains in opposite directions (Fig. 8.4).

8.2.2.2. Dynein:

Dyneins are microtubule motors capable of a sliding movement. Dynein complexes are much larger and more complex than kinesin and myosin motors. They walk towards the minus end of MTs. They facilitate the movement of cilia and flagella. Dynein includes 2 or 3 heavy chains (Fig. 8.4).

Cytoplasmic dynein: It is basically known for its ability to transport cargoes towards the '-' end of the MTs (toward the centrosome). It contains two or three heavy chains. Each is about 4600 amino acid residues long and includes a large globular motor domain and a small stalked globular domain at its tip that mediates its attachment to MTs. From the motor domain a coiled coil extends that is considered as *tail*. A multisubunit complex *dynactin* interacts with cytoplasmic dynein playing a regulatory role. Dynactin shows an elongated 150-kDa dimeric protein showing two microtubule-binding globular heads at its distal end. Other domains of dynactin may bind to dyneins as well as kinesins. Dynein may bind to membranes or some cargo proteins directly via its light chains, but interactions with cargo are often mediated by dynactin.

Functions:

i) It mediate ATP-dependent retrograde movements of vesicles and organelles along microtubules toward the centrosome

ii) It mediates movement and acquiring positions of nuclei. in fungi, cortical dyneins pull the astral MTs and help in the positioning of the spindles.

iii) It transports IFs, centrosomal proteins, mRNA, nuclear envelope breakdown products, movement of signaling and spindle checkpoint proteins etc.

iv) Dynein is often found in the cell cortex and remains associated with the plasma membrane via some accessory proteins. Arp1, binds to dynactin via dynamitin and binds to spectrin, an actin-binding protein of the cortical cytoskeleton. Spectrin in turn binds to ankyryn, which binds to integral membrane proteins. In this way dynactin complex anchors dynein to the plasma membrane.

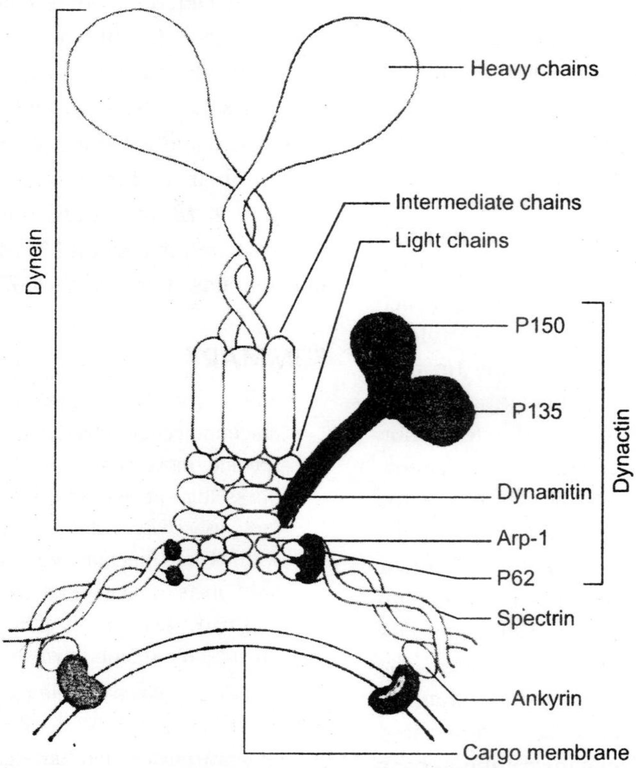

Fig. 8.5 : Diagramatic structure of Dynein-Dynactin complex.

v) Dynein and dynactin are associated with Golgi membranes, which also have a spectrin network on their surface. Its actin related Arp-1 subunit bind to Golgi associated spectrin protein and assist in their transport. Location of the Golgi apparatus near the centrosome is thought to be due to its movement along microtubules toward their '-' ends by dynein mediated by dynein and dynactin.

vi) Early and late endocytic vesicles have associated dynein and dynactin, which may be responsible for moving these vesicles inward from the cell surface.

vii) Interaction of cortical dynein with astral microtubules is essential for the orientation of the mitotic spindle and separation of poles during mitosis.

viii) Another important dynein regulatory protein is lissencephaly-1 which is required for dynein's role in mitosis and nuclear migration.

ix) Some mammalian cell types have motile cilia or flagella, including some respiratory epithelial cells and sperm cells. Many cells have a single short non-motile primary cilium. The photoreceptor structure of each retinal rod and cone cell develops from a non-motile cilium. In addition to their role in ciliary / flagellar movement, MTs of the axoneme provide pathways along which cytosolic and plasma membrane proteins are transported to and from the tip of a cilium or flagellum. This intraflagellar transport is important for the formation and maintenance of cilia and flagella, which grow by addition of subunits at the distal tip where plus ends of axonemal MTs are located. Some axonemal precursor proteins are transported in association with particles. Kinesins transport the particles along axonemal microtubules toward the ciliary/flagellar tip. Cytosolic dyneins transport the particles with associated proteins along axonemal microtubules back toward the cytosol.

Chlamydomonas possesses another cargo transporting cytoplasmic dynein related motor that moves the IFT particles in retrograde direction from the tip of the axoneme to the basal body. This motor is referred as cytoplasmic dynein 1 or 2.

8.3. PLANT-SPECIFIC MOTORS

In contrast to animals, fungi and lower plants, the cells of flowering plants lack dynein motors. However, they contain different types of kinesins. Many of these plant-specific kinesin groups are specialized for functions during plant cell mitosis. During mitosis, the new cell wall is built by the formation of a cell plate starting in the center of the cell. This process is facilitated by phragmoplast, a microtubule array unique to plant cell mitosis. The building of cell plate and ultimately the new cell wall requires kinesin-like motor proteins. Another motor protein essential for plant cell division is *kinesin-like calmodulin-binding protein* (KCBP), which is unique to plants and is partly kinesin, partly myosin like motor protein. Kine sins, as a kind of MT-based motor proteins, usually have a conserved MT-binding site in their motor domain. Two homologous kinesins *KatB* and *KatC* found in *Arabidopsis thaliana,* contain a second MT-binding site in their tail domains (Jiang *et al*, 2007).

SUMMARY

1. The complex interorganelle or simple macromolecular or vesicular itinerary involves their specific movement along a definite cytoskeletal tracks that are regulated by some complex motor proteins. These consist of a motor head portion with ATPase function and a tail part that can either form fibers or attach to a cargo. Motor proteins fall into two categories - Actin motors (myosins) and Microtubule motors (dynein and kinesin)
2. Myosin generates muscle contraction. These are of 10 sub types, I-III, V-VII, IX, XI, XIII and XIV. They show two heavy and four light chains. Each heavy chain shows head and neck regions. Two light chains are associated with each head domain. The

neck is followed backwards as coiled tail. Most myosins move towards the '+' end of the actin filament.

3. Kinesins use a MT track to "walk" towards the '+' or '-' end, depending on the type of kinesin. They facilitate movement of chromosomes, mitochondria, Golgi bodies, and vesicles within the cell. Grossly these are divided into three types – Kinesin-I, Kinesin – II and Unc104/KIF1.

4. Kinesin I shows a motor, a coiled stalk and a globular tail domain. Two light chains associate with two heavy chains near the tail. Kinesin-I assists in the movement of vesicles and lysoso mes along MTs within the axon to the plus ends. Kinesin II contains two polypeptide chains and a sub unit - KAP is tightly bound to its tail domain. It helps in intra flagellar transport of tubulin that is impor tant for the genesis of flagella. Unc104/KIF1 kinesins are monomeric and lacks the coiled coil domain in contrast to other kinesins. These motors transport synaptic vesicles to the presynap tic terminals. Some kinesins mediate different structural alterations of MTs in mitosis.

5. Dyneins are MT motors capable of a sliding movement. They walk towards the minus end of MTs. They facilitate the movement of cilia and flagella. Cytoplasmic dynein contains 2-3 heavy chains, which includes a large motor domain with a stalked domain that mediates its attachment to MTs. From the motor domain a tail extends. These mediate ATP-dependent movements of vesicles and organelles along MTs. Cortical dyneins pull the astral MTs to maintain position of the spindles. It also transports IFs, centrosomal proteins, mRNA, nuclear envelope breakdown products, signaling and spindle checkpoint proteins etc.

6. Dynactin interacts with cytoplasmic dynein playing a regulatory role. Dynactin shows two MT- binding heads, other domains may bind to dyneins as well as kinesins. Dynactin complex anchors dynein to the plasma membrane.

7. Fungi and lower or higher plants lack dyneins but a larger number of different kinesins that assist in mitosis. During cytokinesis during the formation of cell plate microtubules and kinesins .play decessive roles. Another motor protein kinesin-like calmodulin-binding protein (KCBP) is essential for plant cell division.

8. Deficiencies in motor proteins may lead to Charcot-Marie-Tooth disease, some kidney diseases, chronic infections of the respiratory tract, myopathies, Usher syndrome, lissencephaly, deafness etc.

QUESTIONS

Long questions:

1. Describe the structure of kinesins and their functions.
2. Describe the structure and role of dyneins to maintain various physiological activities.
3. Describe the structure of myosin and their role in muscle contraction.

Short questions:

1. Describe the general principle of association of motors with the cargoes.
2. Which motor is associated with intraflagellar transport of tubulins?
3. Which motor protein is related with anaphase-II?
4. Which motor is associated with the transport of different vesicles or organelles?
5. How association of dynein with plasma membrane is maintained?
6. Which motor participates in the movement of endocytic vesicles inwards?
7. Describe the structure and functions of plant specific motors.

Fill up the blanks:

1. ATPase domain of motor proteins occurs in the ———.
2. The tail domain mainly binds firmly with the ———.
3. Myosin — is mainly responsible for muscle contraction.
4. Myosins consist of ——— heavy and ——— light chains.
5. Most of the myosins move towards the ——— end of the actin filament.
6. The only kinesin that is monomeric is ———.
7. Dyneins move towards the ——— end of the MTs.

8. Signaling protein movement is done by ———.

MCQs:

1. **Muscle contraction is mainly caused by**
(a) myosin-I, (b) myosin-II, (c) myosin-III.
2. **The kinesin** (a) Ncd, (b) BimC, (c) MCAK **remains bound to the kinetochore.**
3. **The kinesin** (a) MCAK, (b) BimC, (c) KLP10A **is associated with the '-' end of kMTs.**
4. **Retrograde transport of vesicles is done by**
(a) kinesin I, (b) cytoplasmic dyneins, (c) KIF1.
5. (a) Dynactin, (b) kinesin-I, (c) KIF1 **interacts with dynein regulating its action.**
6. **Charcot Marie tooth disease is caused due to deficiency in**
(a) myosin-I, (b) kinesin, (c) dynein.
7. **Respiratory tract disease is caused due to defects in**
(a) myosin-II, (b) dynein, (c) kinesin.
8. **Myopathy is caused due to defects in**
(a) myosin-II, b, dynein, (c) kinesin.
9. **Neurodegenerative disease is caused due to defects in**
(a) myosin-I, (b) kinesin, (c) dynein.

Answers to MCQs:

1. b, 2. c, 3. c, 4. b, 5. a, 6. b, 7. b, 8. a, 9. b.

Chapter 9

Intracellular Vesicular Trafficking

In eukaryotic cells, any macromolecule whether protein, lipid, polysaccharides or even conjugated proteins or lipids, shortly after synthesis can not take part in any metabolic activity with out proper biochemical transformation. Most of these biochemical transformations occur either in ER or in Golgi complex and most of these macromolecules are transported between the organelles through some vesicles. Thus inter organelle vesicular tranport is the key phenomenon in the post secretion transformation of the macromolecules.

The vesicular transport hypothesis was first deduced by Palade (1975) during his investigation on protein synthesis. He put forward the hypothesis that nascent proteins pass through several vesicles on their way toward maturation and incorporation to different organelles or secretion to extracellular space.

The routes of vesicular transport exhibit passage of protein from ER to Golgi to the outside of the cell by exocytosis or from TGN to lysosomes or from the cell membrane to endosomes to lysosome by enodocytosis. In plants, the secretory pathway comprises a network of organelles that connects three large membranes, the plasma membrane, the vacuole and the endoplasmic reticulum. The Golgi apparatus, various post-Golgi organelles and prevacuolar compartments (PVC) or multivesicular endosomes that control vacuolar sorting, secretion and endocytosis. They can be regarded as intermediate organelles of the endocytic and biosynthetic routes for sorting proteins from the Golgi apparatus to vacuoles, sending missorted proteins back to the Golgi. (Foresti and Denecke, 2008).

9.1. ER-GOLGI VESICULAR TRANSPORT

Proteins after synthesis are transported into the lumen of the ER and are variously modified for their final destination. These properly folded proteins leave the ER from a specialized region of the ER that is devoid of ribosomes called *ER exit site*. In most ERs, the ER exit site is randomly dispersed. The correctly folded cargo proteins display *exit signals* on their surface that are recognized by the coats of the vesicles. Some proteins without such exit signals are also packaged into the vesicles so that the ER resident proteins slowly leak out of the ER. However all the vesicles are transported into an *ER-Golgi*

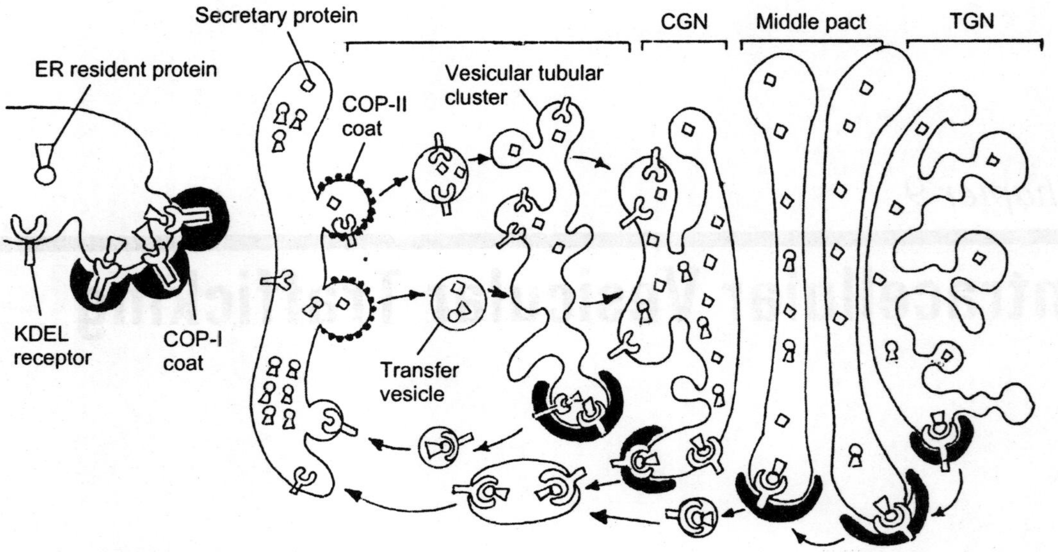

Fig. 9.1 : Anterograde and retrograde vesicular transport pathway.

intermediate compartment (ERGIC) via *transport vesicles* (70–80-nm in diam.). As the transport vesicles move towards the cis Golgi network (CGN), the vesicles shed their coats and fuse with each other to form a convoluted structure called *vesicular tubular clusters* (Fig. 9.2). These are very short lived and move along the microtubule to the CGN.

Only properly folded proteins leave the ER: The vesicles that travel between the ER and Golgi are coated with proteins that have *'addressing tags'* on them. For example, the vesicle that buds

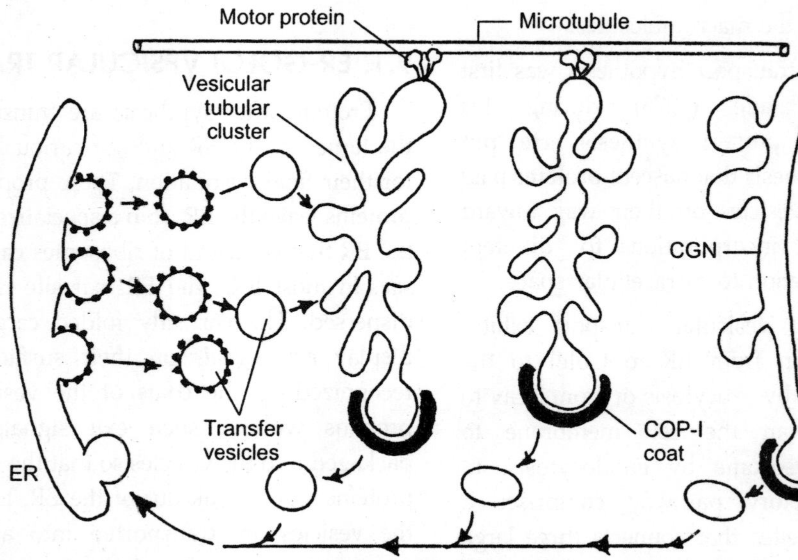

Fig. 9.2 : Transport of transfer vesicles to the vesicular tubular clusters.

off from the ER has coat proteins that specifically direct them to the cis Golgi instead of the medial Golgi or some other organelle. Proteins which are correctly folded do not require a signal for anterograde transport but the misfolded proteins are tagged with a retention signal. The resident ER membrane proteins exhibit two lysines followed by other two amino acids at their C terminal end. It is called *KKXX* sequence (K = lysine and X = any animo acid). Soluble ER resident proteins show 4 amino acids identified as Lys-Asp-Glu-Leu. This is called *KDEL* sequence (K = Lysine, D = Aspartic acid, E = Glutamic acid, L = Leucine). However, the retention signal does not work by holding the proteins back in the ER, all the proteins are transported to the Golgi where they are selectively retrieved by binding to specific membrane bound receptors and are then transported back to the ER. At this point only the correctly folded secretory proteins follow the anterograde direction but the ER resident proteins or the improperly folded proteins are transported back to the ER where they are bound to the chaperone proteins like Bip or Calnexin. These proteins are ultimately transported back to the cytosol where they are degraded by the proteasomes. Thus the ER acts as a *'monitoring point'*.

These vesicles with properly folded proteins gradually merge with the wall of the Golgi cisternae following *anterograde pathway*. These proteins are gradually transformed during their journey through the Golgi. But after transportation of proteins from the ER to the cis Golgi, some proteins are recycled back to the ER following *retrograde pathway* while others are transported to the TGN where the last stages of processing are finished. The TGN thus acts as a *"distribution center"* for the completed product targeted for plasma membrane or exterior of the cell or to the lysosome (Fig. 9.1).

This retrograde transport not only takes place from Golgi to ER but also from one Golgi stack back to previous Golgi stack. Therefore, the Golgi has two sets of vesicles flowing in opposite directions. The first set of vesicles is filled with newly made proteins awaiting further modifications traveling in an anterograde direction. The second set of vesicles is filled with enzymes and mature proteins seeking their resident organelle traveling in the retrograde direction. The tyrosine kinase Src present on the Golgi membranes and Golgi *cis*-medial matrix proteins are reported to be involved in retrograde protein transport from Golgi apparatus to the ER. Large macromolecular cargoes (e.g. 300-nm sized procollagen) are unable to interact directly with the coat proteins and are carried by tubular structures aided by dynactin and kinesin motor proteins.

9.2. TRANSPORT OF SECRETORY SUBSTANCES FROM THE TGN TO THE CELL EXTERIOR–EXOCYTOSIS

Activity of the secretory cells is also directly related with the trafficking of secretory vesicles and their subsequent liberation from the cell. There are two types of secretory processes. As per *regulated secretory pathway* the secretory vesicles only in response to the correct *"signal"*, fuse with the plasma membrane and are released outside the cell and transported to far off places. Hormones, neurotransmitters and digestive enzymes are secreted in this process.

As per *constitutive secretory pathway* the proteins are secreted in a constitutive or constant manner. Irrespective of the signal for release these protein containing vesicles are budded off from the TGN and exported to the plasma membrane for exocytosis. Proteoglycans and glycol proteins are secreted in this way. Such proteins are used just outside the cell in the ECM. Some proteins however need to be delivered immediately and directly to the cell surface without being packaged into the vesicles following a *non selective constitutive pathway or default pathway* (Fig. 9.3).

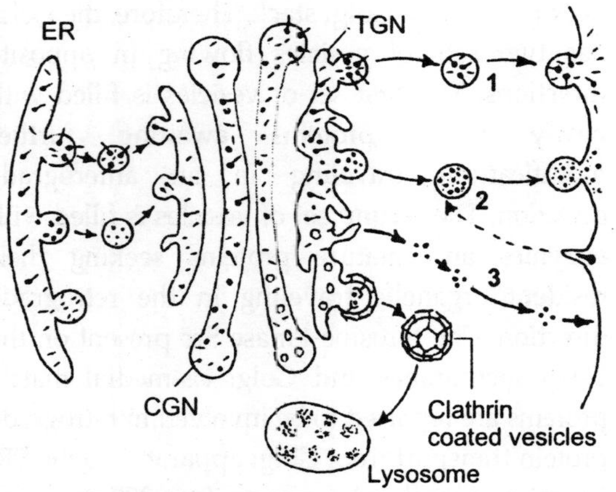

Fig. 9.3 : Different types of exocytic secretion. 1- constitutive secretion, 2- regulated secretion, 3- non selective constitutive secretion.

Once loaded with secretory proteins, the motor proteins propel the vesicles to the cell membrane along the microtubules. Once the vesicles containing the materials for constitutive secretion arrive the plasma membrane they fuse with the membrane. Proteins for regulated secretion first aggregate in the TGN where their ionic features are adjusted. Next they are accumulated in immature vesicles where only secretory proteins are segregated from excess membrane and luminal contents and are

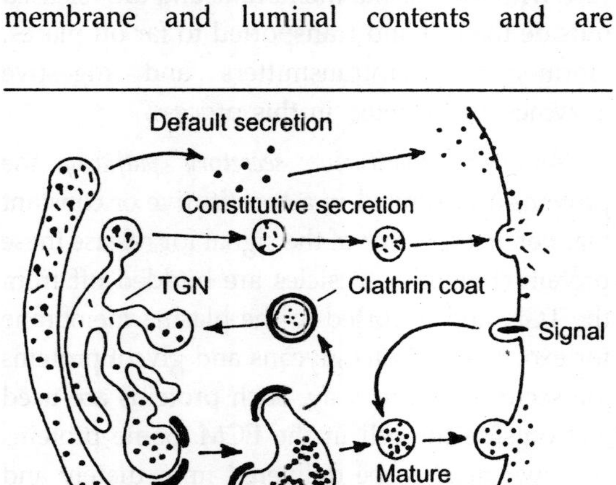

Fig. 9.4 : Formation of secretory materials.

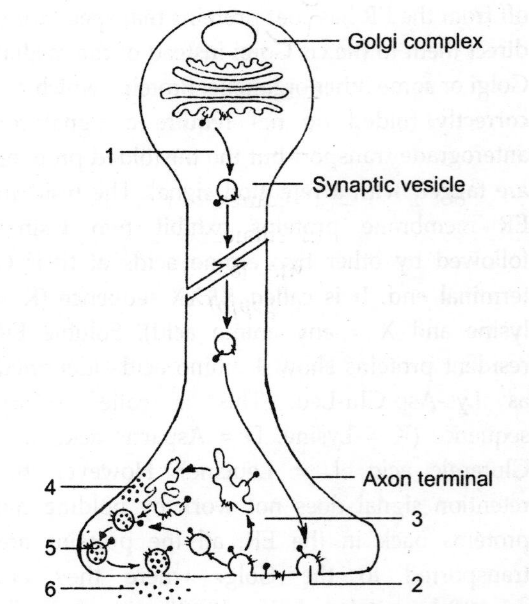

Fig. 9.5 : Release of synaptic vesicles. 1- Fusion of synaptic vesicles with the presynaptic membrane; 2- Endocytosis of the vesicle components; 3 -Delivery of the components to endo-somes; 4- Formation of new vesicles;5 – Loading of neurotransmitters tothe new vesicle; 6 - Exocytosis of neurotransmitters.

retrieved to the TGN via clathrin coated vesicles and finally the secretory proteins are accumulated in definite mature vesicles but the vesicles have to wait for signal to fuse with the membrane (Fig. 9.4). The signal leads to an increase in the concentration Ca^{++} in the cytosol that regulates the fusion of vesicles with the membrane. TNF-α is a potent proinflammatory cytokine, essential for initiating innate immune responses against invading microbes and a key mediator involved in the pathogenesis of acute and chronic inflammatory diseases. Cathepsin B, a lysosomal cysteine proteinase is also involved in the TNF-α-containing vesicle trafficking to the plasma membrane (Ha *et al*, 2008).

9.2.1. Exocytosis of Neurotransmitter:

The exocytosis of neurotransmitter – acetylcholine filled synaptic vesicles involves targeting and fusion of vesicles to a specialized area of the presynaptic plasma membrane the active zone in response to the action potential at the presynaptic axon terminal. Action potential causes opening of voltage gated Ca^{++} channels that allow influx of Ca^{++} from the ECM. The membrane of the synaptic vesicles contains a Ca^{++} binding protein – *synaptotagmin* that senses the rise of Ca^{++} in the Cytosol (Fig. 9.5). The vesicles are linked with each other by a fibrous protein *synapsin.* Rab3A, located in the membrane of the vesicles is needed for targeting the vesicles to the active zone. Ca^{++} induces conformational change in the synaptotagmin that in association with another protein *complexin* mediates docking of the vesicles with the plasma membrane. The fusion of vesicles involves binding of v- and t-SNAREs. The principal v-SNARE in synaptic vesicle is *VAMP (synaptobrevin)* that binds with the t-SNAREs at the plasma membıane of the axon terminal – *syntaxin and SNAP-25.* SNARE complex assembly is regulated by Munc13 and Munc14 proteins. The speed with which the synaptic vesicles fuse with the presynaptic membrane in response to sudden increase in the cytosolic Ca^{++} indicates that the fusion machinery remains pre assembled and only undergoes conformational change leading to exocytosis. At least two proteins *synaptotagmin-1 and VAMP-2* occur in large quantity in the axonal plasma membrane and interchange with recently exocytosed proteins. It suggests that plasma membrane serves as a reservoir of synaptic vesicle proteins particularly synaptobrevin. Thus neurotransmitter exocytosis is a meticulously controlled interaction of Ca^{++}, synaptotagmin, VAMP, SNAP-25 and syntaxin through exo- and endocytosis (Fig. 9.6).

Action potential
↓
Voltage gated Ca^{++} channels open
↓
Influx of Ca^{++} from ECM
↓
Synaptotagmin senses this rapid Ca^{++} influx
↓
Ca^{++} causes conformational changes of Synaptotagmin
↓
Fusion of vesicles with plasma membrane by synaptobrevin
↓
Exocytosis of neurotransmitters

Fig. 9.6 : Flow chart of the main steps of neurotransmitter release.

9.3. VESICULAR TRANSPORT FROM THE TGN TO THE LYSOSOMES

Another pathway of vesicular transport is to provide the acid hydrolases for the lysosomes and substrates for the lysosomes. The substrates delivered to the lysosomes are derived from two sources; one is *heterophagy* involving uptake of foreign bodies by endocytosis and other is *autophagy* i.e. encapsulated deformed or dead organelles that are delivered to the lysosome for degradation.

9.4. TRANSPORT OF EXOGENOUS MATERIALS INTO THE CELL-ENDOCYTOSIS

Endocytosis is the intake of macromolecules through the plasma membrane into the cytosol.

Endocytosis actually occurs in some specialized part of the plasma membrane called *clathrin coated pits* which occupy about 2% of the whole cell membrane area. Clathrin coated pits contain definite transmembrane receptors that receives specific macromolecules for internalization. This type of endocytosis is therefore called *receptor mediated endocytosis.* More than 25 receptors accumulate in these pits and participate in endocytosis. Cholesterol in the

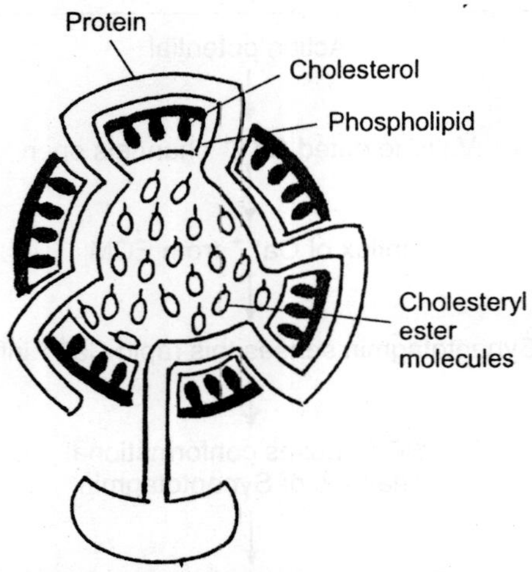

Fig. 9.7 : A low density lipo protein particle.

blood is transported as low density lipoprotein particle (LDL) whose receptors accumulates in the clathrin coated pits and receive LDL particle. In fibroblasts, these areas form about 2500 clathrin coated vesicles every minute (features of clathrin coated vesicles is discussed later on).

Endocytosed materials containing clathrin containing vesicles are pinched off into the cytoplasm and within seconds they shed off their clathrin coats and are at first delivered to small irregularly shaped organelle known as *early endosomes.* Some of the materials they contain, are selectively retrieved and returned to the same domain of the plasma membrane from where they were internalized. The LDL particle (Fig. 9. 7) with receptors follow this pathway, they bind to early endosome from where receptors again recycle back to the membrane and LDL particles deliver cholesterol to the cytosol (Fig. 9.8). Epidermal growth factor (EGF) internalization does follow this route. After its internalization by EGF receptors intracellular signaling pathway is activated, then it down regulates EGF receptors on the cell surface and ultimately the receptors are degraded in the lysosomes. In some cases, the macromolecules are recycled back to the plasma membrane domain that is non homologous to membrane domain from where it was internalized. It is called *trancytosis.* In some cases where retrieval of receptors does not happen, there the contents are directly delivered to the lysosomes for degradation. Transformation of endosomes to lysosomes involves migration of early endosomes along with microtubules.

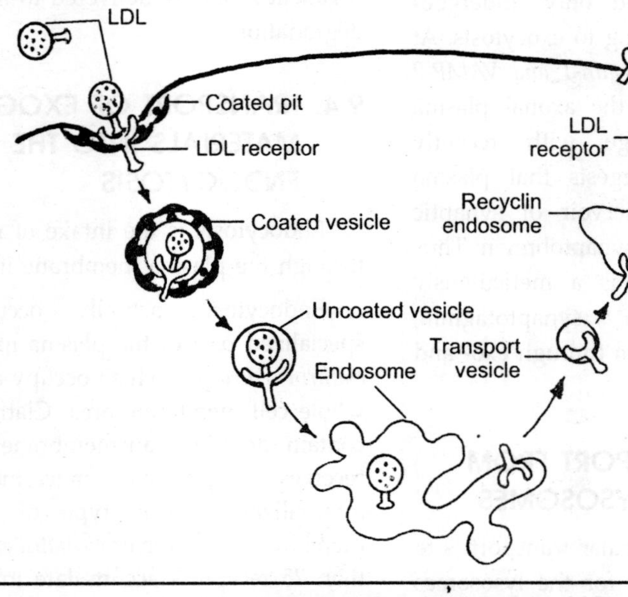

Fig. 9.8 : Receptor mediated endocytosis of LDL.

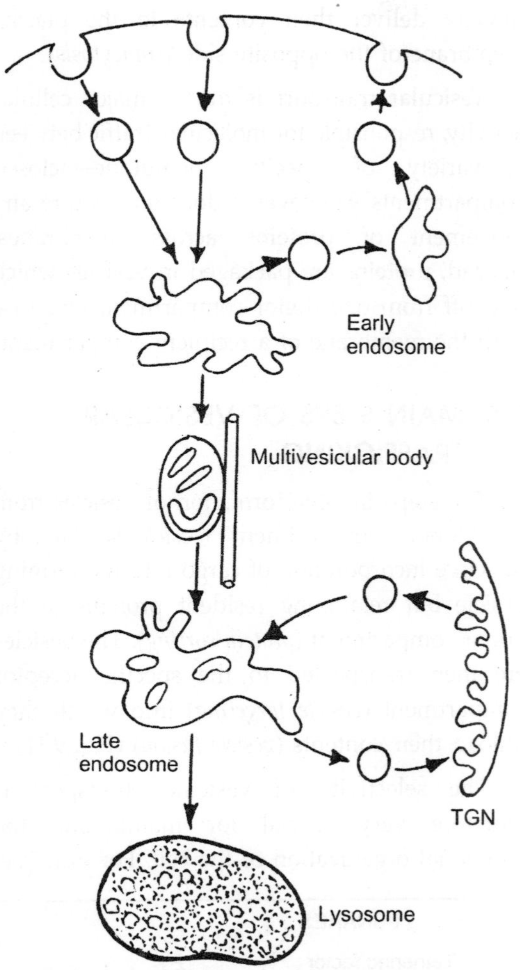

Fig. 9.9 : Endocytotic compartments.

Endocytotic compartments: The endosomes contain large amount of invaginated membranes and internally pinched off vesicles. Therefore these migrating endosomes are called *multivesicular bodies.* They contain membrane proteins that are needed to be degraded but do not contain proteins to be recycled. It is uncertain whether multivesicular bodies fuse to form *late endosomes* or they fuse directly with late endosomes. The endocytosed materials from apical and basolateral domains of plasma membrane communicate with separate early endosomes but if the material do not receive any signal for recycling or trancytosis, they both meet in a common late endosomal compartment before delivery to lysosomes (Fig. 9.9). The contents in the late endosomes are first mixed with the hydrolases (its interior is mildly acidic acidic pH 6.0) since hydrolytic enzymes the vesicles carrying from the TGN fuse with the late endosomes. Mature lysosomes are formed from the fusion of late endosomes, when the acidity of the lysosomes is increased to 4-5.0 which results in full activation of all the acid hydrolases. Also during the transformation to the lysomome, distinct endosomal proteins are lost, which presumably are recycled back to the plasma membrane. Although this is the most popular view, an alternative process the early and late endosomes and lysosomes as being distinct compartments, recognizable by different membrane protein profiles and that some type of vesicular transport occurs between them (Fig. 9.10). The endocytosed membrane proteins are subsequently degraded by three endosomal sorting complexes required for transport which also causes multivesicular body biogenesis.

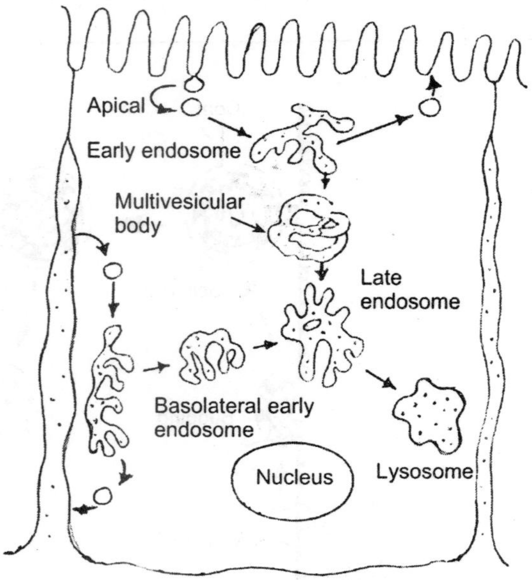

Fig. 9.10 : Differential processing of materials in endosomes from different domains of plasma membrane.

In *autophagy*, the dead or senile cell organelle to be degraded becomes surrounded by membrane of unknown origin to form *autophagosome* that ultimately fuses with late endosome or lysosome.

In *phagocytosis*, the phagosome formed the cell membrane flowing around the particle undergoes lysosomal fusion and the contents are digested.

All eukaryotic cells continually ingest bits of their plasma membrane along with extracellular fluid in pinocytic vesicles. This process is called *pinocytosis*. A macrophage in this way ingests about 3% of its plasma membrane every minute but as the area of plasma membrane remains unchanged therefore same amount of membrane material must be incorporated to the cell membrane by exocytosis. Therefore both exo- and endocytosis are interlinked.

In addition to clathrin derived pinocytosis, caveolae also invaginate along with the cargo materials and form vesicles whose wall is derived from the lipid contents of the caveolae. These vesicles may bind to the early endosomes or may directly deliver their contents to the plasma membrane of the opposite side (*trancytosis*).

Vesicular transport is thus a major cellular activity, responsible for molecular traffic between a variety of specific membrane-enclosed compartments. However it does not require any movement of proteins across membranes. Instead, proteins are packaged in vesicles which bud off from one donor compartment and fuse with the membrane of a recipient compartment.

9.5. MAIN STEPS OF VESICULAR TRAFFICKING

The steps involve formation of vesicles from the 'donor com partment' (*vesicle budding*) by selective incorporation of cargo into the forming vesicle but retainning resident proteins in the donor compartment (*protein sorting*). The vesicles are then transported to the specific acceptor compartment (*vesicle targeting*) into which they unload their contents (*vesicle fusion*) (Fig. 9.11).

The selectivity of vesicular transport is therefore very crucial for maintaining the functional organization of the cell. For example,

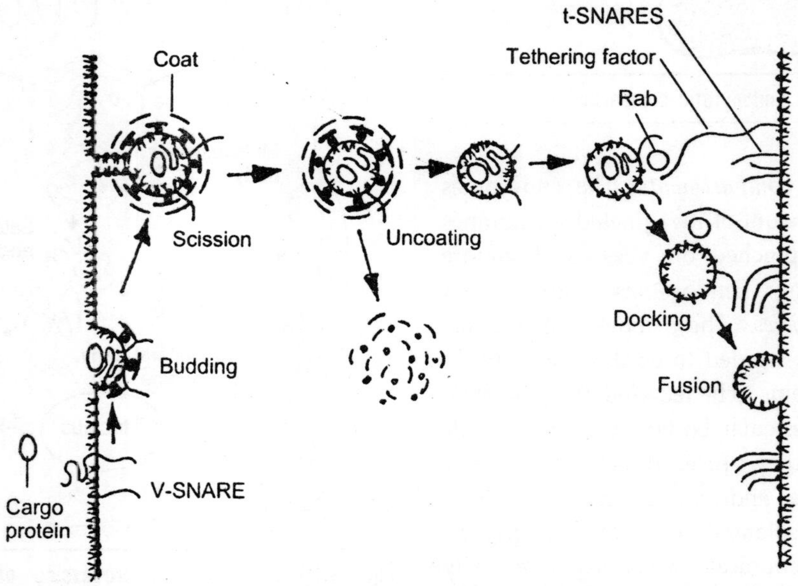

Fig. 19.11 : Steps of vesicle budding and fusion.

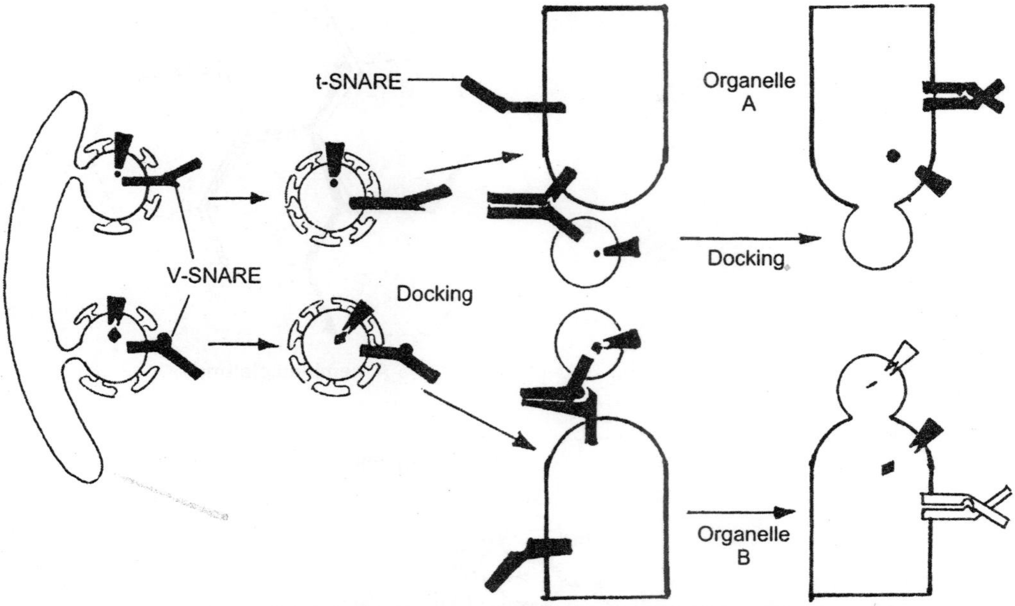

Fig. 9.12 : Role of SNAREs in the selectivity of vesicular transport.

lysosomal enzymes must be transportted specifically from the TGN to lysosomes—not to any other compartment. The specificity of transport is based on the selective packaging of the intended cargo into the vesicles that recognize and fuse only with the appropriate target membrane. In fact, the selectivity actually is regulated by the interaction of specific SNAREs. The v-SNAREs present in different vesicles interact with specific t-SNAREs present in receiving compartment. Therefore SNAREs indirectly regulates the selectivity of the vesicular transport (Fig. 9.12). The whole process involves vesicle packaging, budding, and fusion with other membrane. In addition to simple transport of these molecules, it also presents the opportunity for modification of the proteins.

9.5.1. Types of vesicle coats:

The first step in vesicular transport is the formation of a vesicle by budding from the donor membrane. The cytoplasmic surfaces of transport vesicles are coated with proteins, and it appears to be the assembly of these protein coats that drives vesicle budding by distorting membrane conformation. Three kinds of coated vesicles function in different types of vesicular transport.

9.5.1.1. Clathrin-coated vesicles:

These were initially supposed to participate in most vesicular transport but later studies confirm their localization to post Golgi locations. Clathrin coated vesicles are found trafficking between the Golgi and plasma membrane, the Golgi and endosomes, and the plasma membrane are endosomes and endosomes. They receive extra cellular molecules by the plasma membrane during endocytosis as well as the transport of molecules from TGN to plasma membrane or lysosomes.The coats of clathrin-coated vesicles are composed of two types of protein complexes, *clathrin* and *adaptor proteins,* which assemble on the cytosolic side of the membranes. Clathrin plays a structural role by assembling into a basket like lattice structure that distorts the membrane and drives vesicle budding. Each clathrin subunit consists of three large and three small

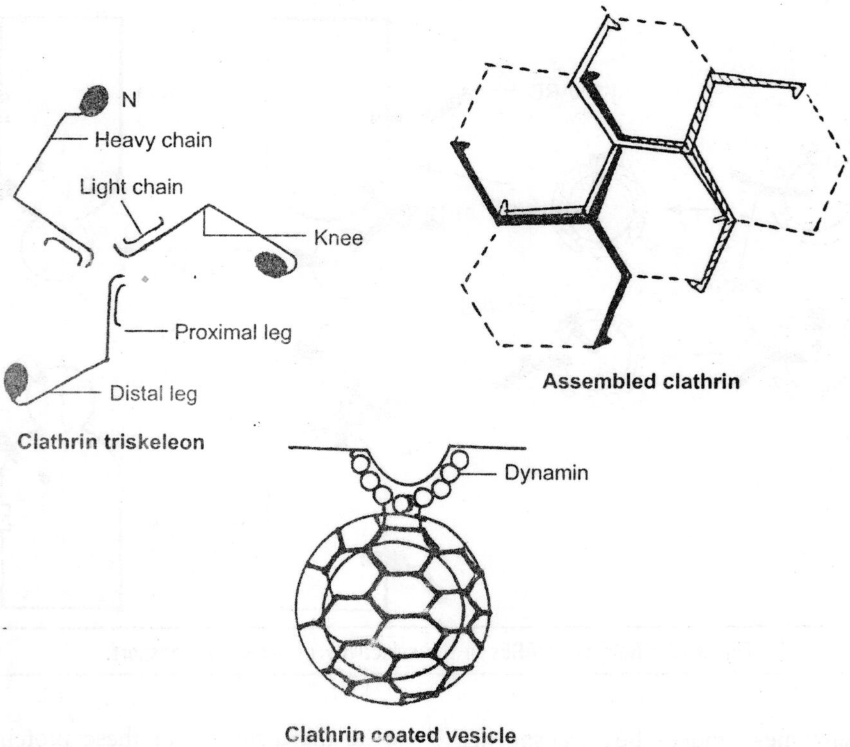

Fig. 9.13 : Structure of triskelion and clathrin coats.

polypeptide chains that collectively form a three legged structure called *triskelion* (Fig. 9.13). These assemble into a basket like framework of hexagons and pentagons to form coated pits on the cytosolic surface of the membranes. The binding of clathrin to membranes is mediated by adaptor proteins called *adaptins.* Four types of adaptins are responsible for the assembly of clathrin-coated vesicles and additionally they select the specific molecules to be incorporated into the vesicles. For example, the AP-1 adaptor protein involved in budding from the TGN binds to the cytosolic portion of the mannose-6-phosphate receptor, thereby directing proteins destined for lysosomes into clathrin-coated vesicles.

9.5.1.2. Non clathrin-coated or COP-coated vesicles:

(COP stands for coat protein) – these vesicles bud from the ER and Golgi complex. These are of two types – ***COP-II-coated vesicles*** that bud from the ER and carry their cargo along the secretory pathway to the ERGIC or Golgi apparatus. In contrast, ***COP-I coated vesicles*** are involved in intra-Golgi transport or bud from the ER-Golgi intermediate compartment (ERGIC) or from the Golgi apparatus and follow the *retrieval pathway* that serve to retain resident proteins in the Golgi or ER. These vesicles transport resident ER proteins marked by the KDEL or KKXX retrieval signals back to the ER from the ERGIC or the *cis* Golgi network. The coats of COP-I and COP-II coated vesicles show distinct protein complexes, which function analogously to clathrin and adaptor proteins in vesicle budding.

The core COP-II components are the small Ras like GTPase Sar1p-Sec23p-Sec24p and Sec13p-Sec31p sub complex. Sar1p-Sec23p-Sec24p forms the membrane proximal layer while Sec13p-Sec31p forms the distal layer of the COP-

II. Sec16p and Sec12p act as additional regulatory proteins. COP-I coat proteins also show seven subunits. Sec23p is important in linking COPII vesicle formation to anterograde transport events along microtubules (Fromme *et al*, 2008).

9.5.1.3. Synaptic vesicles:

These store neurotransmitters and are located at presynaptic terminals in axons.

9.5.1.4. Matrix vesicles:

These are specialized to initiate biomineralization of the matrix in bone, cartilage, and dentin. These are located within the ECM. During normal calcification, a major influx of Ca^{++} and PO_4^{++} into the cell accompanies cell death and formation of matrix vesicles where phosphatidylserine; calcium phosphate complexes is formed in the plasma membrane mediated by a protein - called *annexins*. Matrix vesicles bud from the plasma membrane at the sites of interaction with the ECM and convey calcium, phosphate, lipids and annexins to the ECM which participate in mineral formation. These processes are precisely coordinated to bring about mineralization at the proper place and time during bone development.

9.5.2. Coat assembly:

The budding of transport vesicles and the selective incorporation of cargo into the forming vesicles are both mediated by protein coat. Coats are supramolecular assemblies of proteins that are recruited from the cytosol to the nascent vesicles. The assembly of vesicle coats requires GTP-binding proteins, which regulate the binding of coat proteins to the membrane.

9.5.2.1. COP-I coat assembly:

The budding of both clathrin-coated and COP-I coated vesicles from the Golgi complex requires a GTP-binding protein called ARF (ADP-ribosylation factor). ARF proteins are responsible for both COPI and clathrin coat assembly. The first step in vesicle formation is the association of ARF bound to GDP with the Golgi membrane. Proteins in the Golgi membrane then stimulate the exchange of the GDP bound to ARF for GTP and the COP-I coat proteins bind to the ARF/GTP complex. It then recruits the membrane proximal bgdz and the membrane distal abe sub complexes. COP-I recognizes specific signals in the domains of the transmembrane cargo proteins. However in this case the signals mainly help to retrieve the proteins from the ERGIC or the Golgi complex to the ER. Assembly of the coat is then followed by deformation of the membrane and vesicle budding. ARF then hydrolyzes its bound GTP, leading to the conversion of ARF to the GDP-bound state and the dissociation of coat proteins from the vesicle membrane.

9.5.2.2. COP-II assembly:

Assembly of COP-II coated vesicles from the ER requires a distinct GTP-binding protein called Sar1. COP-II coat assembles by the stepwise

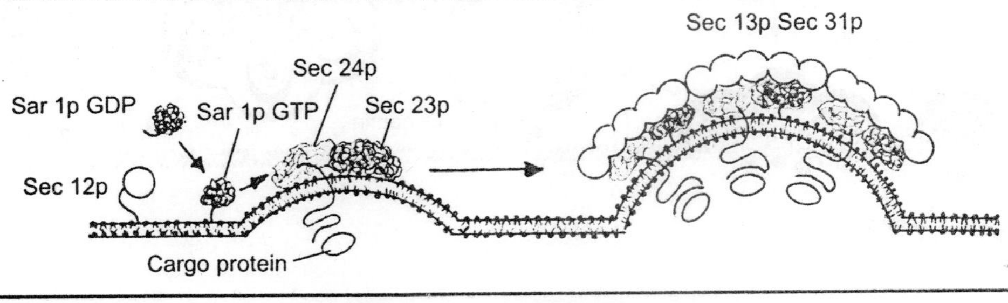

Fig. 9.14 : Steps of COPII assembly.

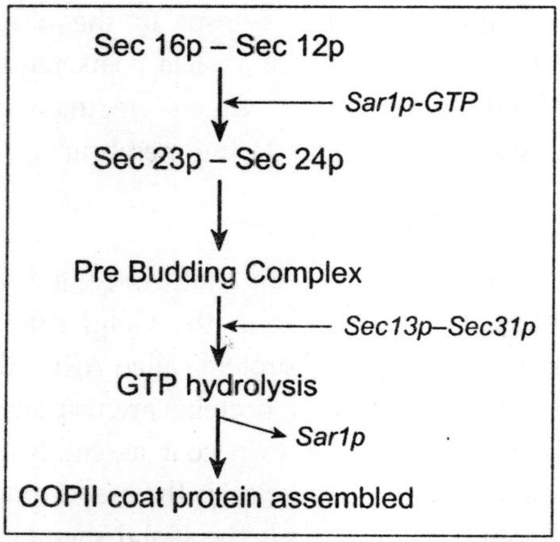

Fig. 9.15 : Flow chart of the COPII coat assembly.

deposition of Sar1p-GTP, Sec23p-Sec24p and Sec13p-Sec31p onto the sites where newly synthesized proteins exit from the ER. These ER exit sites are devoid of ribosomes and range in morphology from a small bud to convoluted network. Sec16p a peripheral ER protein interacts with Sec23p, Sec24P and Sec31P and serve as a scaffold for the nucleation of the assembling coat. Sec16p perhaps acts in conjunction with Sec12p to recruit Sar1p to the ER membrane (Fig. 9.14).

Sar1p along with Sec23p-Sec24p constitute the so called *'pre budding complex'*. It resembles slightly bent structure attaching to both Sec23p and Sec24p. Sec23p directly is communicated with 23p while Sec24p participates in cargo recognition. Once assembled into the membrane the pre budding complex recruits two Sec13p and two Sec31p sub complex (Fig. 9.15).

9.5.2.3. Clathrin coat assembly:

Clathrin coat assembly involves recruitment of a variety of adapters (AP-1, AP-2, AP-3, GGA, Hrs, Epsin1 and ARH proteins) from the cytosol to the membranes onto which clathrin is subsequently deposited. Clathrin and clathrin adaptor complexes can polymerize into spherical cage like structure (Fig. 10.16). Clathrin vesicle assembly is regulated by many kinases, phosphatases and other accessory proteins. The adapters also bind to cargo proteins by recognizing cytosolic sorting signals exhibiting tyrosine or di-leucine amino acids.

9.5.3. Protein sorting or Cargo selection:

Majority of the transmembrane cargo proteins is concentrated in COP-II coated buds prior to export from the ER and they exit the ER by directly binding COP-II protein subunits. Most soluble cargo proteins however bind indirectly to COP-II coated vesicles through some export receptors that leave the ER together with their

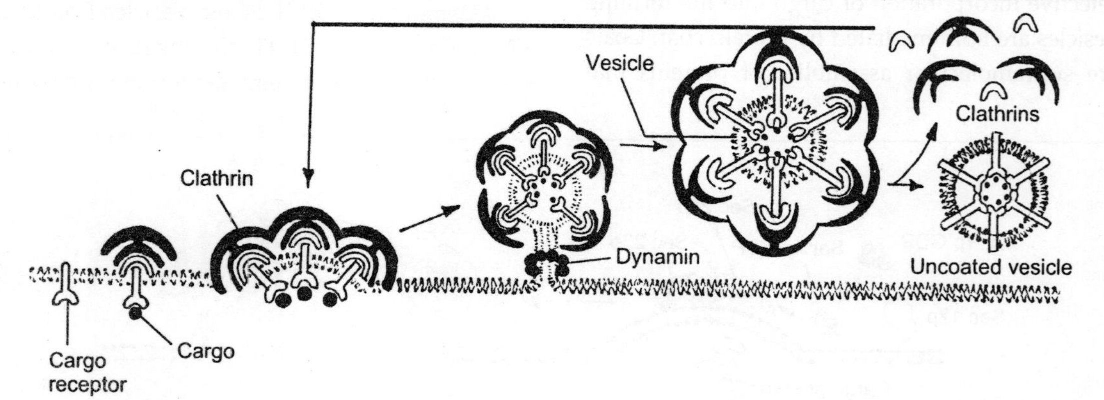

Fig. 9.16 : Clathrin coat assembly.

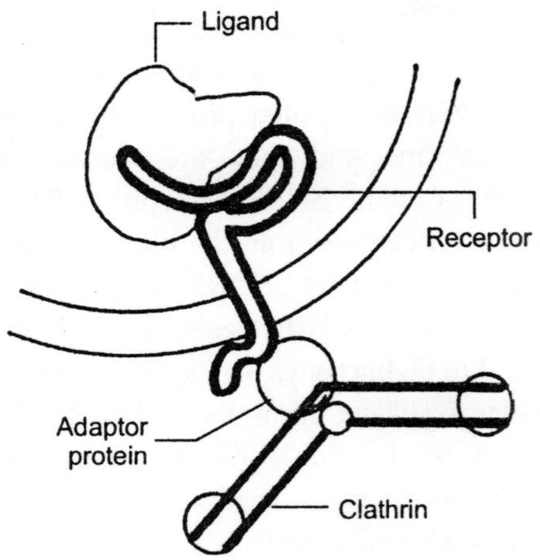

Fig. 9.17 : Association of cargo and clathrin coat.

ligands and unload their cargo into the acceptor compartment and recycle back to ER. In fact COPII coats recognize some sorting signals which are specific amino acids attached to the terminal portion of the cargo polypeptide. These may be represented by [DE] X [DE] or [RK] X [RK] or FF, YYM, FY, LL or IL where D = aspartate, X is any amino acid E = glutamate, R = arginine, K = lysine, F = phenylalanine, Y = tyrosine, M = methionine, L = leucine and IL = isoleucine. Most of these signals bind to Sec24p that displays at least three distinct binding sites termed 'A', 'B' and 'Arg342'.

9.5.4. Vesicle budding:

Sec23p-24p, Sec13p-31p and GTP locked Sar1p are sufficient for deforming the membrane and pinching off a vesicle. The role of Sec13p-31p in causing deformation of the membrane is not clear but they certainly stabilize the deformed membrane. The final stage of vesicle budding is the scission of the neck of the bud. As a clathrin coated vesicle grows soluble cytoplasmic protein *dynamin* assemble as a ring around each bud. Dynamin recruits other proteins that draws the apposing leaflets of the plasma membrane very close together and distort the lipid bilayers leading to scission of the vesicle.

9.5.5. Transport of vesicles:

Vesicular transport or passage of vesicle to definite organelle is regulated by motor proteins which change conformation, as a result move the vesicles along a set 'track' consisting of microtubules that regulate this trafficking from ER to Golgi complex, TGN to lysosomes or plasma membrane and plasma membrane to endosomes or lysosomes. In the majority of animal cells vesicular trafficking switches between anterograde (microtubule '+' end directed) and retrograde (microtubule '-' end directed) movement. Anterograde direction is regula ted by kinesin-II while retrograde direction is regulated by dynein motors. Usually the trafficking occurs in either of these directions but in the flagella and axon, trafficking occurs in both directions. Rho GTPases activate the actin nucleators and mediate their involvement in vesicular trafficking. Intermediate filament – vimentin is also required for proper protein trafficking.

9.5.6. Uncoating:

During transportation of the vesicles shortly after scission from the donor compartment, the vesicles loses its coat due to the action of various events including inactivation of the small GTPase, phosphoinositide hydrolysis and the action of uncoating enzymes. Cytosolic coat proteins are then recycled for additional round of vesicle budding. In case of clathrin coats, a protein *auxillin* activates a chaperone protein hsp70 that acts as uncoating ATPase which peels off the coats.

9.5.7. Vesicle fusion:

The fusion of the naked transport vesicle with its target compartment involves two types of events. First, the transport vesicle must specifically recognize the correct target membrane; for example, a vesicle carrying lysosomal enzymes has to deliver its cargo only

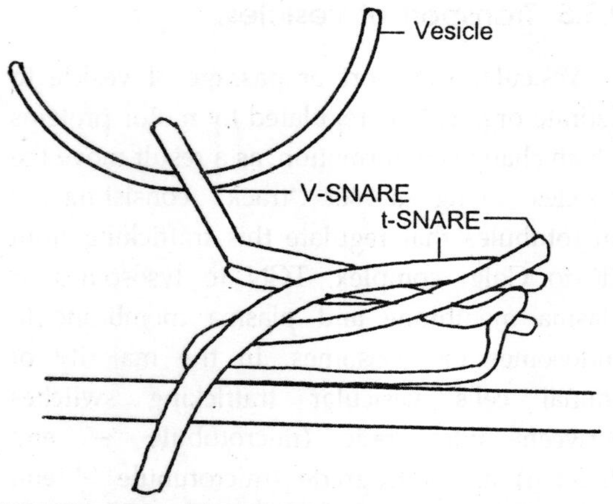

Fig. 9.18 : Formation of t-SNARE complex.

to lysosomes. Second, the vesicle and target mem branes must fuse, thereby delivering the contents of the vesicle to the target organelle. Specific recognition between a vesicle and its target is mediated by interactions between unique pairs of transmembrane proteins, followed by fusion between the phospholipid bilayers of the vesicle and target membranes. The whole process is divided into three phases – *tethering, docking and fusion.*

9.5.7.1. Tethering:

This phase involves definite recognition of vesicle to the membrane of definite compartment. It is done by the combination of a GTP bound Rab and a tethering factor. The tethering action is mediated by N-ethyl maleimide sensitive factor (NSF) that exists in the cytosol or as membrane bound form. A partner protein of NSF, α-SNAP (Soluble NSF asso ciated protein) binds NSF to the membrane and they are received by a receptor called *SNAP receptor* (SNARE). Transport of cargoes from the ER to the Golgi is regulated by specific SNARE recognition. In *S. cerevisiae*, three SNAREs Sed5p, Bet1p and Sec - 22p bind to distinct sites on Sec24p subunit of the vesicular wall and form the v-SNAREs of the transport vesicles. Rab27 also occurs on secretory granules and melanosomes in melanocytes and controls their trafficking.

Rab27-binding proteins form a bridge between Rab27 bound granules and plasma membrane promoting docking of Rab27-bound granules to the plasma membrane.

9.5.7.2. Docking:

In this step the two membranes come close enough together to facilitate protrusion of proteins from the membrane so that the lipid bilayers interact with each other. The vesicle from carries GTP bound Rab and a specific 'v-SNARE' that binds to a cognate 't-SNARE' on the target membrane forming a *t*-SNARE complex (also called 'SNAREpin') (Fig. 9.19). The t-SNARE complex is a four helix bundle (one α-helix contributed by the v-SNARE and three α-helices are contributed by t-SNARE). Assembly of this four helix complex seems to supply the free energy needed to bring apposing membranes

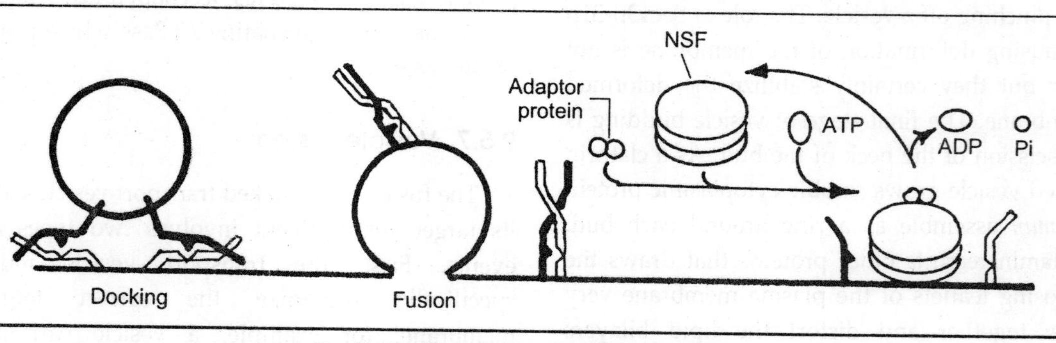

Fig. 9.19 : Dissociation of SNAREs by NSF.

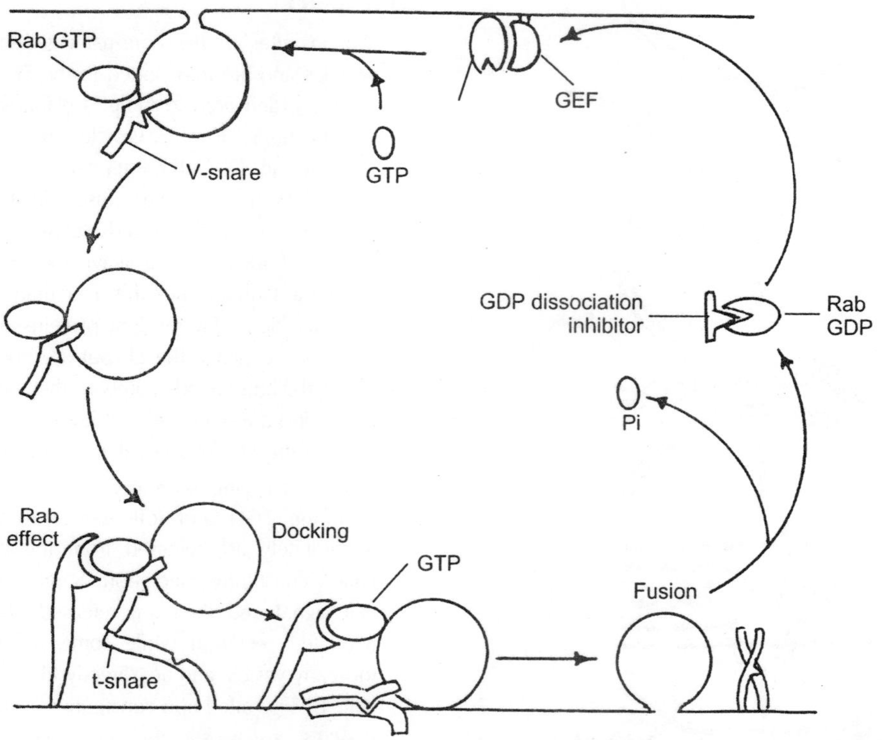

Fig. 9.20 : Role of Rab protein and SNAREs in the docking and fusion of vesicles.

close enough to fuse. The v-t-SNARE interaction is so specific that only specific SNARES bind to form t-SNARE complex.

9.5.7.3. Fusion:

The formation of four helix stable complex is the primary step for fusion. It brings both membranes so close together that the lipid bilayers fuse and lipids flow one membrane to other accompanied from by displacement of water molecules. α-SNAP binds to the four helix complex and recruits NSF which hydrolyzes ATP to dissociate the complex. NSF actually forms a hexameric ring that binds to the end of the SNARE complex and promotes fusion of both the membranes (Fig. 9.20). Binding of both membrane by the assemblage of SNAREs constitutes *stalk.* Fusion of apposing leaflets leads to *hemifusion* which follows *complete fusion* of both the membranes. This model of fusion is called stalk model of fusion (Fig. 9.21). When vesicles of unrelated compatments bind to each it is called t-SNARE complex but when vesicles generated from the same compartment bind to each other other it is called cis-SNARE complex.

At the last step, of vesicular fusion, a *fusion pore* is formed and the contents are released. Lipid rafts the microdomains of plasma membrane rich in cholesterol and glycosphingolipids remain closely associated with the protein SNAP-23 and plays a crucial role in membrane fusion. After the fusion of vesicle with the acceptor compartment has completed the Rab protein hydrolyzes its bound GTP releasing GDP into the cytosol. After the formation of SNARE complex the SNAREs must be disassembled up to the next round of fusion. NSF cycles between membranes and cytosol and

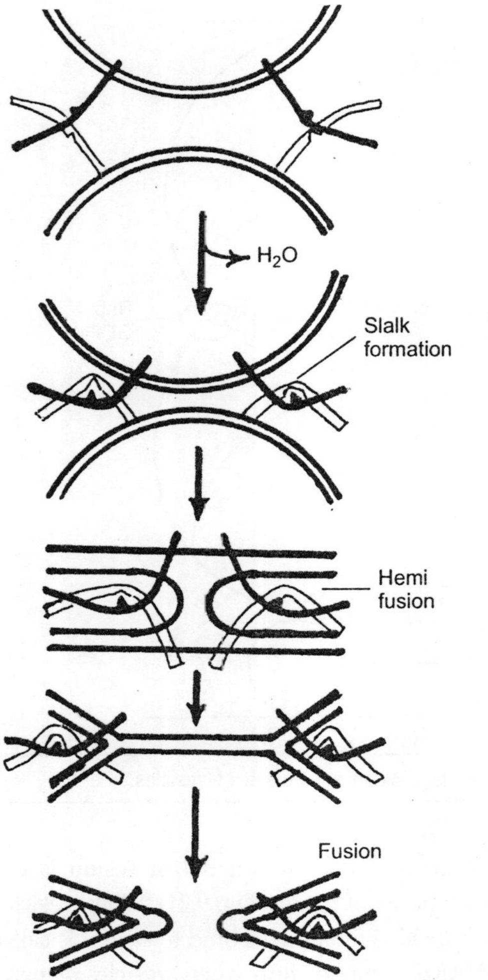

Fig. 9.21 : Different steps of Stalk model of fusion.

catalyzes the disassembly process. After the v- and t-SNAREs mediate the fusion of vesicles the NSF binds to the SNAREs complex via adaptor proteins and hydrolyzes ATP to keep the SNAREs apart. The cytosolic factors GATE-16 and LMA1 additionally bind to v- and t-SNAREs and keep them separate. In some cases, t-SNARE complex formation is arrested at an intermediate stage by an accessory Ca^{++} sensor protein synaptotagmin that promotes synaptic vesicle fusion in response to Ca^{++} influx in response to a fusion signal.

SUMMARY

1. Proteins after synthesis are packaged into transport vesicles and are transported to the ER from where these are delivered to Golgi apparatus for various modifications. However vesicles are at first transpor ted into an ER-Golgi intermediate compartment (ERGIC) from where these fuse with the cis Golgi portions and then proceeds towards TGN. Only properly folded proteins leave the ER and follow the intra Golgi route; this is called anterograde pathway. Some ER resident proteins or misfolded proteins also leave the ER but are not allowed to follow the anterograde pathway, they bear retention signal and are selectively retrieved and transported back to the ER. This is called retrograde pathway.
2. Vesicles carrying secretory proteins are escaped from the TGN and follow four routes – some immediately are released from the cell. This is called constitutive secretion. Some proteins need exit signal and then are released, this is called regulated secretion while some proteins do not form any vesicle and are directly liberated from the cell. This is called non constitutive secretion. Some vesicles carry specific enzymes needed for lysosome. They are delivered to the cytosol and are carried to the late endosome to form lysosome. Liberation of the protein containing vesicles to the outside is called exocytosis.
3. Once loaded with secretory proteins, the motor proteins propel the vesicles to the cell membrane along the microtubules. The membrane of the synaptic vesicles contains a Ca^{++} binding protein – synaptotagmin that senses the rise of Ca^{++} in the cytosol. The vesicles are linked with each other by a fibrous protein synapsin. Rab3A, located in the membrane of the vesicles is needed for targeting the vesicles with the presynaptic membrane. Ca^{++} induces conformational change in the synaptotagmin that in association with another protein complexin mediates docking of the vesicles with the plasma membrane. The fusion of vesicles involves binding of v- and t-SNAREs. SNARE complex assembly is regulated by Munc13 and Munc14 proteins
4. Another route for vesicular transport is the endocytosis of macromolecules that actually occurs in some specialized part of the plasma membrane called clathrin coated pits which occupy about 2%

of the whole cell membrane area. The endocytosed materials containing vesicles first fuse to form early endosomes whin merge to form multivesicular bodies and ultimately late endosomes. Hydrolase containing vesicles then fuse with the late endosomes to form lysosomes that is the main site of degradation.

5. The main steps of vesicular trafficking involve formation of vesicles from the 'donor compart ment' (vesicle budding) by selective incorporation of cargo into the forming vesicle but retainning resident proteins in the donor compartment (protein sorting). The vesicles are then transported to the specific acceptor compartment (vesicle targeting) into which they unload their contents (vesicle fusion).
6. The cytoplasmic surfaces of transport vesicles are coated with proteins, and it appears to be the assembly of these protein coats that drives vesicle budding by distorting membrane conformation. The budding of transport vesicles and the selective incorporation of cargo into the forming vesicles are both mediated by protein coats Three kinds of coated vesicles function in different types of vesicular transport.
7. Clathrin coated vesicles are found trafficking between the Golgi and plasma membrane, the Golgi and endosomes, and the plasma membrane and endosomes. The coats of clathrin-coated vesicles are composed of two types of protein complexes, clathrin and adaptor proteins, which assemble on the cytosolic side of the membranes. The binding of clathrin to membranes is mediated by adaptor proteins called adaptins.
8. Non clathrin-coated or COP-coated vesicles bud from the ER or Golgi complex. The coats of COPI- and COPII-coated vesicles show distinct protein complexes, which function analogously to clathrin and adaptor proteins in vesicle budding. Sar1p-Sec23p-Sec24p forms the membrane proximal layer while Sec13p-Sec31p forms the distal layer of the COPII. COPI coat proteins also show seven subunits.
9. The assembly of vesicle coats requires GTP-binding proteins, which regulates the binding of coat proteins to the membrane. The budding of both clathrin-coated and COPI-coated vesicles from the Golgi complex requires a GTP-binding protein called ARF which are responsible for both COPI and clathrin coat assembly. The first step in vesicle formation is the association of ARF bound to GDP with the Golgi membrane. Proteins in the Golgi membrane then stimulate the exchange of the GDP bound to ARF for GTP and the COPI coat proteins bind to the ARF/GTP complex. Assembly of the coat is then followed by deformation of the membrane and vesicle budding. Assembly of COPII-coated vesicles from the ER requires a distinct GTP-binding protein called Sar1. Sec16p perhaps acts in conjunction with Sec12p to recruit Sar1p to the ER membrane. Sec23p-24p, Sec13p-31p and GTP locked Sar1p are sufficient for deforming the membrane and pinching off a vesicle.
10. Majority of the transmembrane cargo proteins is concentrated in COPII coated buds prior to export from the ER and they exit the ER by directly binding COPII protein subunits. As a clathrin coated vesicle grows soluble cytoplasmic protein dynamin assemble as a ring around each budto pinch it off into the cytosol.
11. Vesicular transport or passage of vesicle to definite organelle is regulated by motor proteins, as a result they move the vesicles along a set 'track' consisting of microtubules that regulate this traffic cking from ER to Golgi complex, TGN to lysosomes or plasma membrane and plasma membrane to endosomes or lysosomes.
12. The fusion of the naked transport vesicle with its target compartment involves two types of events. First, the transport vesicle must specifically recognize the correct target membrane. A partner protein of NSF, α-SNAP (Soluble NSF associated protein) binds NSF to the membrane and they are received by a receptor called SNAP receptor (SNARE). Rab27-binding proteins form a bridge between Rab27 bound granules and plasma membrane promoting docking of Rab27-bound granules to the plasma membrane.

QUESTIONS:

Long questions:

1. Describe ER-Golgi transport of proteins.
2. Describe the process of exocytosis.
3. Describe the process of endocytosis.
4. Describe the processes of vesicle budding and fusion.

Short questions:

1. What are the main sites of vesicular transport?
2. What are the main significances of vesicular transport?
3. What do you mean by ER exit site? What is their importance?
4. What do you mean by exit signal?
5. What is ERGIC?
6. What are trans port vesicles?
7. What do you mean by vesicular tubular clusters?
8. What do you mean by retention signal?
9. What are the basic differences between KKXX and KDEL sequences?
10. What are the differences between anterograde and retrograde transport?
11. Describe how ER acts as monitoring point of protein transport?
12. Which proteins of Golgi are involved in retrograde transport?
13. What are the basic differences between regulated and constitutive secretion?
14. What is default pathway of secretion? How it differs from constitutive secretory pathway?
15. Differentiate autophagy and heterophagy.
16. What do you mean by receptor mediated endocytosis?
17. What is trancytosis? 18. How early endosomes differ from multi vesicular bodies?
19. How multivesicular bodies differ from late endosomes.
20. How exocytosis and endocytosis are inter linked?
21. What do you mean by selectivity of vesicular transport?
22. What do you mean by v- and t-SNAREs?
23. What are the main structural subunits of COPII coats?
24. What is prebudding complex in the COPII coat assembly?
25. What do you mean by sorting signals?
26. What is the role of dynamin in vesicle budding?
27. How SNAREs mediate vesicle fusion?
28. Differentiate cis and trans SNARE complex.
29. Describe the stalk model of vesicle fusion.

Fill up the blanks:

1. ER exit sites are devoid of ———.
2. ER resident proteins after cycling back to ER binds to ———.
3. Synaptic vesicles are linked up by a fibrous protein ———.
4. Docking of synaptic vesicles to the presynaptic active zone is mediated by ———.
5. The pH of late endosome is ———.
6. The unit of clathrin sub unit is called ———.
7. The GTP binding protein for the assembly of clathrin coat is ———.
8. The GTP binding protein for the assembly of COPII coat is ———.
9. Anterograde trafficking is regulated by ——— while retrograde trafficking is regulated by ——— motors.
10. Uncoating of clathrin coat is mediated by ———.

MCQs:

1. **In KDEL sequence K stands for**
   (a) lysine, (b) leucine (c) isoleucine.
2. **Proteoglycan secretion follow**
   (a) regulated;
   (b) constitutive
   (c) default pathway of secretion.
3. **Chymotrypsin secretion follows**
   (a) regulated,
   (b) constitutive
   (c) default pathway of secretion.
4. (a) Mg^{++}; (b) Ca^{++}; (c) Na^{+} **regulate fusion of exocytotic vesicles.**
5. **Synaptotagmin senses the rise of**
   (a) Ca^{++}, (b) Na^{+}, (c) K^{+} concentration in the axoplasm.
6. **COPI coated vesicles transit from**
   (a) ER to Golgi;
   (b) Golgi to ER;
   (c) TGN to lyso somes.
7. **COPII coated vesicles transit from**
   (a) ER to Golgi; (b) Golgi to ER; (c) TGN to lysosomes.
8. **Clathrin coated vesicles transit from**
   (a) ER to Golgi; (b) Golgi to ER; (c) TGN to lysosomes.

Answers to MCQs:

1. a, 2. b, 3. a; 4. b, 5. a, 6. b, 7. a, 8. c.

Chapter 10

Ageing

Ageing is generally characterized by the declining ability to respond to stress, increasing homeostatic imbalance and increased risk of disease. It is the result of the progressive and irreversible decline of the capacity of an organism to adapt to its ever-changing environment . The cellular and molecular mechanisms of ageing is mainly concerned with age related damages in molecules (DNA, proteins, lipids) leading to pathologies like osteoarthritis, cardiovascular diseases, emphysema, malignant tumors, dementias, osteoporosis, heart disease, cancer, Alzheimer's Disease, etc. The mechanisms recognized in cell and tissue ageing mainly involve post-synthetic modifications of the macromolecules of intercellular communication as a result of receptor decay. Some researchers consider the role of genes on ageing. Diet has also been considered as a prime factor for ageing specifically, caloric restriction (i.e taking 30-50% less calory than a man normally consumes), has been shown to increase lifespan.

10.1. SYMPTOMS OF AGEING

These include:

1. Loss of hearing ability.
2. There is a decline in the ability to taste salt and bitter.
3. There is a reduction of the thymus gland to 5"10% of its original mass by age 50.
4. Levels of antibodies increase with ageing.
5. One third of men and half of women over 65 report some form of arthritis.
6. About half of the aged persons above 65 have lost all teeth.
7. The elderly require twice as much insulin to achieve the glucose uptake of the young.
8. There is reduced sensitivity to growth factors and hormones due to fewer receptors and dysfunctional post-receptor pathways.
9. The temperature needed to separate DNA strands increases with age.
10. Weight declines after age 55 due to loss of lean tissue, water and bone.
11. Body fat increases to age 60.
12. Muscle strength for men declines 30-40% from age 30 to age 80.
13. Elderly people tend to sleep more lightly, more frequently and for shorter periods with a reduction in rapid eye-movement (REM) sleep.

14. Degree of saturation of fats drops by 26% in the brains of old animals.
15. Presbyopia (reduced ability to focus on close-up objects) occurs in 42% of people aged 52-64, 73% of those 65-74 and 92% of those over 75. Most people over age of 75 have cataracts.
16. About half of those over 85 are disabled.
17. Over 75% of people over 85 have 3-9 pathological conditions those may lead to death.

Usually ageing eventually progress towards death but this does not mean any type of ageing symptom may lead towards death. For example, graying of hair is a symptom of ageing, but graying does not increase likelihood of mortality. Ageing changes which are not associated with a specific disease, but are associated with a generalized increase in mortality would qualify as biomarkers of ageing and would distinguish biological age from chronological age. Biomarkers would be better predictors of the increased likelihood of mortality than the chronological age. No biomarkers of ageing have been universally accepted.

10.2. THEORIES OF AGEING

Irrespective of the causes, most workers are at one that ageing is nothing but dysfunction at the cellular level. Differences in life span between species raise critical questions? Why is a rodent old at 3 years, a horse old at 35 years and a human old at 80 years? Don't they have cells of similar biochemical and molecular feature? Why at age 3 about 30% of rodents show cancer, whereas at age 85, about 30% of humans have had cancer? Why cancer cells, stem cells and human germ cells seem "immortal" when compared to other cells.

When discussing ageing it is important to distinguish two points on survival curves - ***mean life span*** (average life span) and ***maximum life span*** corresponds to the age at which the oldest known member of the species has died. Today, people in the most developed countries have an average lifespan of about 80. Maximum lifespan for humans, however, has remained about 115"120 all through known history.

Considering various aspects of life expectancy different theories have been put forward which are mainly categorized into two major classes - programmed ageing and wear and tear ageing.

Programmed ageing concerns ageing due to something inside an organism's control mechanisms that forces elderliness and deterioration. It seems to be controlled by genes.

Wear and tear ageing is mainly concerned with the ageing due to radiation, chemical toxins, metal ions, free radicals, glycation, disulfide-bond cross-linking, etc. Such damage can affect genes, proteins, cell membranes, enzyme function, blood vessels, etc.

As per ***Glycation theory of ageing*** proteins can be damaged both by free radicals and by glycation (also called the *Maillard reaction*, or *non-enzymatic glycosylation*) that is a reaction by which reducing sugars become attached to proteins without the assistance of an enzyme. This attachment occurs at the free amine group of lysine or arginine, which is not involved in the peptide bond.

The ***"error catastrophe" theory of ageing*** proposed that accumulation of damaged proteins damage to the machinery of protein synthesis itself, leading to an escalating viscious circle of malfunctioning cellular components. Rate of both protein synthesis and protein degradation decline with age, and the inability to eliminate damaged macromolecules may be more catastropic than the synthesis of new defective ones.

According to ***"viscious cycle" theory*** the free radical damage to mitochondrial DNA leads to mitochondria that produce more superoxide.

According to ***"mitochondrial theory of ageing"*** damage to mtDNA and organelles by free radicals leads to loss of mitochondrial function and loss of cellular energy.

10.3. PROTEIN QUALITY DEGRADATION AND AGEING

There is a close relationship between oxidative damage to protein and ageing. Cysteine and methionine are by far more vulnerable to oxidation by reactive oxygen species than any other amino acids because of their sulfhydryl groups. Oxidation of cysteine sulfhydryls can result in disulfides that cause protein aggregation and lipofuscin. Proteins are continually being degraded (hydrolyzed) within cells by proteolytic enzymes, both for regulation of cellular processes and for "quality control" of proteins (eliminating malformed or malfunctional ones). Normally damaged or misfolded proteins are eliminated by proteasomes, found both in the nucleus and in the cytoplasm. Proteasome activity declines with age.

Degradation of cellular organelles, proteins and other materials are mainly carried out by lysosomes. Proteins are brought into the lysosome for degradation by *chaperone-mediated autophagy* in which members of the hsp70 heat shock protein family attach to a target protein and then bind to a lysosome receptor protein mediated by the signal - KFERQ. (K=Lysine, F=Phenylalanine, E=Glutamate, R=Arginine, D=Glutamine. Proteins with short half-lives tend to be broken-down by the proteasome, whereas proteins with half-lives in excess of ten hours tend to be degraded by autophagy. With ageing, lysosomes become increasingly bloated with aggregates of oxidized, glycated, cross-linked proteins that are resistant to enzymatic degradation forming *lipofuscin.*

Blockage of proteasomes by protein aggregates would result in cell dysfunction due to the inability to degrade short-lived regulatory proteins. Impaired degradation of p53 protein can result in excessive cell senescence or apoptosis. Impaired proteasome degradation of immune system regulators can result in immune deficiency. Protein aggregation can also impair chaperone-mediated autophagy.

10.4. SEX AND AGEING

A US National Institute of Ageing study showed that 44% of men over age 80 are "robust and independent" compared with only 28% of women. And the percentage of surviving males increases from 15% at age 100 to 40% at age 105 in the United States.

The reproductive organs of the human female exhibits a rate of ageing that is among the most rapid of body systems. The complete shut down of female fertility at menopause may be of value in preventing the birth of deformed children or death of newborn babies.

The most serious complications of menopause are osteoporosis and a decline in cardiovascular health. There is no sudden "andropause" event in males that is comparable to the menopause event of females. Instead, testosterone levels tend to decline gradually with age.

10.5. CYTOLOGICAL CHANGES AND AGEING

Ageing is no more than an accumulation of damage in biochemical systems. After a limited number of cell divisions, somatic cells lose the capacity for proliferation, called *cellular replicative senescence* or *Hayflick limit*. Senescence, which is triggered by the loss of DNA sequences at the telomeres, is often seen as an example of a regular "biological clock."

Telomerase, the enzyme capable of resynthesising the telomeres on each end of chromosomes is repressed in most somatic cells but reexpressed in the large majority of transformed, malignant cells. The loss of telomeres was shown to be more rapid in lymphocytes isolated from the blood of trisomic patients in accord with their more rapid ageing. Telomere structure and length is maintained by *Telomeric Repeat-binding Factors (TRFs)* - TRF1 and TRF2. TRF2 stimulates the helicase activity that may play a role in telomere maintenance (Opresko, 2005). Loss of TRF2 due to gradual cell divisions directly signals apoptosis. Delaying mitochondrial dysfunction led to slowing of telomere shortening.

10.6. AGEING OF THE EXTRACELLULAR MATRIX

The two other major families of matrix macromolecules, proteoglycans and structural glycoproteins exhibit age-dependent variations, both at the synthetic and postsynthetic levels. The rate of biosynthesis of matrix macromolecules changes with age. Some decline, others increase. The biosynthesis of hyaluronan, a glycosaminoglycan, was shown to decrease with ageing of human skin fibroblasts. Post synthetic modifications of proteoglycans and glycosaminoglycans consist of both enzymatic and non-enzymatic degradation. This process is directly involved in the development of osteoarthritis. Ageing cells increasingly synthesize Fibronectin, the most abundant structural glycoprotein. As fibronectin is very sensitive to proteolytic degradation, its fragmentation is also increasing with age as well as in some pathology where its contact with proteases and especially elastases is increased as in cystic fibrosis of the pancreas.

10.7. ROLE OF MITOCHONDRIA IN AGEING

Among the cellular organelles mitochondria play the most crucial role in ageing. However its following features are affected by ageing :

Oxidative damage: An estimated 1"2% of oxygen used by mitochondria will normally "leak" from the respiratory chain to form superoxide. Ageing is associated with decreased oxidative phosphorylation coupling efficiency and increased superoxide production. Free radicals can damage the mitochondrial inner membrane, creating a positive feedback-loop for increased free-radical creation. Among the byproducts of mitochondrial energy production are "reactive oxygen species" (ROS) that include hydrogen peroxide. Many of these ROS are free radicals. Unless these are quickly neutralized by antioxidants, these can cause considerable damage to the membranes of mitochondria and to mitochondrial DNA. The affected mitochondria become so inefficient that they are unable to generate sufficient energy to meet cellular demands. However certain signaling molecules like nitric oxide (NO) can slow the age-related decline of mitochondrial function and boost up mitochondrial function.

Mitochondrial DNA - an easy target: Mitochondrial DNA is not as well protected as nuclear DNA, which is coated by proteins. The "naked" mitochondrial DNA becomes an easy target for reactive oxygen species. When cultured animal cells were exposed to various types of ROS, their mitochondrial DNA was more severely damaged than nuclear DNA. Moreover mitochondrial DNA damage was more extensive and persisted longer than nuclear DNA damage in human cells following oxidative stress. In general as cells age, the number of gaps and errors in their mitochondrial DNA tends to increase, and oxidant exposure is the likely cause. Recently several "hot-spots" in mitochondrial DNA-regions have been located where defects and mutations tend to cluster. Mutations are not widely distributed in mtDNA, but appeared to be clustered in so-called these "control" regions of the DNA that regulate its replication. [*Homoplasmy* describes the original condition of all of a person's mtDNA being the same, but as mtDNA mutations occur and the mutated mtDNA replicates, cells, tissues and even mitochondria can have a mixture of mtDNA types, a condition known as *heteroplasmy*.] Most strikingly, a DNA sequence rearrangement was found in high proportion of mitochondrial DNA molecules in individuals above 65 years of age. A significant portion of *"photoageing"* of the skin may be due to mtDNA deletions induced by ultraviolet light.

Role of Cardiolipin, Carnitine and CoQ: As the body ages, we absorb nutrients less efficiently and this affects the efficiency of mitochondrial function.

Cardiolipin, a component of the energy-producing process is found exclusively in mitochondria. Its levels decline with age. Lipid peroxidation, a type of oxidant damage more common in older cells, leads to a decrease in cardiolipin. Progressive accumulation of crippled cardiolipin molecules is another way in which oxidant damage can jeopardize the efficiency of energy production.

Carnitine, an amino acid, is also important to mitochondrial metabolism because it helps fatty acids metabolism in mitochondria. Carnitine deficiency in aged persons leads to an inability to harvest the energy stored in fatty acids and to a build-up of fatty intermediates that can prove toxic to cells.

Coenzyme Q10, also known as CoQ10 or ubiquinone, is another factor necessary for energy production. It is available in the diet and it can be manufactured from simpler precursors. CoQ10 deficiency can affect muscle, brain and nerve functions.

Thus molecular damage to mitochondria and mitochondrial DNA (mtDNA) leads to ageing of humans. Age-related impairment in the respiratory enzymes not only decreases ATP synthesis but also enhances production of ROS through increased electron leakage in the respiratory chain. Human mtDNA, which is not protected by histones and yet is exposed to high levels of ROS and free radicals in the matrix of mitochondria, is susceptible to oxidative damage. Moreover, the activities of Cu, Zn-SOD, catalase and glutathione peroxidase decrease with age and intracellular levels of H_2O_2 and oxidative damage to DNA and lipids are significantly increased with age. Thus mtDNA mutations, oxidative damage altered expression of a few clusters of genes, and apoptosis are increased in ageing human tissues suggest that mitochondrial theory of ageing is mature (Lee and Wei, 2007).

10.8. ROLE OF LYSOSOME IN AGEING

The most damaged mitochondria are consumed by lysosomes whereas the more defective mitochondria (which produce less ATP as well as less superoxide) remain to reproduce themselves. But the efficiency of lysosomes to consume malfunctioning mitochondria declines with age, resulting in more mitochondria producing higher levels of superoxide. Mitochondria of older organisms are fewer in number, larger in size and less efficient (produce less energy and more superoxide). Ageing mitochondria become enlarged and, if they can be engulfed by lysosomes, are resistant to degradation and contribute to lipofuscin formation. Due to insufficient digestion of oxidatively damaged macromolecules and organelles by autophagy and other degradative systems, long-lived post mitotic cells, such as cardiac myocytes, neurons and retinal pigment epithelial cells, progressively accumulate biological 'garbage' which include lipofuscin, defective mitochondria, other organelles and aberrant proteins, often forming aggresomes. An interaction between lipofuscin-loaded lysosomes and mitochondria seems to play a pivotal role in the progress of cellular ageing. Lipofuscin deposition hampers autophagic mitochondrial turnover, promoting the accumulation of senescent mitochondria, which are deficient in ATP production but produce-increased amounts of reactive oxygen species. Increased oxidative stress, in turn enhances damage to mitochondria and lysosomes, triggering mitochondrial and lysosomal pro-apoptotic pathways culminating in cell death (Terman *et al*, 2007).

10.9. EFFECT OF AGEING ON DIFFERENT ORGANS

1. Ageing in the female reproductive system provides the best example of programmed ageing in mammals. Ageing also has profound effects on heart, brain, lung and kidney. Most persons of 65 years of age have at least some symptoms of sub clinical diabetes and most of the symptoms of ageing are accelerated in diabetes.

2. Skin, lungs, muscles, blood vessels and organ-function in general is adversely affected by protein cross-linking.

3. Generalized reduction in blood flow due to atherosclerosis also has an adverse effect on most organ systems.

4. The kidney provides perhaps the most striking example of individual variation in the effects of ageing. On average, kidney weight declines about 15% between ages 40 and 80. Its filtering capacity declines drastically.

5. Cardiovascular disease is the most frequent cause of death among those over age 85. The left ventricle of the heart increases in size with age (hypertropy) due to an increase in size of the heart muscle cells that work harder to pump blood through a circulatory system that has narrower channels and reduced elasticity. Lipofuscin content of aged heart muscle cells increases. Arteries thicken with age so that about three-quarters of elderly people have increased blood pressure (both systolic and diastolic).

6. 2% of neurons are lost, on average, between ages 20 and 90 (up to 40% of this loss in the frontal cortex). Those over age 86 show an average 10% decline in brain weight from age 20. Dramatic reduction in cerebral blood flow and in brain oxygen and glucose utilization is frequently seen after the 8th decade of life.

7. Ageing results in greater loss of fast-twitch than slow-twitch muscle. Fast-twitch fibers atrophy due to loss of the nerves that innervate them. Muscle fibers are replaced by fat and connective tissue.

8. Muscles in the iris of the eye, atrophy and pupil size reduces, with age increasing the need for illumination. The lens thickens and becomes yellowish, reducing green-blue-violet discrimination.

9. Collagen and elastin in tendons and ligaments become less resilient and more fragmented as a person grows older, particularly due to glycation (cross-linking of proteins by sugar). Articular cartilage becomes frayed and the synovial fluid between joints becomes "thinner". Decline in circulatory function contributes to this process.

10. Ageing of skin is commonly divided into "chronological ageing" and "photoageing", with up to 80% of skin ageing attributed to photoageing in non-smokers. Photoageing is due to ultraviolet (UV) light, which activates inflammatory cytokines and metalloprotein collagenases as well as inducing free radicals. UV radiation generates singlet oxygen that both activates metalloproteinases and causes large-scale deletions of mitochondrial DNA in the skin cells.

11. Collagen and elastin also cross-link in skin, resulting in a loss of elasticity. The epidermis thins with age, leading to wrinkles. Decreased secretion by sweat glands increases vulnerability to heat stroke. When the melanocytes associated with hair follicles cease functioning, hair turns white. Partial reduction of melanocyte function results in hair that appears "gray".

12. Loss of flexibility of the collagen and elastin in the lung results in loss of elastic recoil. It becomes too difficult to fully exhale, which reduces air exchange, reducing the capacity to do work. Oxygen-to-tissue transfer rate is often halved by age 70.

13. Bone is typically 25% water, 30% soft tissue and 45% mineral deposits (mostly calcium). Most of the white ash remaining after cremation is calcium, lead, zinc and potassium from bone. Both men and women lose bone mass between the ages of 39 and 70 (osteoporosis), but

post-meno pausal women lose bone mass at twice the rate as men. Decreased growth hormone causes bone loss in both sexes. The physical inactivity and malnutrition also worsens bone loss.

14. Joints in the bones of the inner ear calcify, contributing to a loss in the ability to hear higher tones. Loss of sweat glands in the ear causes earwax to become drier and crustier. Wax obstruction reduces the ability to hear low frequencies.

15. Ageing reduces salivary secretion resulting in a drier mouth and decreased protection from bacterial infection of the mouth. Gastric juice volume is reduced 25% by age 60 and there is a 60% decline in pepsin activity. But this does not noticeably affect digestion except in the case of heavy meats. Absorption of Vitamin D, Vitamin B_{12} and folic acid decline with age.

16.10. MEMORY DECLINE IN AGEING

One of the key concerns of older adults is the memory loss, especially as it is one of the hallmark symptoms of Alzheimer's disease. However, memory loss is qualitatively different in normal ageing from the kind of memory loss associated with Alzheimer's disease. Studies comparing the effects of ageing on episodic memory, semantic memory, short-term memory and priming find that episodic memory is especially impaired in normal ageing. These deficits may be related to impairments seen in the ability to refresh recently processed information. In addition, even when equated in memory for a particular item or fact, older adults tend to be worse at remembering the source of their information, a deficit that may be related to decline in the ability to bind information together in memory.

10.11. BRAIN AGEING AND NEURODEGENERATIVE DISORDERS

Multiple molecular, cellular, structural, and functional changes occur in the brain during ageing. Neural cells may respond to these changes adaptively, or they may succumb to neurodegenerative cascades that result in disorders such as Alzheimer's and Parkinson's diseases. Multiple mechanisms are employed to maintain the integrity of nerve cell circuits and to facilitate responses to environmental demands and promote recovery of function after injury. The mechanisms include production of neurotrophic factors and cytokines, expression of various cell survival-promoting proteins (e.g., protein chaperones, antioxidant enzymes, Bcl-2 and inhibitor of apoptosis proteins), preservation of genomic integrity by telomerase and DNA repair proteins, and mobilization of neural stem cells to replace damaged neurons and glia. The ageing process challenges such neuroprotective and neurorestorative mechanisms. At the cellular and molecular levels, neurons increase production of neurotrophic factors and stress proteins can facilitate successful brain ageing though brain cells have remarkable capacity for adaptation to ageing and resistance to disease.

The major classes of signaling molecules important in brain ageing include neurotrophic factors, [neurotrophins, nerve growth factor (NGF), neurotensin (NT)-3, and NT-4], fibroblast growth factors, glial-derived neurotrophic factor (GDNF), and ciliary neurotrophic factor (CNTF)] neurotransmitters, cytokines and steroids. Neurotrophic factors have been shown to promote the survival of specific populations of brain neurons. In addition to their roles as mediators of synaptic transmission, neurotransmitters such as glutamate, acetylcholine, and dopamine also play important roles in regulating the formation of neuronal circuits during development and in influencing the neurodegenerative process in brain disorders of ageing. Sex steroids and stress steroids (glucocorticoids) have been shown to have quite striking effects on brain function and structure, and alterations in regulation of their production and signaling mechanisms have been reported to occur during ageing. The status of such neurotransmitter, neurotrophic factor, cytokine, and steroid hormone signaling systems have a major influence on the outcome of brain ageing.

As in other organ systems, cells in the brain encounter a cumulative burden of oxidative and metabolic stress that may be a universal feature of the ageing process. Each of the major classes of cellular molecules, including proteins, nucleic acids and lipids, is oxidatively modified during brain ageing. These modifications of proteins, nucleic acids, and lipids are greatly exacerbated in neurodegenerative disorders such as Alzheimer's disease (AD) and Parkinson's disease (PD) consistent with a major role for oxidative stress of ageing in the pathogenesis of those disorders. ***Dietary factors in Brain ageing and neurodegenerative disorders:*** There are numerous dietary factors that have been reported to affect brain physiology that could modify brain ageing and the pathogenesis of neurodegenerative disorders. These range from amino acids such as tryptophan to caffeine and related stimulants to omega-3 fatty acids.

Many such dietary factors include caloric intake folic acid, and antioxidants (Fig. 10.1).

Folic Acid: Humans cannot synthesize folic acid and therefore must obtain it in the diet; major sources of folate are green vegetables, citrus fruits, liver, and whole grains. Neural cells may be particularly sensitive to folic acid. Folic acid deficiency causes abnormalities in cell proliferation, differentiation, and survival. The important role for folic acid in development of the nervous system is further demonstrated by genetic alterations in methyl tetrahydrofolate reductase (MTHFR), which decreases enzyme activity, resulting in decreased folic acid levels. Folic acid deficiency results in depletion of S-adenosylmethionine (SAM), which is the major methyl donor in most biochemical reactions. A reduction in the methylation of cytosine in DNA may enhance gene transcription and DNA strand breakage that can trigger malignant transformation.

10.12. DIETARY RESTRICTION (DR) REGULATING THE MOLECULAR MECHANISM OF AGEING

Low-calorie diet decreases the risk of the most prominent age-related diseases in humans including cardio vascular disease, diabetes, and cancers. It can also delay age-related functional deficits in the brain and may reduce the risk of major neurodegenerative disorders. DR counter acts age-related changes in the expression of genes that encode proteins involved in innate immune responses, oxidative stress, and energy

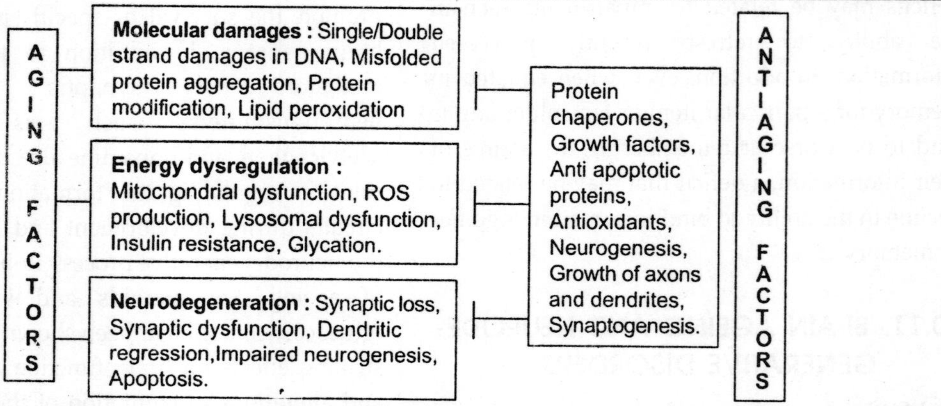

Fig. 10.1 : Balance of aging and antiaging factors.

metabolism. This retardation of brain ageing at the molecular level may underlie the preservation of brain function during ageing in animals maintained on DR. DR can protect against neurodegenerative disorders (Fig. 10.2). DR modifies shared bio chemical cascades that lead to cell dysfunction and disease. In the case of neurodegenerative disorders, it is clear that while different genetic and environmental factors may initiate neurodegenerative process in different disorders. Increased oxidative stress, perturbed cellular calcium homeo stasis, and impaired energy metabolism occur in every neurodegenerative disorder studied to date. These alterations render neurons vulnerable to apoptosis involving the proteins - Par-4, Bcl-2 family members, and caspases. DR can stabilize mitochondrial function and reduce oxidative stress in brain cells of rodents and this may increase the resistance of neurons to many types of genetic and environmental factors. DR can induce the expression of genes that encode proteins that promote neuronal survival and plasticity. Levels of heat shock protein-70 (HSP-70) and glucose- regulated protein-78 (GRP-78) are increased increased in different parts of the brain of rats and mice maintained for several months on a restricted diet. HSP-70 and GRP-78 can protect neurons against excitotoxic and oxidative injury suggesting that their increased levels contribute to the neuroprotective effect of dietary restriction. DR can induce the expression of several different neurotrophic factors in brain cells. Levels of brain-derived neurotrophic factor - BDNF are increased in neurons maintained on dietary restriction. BDNF can protect neurons and against excitotoxic, metabolic and apoptotic processes. Neurotrophic factors may protect neurons by stimulating the production of proteins that suppress oxidative stress and stabilize cellular calcium homeostasis. Protein chaperones and neurotrophic factors are known to be induced by cellular stress, and it is therefore very likely that DR elicits a cellular stress response; this might result from decreased glucose availability to the cells. The adult brain contains populations of cells that are capable of dividing and then differentiating into neurons (neurogenesis) or glial cells (gliogenesis). Caloric restriction can increase neurogenesis in the brains. As the insulin-signaling pathway, are believed to play major roles in the ageing process, they are likely to have an important influence on brain ageing.

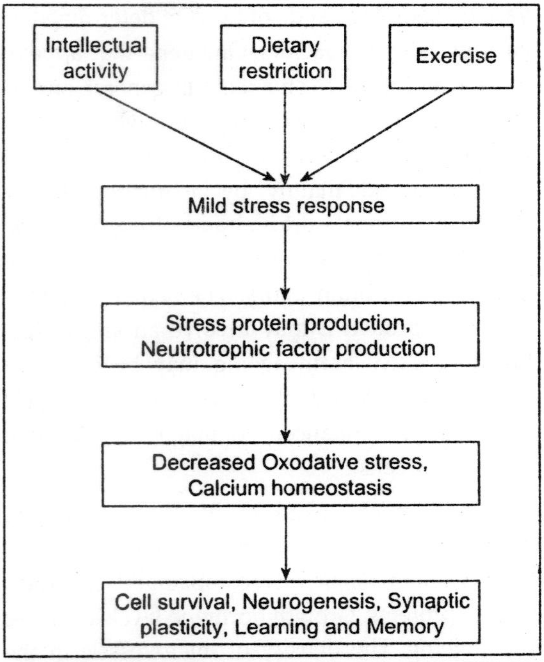

Fig. 10.2 : Model of the mechanisms whereby dietary restriction, intellectual activity and exercise promote neuronal survival and plasticity.

10.13. GENES THAT INFLUENCE LIFE SPAN AND BRAIN HEALTH SPAN

The probability of living a long life with preservation of a high level of brain function is strongly influenced by the genes one inherits. Accordingly, genetic factors also play important roles in determining one's risk of age-related neurodegenerative disorders. There is ample evidence that life span, intelligence, and risk of various neurological disorders are determined, in part, by heritable factors. One gene that appears to have an influence on ageing in general, and on risk of age-related neurodegenerative disorders, is apolipoprotein E. In the brain, apolipoprotein E is produced mainly by astrocytes. It can promote neuronal survival and out growth and may play important roles in adaptive responses to ageing and brain injury. The beneficial effects of apolipoprotein E may involve an antioxidant function. Early brain ageing may be due to a diminished antioxidant activity of this protein resulting in enhanced damage to vascular endothelial cells.

Other genes linked to life span that may influence brain ageing are those encoding growth hormone and major histocompatability complex (MHC) proteins. Growth hormone levels decrease with ageing. The density of microvessels in the brain decreases during ageing and treating the animals with growth hormone can reverse this.

Inherited variability of mitochondrial genes encoding proteins involved in oxidative phosphorylation and other aspects of mitochondrial function may also contribute to ageing and the pathogenesis of neurodegenerative disorders. Mitochondrial DNA damage has been shown to increase in brain cells during ageing, and it has been proposed that accumulation of mitochondrial DNA mutations is a major factor in the ageing process itself.

10.14. CHANGES IN GENE EXPRESSION DURING AGEING

Ageing is associated with characteristic changes in gene expression. During ageing, the expression of certain stress response genes is increased in age-specific and tissue-specific patterns. At least part of this increase appears to be a response to oxidative stress. Normal ageing results in changes in the brain that contribute to the decline of various functions, including learning and memory. Mechanisms causing this decline have not been clearly established. Increases in chemokine gene expression that occur in conjunction with increased microglial activation suggest that chemokines may contribute to the decreased brain function that occurs during normal ageing. Increase in RPS25 (ribosomal protein S25) expression is further evidence that the composition of ribosomes may alter with age and as a result could have significant effects on protein synthesis.

Ageing and oxidative stress responses shared the up-regulation of purine biosynthesis, heat shock protein, antioxidant, and innate immune response genes. As we age, genes are expressed differently throughout our body. The most obvious examples are the hormonal changes seen in adolescence and in the menopause. In many models of how genes are expressed during older age, one of the prevalent models - the *programmed ageing* model - claims that ageing is caused by genetically programmed cell death. In other words, ageing is programmed into our cells. Opposed to this model we have the normally less favoured *stochastic model*, which claims that random biological events play a significant role in ageing.

10.15. HORMONES AND AGEING

Hormones alter the gene expression of DNA throughout the body. The hormones DHEA, melatonin, thyroid, and somatotropin (Growth Hormone, GH) decline with age. According to the glucocorticoid cascade hypothesis, glucocorticoid steroid hormones show rising blood levels with age, which increasingly damages feed back inhibition resulting greater increases of blood glucocorticoid and a destructive feed back loop. Although the glucocorticoid cascade hypothesis is not a theory of ageing in the sense of reducing lifespan, but it is related with brain ageing. About half of Alzheimer's disease patients show significantly elevated glucocorticoid level. Growth Hormone

(GH) also declines with age (about 14% per decade after age 25), which is blamed for increased fat deposition, loss of muscle mass and bone demineralization.

10.16. IMMUNE SYSTEM AND AGEING

According to the "immune system theory of ageing", many ageing effects are due to the declining ability of the immune system to differentiate "foreign" from "self" proteins. Not only does the immune system become less capable of resisting infection and cancer, but declining cell function could be due to attacks by the immune system against native tissues. Arthritis, psoriasis and other autoimmune diseases increase with age. The thymus shrinks gradually after 50 years age and thereafter-lymphoid tissues are replaced by fat.

10.17. COMMON DISEASES ASSOCIATED WITH AGEING

No disease condition displays all symptoms of accelerated ageing. Diseases that resemble certain aspects of accelerated ageing are known as ***segmental progerias.*** If ageing is due to a variety of cellular and molecular damages, segmental progerias may represent subsets of those damages. These are primarily diseases of defective DNA-repair.

Werner's syndrome is associated with early onset of many age-related diseases. Werner's Syndrome is due to a defect or deletion of a single gene (*WRN*), resulting in defects of both telomeres and DNA repair. The WRN gene is a member of the helicase family that causes DNA to unwind, which is a requirement for most forms of DNA repair. Defective WRN protein results in a reduction of p53-mediated apoptosis.

In ***Hutchinson-Gilford Progeria Syndrome*** (also called "*childhood progeria*") a child is born with abnormally short telomeres. The disease is caused by a point mutation in the gene for lamin A, (converting it to prolamin A) a filament protein in the nuclear matrix and nuclear lamina that is required for DNA replication and nuclear organization. At age 5 the telomeres of a Hutchinson-Gilford syndrome child are as short as those of a very elderly person including down-regulation of some nuclear proteins, suggesting that lamin A defects contribute to normal ageing.

Bloom's syndrome is due to a defective helicase-type protein BLM leading to chromosome aberrations. Bloom's victims usually die of solid tumors at an average age of 35. BLM protein deficiency renders cells highly vulnerable to apoptosis.

SUMMARY

1. Ageing is generally characterized by the declining ability to respond to stress, increasing homeostatic imbalance and increased risk of disease. The cellular and molecular mechanisms of ageing is mainly concerned with age related damages in DNA, proteins, lipids etc. leading to diseases like osteoarthritis, cardiovascular diseases, emphysema, malignant tumors, osteoporosis, heart disease, cancer, Alzheimer's Disease, etc. Some researchers consider the role of genes on ageing.
2. Symptoms of ageing include increased levels of antibodies, reduction of weight, muscle strength and thymus gland, loss of teeth, hearing, increase of body fat, arthritis, diabetes, decline of, increase production of ROS, presbyopia etc. Usually ageing eventually progress towards death. However no biomarkers of ageing have been universally accepted.
3. Programmed ageing concerns with the defects of organism's internal control mechanisms. According to "mitochondrial theory of ageing" damage to mtDNA and organelles by free radicals leads to loss of mitochondrial function and loss of cellular energy. Wear and tear ageing is mainly concerned with environmental hazards accompanied with damaged genes, proteins, cell membranes, enzyme function etc. Glycation theory of ageing concerns with reducing sugars attached to proteins without the assistance of an enzyme. Error catastrophe theory of ageing concerns with accumulation of damaged proteins. According to "mitochondrial

theory of ageing" damage to mtDNA and organelles by free radicals leads to loss of mitochondrial function and loss of cellular energy.

4. Rate of both protein synthesis and protein degradation decline with age, and the inability to eliminate damaged proteins may be more catastrophic than the synthesis of new defective ones. There is a close relationship between oxidative damage to protein and ageing. Proteasome activity declines with age. Impaired degradation of p53 protein can result in excessive cell senescence or apoptosis.
5. Ageing is no more than an accumulation of damage in biochemical systems. The rate of biosynthesis of matrix macromolecules changes with age. Ageing cells increasingly synthesize fibronectin, the most abundant structural glycoprotein.
6. Among the cellular organelles, mitochondria play the most crucial role in ageing. However certain signaling molecules like nitric oxide (NO) can slow the age-related decline of mitochondrial function. Moreover mtDNA damage was more extensive and persisted longer than nuclear DNA damage in human cells following oxidative stress. Moreover, the activities of Cu, Zn-SOD, catalase and glutathione peroxidase decrease with age and intracellular levels of H_2O_2 and oxidative damage to DNA and lipids are significantly increased with age. Thus mtDNA mutations, oxidative damage altered expression of a few clusters of genes, and apoptosis are increased in ageing human tissues suggest that mitochondrial theory of ageing is mature.
7. Ageing in the female reproductive system provides the best example of programmed ageing in mammals. Ageing also has profound effects on heart, brain, lung and kidney. On average, kidney weight declines about 15% between ages 40 and 80. Lipofuscin content of aged heart muscle cells increases. Ageing results in greater loss of fast-twitch than slow-twitch muscle. Ageing of skin is commonly divided into "chronological ageing" and "photoageing", with up to 80% of skin ageing attributed to photoageing in non-smokers. The epidermis thins with age, leading to wrinkles. Oxygen-to-tissue transfer rate is often halved by age 70. Absorption of Vitamin D, Vitamin B_{12} and folic acid decline with age.
8. Amyloidosis is a heterogeneous group of extra cellular protein folding and deposition.
9. Multiple molecular, cellular, structural, and functional changes occur in the brain during ageing. Each of the major classes of cellular molecules, including proteins, nucleic acids, and lipids, is oxidatively modified during brain ageing. The mechanisms include production of neurotrophic factors and cytokines, expression of various cell survival-promoting proteins (e.g., protein chaperones, antioxidant enzymes, Bcl-2 and inhibitor of apoptosis proteins), preservation of genomic integrity by telomerase and DNA repair proteins, and mobilization of neural stem cells to replace damaged neurons and glia. The ageing process challenges such neuroprotective and neurorestorative mechanisms. At the cellular and molecular levels, neurons increase production of neurotrophic factors and stress proteins can facilitate successful brain ageing though brain cells have remarkable capacity for adaptation to ageing and resistance to disease. There are numerous dietary factors that have been reported to affect brain physiology that could modify brain ageing and the pathogenesis of neurodegenerative disorders.
10. Low-calorie diet decreases the risk of the most prominent age-related diseases in humans including cardiovascular disease, diabetes, and cancers. Restricted diet counteracts age-related changes in the expression of genes that encode proteins involved in innate immune responses, oxidative stress, and energy metabolism. This retardation of brain ageing at the molecular level may underlie the preservation of brain function during ageing in animals maintained on such diet.

    Levels of brain-derived neurotrophic factor - BDNF are increased in neurons maintained on dietary restriction. As the insulin-signaling pathway, are believed to play major roles in the ageing process, they are likely to have an important influence on brain ageing.
11. One gene that appears to have an influence on ageing in general, and on risk of age-related neurodegenerative disorders, is apolipoprotein E. Early brain ageing may be due to a diminished antioxidant activity of this protein resulting in enhanced damage to vascular endothelial cells.

Other genes that may influence brain ageing are those encoding growth hormone and major histocompatability complex (MHC) proteins. Growth hormone levels decrease with ageing. Inherited variability of mitochondrial genes encoding proteins involved in oxidative phosphorylation and other aspects of mitochondrial function may also contribute to ageing and the pathogenesis of neurodegenerative disorders. Mitochondrial DNA damage has been shown to increase in brain cells during ageing, and it has been proposed that accumulation of mitochondrial DNA mutations is a major factor in the ageing process itself. Creatine increases phosphocreatine levels in muscle and brain cells and may thereby improve cellular energetics and reduce oxyradical production that significantly reduces brain damage.

12. With respect to ageing, DNA repair declines; the initiation of translation appears unaltered. During ageing, the expression of certain stress response genes is increased in age-specific and tissue-specific patterns. Normal ageing results in changes in the brain that contribute to the decline of various functions, including learning and memory. Increases in chemokine gene expression that occur in conjunction with increased microglial activation suggest that chemokines may contribute to the decreased brain function that occurs during normal ageing. Increase in RPS25 (ribosomal protein S25) expression is further evidence that the composition of ribosomes may alter with age and as a result could have significant effects on protein synthesis. Ageing and oxidative stress responses shared the up-regulation of purine biosynthesis, heat shock protein, antioxidant, and innate immune response genes. As we age, genes are expressed differently throughout our body. In many models of how genes are expressed during older age, one of the prevalent models - the programmed ageing model - claims that ageing is caused by genetically programmed cell death. In other words, ageing is programmed into our cells.
13. Hormones alter the gene expression of DNA throughout the body. The hormones melatonin, thyroxine and growth hormone decline with age. Growth hormone level declines with age that is blamed for increased fat deposition, loss of muscle mass and bone demineralization. Glucocorticoid level is also related with brain ageing.
14. According to the "immune system theory of ageing", many ageing effects are due to the declining ability of the immune system to differentiate "foreign" from "self" proteins. Arthritis, psoriasis and other autoimmune diseases increase with age.
15. Diseases that resemble certain aspects of accelerated ageing are Werner's syndrome (associated with early onset of many age-related diseases); Hutchinson-Gilford Progeria Syndrome (a child is born with abnormally short Telomeres) and Bloom's syndrome (is due to a defective helicase-type protein BLM leading to chromosome aberrations).

QUESTIONS

Long questions:

1. What do you mean by ageing? Point out the main symptoms of ageing. Discuss the theories of ageing. Describe the role of mitochondria in ageing.
2. What do you mean by brain ageing? Discuss the factors responsible for brain ageing.

Short questions:

1. Name some diseases associated with ageing.
2. Differentiate biological ageing and chronological ageing.
3. What do you mean by 'mean life span' and maximum life span'?
4. What do you mean by 'programmed ageing'and 'wear and tear ageing'?
5. What is Glycation?
6. What is Error catastrophe theory of ageing?
7. Describe mitochondrial theory of ageing?
8. Describe chaperone medited autophagy.
9. What are the main complications of menopause.
10. What is Hayflick limit?
11. How telomere shortening is related with ageing?
12. Which ECM show drastic changes due to ageing?
13. How ROS is related with ageing?
14. Differentiate homoplasmy and heteroplasmy?
15. What do you mean by biological garbage?
16. What is amyloidosis?

17. Name a gene which delays ageing?
18. Name some antioxidants.
19. How hormone changes are related with ageing?
20. What are the common symdromes associatd with ageing?

MCQs:

1. **The cellular organelle that produces more ROS is**
(a) lysosome, (b) ER, (c) mitochondri(a)
2. **Dysfunctional organelle are phagocytosed by**
(a) Golgi, (b) lysosome (c)ER.
3. **The age related functional decline of mitochondria can be delayed by**
(a) Ca^{++}, (b) NO, (c) Mg^{++}.
4. **Photoageing of skin is related with dysfunctional**
(a) mt DNA, (b) ROS, (c) telomere shortening.
5. **In glycation, reducing sugars become attached to**
(a) proteins, (b) fats, (c)nucleic acids.
6. **Cardiolipin content in mitochondria**
(a) declines,
(b) increases,
(c) remains unchanged in ageing.
7. **The proteins affected in ageing skin are**
(a) fibronectin. (b) collagen, (c) proteoglycan.
8. **Parkinson's disease make use of the neurotransmitter**
(a) GABA, (b) acetylcholine, (c) dopamine.

Answers to MCQs:

1. c, 2. b, 3. b, 4. a, 5. a, 6.a, 7. b, 8. c.

Chapter 11

Cellular Organelles

The organelles described in this section are :

Chapter 11.1

Endoplasmic Reticulum

In the endoplasm of eukaryotic cells, in between nucleus and plasma membrane shows a network of interconnected tubules and sacs called endoplasmic reticulum (ER). ER is a fundamental organelle required for assembly and remodeling of proteins, lipid biosynthesis, storage and production of glycogen. In fact it is central organelle that participate in the secretion, membrane trafficking, protein remodeling, Ca^{++} signaling etc. Membranes of ER, Golgi, endosomes and lysosomes are functionally linked to carry out integrated role in maintaining the physiological balance of the cell.

Network like structure invested by single layered membrane, occurring in the endoplasm that play vital roles in the remodeling of different macromolecules.

11.1.1. MORPHOLOGY (Fig. 11.1.1):

A. On the basis of shape:

The endoplasmic reticulum (ER) is a large membrane-bound compartment spread throughout the cytoplasm of eukaryotic cells. It is divided into three major morphologies that include the nuclear envelope (NE), peripheral ER cisternae, and an interconnected tubular network. It has diverse structural domains - flat, cisternae, highly curved tubules (*tubular*), round or oval (*vesicular*) (Fig. 11.1.1) and other domains maintaining contacts with nearly every organelle and with the plasma membrane. NE is the most visually obvious domain made from ER membrane. It surrounds the nucleus as a double membrane bilayer and acts as a barrier to selectively control transport of molecules into and out of the nucleus. The peripheral ER branches out from the NE as an interconnected network comprised of cisternae and tubules. ER cisternae tend to be localized closer to the NE whereas small cisternae and tubules predominate in the periphery that pervade the cytoplasm, but they form polygonal network at the cortex of the cell which is anchored to the plasma membrane (Puhka *et al.* 2007). Structure of ER cisternae is regulated by an integral ER membrane protein – Climp 63 with interaction with cytoskeleton (Friedman and Voeltz, 2011). The morphology of the tubular peripheral ER is maintained by the integral membrane proteins Reticulon (Rtn) and DP1/Yop1 (West *et al.* 2011).

ER shows a number of domains associated with specific functions like protein storage, or direct interaction with other organelles (Griffing,

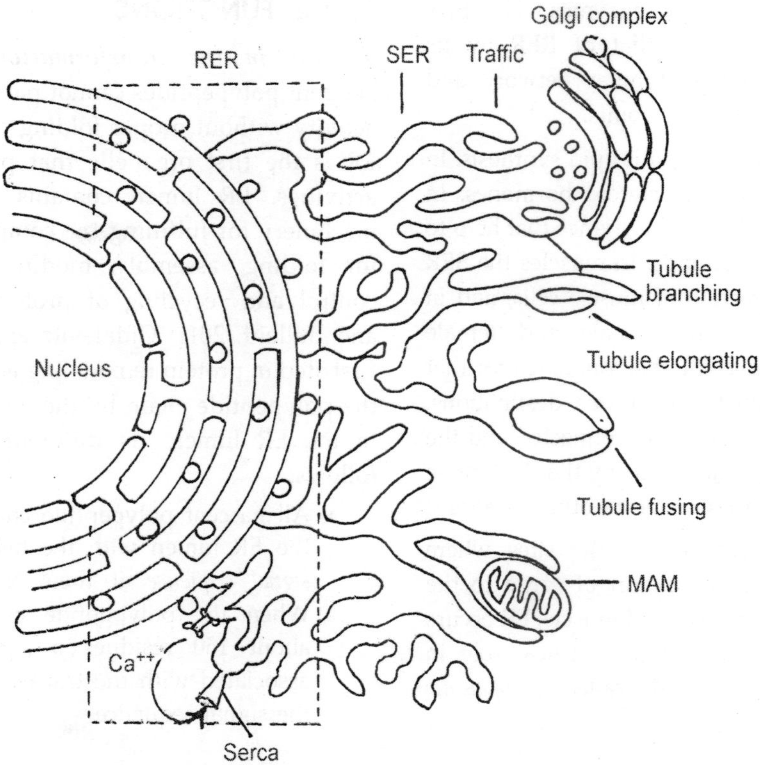

Fig. 11.1.1 : Different morphological structures of ER.

2010; Bravo *et al*, 2013). The process by which ER expands or changes is not properly understood. Fusion of small ER vesicles results into large interconnected network. Moreover weak interaction between cisternal compartments leads to separate cisternae.

Endocrine cells show dramatic expansion of the RER. Prior to this formation of RER stacks called ***karmellae*** are noted. During the differentiation of B-cells into antibody secreting cells, UPR is activated and expansion of ER membranes occurs prior to immunoglobulin systhesis.

B. On the basis of association with ribosomes (Fig. 11.1.2) :

ER maintains an intimate relation with ribosomes. On this basis, it is of two types - ***Rough ER (RER)*** – showing ribosomes on their surface and ***Smooth ER (SER)*** – surface ribosomes are absent. In between RER and SER an intermediate form is noted, called ***transitional element*** RER and SER occur side by side. In many cells, RER and SER do not occupy spatially segregated regions; however, in some cells such as hepatocytes and neurons, the smooth and RER occupy different cellular areas (Borgese *et al.*, 2006). Cisternae are better suited for ribosome binding or ribosome binding stabilizes cisternal structure (Hetzer, 2010).

RER is mainly concerned with synthesis and storage of proteins; therefore they are very prevalent in protein synthesizing cells. In lymphocytes, they produce antibodies, in pancreatic cells, they produce insulin.

SER is found in a variety of cell types serving different functions. It consists of tubules and vesicles that increases the surface area for the action or storage of key enzymes. The SER morphology differs from that of RER by its typically more complex, tubular network and greater numbers of branch points. .

SER is also associated with lipid synthesis. In adrenal cortex, they store steroid hormones. In liver cells the SER produces enzymes that help to detoxify certain compounds. In muscles the SER assists in the contraction of muscle cells, and in brain cells it synthesizes male and female hormones. It serves as a transitional area for that transport ER products to various destinations. RER and SER are usually interconnected and the proteins and membranes made by the RER move into the SER to be transferred to other locations.

Transitional elements are the sites where vesicles bud during the export of cargo to the Golgi apparatus. In the striated muscle, ER occurs mainly stores Ca^{++} and play a critical role in muscle contraction these are called *sarcoplasmic reticulum* (SR).

11.1.2. FUNCTIONS

Role in the transformation of protein: Nascent polypeptides cannot participate in any activity without proper folding or remodeling. ER is the first organelle that performs these activities. ER lumen contains much of the machinery for fulfilling the complex pathways for folding, assembly, modification, quality control and recycling of proteins (Braakman and Bulleid, 2011; Gidalevitz *et al*, 2013). The first step in protein remodeling is the folding of the polypeptide chain by the enzymes present in the ER lumen. Its different steps are as follows:

i) All nascent polypeptide chains enter into the ER lumen with the help of *leader or signal sequence* at their N-terminal end. When the polypeptide chain grows to about 150 residues a signal peptidase associated with the translocon cleaves off the signal sequence.

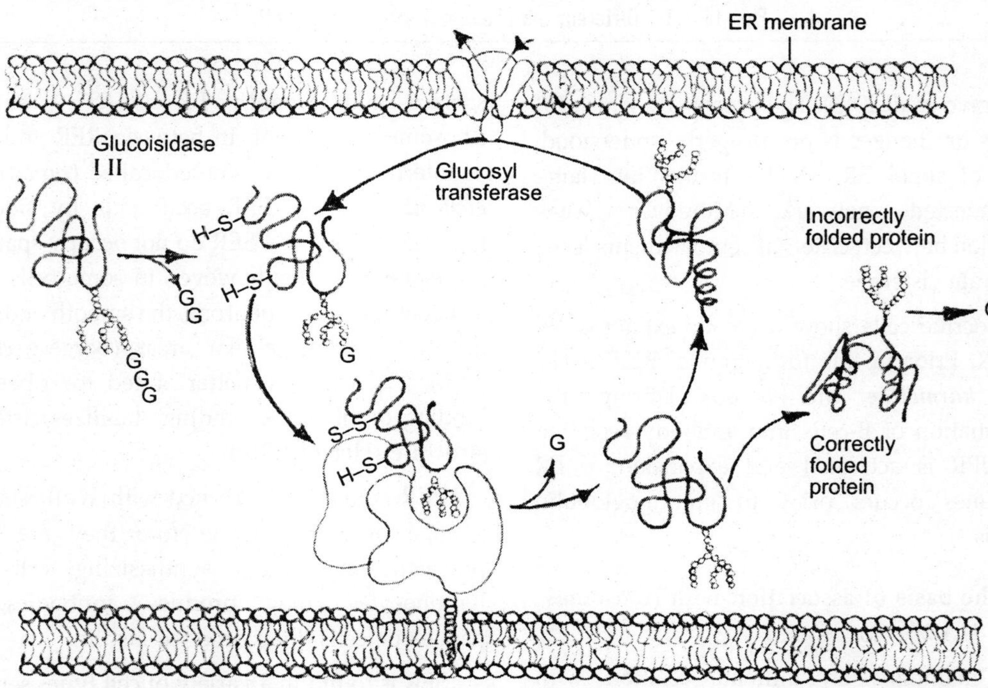

Fig. 11.1.2 : Protein modification in the ER lumen.

ii) Within the ER lumen, proteins undergo four types of modifications – *glycosylation* (addition of carbohydrates), *disulfide bond formation, oligomerization* (folding of the protein) and *proteolytic cleavage.* Proteins become degraded if these are misfolded.

ER has a *quality-control system* for *'proof-reading'* newly synthesized proteins, so that only properly folded proteins reach their final desti nations. Proteins that fail to fold correctly are translocated back to the cytosol where they are degra ded (Swanton and Bulleid, 2013). Disruption of ER homeostasis causes accumulation of unfolded and misfolded proteins in the ER. This condition is called ***ER stress.***

There are two types of ER stress conditions - tolerable and unresolvable. Tolerable ER stress can be physiological, acute and/or mild. Under tolerable ER stress conditions, the unfolded protein response (UPR) is properly activated. In contrast, unresolvable ER stress conditions are pathological or chronic. Under unresolvable ER stress conditions, the UPR is hyperactivated inducing cell dysfunction and death. In response to tolerable ER stress, the UPR triggers transcription of adaptive and survival genes. In response to unresolvable ER stress, the UPR is hyperactivated triggering the activation of apoptotic pathways and upregulating apoptotic genes (Oslowski, and Urano, 2011). Different stressful conditions such as hypoxia, nutrient deprivation, pH changes, or reduced vascularization, potentially able to act as growth-limiting factors for tumor cells, activate the unfolded protein response (UPR). UPR is therefore involved in tumor growth and adaptation to severe environments and is generally cytoprotective in cancer (Giampietri *et al*, 2015).

The UPR has a dualistic function; it exists to alleviate damage associated with ER stress, and if this is not possible, then it signals for cell death through apoptosis. Cancer cells are largely due to an altered state of the UPR (Wang *et al,* 2014). How UPR exerts its action is not properly understood but it is suggested that Mitogen-Activated Protein Kinase (MAPK) signaling pathways may contribute to the consequences of ER stress. Indeed, MAPK signalling pathway is known to regulate cell cycle progression and cell survival or death responses following a variety of stresses (Darling and Cook, 2014).

Even though most secretory proteins are co-translationally translocated, some are translated in the cytosol and later transported to the ER / plasma membrane by a post-translational system. In prokaryotes this requires certain cofactors such as SecA and SecB. This pathway is poorly understood in eukaryotes, but is facilitated by Sec62 and Sec63, two membrane-bound proteins. In addition, proteins targeted to mitochondria, chloroplasts, or peroxisomes, use specialized post-translational pathways. Also, proteins targeted for the nucleus are translocated post-translation.

Trafficking of proteins from ER to Golgi apparatus: Only correctly folded proteins are sent to their final destinations. Proteins from ER are enclosed in minute *transitional vesicles,* coated with a protein called COPII, that accumulate in *ER-Golgi intermediate compartment (ERGIC).* In the most popular current view, the ERGIC is the most dynamic part showing mobile transport complexes that deliver secretory cargo from ER-exit sites to the Golgi. Recent data suggest a stable compartment model in which ER-derived cargo is first shuttled from ER-exit sites to stationary ERGIC clusters in a COPII-dependent step and subsequently to the Golgi in a second vesicular transport step (Appenzeller-Herzog and Hauri, 2006). Many proteins cycle back to the ER from pre-Golgi intermediates or the Golgi following *retrograde transport process.* These are enclosed in COPI-coated vesicles and transit via tubular structures.

Calcium balance in the cell: ER acts as the storage site of intracellular Ca^{++}, that is released

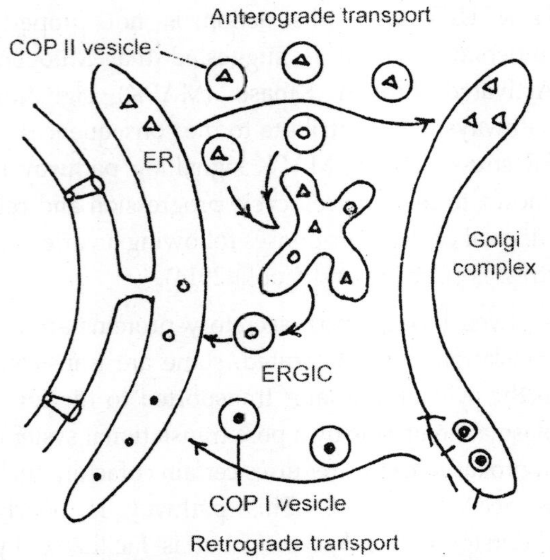

Fig. 11.1.3 : Protein trafficking from ER to Golgi complex.

at the time of emergency. Release of Ca^{++} from ER is mediated by an intraluminal ER-Ca^{++} binding protein – *calreticulin,* an ER-Ca^{++} binding chaperone *calnexin,* two ER Ca^{++} release channel proteins – *IP3R* and *ryanodine receptor.* Ca^{++} release depends upon various intracellular signaling. In SR, depolarization of sarcolemma and 'T' tubules activates the release of Ca^{++} during muscle contraction. Release of Ca^{++} is known to affect actin polymerization. Ca++ release also plays a vital role in the formation of fertilization membrane. Storage of Ca^{++} is important for folding of the newly synthesized proteins. Disruption of ER- Ca^{++} homeostasis results in the development of *ER stress response* that in turn may lead to neurodegenerative diseases.

Biogenesis of peroxisome:

Peroxisomes are formed *de novo* from the ER through budding and fusion of two vesicle types, V1 and V2. Peroxin 1 and 6 (Pex1 and Pex6) are found in separate pre peroxisomal vesicles and are necessary for the fusion of V1 and V2 vesicles and the formation of mature peroxisomes. Pex3 is required for egression of vesicles from the ER. Mature peroxisomes can multiply by growth, with the help of proteins and membranes synthesized from the ER and fission, mediated by Pex11 (Nuttall *et al,* 2011).

As per *semiautonomous model of peroxisomal biogenesis* peroxisomal membrane proteins (PMPs) traffic from the ER to the peroxisome by

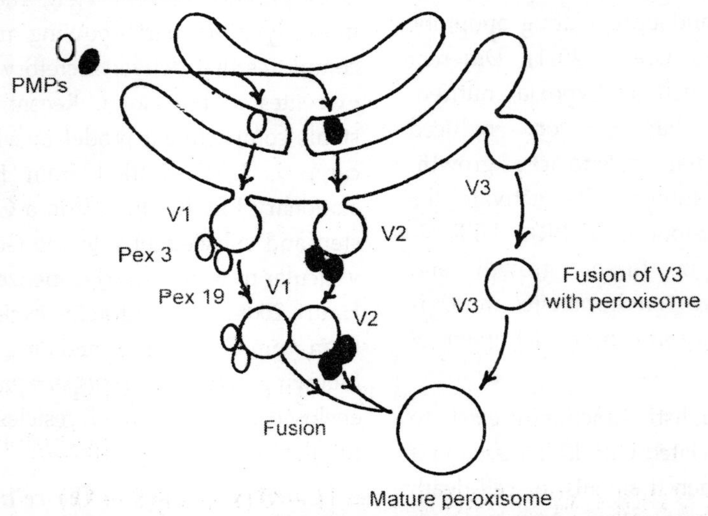

Fig. 11.1.4 : Role of ER in the biogenesis of Peroxisome.

a vesicular budding, targeting, and fusion process while peroxisomal matrix proteins are imported into the organelle by an autonomous, post translational mechanism. (Dimitrov *et al*, 2013). However the enzymes, catalase and thiolase, that are essential for structure and function of peroxisome are synthesized by ER and later on transported to peroxisome.

Role in apoptosis: Apoptotic pathways are activated by a diverse array of extrinsic and intrinsic signals, most of which are coupled to the activation of effector caspases. It involves an egress of critical proapoptotic regulators to the cytosol. It shows a complex interplay between antiapoptotic members of the BCL-2 family, and the proapoptotic BAX and BAK subgroups. Constituents of all these proteins converge on the ER which contributes in a fundamental way to Fas-mediated apoptosis and to p53-dependent pathways resulting from DNA damage and oncogene expression. Mobilization of ER calcium stores can initiate the activation of cytoplasmic death pathways as well as sensitize mitochondria to direct proapoptotic stimuli ER plays a vital role in this aspect by controlling the release of nucleases causing intranuclear DNA fragmentation (Breckenridge *et al*, 2003).

Glycogen metabolism: ER plays a vital role in the storage and modification of glycogen.

Repair of plasma membrane: ER synthesizes some proteins that are essential for the repair of plasma membrane.

Steroid metabolism: In adrenal cortex, SER metabolizes steroids and participates in the formation of steroid hormones.

Lipid synthesis: Synthesis and sorting of lipids are essential events for membrane biogenesis. A conserved family of proteins, fat storage-inducing transmembrane (FIT) proteins, is required for proper budding of lipid droplets from the ER (Choudhary *et al*, 2015).

Nuclear envelope recycling: Nuclear envelope NE breaks down during prometaphase and is reabsorbed by the ER which again participates in the formation of new NE and nuclear pore complexes (Lu *et al*, 2011).

Quality control: Quality control in the ER prevents the arrival of incorrectly or incomepletely folded proteins at their final destinations and targets permanently misfolded proteins for degradation. Such proteins have a high affinity for the ER chaperone BiP and are finally degraded via retrograde translocation from the ER lumen back to the cytosol. This ER-associated protein degradation (ERAD) is currently thought to constitute the main disposal route. This mechanism of quality control is closely linked to calcium homeostasis whose balance is presumed to be regulated by a specific cellular compartment (Araki and Nagata, 2011).

Mitochondrial division: The ER and mitochondria exhibit extensive contacts. Mitochondrial division occurred at positions where ER tubules remain in contact with mitochondria Thus, ER tubules play an active role in defining the position of mitochondrial division sites (Westermann, 2011).

11.1.3. ER BIOGENESIS

Mostly ER is maintained as a continuous network throughout mitosis and is pinched off into roughly two equal partsTwo proteins, reticulon and DP1/Yop1p families can generate tubular membrane networks from phospholipid vesicles suggesting a new process of ER biogenesis.

SUMMARY

1. Endoplasm of eukaryotic cells, in between nucleus and plasma membrane shows a network of interconnected tubules and sacs called endoplasmic reticulum (ER). Two membrane proteins, Rtn4a/NogoA and DP1/Yop1p, are responsible for the generation of tubular morphology of the ER.

The lumen of ER differs from cytoplasm in ionic composition. It maintains a close relation with ribosomes. On this basis, ER is of two types - Rough ER (RER) - bearing ribosomes on their surface and Smooth ER (SER) - having no ribosome on the surface.

2. ER is required for assembly and remodeling of proteins, lipid biosynthesis, storage and production of glycogen.

   From ER, each protein follows a complex route for delivery to its final destination. Translocation of the nascent polypeptide chain into the ER is regulated by the specific amino acid sequence called leader or signal sequence on the growing polypeptide that is the first synthesized part of the polypeptide chain with 13-36 extra amino acids at their N-terminal end. Ribosome with its nascent polypeptide and the signal recognition particle (SRP) binds to the SRP receptor on the surface of the ER followed by dissociation of SRP and SRP receptor. It permits elongation and polypeptide translocation. The synthesized protein is introduced into the lumen of the ER. After entrance into the ER lumen, signal peptide is cleaved and is immediately proteolyzed.

3. As the elongating polypeptide chain enters into the ER lumen it binds with a chaperone – BiP. In eukaryotes, cotranstational translocation of secretory proteins into the ER occurs but in yeast secretory proteins enter into the ER lumen after the completion of the translation without the involvement of SRP and SRP receptor but with the assistance of transmembrane Sec63 complex and a luminal chaperone BiP. As the polypeptide elongates more and more BiP-ADP binds with the polypeptide and draws the entire polypeptide into the ER lumen.

4. Integral membrane proteins have been classified into four topological classes. Type I proteins (e.g. Glycoproteins, Insulin receptors, Growth hormone receptors etc.) possess an N terminal signal sequence and an internal hydrophobic amino acid sequence that acts as stop transfer sequence. Once the polypeptide enters the ER lumen, the signal sequence is cleaved but the chain continues to elongate until the stop transfer sequence is synthesized which anchors the chain with the wall of the translocon. Type IV proteins fall into two major categories – type IV-A proteins with N terminus in the cytosol (e.g. various glucose transporters and ion transporters) and type IV-B proteins with N terminus in the ER lumen (e.g. G protein coupled receptors). Type IV-B proteins show initial C terminus in the cytosol followed by same sequences eventually showing N terminus in the ER lumen.

5. Within the ER lumen, after removal of the signal sequence, proteins undergo four types of modifications – glycosylation (addition and processing of carbohydrates), disulfide bond formation, oligomerization and proteolytic cleavage.

6. The oxidizing environment of the ER lumen favours disulphide bond formation by the enzyme protein disulphide isomerase (PDI). Proteins that contain a signal peptide are called preproteins Most proteins undergo proteolytic cleavage following translation. Inactive precursor proteins are called proproteins. This class of proteins is termed preproproteins. Still other proteins are synthesized as inactive precursors called zymogens.

7. Retention of unfolded and improperly folded proteins in the ER lumen increase the overall efficiency of folding machinery. A dimeric ER membrane protein Ire1 promotes synthesis of a transcription factor – Hac1 that activates the expression of genes in response to unfolded proteins. Misfolded and unfolded proteins in the ER lumen are degraded rather than being exported from the ER but recent findings confirm that misfolded proteins are retrograded to the cytosol from the ER where they are degraded by the ubiquitin mediated proteolytic pathway. Protein misfolding in the ER triggers following two signaling pathways -

   Unfolded protein response pathway: It involves misfolding of the mutant proteins, inhibition of glycosylation, disulphide bond formation or over production of normal proteins.

   ER overload response pathway: It involves apoptosis of the cells containing misfolded proteins.

8. ER has a quality-control system for 'proof-reading' newly synthesized proteins, so that only properly folded proteins reach their final destinations. Protein translocation in the ER and N-glycosylation are the initial steps in the protein transformation in

the intracellular pathway. Eventually it forms secretory proteins and membrane proteins of the endomembrane system and plasma membrane.

9. Trafficking of proteins from ER to the Golgi apparatus is noted to be highly specific for selective retention and retrieval of ER resident proteins and export of cargo proteins. Proteins released from ER are enclosed in minute transitional vesicles, coated with a protein called COPII, that accumulate in ER-Golgi intermediate compartments (ERGIC). ER membrane cholesterol seems to be essential ER-to-Golgi transport of secretory membrane proteins.
10. ER plays a vital role in the storage and modification of glycogen. It regulates Ca^{++} storage and release depending upon long range signaling through intra-ER tunnels. BCL-2 family of proteins plays a central role in regulating apoptosis mediated by the ER. ER synthesizes some proteins that are essential for the repair of plasma membrane. An ER membrane protein with reductase activity catalyses the hydroxylation of endogenous steroids. Quality control in the ER prevents the arrival of incorrectly or incomepletely folded proteins at their final destinations and targets permanently misfolded proteins for degradation. This ER-associated protein degradation (ERAD) is currently thought to constitute the main disposal route.

QUESTIONS

Long type questions:

1. Describe the morphological and molecular structure of endoplasmic reticulum.
2. Describe the process of protein translocation in to the ER lumen.
3. Describe the processes of protein remodeling in the ER lumen.

Short type questions:

1. What do you mean by transitional elements?
2. What do you mean by microsomes?
3. What is signal sequence?
4. Distinguish among preprotein, proproteins and preproprotein. Give examples.
5. What is SRP? Discuss its role in protein translocation in to ER lumen.
6. Differentiate SRP and SRP receptor.
7. What do you mean by glypiated proteins?
8. What is dolichol? How it influences protein remodelling?
9. What is GPI linkage? Name some GPI linked proteins.
10. What is chaperone? How they help in protein folding?
11. What do you mean by quality control system of the ER? What is its practical significance?
12. Name some diseases due to misfolding of proteins.
13. What do you mean by retrograde transport?
14. Describe the role of ER in the calcium homeostasis of the cell.

MCQs:

1. **RER is mainly found in**
   (a) carbohydrate processing cells;
   (b) lipid synthesizing cells and
   (c) protein synthesizing cells.
2. **In the formation of glycoprotein**
   (a) proline,
   (b) methionine and
   (c) glycine does not participate.
3. **Disulphide bond formation is noted in**
   (a) endocytic proteins,
   (b) exocytic proteins,
   (c) secretory proteins.
4. **SRP**
   (a) carries the signal sequence to the translocon,
   (b) cleaves the signal sequence,
   (c) helps in the transfer of the nascent polypeptide into the ER lumen.
5. **Peptide exit site is the**
   (a) pore of translocon,
   (b) pore of ER through which folded proteins are transferred to the Golgi,
   (c) the pore of large subunit of ribosome through which nascent polypeptide emerges.
6. **The main role of ER is**
   (a) processing of proteins,
   (b) breakdown of proteins,
   (c) autophagy.
7. **ER originates from**
   (a) Golgi complex, (b) Mitochondria, (c) Nucleus.

8. **Sarcoplasmic reticulum is the ER of**
   (a) Neuron, (b) Adipose tissue, (c) Muscle.
9. **What is more abundant in SER?**
   (a) Tubules, (b) Vesicles, (c) Tubules and Vesicles.
10. **What is more abundant in RER?**
    (a) Cisternae, (b)Tubules, (c) Vesicles.
11. **RER is well developed in cells engaged in**
    (a) Lipid synthesis,
    (b) Protein synthesis,
    (c) Detoxyfication.
12. **SER is well developed in cells engaged in**
    (a) Lipid synthesis,
    (b) Protein synthesis,
    (c) Detoxyfication.
13. **In rapidly dividing cells ER is**
    (a) Absent, (b) Non-functional, (c) Poorly developed.
14. **In hepatocytes, the detoxification site is**
    (a) RER, (b) SER, (c) Lysosome.

Answer to MCQs:

1. c; 2. a; 3. b; 4. a; 5. c; 6. a; 7. c; 8. c; 9. c; 10. a; 11. b; 12. a; 13.c; 14. b.

Chapter 11.2

Golgi Complex

In eukaryotic cells, it is a part of the cellular endo membrane system, that packages proteins inside the cell after liberating from The ER before they are sent to their destination; it is particularly important in the processing of proteins for secretion. In fact, Golgi apparatus is the central processing and sorting station of the secretory pathway of the cell.

Definition: Golgi apparatus is a heterogeneous assemblage of different structural components, that are involved in the secretion of the cell.

In non mammalian cells, Golgi complex remains randomly distributed throughout the cytosol but in mammalian cells these are anchored to the pericentriolar region. Thus a close relationship between Golgi and MTOC are conjectured because microtubule motor proteins dynein and kinesin bind to Golgi membrane and help in the vesicular transport to and from Golgi. The trans faces of Golgi complex remain intimately associated with ER and lysosomes forming a large complex called *GERL (Golgi-ER-Lysosome).*

11.2.1. MORPHOLOGY (Fig. 11.2.1):

The whole organelle is roughly semi circular. Around the whole organelle there is a clear zone, devoid of any structure, called *Zone of Exclusion.* Its nuclear facing side is convex while plasma membrane facing side is concave. The convex side is called *Forming or CIS face* while the concave side is called *Maturing or Trans face.* The middle part is called *Medial region.* The whole organelle consists of mainly 3 types of structural components - *Cisternae, Vacuoles and vesicles.*

Cisternae are flat saccular structures, arranged parallel to each other. Each stack consists of closely aligned, flattened membrane-limited cisternae. The number of cisternae in a Golgi stack varies from 2-30. Each cisterna is bounded by single layered membrane enclosing the lumen. The cisternae remain connected with each other by protein scaffold. The structural integrity of Golgi cisternae is maintained by a protein called *golgin* which maintains a dynamic exchange between the membrane and the cytoplasm.

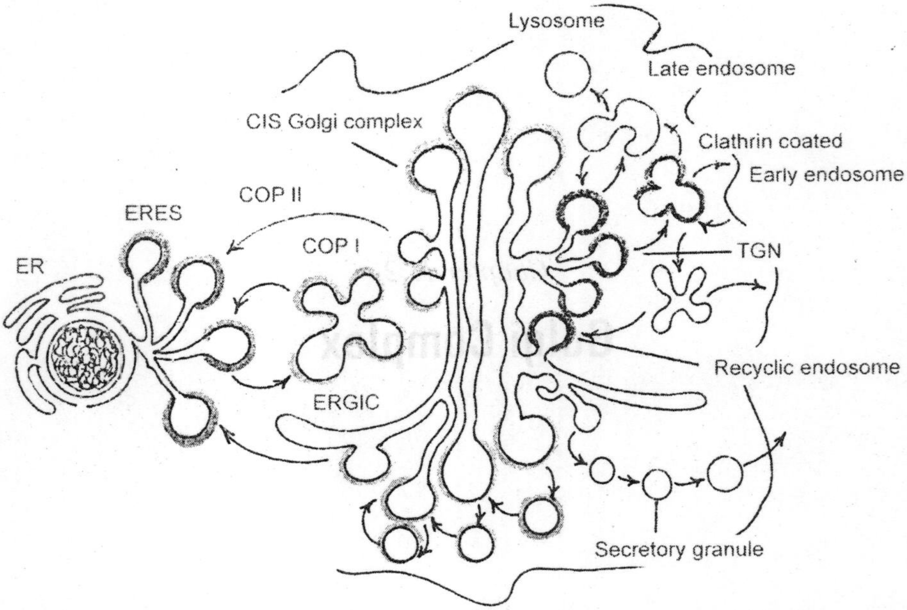

Fig. 11.2.1 : GERL system.

Vacuoles are large sized globular structures present at the periphery of the Golgi complex. Vesicles are minute structures present at the trans face and due to the presence of these structures this part looks fenestrated.

The mammalian Golgi complex is a highly dynamic organelle that contribute to cargo import and export, intra-cisternal trafficking, and overall Golgi architecture. At the morphological level, all these structures are continuously remodelled to carry out trafficking functions. Recent advances have shown that continual phospholipid remodeling by phospholipase A (PLA) and lysophospholipid acyltransferase (LPAT)

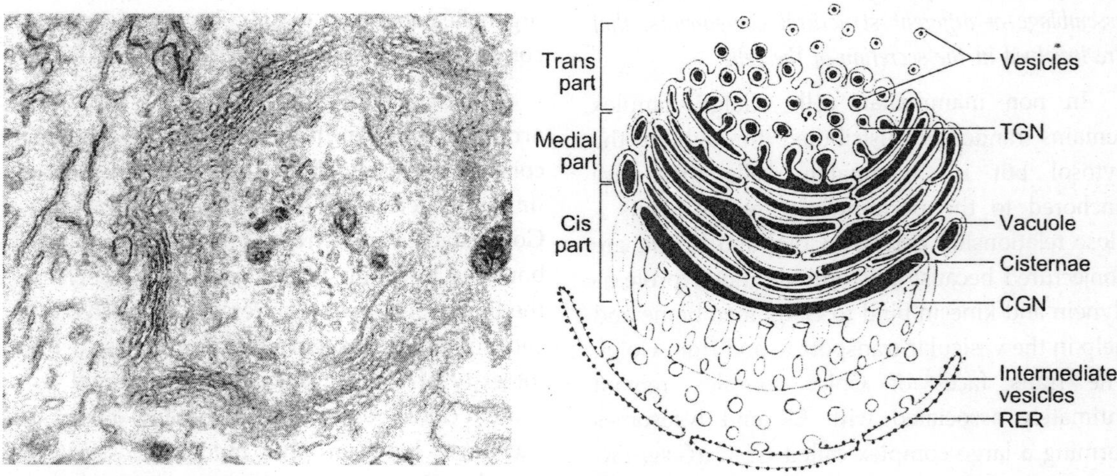

Fig. 11.2.2 : Morphology of Golgi complex.

Fig. 11.2.3 : Schematic representation of inter golgi connection and TGN.

enzymes, which deacylate and reacylate Golgi phospholipids, respectively, contributes to this morphological remodeling (Ha *et al*, 2012).

Maturational Model (Bannykh and Balch, 1997): As per this model all classes of Golgi components are dynamically associated and a flow of structural components are noted from Cis to Trans part. enzymes and recycling components are continuously exiting and reentering the Golgi apparatus by membrane trafficking pathways to and from the ER, whereas Golgi matrix proteins and coatomer undergo constant, rapid exchange between membrane and cytoplasm.

Mechanical model of the Golgi stack (Derganc et al, 2006): As per this model: (i) Golgi membrane bend ing elasticity is related to the mem brane curvature and inter cisternae adhesion. (ii) Number of cisternae is proportional to the propagating secretory wave. In the cis part, the intermediate vesicles, tubules released from the ER form a network like structure called *Cis Golgi Network (CGN)*. In the trans region also a network of tubular structures originate and form an anastomosing structure, called *Trans Golgi Network (TGN)*. They remain united with other similar structures of neighboring Golgi (Fig. 6.2.3). TGN is a tubular-vesicular organelle that matures from the trans cisternae of the Golgi apparatus. In plants, it functions as a central hub for three pathways: the secretory pathway, the vacuolar trafficking pathway and the endocytic pathway. A novel TGN-localized membrane protein, *continuous vascular ring* (COV1) is crucial for Golgi morphology and vacuolar trafficking in *Arabidopsis* (Shirakawa *et al*, 2014).

Physiological significance of TGN: TGN is an extension of the trans Golgi portion where different types of vesicles are formed. It is a major protein sorting station where mature proteins are sorted prior to transport to several locations inside and outside the cell.

11.2.2. CHEMICAL COMPOSITION

Golgi complex shows specific enzymes for the formation of conjugated protein / lipid. The *cis* part shows lipid rich membranes while *trans* part, shows high acid phosphatases for the liberation of lysosomes and middle part shows enzymes for glycolipids and glycoprotein formation.

11.2.3. FUNCTIONS

Secretion: The main function of Golgi complex is to control the secretory activity of the cell. For the production of secretory droplets, proteins enclosed in *transport vesicles* are transported in *anterograde (CGN to TGN) directions*. These vesicles merge with the wall of the CGN and their contents transformed during their journey through the Golgi. The vesicles containing the secretory droplets merge with the plasma membrane with the help of Ca^{++}. Some times some vital proteins are transport ted to the Golgi but the cisternae of Golgi can identify and immediately sent them back to the RER following *retrograde direction*. Sometimes cisternal progress is accompanied by anterograde as well retrograde transport packaging and sorting of secretory

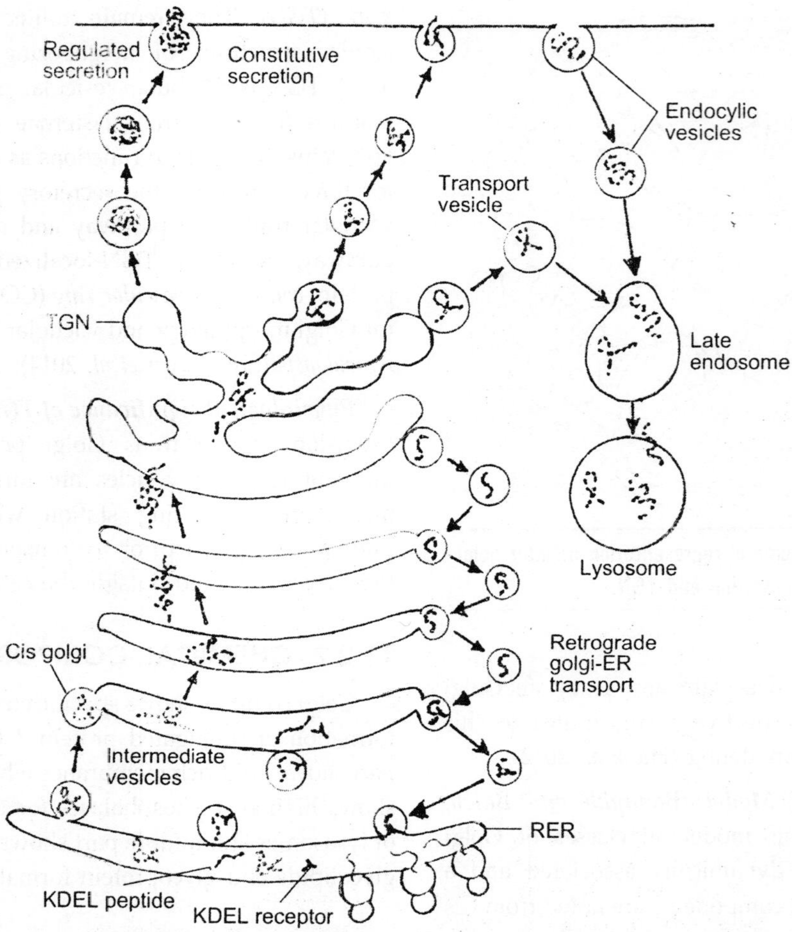

Fig. 11.2.4 : Types of secretion through the Golgi complex.

substances may follow one the following pathways.

As per ***regulated secretory pathway*** the secretory vesicles may contain a high concentration of specific protein. When they receive the correct "signal", they fuse with the plasma membrane and are released into the blood and transported to far off places. Endocrine cells, neurons and digestive enzyme forming cells follow this pathway.

As per ***constitutive secretory pathway*** some proteins are produced and secreted constantly. Irrespective of getting signal for release, these are budded off from the TGN and exported directly to the plasma membrane for exocytosis. Such proteins are used for extracellular matrix. Some proteins are modified to attain lytic action and are stored in special vesicles that go back to the interior of the cytosol and act as primary lysosomes. These are bound to mannose 6-phosphate receptor. These proteins then buds off in a vesicle that fuses with lysosome. The low pH in the lysosome causes the mannose 6-phosphate bearing protein to dissociate from the receptor. The released receptor is recycled to the TGN. Some proteins are not packaged into vesicles and are delivered directly to the cell surface following a *non selective constitutive pathway or default pathway* (Fig. 11.2.4).

Formation of glycoproteins by the Golgi complex: During the processing of different types of conjugated proteins, *cis* region adds lipids to membranes and performs phosphorylation; the *middle* region adds sugars or other groups to the product while final sorting occurs in the trans part. Protein of all glycoproteins is synthesized in RER; processing of the sugar groups occurs in the ER lumen but O-linked and N-linked attachment of sugars in glycoproteins occurs in the Golgi apparatus.

Post translational processing of proteins by Proteolytic Cleavage: Cleavage of signal peptide from *preproteins*, proteolytic cleavage following translation, formation of active protein from inactive precursors (*proprotein*), proteolysis of large *preproproteins* all occurs in the Golgi complex.

Lipid synthesis: ER and Golgi complex synthesize most of the cellular lipids. Phosphoglycerides are largely synthesized by the ER. These are later on transported by intermediate vesicles to Golgi complex. Sphyngolipid synthesis largely occurs in the Golgi. Ceramide is the backbone of sphyngolipids. Its synthesis begins in the cytosolic face of ER but a subsequent step of synthesis occurs in the Golgi complex.

Sorting of lysosomal enzymes: The proteins destined for lysosome are received from the RER in the *cis* region of Golgi complex. There it has a phosphate radical attached to the mannose residue This mannose-6-phosphate forms a sorting signal that moves through the cisternae to the trans region where it binds to a specific receptor and begins to bud. A "cage" or "coat" made up of clathrin is formed around the bud (to strengthen it). It moves away to fuse with a developing lysosome. This lysosome contains a H+ pump that works to acidify the interior of the lyso some. This removes the phosphate and

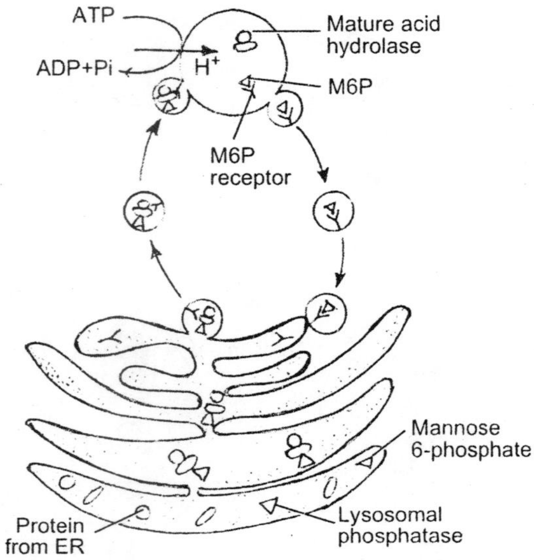

Fig. 11.2.5 : Sorting of lysosomal enzymes through Golgi complex.

dissociates the hydrolase from the receptor. The receptor is then recycled back to the Golgi complex (Fig. 11.2.5).

Formation of extra cellular matrix protein: Many extracellular proteins (like fibronectin) are specially modified in the lumen of the Golgi cisternae. Vacuoles and vesicles become filled with protein-lipoid material for storage. These stored products help in secretory action.

Synthesis of membrane proteins: Golgi complex synthesizes proteins needed to repair the plasma membrane. Moreover the plant-specific, nuclear envelope (NE)-associated proteins are localized at the Golgi and are conveyed to the NE.

Formation of acrosome: Acrosome is the lytic enzyme laden part at the tip of the sperm head. During the transformation of spermatids into sperms, Golgi complex is transformed into acro some which occupies the apical position on the head of the sperms (Fig. 11.2.6).

Formation of Cortical sacs: Cortical granules are membrane bound organelles located in the

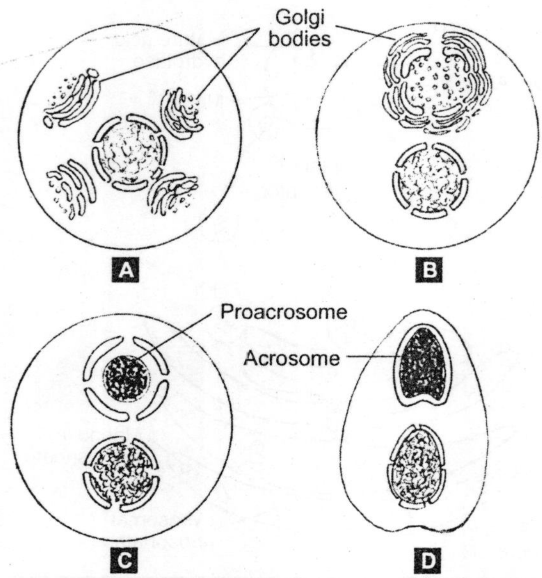

Fig. 11.2.6 : Formation of acrosome from Golgi complex.

cortex of unfertilized oocytes. Following fertilization, they undergo exocytosis to release their contents into the perivitelline space. Enzyme laden cortical sacs are derived from the Golgi complex of the ootids and occur just beneath their plasma membrane. They form fertilization membrane just after penetration of the sperm head.

Formation of Cell wall: Phragmoplast is formed during late cytokinesis of plant cell division. It serves as a scaffold for cell plate assembly and formation of a new cell wall between two daughter cells. It is a complex assembly of microtubules (MTs), microfilaments (MFs), and ER elements, that assemble in two opposing sets, perpendicular to the plane of the future cell plate. The MTs and actin filaments of the phragmoplast serve to guide vesicles with cell wall material to the growing cell plate. Actin filaments are possibly involved in guiding the phragmoplast to the site of the former preprophase band location at the parent cell wall. In the growing cell plate, segments of SER are

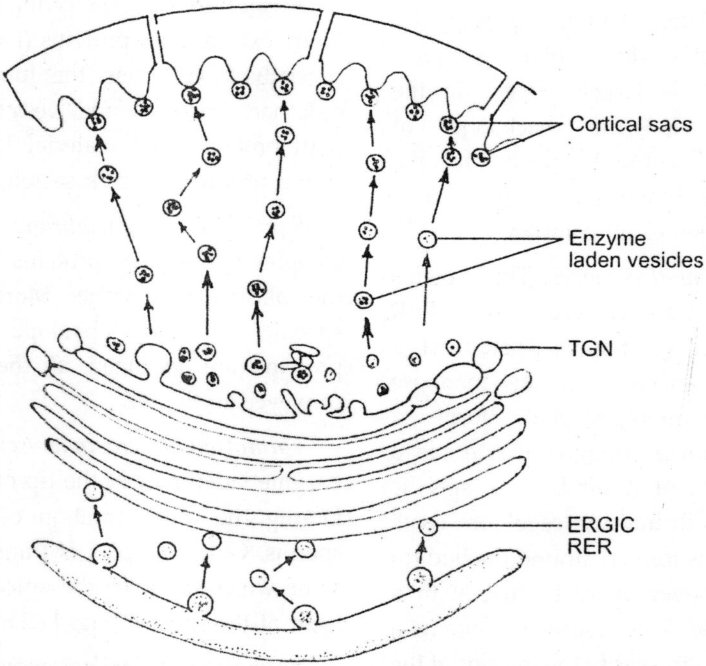

Fig. 11.2.7 : Role of Golgi in the formation of yolk materials and cortical granules.

trapped that form plasmodesmata between the two daughter cells.

Plant cells undergoing cytokinesis (cytoplasmic division) form a cell plate in the center of the phragmoplast. The phragmoplast consists of both cytoskeletal and membranous elements. The cytoskeletal elements are required for the delivery of vesicles to the equatorial region of the cell and provide guidance to the cell plate. Vesicles, originating from the Golgi apparatus, travel along microtubules and fuse in a unique manner at the center of the phragmoplast to give rise to the new plasma membranes of the two daughter cells.

Role in apoptosis: The Golgi complex undergoes fragmentation during apoptosis, as a result of caspase mediated cleavage of several Golgi-associated proteins. Golgi structure is also regulated by the cytoskeleton and cytoskeletal changes have been implicated in inducing apoptosis. However breakdown of the Golgi is an early event during apoptosis that occurs independently of major changes to the actin and tubulin cytoskeleton. Golgi is involved in storing Caspase-2 which plays vital role in the cleavage of Golgin-160, a protein that has crucial role in cell survival (Mukherjee *et al*, 2007).

Role in the synthesis of yolk materials: In reptiles and birds, eggs contain a large amount of yolk that are utilized by the developing embryo. These are synthesized in the liver and are carried by the blood to the developing oocytes. Inside the oocytes these are incorporated in the ER and Golgi complex and are enzymatically transformed into proteinous and lipoid yolk which are stored in the cytoplasm (Fig. 11.2.8). It also helps in the production of follicular fluid from granulosa cells of ovary.

Absorption of compounds: Golgi complex can absorb Fe, Cu, Au, Bismutose (compound of albumin and bismuth) and protargol (compound of albumin and silver) and is concerned with the storage and absorption of lipids. Golgi membranes may remove water from the products of synthesis during the formation of secretory granules.

Enzyme formation: It acts as an intracellular centre of enzyme formation. It may release zymogen granules which arise from cisternae into secretory vesicles. Secretory vesicles containing granules migrate to cell surface where the membrane of the secretory vesicle and the cell membrane merge, releasing the contents into pancreatic ducts from where they pass into intestine as active digestive enzymes.

Production of hormones: Golgi body in endocrine cells helps in secretion of hormones. Any harm to Golgi apparatus in thyroid gland cells results in the decline in secretion of its hormone.

11.2.4. EXISTANCE OF GOLGI COMPLEX DURING CELL DIVISION

In mammalian cells, at the onset of cell division, Golgi complex in the pericentriolar

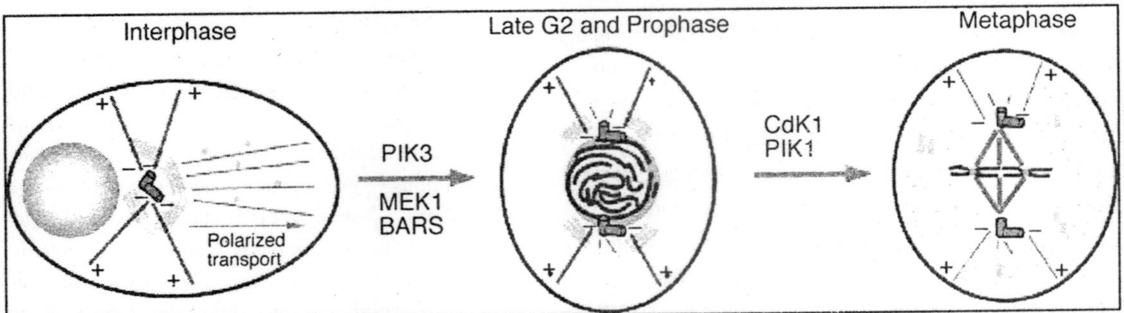

Fig. 11.2.8 : Position of Golgi during different phases of cell division.

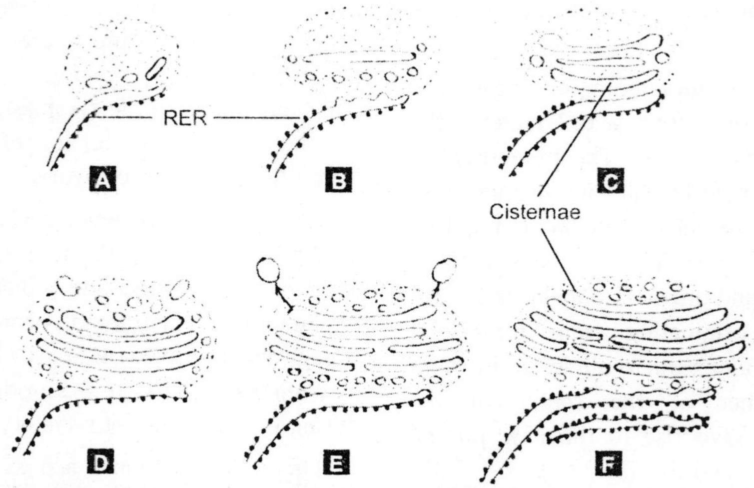

Fig. 11.2.9 : Schematic representation of the origin of Golgi from ER.

region, are fragmented into smaller elements called *'Golgi blobs'* which undergo further - fragmentation in later stages of division. In the absence of such fragmentation, cell gets arrested in G_2. Moreover at the onset of cell division Golgi complex is resorbed into the ER so that division of Golgi is dependent upon division of ER. However after telophase, the smaller fragments aggregate to form larger fragments that form Golgi complex in the pericentriolar region.

11.2.5. BIOGENESIS OF GOLGI COMPLEX (Fig. 11.2.9)

Golgi complex undergo rapid disassembly and reassembly during mitosis and in response to perturbations in membrane trafficking pathways. Recent works suggest that membrane recycling pathways operating between the Golgi and ER play a crucial role in Golgi maintenance and biogenesis. Two processes occur during Golgi disassembly: unstacking of the Golgi cisternae mediated by phosphorylation of the Golgi stacking proteins by mitotic kinases and fragmentation of the Golgi cisternae by COPI vesicle budding. Reassembly of the Golgi fragments into Golgi stacks involves membrane fusion and restacking.

SUMMARY

1. In between nucleus and plasma membrane cisternae, vacuoles and veicles containing Golgi complex is primarily concenrned with secretion but additionally performs many important roles. A Conserved Oligomeric Golgi (COG) complex is a peripheral Golgi protein which is associated with the cytoplasmic surface of the Golgi and is required for normal Golgi morphology and function..
2. Proteins, after synthesis are primarily modified in the ER from where they are delivered to the Golgi for final modification. The trans faces of Golgi complex remain intimately associated with ER and lysosomes forming a large complex called GERL (Golgi-ER-Lysosome).
3. As per Maturational model no part of Golgi is static and all the components actually mature gadually keeping pace with their translocation fron cis to trans face. Enzymes and recycling components are continuously exiting and reentering the Golgi apparatus by membrane trafficking pathways to and from the ER, whereas Golgi matrix proteins and coatomer undergo constant, rapid exchange between membrane and cytoplasm. When ER to Golgi transport is inhibited, the Golgi structure

disassembles and all Golgi components redistribute into the ER or cytoplasm. As per Mechanical model Golgi membrane bending elasticity is related to the membrane curvature and inter cisternae adhesion. In the cis part the intermediate vesicles, tubules released from the ER form a network like structure called Cis Golgi Network (CGN). It forms an anastomosing structure, called Trans Golgi Net work (TGN). They remain united with other similar structures of neighboring Golgi.

4. In mammalian cells, at the onset of cell division, Golgi complex in the pericentriolar region, are fragmented into smaller elements called 'Golgi blobs' which undergo further fragmentation in latter stages of division. These are resorbed into the ER so that partitioning of Golgi is dependent upon partitioning of ER.

5. Chemically Golgi complex shows specific enzymes for modification of proteins forming glyco or lipoproteins.

   Proteins after synthesis are transported into the lumen of the ER and are variously folded for their final destination. As per anterograde or retrograde directional model proteins enclosed in vesicles are transported in anterograde (cis to trans Golgi) or retrograde (cis Golgi to ER) directions.

6. The proteins leave the ER in an ER-Golgi intermediate compartment (ERGIC) via some vesicles called transport vesicles. These vesicles are specifically directed to the CGN. This is called anterograde transport. The modified carbohydrates, proteins and lipids are sent to the trans Golgi by the transport vesicles where the last stages of processing are finished. Sometimes vital proteins needed for the RER are transported along with the other proteins in the Golgi complex. The Golgi complex identifies these default proteins and send them back to the RER. This is called Retrograde Transport, which not only takes place from Golgi to ER, but it may also occur from one Golgi stack back to their previous Golgi stack. The ER receptor protein captures all of the proteins that carry the ER residency signal. The tyrosine kinase Src present on the Golgi membranes has an appreciable role in the organization of the Golgi apparatus. It may be linked to its involvement in protein transport from Golgi apparatus to the ER. Golgi cis-medial matrix proteins also participate in Golgi-to-ER retrograde traffic. The vesicles merge with the cis cisternae of the Golgi and they move gradually towards the trans region but simultaneously retrograde transport is also noted.

7. During the processing of different types of conjugated proteins different parts of Golgi performs different roles. Further modifications to the carbohydrates are completed in the trans Golgi. Proteins that contain a signal peptide are called preproteins Most proteins undergo proteolytic cleavage following translation. Inactive precursor proteins are termed, proproteins.

8. ER and Golgi complex synthesize most of the cellular lipids. Sphyngolipid synthesis largely occurs in the Golgi. The proteins destined for lysosome are received from the RER in the cis region of Golgi complex.. The receptor is then recycled back to the Golgi complex.

   The main function of Golgi is to control the secretory activity of the cell. Golgi complex also synthesizes proteins needed to repair the plasma membrane.

   Wall fragments are synthesized within the cisternae of the Golgi apparatus. Galactans accumulate in specialized Golgi-derived vesicles during cell-wall thickening. Recent works suggest that membrane-recycling pathways operating between the Golgi and ER play an indispensable role in Golgi maintenance and biogenesis.

QUESTIONS

Long Questions:

1. Describe the morphological structure of Golgi complex.
2. How secretory vesicles move from ER to TGN and what are the biochemical modifications during this passage.

Short Questions:

1. Name different components of Golgi complex.
2. How would you justify that Golgi components are dynamic in nature.
3. What do you mean by CGN and TGN?
4. What is Zone of exclusion?
5. What happens when ER to Golgi trafficking is inhibited?

6. What is GERL?
7. Which proteins are known to be associated with trafficking of Golgi vesicles?
8. What do you mean by Golgi blobs?
9. What is the basic difference between constitutive and regulatory secretory pathway?
10. Clarify the role of addressing tags for the trafficking of vesicles through Golgi.
11. What is the basic difference between transitional vesicles and secretory vesicles.

MCQs:

1. **The vesicles carrying proteins from ER to Golgi are called**
   (a) secretory vesicles,
   (b) transport vesicles,
   (c) primary lysosome.
2. **Phosphorylation of the substances in the Golgi occurs in the**
   (a) TGN, (b) Middle part (c) CGN.
3. **Addition of various sugars in the Golgi occurs in the**
   (a) CGN, (b) middle part, (c) CGN.
4. **Addition of acid phosphatase to the protein occurs in the**
   (a) TGN, (b) CGN (c) middle part.
5. **TGN produces**
   (a) Vesicles, (b) Primary lysosomes, (c) Both.
6. **Zone of Exclusion is associated with**
   (a) ER, (b) Lysosome, (c) Golgi.
7. **Acrosome of sperm is formed from**
   (a) Peroxisome, (b) Golgi comples, (c) ER.
8. **Cortical sacs of ovum are formed from**
   (a) Peroxisome, (b) Golgi comples, (c) ER.
9. **Phragmoplasts are derived from**
   (a) ER. (b) Lysosome, (c) Golgi complex.

Answers to MCQs:

1. b; 2. c; 3. b; 4. a; 5. c; 6. c; 7. b; 8. b; 9. c .

Chapter 11.3

Mitochondria

The most important eukaryotic cellular organelle next to nucleus is the mitochondria but according to ***endosymbiotic theory*** (Margulis, 1960) during the early period of the origin of cell, a prokaryotic cell phagocytosed a heterotrophic bacterium which survived and subsequently over billion of years have been maintaining a symbiotic relationship with the host prokaryote cell citing an example of ***primary endosymbiosis.*** This theory led to the establishment of ***Serial Endosymbiotic Theory (SET)*** which proposed that "symbiotic consortiums" of prokaryotic cells were the ancestors of eukaryotic cells and accordingly the host cell underwent infoldings of plasma membrane that gave rise to membrane coated organelles and nucleus generating a heterotrophic eukaryote. Ultimately, both the prokaryotic host and the endosymbionts developed inter dependence. The ingested bacteria performed oxidative metabolism generating ATP, necessary for the survival of the host cell. Ingestion of photosynthetic bacteria by endosymbiontic prokaryotes or eukaryotes led to the evolution of the ancestors of eukaryotic plants and photo synthetic protists. The prokaryote performing the role of oxidative metabolism evolved into mitochondrion and the photosynthetic bacterium gave rise to chloroplast. Considerable gene transfer took place between the host cell and endosymbionts in the course of time. Margulis' prediction regarding the prokaryotic nature of mitochondrial and chloroplast genome has been confirmed later on. Endosymbiotic prokaryotes are believed to transfer certain genes to the nuclei of their host cells in a process known as ***endosymbiotic gene transfer***. This led mitochondria and chloroplasts fully dependent on their host's DNA to direct synthesis of most of their components.

Mitochondria contains all the enzymes necessary to oxidize the assimilated food to produce energy to be used in all vital activities. In higher plants all mitochondria are collectively called *chondriome*. Due to this potentiality, it is referred as ***'Power house of the cell'***. In addition to the fundamental metabolic processes such as

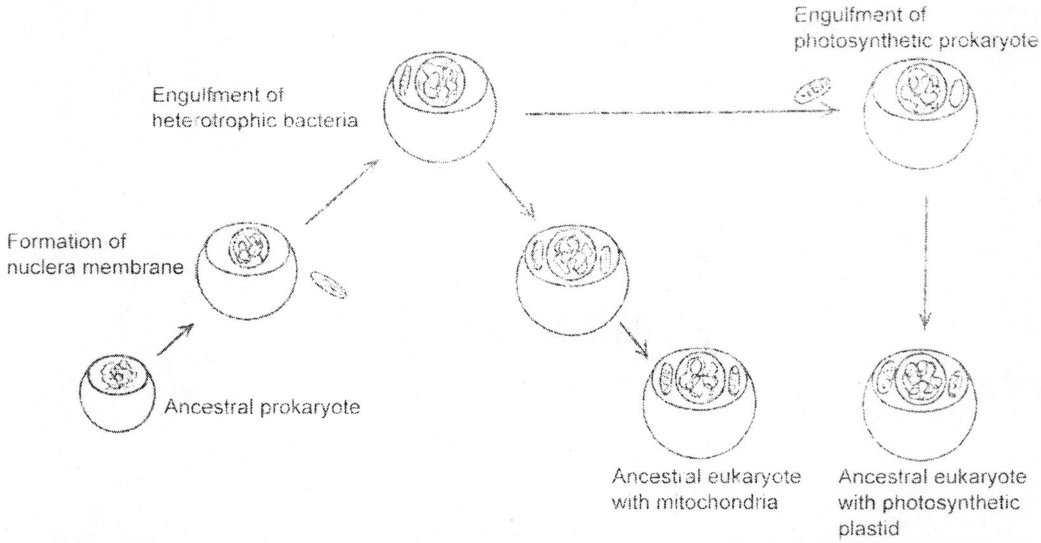

Fig. 11.3.1 : Gradual incorporation of mitochondria and chloroplast forming bacteria in eukaryotes.

electron transport oxidative phosphorylation, β-oxidation of fatty acids, the citric acid cycle, part of the urea cycle and heme synthesis performed by mitochondria very recently its role in apoptosis, aging, signal transduction and cancer has been well documented. They have their own genome necessary for synthesis of specific proteins and auto reproduction. Mitochondrial half-life is thought to be about 5-12 days.

Long, scattered, self replicating organelle invested with double layered membrane that actively participate in internal respiration and many vital activities.

11.3.1. MORPHOLOGY

1. They may be fibrillar, thread like, granular, sac like, club / rod shaped or tennis racket shaped with a stalk and a globular head or may be branched like tubular networks. They may be fragmented during cell division. Inter-mitochondrial fusion often occurs enabling recombination of mtDNA.

2. They measure 7μ / 0.5-1μ.

3. They may be ***dark*** or ***bright*** depending on the protein content of the matrix of mitochondria.

4. Depending on the appearance of the inner membrane, mitochondria are classified as:

- ***crista type*** (most common, with crest-like protrusions),
- ***tubular type*** (with larger tubular protrusions found in steroid synthesising cells),
- ***prismatic type*** (here the tubuli show triangular profiles present only in rat astrocytes)
- ***saccular type*** (shows saccular pearl-like protrusions of the tubuli, present only in cells of the Zona fasciculata of the suprarenal gland).
- ***intermediate forms*** between crista and tubular type are also found.

11.3.2. ULTRA STRUCTURES

1. A double-layered membrane invests the organelle; each layer is 60-70Å thick. In fact the inner membrane is derived from the bacterial plasma membrane. There is a 60-80Å wide *peri*

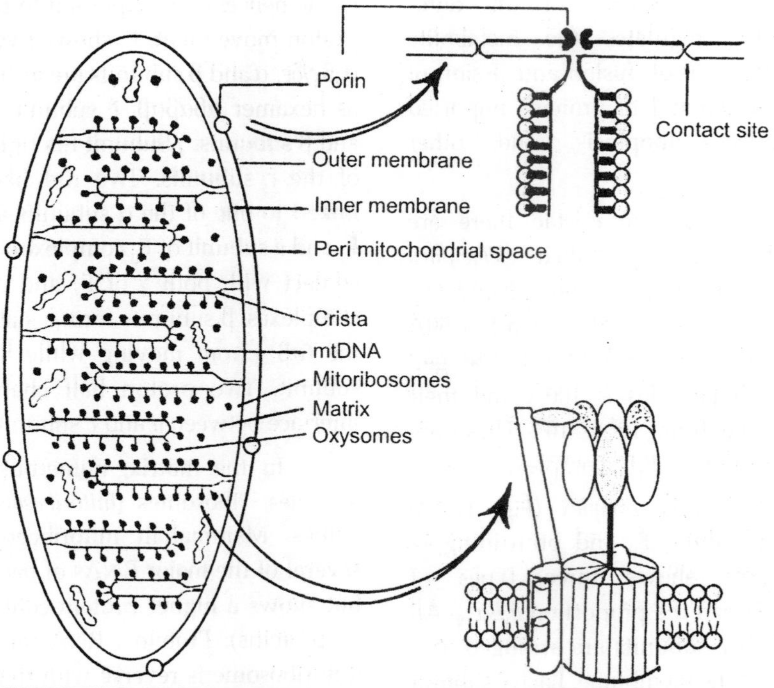

Fig. 11.3.2 : Morphological features of mitochondria.

mitochondrial space / *Outer chamber* between these two layers (Fig. 11.3.2).

2. The portions invested by the inner membrane is called *inner chamber.* The outer membrane can be dissociated containing only the inner membrane and the matrix. This is called "*mitoplasts*" which remains filled up with a jelly like *matrix.*

3. The two membranes show qualitative and quantitative differences. The outer membrane shows 50% phospholipids and contains a variety of enzymes involved in oxidation of epinephrine, degradation of tryptophan, and elongation of fatty acids. It contains numerous integral channeled proteins called *porins,* that allow the passage of ions and small molecules through the outer membrane. The inner membrane, in contrast, contains more than 100 different polypeptides and has a very high protein content. Additionally it is rich in the phospholipid, cardiolipin. The inner membrane is highly impermeable; almost all ions and molecules require special transporters to enter or exit the matrix.

4. The outer and inner boundary membranes are associated at many sites called *contact sites* where both membranes are in direct contact to each other showing channels through which translocation of different substances occur. Contact sites also regulate Ca^{++} homeostasis.

5. The inner membrane is divided into outermost *inner boundary membrane;* occur in close apposition to the outer membrane and *cristae membrane* which is the invagination of the inner boundary membrane that protrudes into the matrix as tubular, lamellar or triangular *cristae.* Cristae are connected to the inner boundary membrane by narrow tubular *cristae junctions* (Fig. 11.3.2). These a) limits the lateral diffusion in and out of the cristae membrane; b) act as a kind of reservoir for inner membrane components that, on demand, become available

to interact with components of the outer membrane; Cyt-c. regulates the metabolite exchange with the cytosol, fusion and fission of mitochondria, the amount of proteins imported into mitochondria, apoptosis, and other processes.

6. On the wall of the cristae there are numerous stalked granules called *elementary particles /Oxysomes / Farnandez Moran sub units/ F1 particles*. These show a base (80Å diam), a 50Å long stalk and a head (100Å diam). The gap between two such particles is 100Å and their number may range from 10^4 - 10^5. These are actually ATP synthese or F_1F_0 complex.

ATP synthase or F_1F_0 complex (Fig. 11.3.3) shows membrane bound F_0 and protruding F_1 complex. F_0 part shows three types of components - *a, b, c* in the proportion $a_1b_2c_{14}$. All the components of *c* subunits are arranged as a ring in the plane of the membrane. Each *c* subunit contains two membrane spanning α-helices. An aspartate residue, Asp61 lies in the centre of one of the helices. It is supposed to participate in the proton move ment. F_1 shows five polypeptides – $\alpha_3\beta_3\gamma\delta\varepsilon$. α and β subunits are arranged alternately as hexamer αβαβαβ. δ subunit is linked with a and β subunits. γ Subunit fits rigidly in the centre of the *c* subunits. Two rod like β subunits is linked to one of the α subunits and δ subunit of F_1 and *a* subunit of F_0 complexes. ε subunit makes contact with both γ of F_1 and *c* subunits of F_0 complexes. β subunits form a stator that prevents the $(\alpha\beta)_3$ from moving while it rests on the γ subunit. Two proton half channels lie at the interface between *a* and *c* subunits of F_0 complex.

7. In the matrix, numerous phospholipoid granules, ribosomes (*mitoribosomes)* having 55S values. Mammalian mitoribosomes (55S) lack several of the major RNAs of bacterial ribosomes but shows a higher protein content (as many as 80 proteins). Protein : RNA ratio (69:33) in the mitoribosome is reverse with that of prokaryotic ribosomes (33:67). They are physically larger than bacterial ribosomes.The mitochondrial ribosome

Table: 11.3.1. Comparative accounts of mitochondrial membranes:

Outer mitochondrial membrane	Inner mitochondrial membrane
1. The membrane appears smooth.	1. The membrane contains many particles.
2. Protein content is roughly equal to that of lipids.	2. Protein content is comparatively very high (80%).
3. It is permeable to metabolites due to the presence of porins.	3. It is selectively impermeable. Porins are absent. Instead carrier and other transport proteins are present.
4. Cholesterol and other lipids are present. Cardiolipin is absent.	4. Cardiolipin is present.
5. Infoldings are absent.	5. It is thrown into folds and tubules called cristae. Cristae may be branched. Cavity of the cristae is continuous with the intermembrane space.
6. ETS is absent.	6. It bears ETS and oxysomes.
7. Enzymes: Monoamine oxidase, NADH cytochrome reductase, Hexokinase-2, Glycerophosphate acyl transferase.	7. Enzymes: ATPase, Succinic dehydrogenase, Cytochrome oxidase, 3-hydroxybyuerase dehydrogenase, Malate and isocitrate dehydrogenase, Fumerase and Aconitase.

possesses inter subunit bridges; it has a gate like structure at its mRNA entrance and it has a poly-peptide exit tunnel suggesting a unique nascent-polypeptide exit mechanism.

8. Mitochondrial DNA molecules vary in size from 25,00000 bp - 5,966 bp [in human, it is 16, 569 bp long]. Each mitochondrion contains 5-11, circular DNA molecules (in *Chlamydomonus* and *Paramoecium* it is linear). Vertebrate mt-genome shows 37 genes, (2 codes for rRNA, 22 for tRNA and 13 for polypeptides involved in oxidative phosphorylation). Invertebrate mtDNA is of same size but have different genetic organization.

ER-mitochondria association: ER and mitochondrial networks are fundamental for the maintenance of cellular homeostasis and for determination of cell fate under stress conditions. Recent studies revealed the interaction of these networks at some specific points called MAM (Mitochondria Associated Membranes) (Pinton *et al*, 2011). It plays vital role in the coordination of calcium transfer, regulation of mitochondrial fission and inflammasome formation showing some apoptotic factors inducing pyroptosis (Marchi *et al*, 2014, Lewis *et al*, 2016).

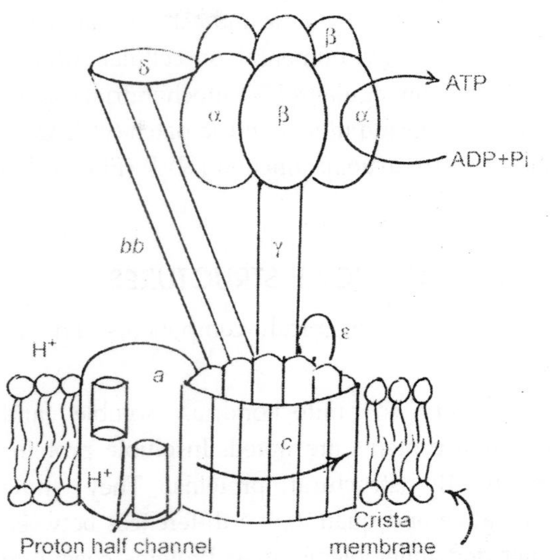

Fig. 11.3.3 : Structure of ATP synthese.

Mitochondrial membrane biogenesis and lipid metabolism require phospholipid transfer from ER to mitochondria at inter organelle contact sites where specific ER membrane protein interacts with the mitochondrial translocase of the outer membrane Tom5 complex which is necessary for phospholipid synthesis and cell growth (Voss *et al.* 2012; Lahiri *et al*, 2014).

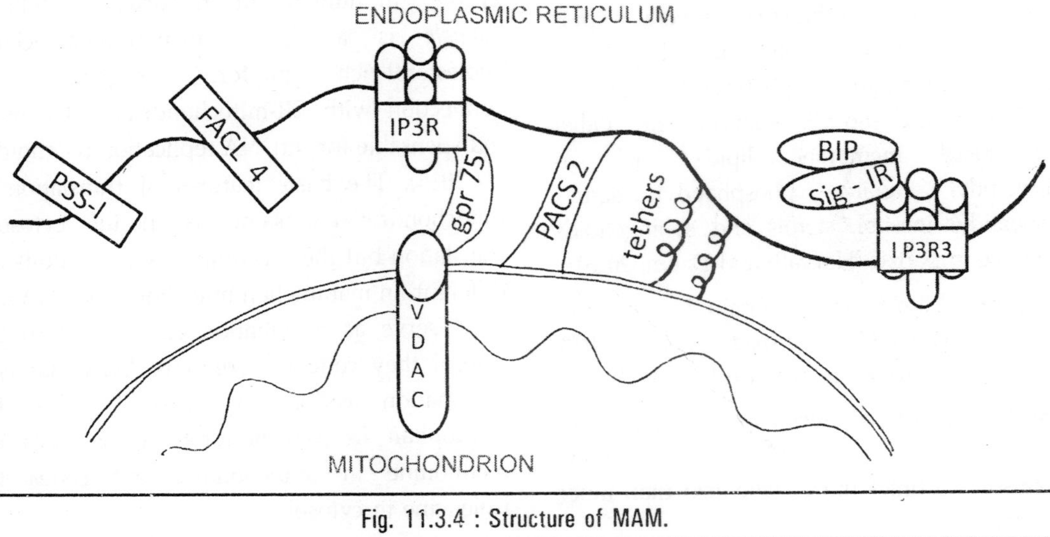

Fig. 11.3.4 : Structure of MAM.

Contact sites are dynamic and can be changed in response to environmental or cellular stimuli. Lam 6 protein regulates ER-mitochondrial contact sites, vacuole and mitochondria patch (vCLAMP) and nuclear vacuolar junction (NVJ) (Elbaz-Alon *et al*, 2015).

11.3.3. CHEMICAL STRUCTURES

The main chemical components are as follows:

Protein: In mitochondria, soluble and insoluble proteins are noted. Insoluble proteins are mostly structural proteins. They show qualitative and quantitative differences between outer and inner membranes. Proteins constitute 50% of the outer and 80% of the inner membranes respectively. In the inner membrane, 14 while in the outer membrane 21 types of proteins are noted. Moreover the outer membrane contains more porin protein molecules. Soluble proteins mainly constitute the enzymes.

Lipid: The main lipid content of the mitochondrial membrane is phospholipid, cholesterol and minute quantity of neutral fat. The outer membrane contains 40% but inner membrane contains 20% lipid. Moreover outer membrane contains more cholesterol, phosphatidyl inositol and minute quantity of cardiolipin. Mitochondria are capable of synthesizing several lipids autonomously such as phosphatidyl glycerol, cardiolipin and in part phosphatidyl ethanolamine, phosphatidic acid and CDP-diacylglycerol. Other mitochondrial membrane lipids such as phosphatidyl choline, phosphatidyl serine, phosphatidyl inositol, sterols and sphingolipids have to be imported (Horvath and Daum, 2013).

Enzyme: Mitochondria contain 120 types of enzymes of which 37% is of oxidoreductase, 10% is of ligase and 0.09% is of hydrolase type.

Sialic acid: Mitochondrial membranes contain much more sialic acid. The outer membrane contains 4 – 5 times more sialic acid than inner membrane.

11.3.4. MITOCHONDRIAL GENOME

Mitochondria have their own genome and it resembles with that of prokaryotes and contains 1% of the total cellular DNA. The genomes consist of a single circular molecule of DNA with more (G) guanosine and cytosine (C) than other nucleosides but there is no histone associated with the DNA. mtDNA varies in size from 6 kb (in *Plasmodium*) to 2500 kb in some flowering plants. Each mitochondrion contains 5-11 DNA mole- cules that are mostly circular (in the alga *Chlamydomonus* and the ciliate *Paramoecium* they are linear). Vertebrate mt-genome shows 37 genes, packed in 16 – 17 kb circle leaving no space between genes. Human mtDNA is 16,659 bp long and contains 37 genes (2 code for rRNA, 22 for tRNA and 13 for polypeptides involved in oxidative phosphorylation). Nuclear genes encode only some of the necessary proteins. Human mtDNA shows one denser strand and accordingly are referred to as H (heavy) and L (light) strands. Unlike nuclear transcription, both strands are transcribed; H strand encodes two rRNAs, 14 tRNAs and 12 polypeptides. The L strand encodes 8 tRNAs and one polypeptides. Only mother transmits mtDNA to the offspring (since sperm contributes no mitochondrion to the zygote). In human mitochondria, the helicase *Twinkle* is a tightly membrane-associated cholesterol-rich platform, organized in connection with ER-mitochondrial junctions is indispensable for mtDNA replication (Gerhold *et al*, 2015). The basic features of translation in mitochondria are same as in the cytosolic translation but the meaning of some codons are different. In mammalian mitochondria, AGA and ACG serve as termination codons but in the cytosol they code for arginine, UGA that is a termination codon in cytosol, codes for tryptophan in mitochondria, AUA codes for methionine in mitochondria but codes for isoleucine in cytosol.

Unlike nuclear RNA synthesis most of the plant mt-mRNAs undergo editing during which some nucleotides are changed (usually C → U but U → C is very rare). During editing small RNA molecules that are partially complementary to the mtDNA transcripts serve as guides for the editing purpose. These RNAs are called *guide RNAs (gRNAs).*

11.3.5. FUNCTIONS

11.3.5.1. Oxidative phosphorylation by Mitochondria:

Oxidative phosphorylation is the process that couples the removal of H^+ from one molecule and giving phosphate molecules to another molecule. Mitochondria use a combination of pumps, carriers, electron transporting proteins and enzymes to convert energy extracted from the nutrients to ATP. Outer membrane contains porins that allow the passage of metabolites required to run Krebs cycle and for the synthesis of ATP. The inner membrane and cristae are selectively permeable but impermeable to H^+ and other molecules in the energy conversion reactions. To gain a thorough knowledge of glucose oxidation, following reaction steps should be properly known.

Glycolysis: The best energy-supplying nutrient is sugar. Other foods like fats can also be broken down and used as fuel. Inside the cell, glucose is broken down to make ATP in two pathways. The first pathway requires no oxygen i.e. anaerobic, it is called ***glycolysis;*** it occurs in the cytoplasm. Cells use the glycolysis pathway to extract energy from sugars (glucose) and store it as ATP and NADH. The end product of glycolysis is pyruvic acid, which can be used in other metabolic pathways to yield additional energy. Pyruvate is carried into the mitochondria and there it is converted into acetyl Co-A that enters the Krebs cycle.

During the initial phases, Glucose is converted into Glyceraldehyde 3 phosphate (3C). Up to this step it is called *Preparatory phase.* The latter phases involving the conversion of glyceraldehyde-3-phosphate to pyruvic acid is called *Pay off phases.* However the main steps of glycolysis are as follows:

i) Glucose is phosphorylated into Glucose-6-phosphate with the help of ATP and Mg^{++} as the co-factor and by the enzyme *hexokinase.* Glycogen in animal cells is broken down into numerous molecules of Glucose 1 phosphate with the help of the enzyme *phosphorylase* that are later converted into glucose-6-phosphate with the help of the enzyme *phosphoglucokinase* and Mg^{++} as the co-factor.

ii) Glucose-6-phosphate is isomerised to Fructose-6-phosphate by the enzyme *phosphohexose isomerase.* The reaction is bidirectional and involves conversion of an aldose to a ketose.

iii) Fructose-6-phosphate is again phosphorylated by ATP to form Fructose 1, 6-bisphosphate by the enzyme *phosphofructokinase* and the co-factor Mg^{++}.

iv) In this phase the six carbon glucose is cleaved into two molecules of 3 carbon components Dihydroxyacetone phosphate. and Glyceraldehyde 3 phosphate by the enzyme *aldolase.* It is also a bidirectional reaction. Both triose phosphates are interconverted by *phosphotriose isomerase* enzyme.

v) After the formation of these two components, the steps of same reactions proceed in two pathways one following the conversion of Dihydroxyacetone phosphate and other following the conversion of glyceraldehyde-3-phosphate.

vi) Glyceraldehyde-3-phosphate is oxidized to 1,3-phosphoglycerol phosphate by the enzyme *3-phosphoglyceroldehyde dehydrogenase.* The hydrogen released in this reaction is accepted by NAD to form NADH + H^+.

vii) 1,3-phosphoglycerol phosphate is converted into 3, phosphoglycerate with the help of the enzyme 3, *phosphoglycerate kinase*. This reaction requires the presence of Mg^{++} as co-factor and it yields one molecule of ATP with the expence of ADP + Pi.

viii) 3-phosphoglycerate is converted into 2-phosphoglyceromutase with the help of the enzyme *phosphoglyceromutase* and Mg^{++} as Co-factor.

ix) 2-phosphoglyceromutase is converted into phosphoenol pyruvate with the help of the enzyme *enolase* and Mg^{++} as the co-factor.

x) Phosphoenol pyruvate is converted into Pyruvic acid (3C) with the help of the enzyme pyruvate kinase. This reaction also produces one molecule of ATP with the release of inorganic phosphate that reacts with ADP.

ATP generation in Glycolysis: In the preparatory phase, one molecule ATP is utilized during the conversion of Glucose to Glucose-6-Ph. and one molecule during the conversion of Fructose-6-phosphate to Fructose,1-6-bis-phosphate. In the Pay-off phase 2 molecules of ATP are produced during the conversion of 1,3,phosphoglycerol phosphate to 3, phosphate and phosphoenol pyruvate to pyruvic acid. Thus 4 molecules of ATP are produced in the Pay-off-phase. Thus altogether two molecules of ATP are directly produced in the whole process. In addition two molecules of NADP are produced that can yield 5 molecules of ATP in the electron transport system. Thus altogether 7 molecules of ATP are produced in the glycolysis process. One molecule of ATP yields 7.3 kcal energy; therefore from glycolysis one molecule of glucose yields 7 × 7.3 Kcal energy.

Regulation of Glycolysis: Glycolysis is regulated by the concentration of glucose in the blood, the relative concentration of critical enzymes, the competition for the intermediate products of glycolysis and the levels of certain hormones in the blood stream. In humans and other mammals high sugar levels stimulate the pancreas to produce insulin, which enhances the entry of glucose into the cell and increases the production of the glycolysis enzymes. These actions stimulate glycolysis and lower blood glucose levels. Alternatively, low blood sugar levels depress insulin production and stimulate the pancreas to release the hormone glucagon, which inhibits glycolysis.

i) At the entry point of glucose is converted into glucose-6-phosphate. The activity of hexokinase is feedback inhibited by its product, so the phosphorylation of glucose will be inhibited if there is a buildup of glucose-6-phosphate. In mammalian cells, the breakdown of glycogen is regulated by covalent modification of glycogen phosphorylase. This regulation reduces the rate of formation of glucose-6-phosphate. It has other metabolic fates than simply leading to pyruvate. It can be used to synthesize ribose for DNA and RNA bases.

   The hexokinase causes conversion of glucose to glucose-6-ph., Hexokinase competes with another enzyme, glucokinase, that turns glucose into glucose-1-ph. This substance forms the glucose polymer glycogen, as well as fat and cholesterol. High levels of G-6-P inhibit hexokinase but have no effect on glucokinase. Therefore, high levels of G-6-P cause the cell to store excess glucose rather than metabolize it in glycolysis. When glucose levels drop, hexokinase activity resumes and causes metabolic utilization of glucose.

ii) The activity of phosphofructokinase (PFK) is regulated by the energy charge of the cell, i.e. the presence of ATP in the cell. ATP inhibits the PFK reaction by raising the K_m for Fructose-6-phosphate. AMP activates the reaction.

   Thus, when energy is required, glycolysis is activated; when energy is plentiful, the reaction is slowed down.

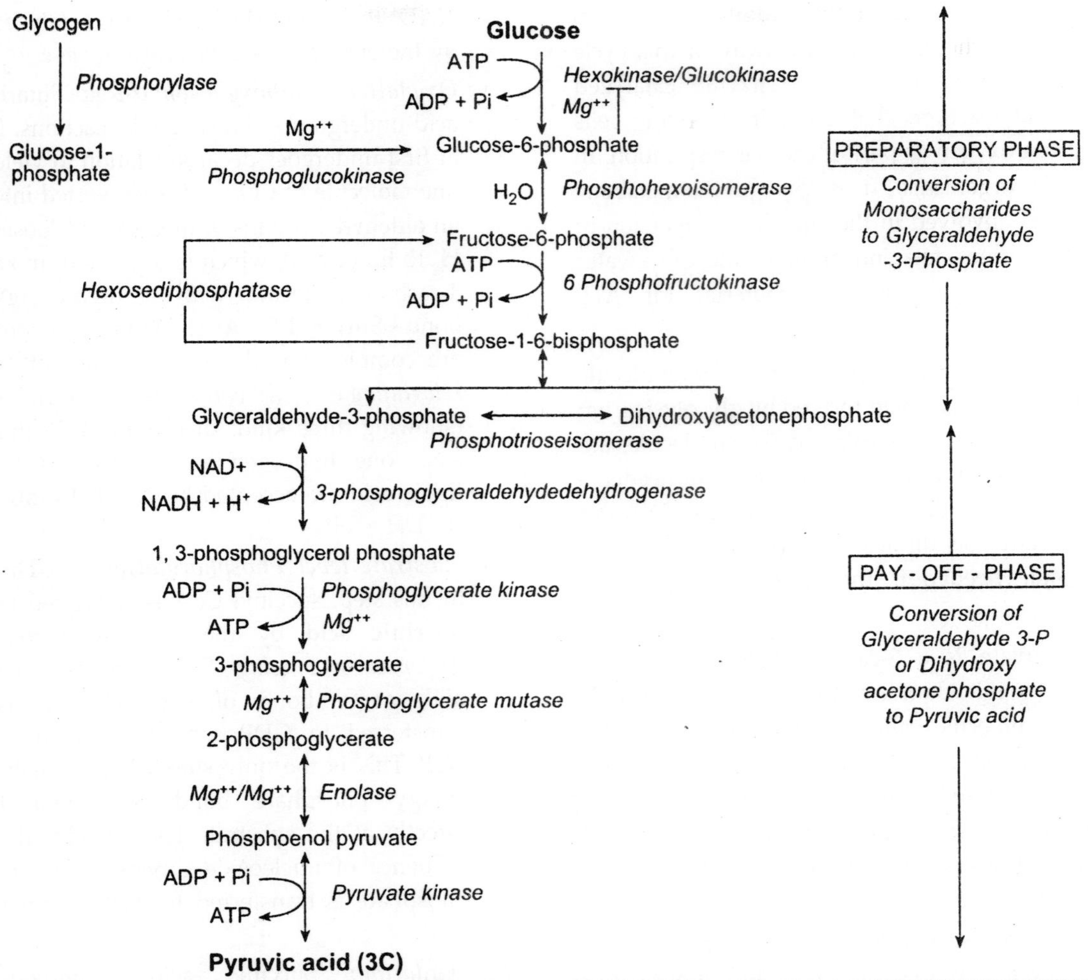

Fig. 11.3.5. Steps of Glycolysis.

PFK is inhibited by high levels of ATP, low pH levels and high levels of citrate, a byproduct of TCA cycle. Conversely, high levels of AMP and Fructose-2-6 bisphosphate (F2-6-BP) activate PFK and increase the rate of glycolysis. Glucagon inhibits production of F2-6-BP, thereby slowing down glycolysis when glucose is scarce.

iii) In the conversion step - phosphoenol pyruvate to pyruvic acid the concentration of ATP in the cell is increased and high ATP levels inhibit PK. ATP, acetyl CoA and fatty acids inhibit pyruvate kinase, serving as an indicator that alternative energy sources are available for the cell. PK is activated by Fructose 1-6-bisphosphate. As glycolytic inter mediates feed into several other pathways, the regulation of glycolysis occurs at more than one point. e.g. dihydroxyacetone phosphate is the precursor to the glycerol component of lipids. Thus glycerol is needed for the formation of triglycerides.

Krebs cycle: The ***citric acid cycle*** (also called ***Tricarboxylic acid cycle*** or *TCA* cycle, or the

Krebs cycle, after Hans Adolf Krebs who discovered the enzymatic pathway of this cycle in 1936) It is a series of enzyme catalyzed chemical reactions that occur in all living cells that use oxygen as part of cellular respiration. In aerobic organisms, It is part of a metabolic pathway involved in the chemical conversion of carbohydrates, fats and proteins into CO_2, water and to supply the raw material for ATP production.

To run Krebs cycle properly the first molecule needed is pyruvic acid formed by glycolysis from glucose. Next some electron carrier molecules are needed. Two types of such carriers are needed - nicotinamide adenine dinucleotide (NAD^+) and flavin adenine dinucleotide (FAD^+). The third molecule needed is the electron acceptor the oxygen.

Pyruvate is a 3-carbon molecule. After it enters the mitochondria, it is broken down to a 2-carbon molecule called **Acetyl CoA** by *pyruvic dehydrogenase.* This releases one molecule of CO_2 and the released H molecule yields NADH + H^+ yielding 2.5 molecules of ATP in Electron Transport System. The main steps of Krebs cycle are as follows:

i) *Condensation:* Acetyl CoA enters the Krebs cycle by joining to a 4-C molecule - oxaloacetate to form a 6-C **citric acid.** It is catalyzed by the enzyme *citrate synthetase.*

ii) *Dehydration:* Citric acid at first undergoes dehydration and is converted into **cis-aconitic acid.** It is catalyzed by the enzyme *aconitase.*

iii) *Hydration I:* With the addition of one molecule of water and by the cooperation of the enzyme – aconitase. Cis-aconitic acid forms **isocitric acid.**

iv) *Decarboxylation I:* This step follows a decarboxylation reaction involving conversion of isocitric acid is converted to a 5-carbon molecule – **α-Ketoglutaric acid** and one molecule of CO_2 and one atom of hydrogen is liberated which is accepted by NAD^+ to form NADH + H^+. It is catalyzed by the enzyme *isocitrate dehydrogenase.*

v) *Oxydative decarboxylation:* α-KetoGlutaric acid undergoes a series of sub reactions. It at first undergoes decarboxylation to yield one molecule of CO_2 and is converted into an aldehyde. It joins with CoA and looses H_2 to lipoic acid, which is converted into a 4-carbon molecule with a high-energy bond - **Succinyl Co-A.** The whole reactions are completed under the influence of *α-ketoglutarate dehydrogenase* complex (showing three kinds of enzymes). In this step, one hydrogen atoms released by lipoic acid are accepted by NAD^+ to form NADH + H^+.

vi) *Substrate level Phosphorylation of GDP:* In this step, succinyl CoA is converted to **Succinic acid** by the enzyme *succinic thiokinase* or *Succinyl CoA synthetase.* The high-energy bond of succinyl CoA is transferred to GDP converting them to GTP. This is the only step where a high-energy phosphate bond is generated directly at the substrate level. Under the influence of nucleoside diphosphokinase, phosphate is transferred to ADP forming ATP.

vii) *Oxidation:* Succinic acid undergoes dehydrogenation and is converted to **Fumaric acid** by the enzyme *succinic dehydrogenase.* FAD directly accepts hydrogen atoms released by this reaction to form $FADH_2$.

viii) *Hydration II:* In this step, fumaric acid is hydrated and is converted to **Malic acid** under the influence of fumarase.

ix) *Oxidation:* Malic acid is dehydrogenated by the enzyme *malic dehydrogenase* to form **Oxaloacetic acid.** The hydrogen atom released in this step is accepted by NAD^+ to form NADH + H^+.

Each NADH molecule contributes enough proton motive force to generate *2.5 molecules of ATP.* Each $FADH_2$ molecule produces *1.5 molecules of ATP*. Altogether 2 molecules of NADH and 2 molecules of ATP are formed from glycolysis; 2 molecules of NADH are produced during conversion of pyruvic acid to acetyl CoA; 6 molecules of NADH are formed from Krebs cycle; 2 molecules of $FADH_2$ and 2 molecules of ATP (direct) molecules are formed during oxidation of one molecule of glucose produce 32 ATP molecules (Berg *et al*, 2002). This ATP yield is the theoretical maximum value; in fact there is some proton leakage across the membrane, resulting in some what lower ATP yields.

Electron transport system: As the Krebs cycle runs, H^+ are donated to two carrier molecules NAD or FAD forming NADH + H^+ and $FADH_2$ respectively and are carried to the electron transport chain found in the mitochondrial cristae. At each site there is a hydrogen (or proton) pump that transfers H^+ from one side of the membrane to the other. The pumps then transport the H^+ to the inter membrane space where they accumulate in high concentration to fuel the ATP pumps. The ATP synthase uses the energy of the H^+ - also called *proton gradient* to

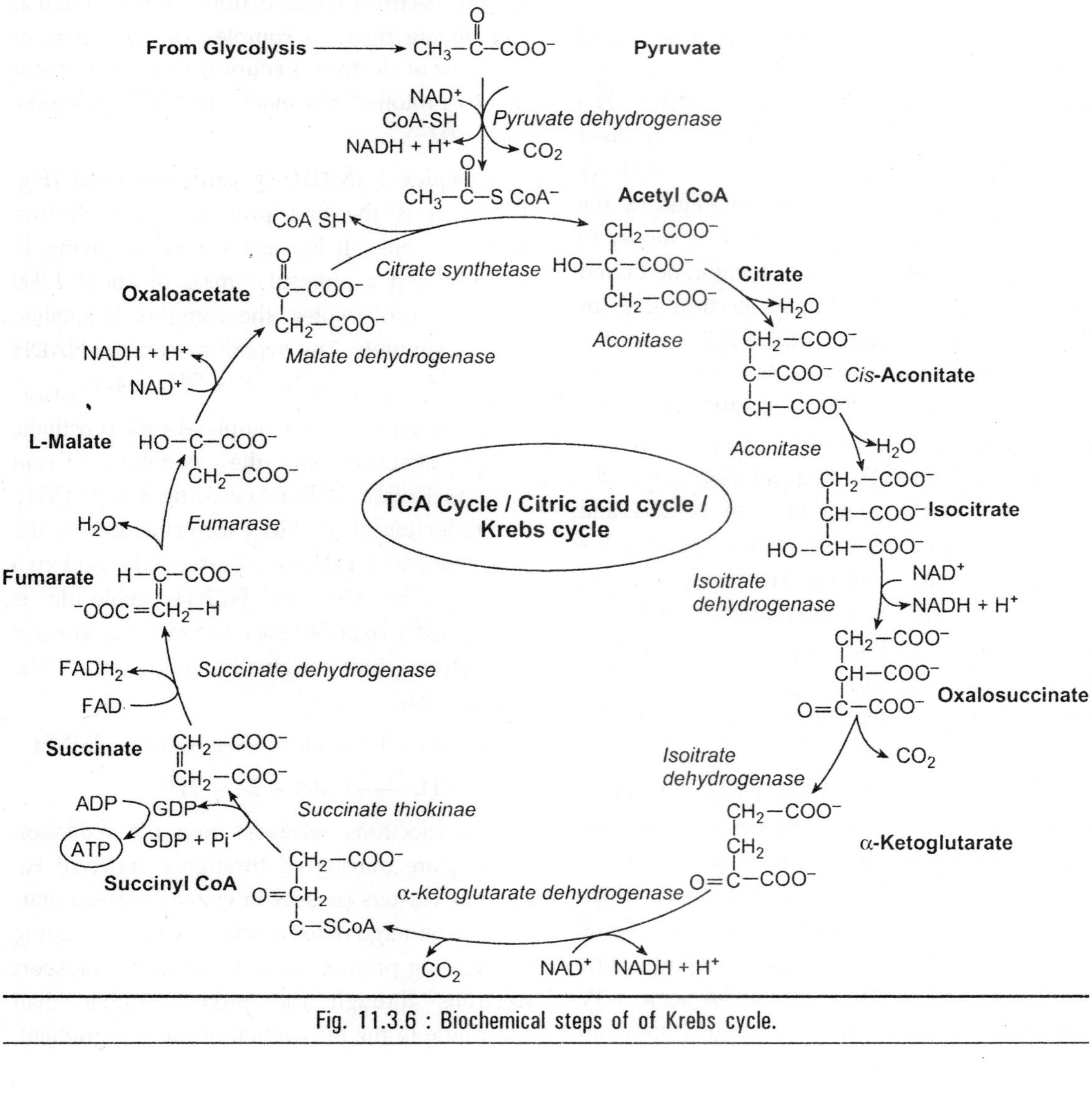

Fig. 11.3.6 : Biochemical steps of of Krebs cycle.

form ATP from ADP and inorganic phosphate (Pi). In this way "oxidation" is coupled to "phosphorylation". The hydrogens that are pumped back into the matrix by the ATP pump then combine with the oxygen to make water. And that is very important because, without oxygen, they will accumulate and the concentration gradient needed to run the ATP pumps will not allow the pumps to work. The whole idea behind this process is to get as much ATP out of glucose (or other food products) as possible. From the Krebs cycle, we get 25 ATP molecules out of one molecule of glucose converted to pyruvate.

Mitochondria show a basic mechanism called ***chemiosmosis*** to generate ATP. This process involves the formation of proton (H^+) concentration gradient and voltage gradient across the membrane. These are collectively called *proton motive force*. It is generated by the stepwise movement of electrons from higher to lower energy states via inner membrane bound electron carriers to the ultimate electron acceptor the oxygen. Mitochondria utilize F_0F_1 complex – *ATP synthase* to synthesize ATP. Protons flow through the ATP synthase from perimitochondrial space to the matrix driven by proton motive force. Synthesis of most of the ATP is coupled to the reoxidation of NADH and $FADH_2$ by O_2 in a stepwise process involving the respiratory chain called ***electron transport chain (ETC)*** that consists of many carriers. During this process protons (H^+) from the matrix are pumped to the perimitochondrial space through these carriers. The pumping causes matrix (-) than the inter membrane space.

Flow of electrons from NADH and $FADH_2$ to O_2 occurs through four multiprotein complexes. Each complex spans through the inner membrane and contains several prosthetic groups, which participate in the passage of electrons. Some carriers perform flow of electrons from one complex to another. Several types of heme, a Fe containing prosthetic groups are tightly bound to mitochondrial proteins and forms cytochromes. Flow of electrons occurs through oxidation and reduction of Fe atom in the center of heme molecule. Electrons move through *Cyt b, c,* c_1 *a,* a_3. All these except *Cyt-c* are components of the multiprotein complex. *Cyt-c* moves freely by diffusion in the peri mitochondrial space.

Fe-S clusters are non-heme Fe containing prosthetic group that accepts and releases electrons one at a time. Co-enzyme Q (ubiquinone) is the only electron carrier in the respiratory chain that is not protein bound prosthetic group. It is a carrier of H atoms. It accepts electrons released from complex-I and II and donate them to complex III. Transport of each pair of electron is coupled to the movement of two protons from matrix to inter membrane space (IMS).

Complex I (NADH-Q oxidoreductase) (Fig. 11.3.7): It is the first protein in the electron transport chain. It is a giant enzyme having 46 subunits and a molecular mass of about 1,000 (kDa). In prokaryotes, the complex is smaller, with 14 subunits. The overall reaction is: NADH + Q + $5H^+_{matrix} \longrightarrow NAD^+ + QH_2 + 4H^+_{cytosol}$

The electrons enter Complex-I via a prosthetic group attached to the complex, flavin mononucleotide (FMN) converting it to $FMNH_2$. The reduction of the FMN molecule, due to the acceptance of 2 electrons, causes it to bind two protons. The resulting $FMNH_2$ molecule is subsequently oxidized back to FMN and donates 2 electrons that originally reduced it. The reactions are:

$$NADH + H^+ + FMN \longrightarrow NAD^+ + FMNH_2;$$

$$FMNH_2 \longrightarrow FMN + 2e^- + 2H^+$$

The electrons released from the oxidation $FMNH_2$ are transferred through a series of Fe-sulphur clusters present in enzymes. Fe-sulphur clusters undergo redox reactions without binding or releasing protons, so serve solely to transport electrons through the protein, rather than contribute to the production of the pH–gradient.

The electrons that pass through the Fe-sulphur clusters are transferred to a coenzyme Q (ubiquinone or UQ) present in the complex. There are three 4Fe-4S clusters present between the $FMNH_2$ and Q groups. The electrons flow through all three on their way from $FMNH_2$ to Q. The basic mechanism involves the uptake of two protons as Q is reduced to QH_2 due to the transfer of 2 electrons from the afore mentioned Fe-S cluster. The two electrons that were released during the oxidation of $FMNH_2$ are passed to a Q molecule from the third Fe- sulphur cluster) – $2H^+ + 2e^- \longrightarrow QH_2$. The two high transfer potential electrons present in the QH_2 molecule are transferred to another Fe-S complex and the two protons are released into the cytosolic side; which contributes to the production of the pH gradient. Finally, these two electrons are transferred to a mobile Q molecule in the membrane that accepts the electrons and is reduced to QH_2, but is then released from the protein complex into the hydrophobic core of the membrane. This QH_2 molecule reattaches to complex III where it is oxidized to release its two high transfer potential electrons.

Complex II [Succinate-Q(ubiquinione) reductase] (Fig. 11.3.7): Succinate dehydrogenase that oxidizes in Krebs cycle is the integral component of complex-II. It is an alternative entry point to the electron transport chain. It shows 4 polypeptides and is made up of

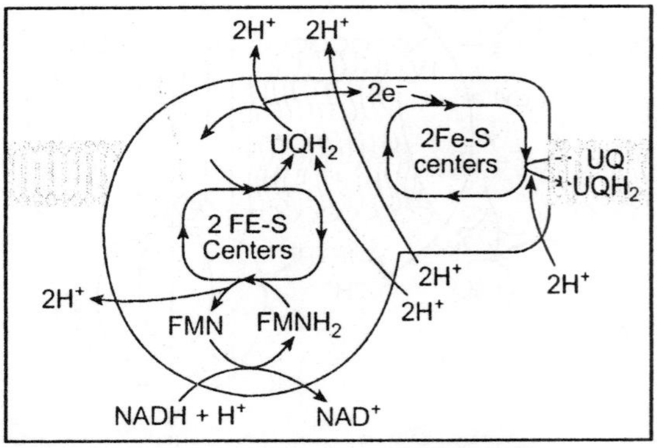

Fig. 11.3.7 : Electron transport in complex I.

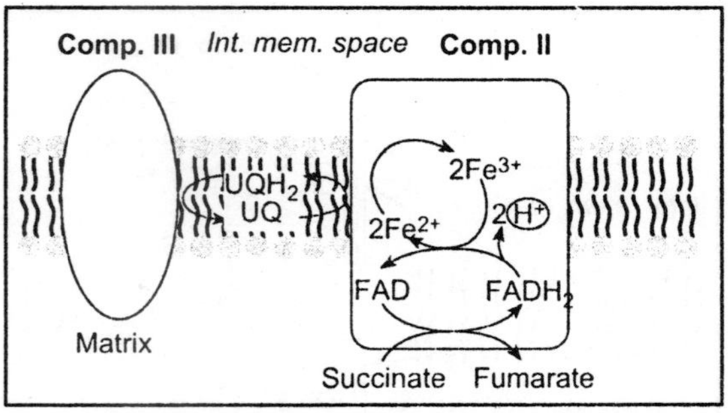

Fig. 11.3.8 : Electron transport in complex II.

coenzyme Q (ubiquinone) reductase. It oxidises succinate to fumerate. The two electrons are at first transferred to FAD and then to Fe-S and finally to CoQ to yield $CoQH_2$. Passage of electrons is not accompanied by proton translocation. The overall reaction is: Succinate + CoQ $\longrightarrow$ Fumerate + $CoQH_2$.

Complex III ($CoQH_2$ – Cytochrome-c reductase) (Fig. 11.3.8): $CoQH_2$ generated by complex – I donates two electrons to complex-III. The electrons are first transferred to Fe-S cluster and then to Cyt-c_1 or b_L and b_H.

Finally the electrons are transferred to the oxidized form of Cyt-c. Along with this passage of electrons, it releases two protons into the IMS from matrix. The overall reaction is - $CoQH_2$ + 2Cyt-c^{3+} $\longrightarrow$ CoQ + 2H+ + 2Cyt-c^{2+} A cytochrome is a kind of electron-transferring protein that contains at least one heme group. The Fe ion inside complex II's heme groups alternate between a reduced ferrous (+2) and oxidized ferric (+3) state as the electrons are transferred through the protein. The overall reaction is: QH_2 + 2Cyt-c_{ox} + $2H^+$matrix $\longrightarrow$ Q + 2Cyt-c_{red} + $4H^+_{cytosol}$

The enzyme contains 3 heme groups that participate in transferring the electron through it. The electron originates from the mobile coenzyme Q which was released from complex I. It terminates at Cyt-c, an electron carrying protein containing a heme group. Unlike Co-Q, which carries 2 electrons, Cyt-c carries only 1 reduced

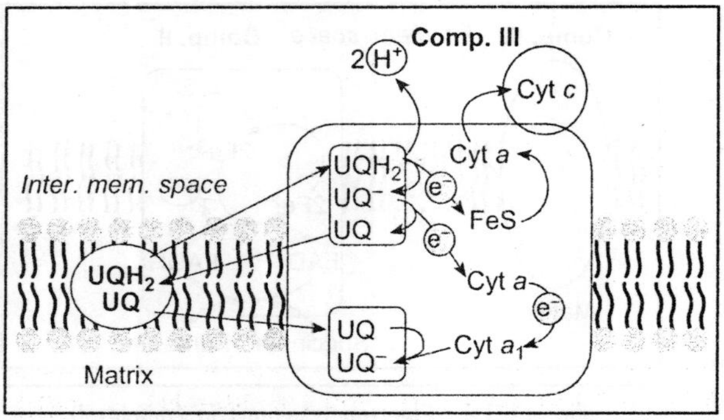

Fig. 11.3.9 : Electron transport in complex III.

electron. The coenzyme Q binds to a site inside complex III and transfers its electrons one at a time. The electron travels first through a 2Fe-2S cluster and then to a heme group held inside complex III. The electron travels from this immobile heme molecule to the heme group inside the Cyt-c protein bound to complex III. The heme group within the Cyt-c protein is now in its reduced form and the enzyme is free to diffuse away from complex III. The H^+ released during the oxidation of QH_2 are transferred to the cytosolic side of the membrane and hence serve to create the proton gradient.

Complex IV (Cytochrome-c oxidase) (Fig. 11.3.9): Two electrons from Cyt-c are transferred to the complex-IV. The electrons are first accepted by a copper ion Cu_a^2 and then to cyt-a and then to Cu_b^2-Cyt-a_3. The electrons are finally transferred to O_2 to produce H_2O. For each pair of electron transferred, two protons are translocated from matrix to IMS. The overall reaction is : $2Cyt.c^2 + 2H^+ + {}^1/{}_2O_2 \longrightarrow 2Cyt.c^3 + H_2O$

Proton gradient across the membrane: The electron transport chain is the flow of electrons from NADH to O_2. This is an exergonic process. The process terminates at the synthesis of ATP from ADP and orthophosphate. The work that drives the reaction is the net flow of protons across the membrane in which the ATP synthase is

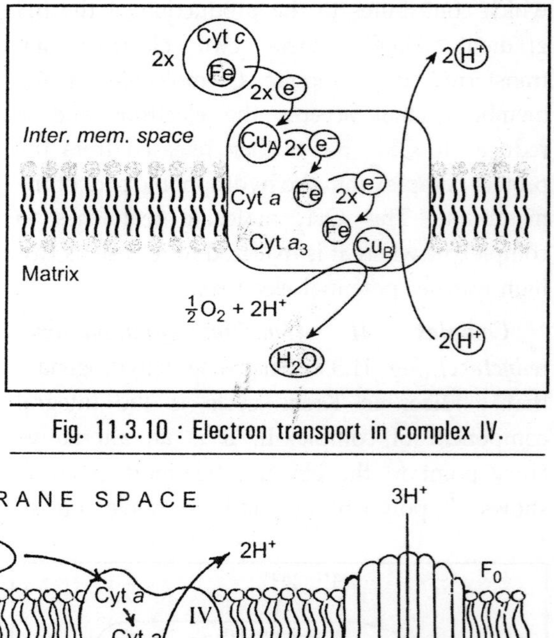

Fig. 11.3.10 : Electron transport in complex IV.

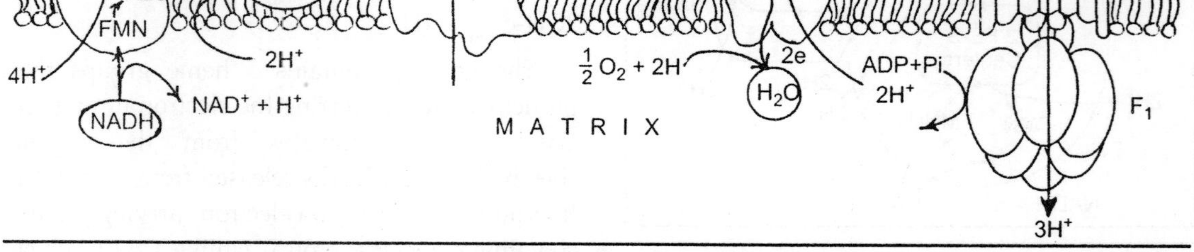

Fig. 11.3.11 : Diagrammatic representation of ETC.

embedded. The reaction creates a net difference of proton concentration across the membrane. The decrease in entropy across the membrane drives the endogonic process of ATP synthesis. This reaction is carried out by one single protein complex known as ATP synthase. As protons cross the membrane through the base of ATP synthase - the central stem rotates inside the α and β subunits. This movement of the α and β subunits cause the movement of their active sites producing of ATP.

Q-cycle (Fig. 11.3.12): Flow of electrons from complex-I to III involves a conserved cycle *Q-cycle*. $CoQH_2$ is generated by complex-I, II and III itself. $CoQH_2$ is accepted by the com plex-III in two sites Qo site, adjacent to the IMS and Qi site adjacent to the matrix. At first, one molecule of $CoQH_2$ binds to the Qo site. Once bound $CoQH_2$ releases two protons to the IMS. It then transfers two electrons to Qo site. One of the electrons is passed to Cyt-c via Fe-S cluster and the other electron moves through Cyt-b_L and b_H to Qi site. After the loss of two electrons and two protons, $CoQH_2$ is oxidized to CoQ at Qo site and dissociates. A second $CoQH_2$ then binds to the Qo site. It again follows the same pathway of proton and electron release. The electrons then move to the Qo site, there addition of two protons from the matrix gene rates $CoQH_2$. It then dissociates from the complex-III. A new CoQ molecule then occupies the void site. This cycle shuttles between CoQ and $CoQH_2$ and simultaneously pumps two protons per two electrons that mean altogether four protons to the IMS from the matrix.

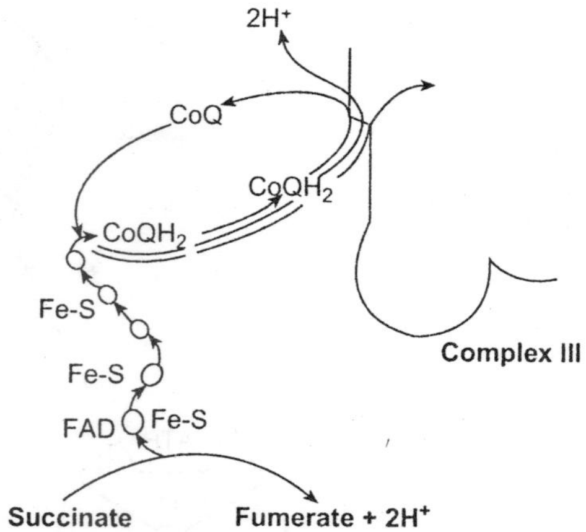

Fig. 11.3.12 : Diagramatic representation of Q cycle.

Production of ATP: The primary function of the mitochondria is the production of ATP. This is done by metabolizing the major products of glycolysis - pyruvate and NADH (glycolysis is performed outside the mitochondria, in the cytosol). Each pyruvate molecule is actively transported across the inner mitochondrial membrane into the matrix where it is combined with CoA to form acetyl CoA. Once formed, acetyl CoA enters into the *Citric acid cycle / Tricarboxylic acid (TCA) cycle / Krebs cycle*. This process creates 12 molecules of NADH and 1 molecule of FADH2, which participate in the *electron transport chain*. This energy from NADH and $FADH_2$ is transferred to oxygen (O_2) in several steps via electron transfer chain. The protein complexes in the inner membrane (NADH dehydrogenase, Cyt c reductase, Cyto-c oxidase) use the released energy to pump protons (H^+) against a gradient (the concentration of protons in the inter membrane space is higher than that in the matrix). As the proton concentration increases in the lumen of cristae, a strong *concentration gradient* is built up and protons from the lumen is again transported back into the matrix through the *ATP synthase* complex (Fig. 11.3.2). Mitochondria use the same basic mechanism called *chemiosmotic coupling or chemiosmosis* to generate ATP from ADP and Pi. It involves a proton (H^+) concentration gradient and an electric potential (voltage gradient) across the membrane collectively called *proton motive force* to drive ATP synthesis. It is generated by stepwise movement of electrons from higher to lower energy states via membrane bound electron carriers. Electrons are

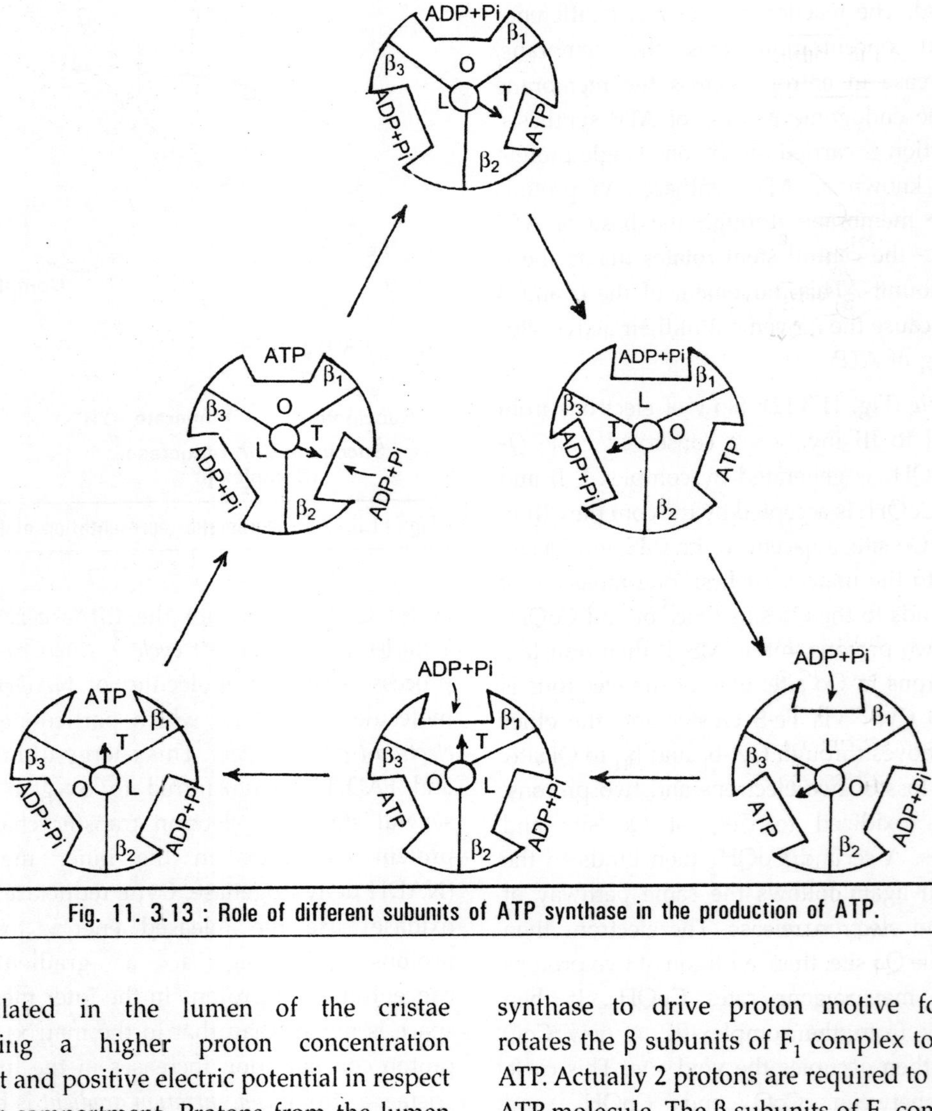

Fig. 11. 3.13 : Role of different subunits of ATP synthase in the production of ATP.

accumulated in the lumen of the cristae generating a higher proton concentration gradient and positive electric potential in respect to inner compartment. Protons from the lumen are forced through the F_0 complex of the ATP synthase to drive proton motive force that rotates the β subunits of F_1 complex to produce ATP. Actually 2 protons are required to form one ATP molecule. The β subunits of F_1 complex can bind ADP and Pi to catalyze ATP synthesis. The

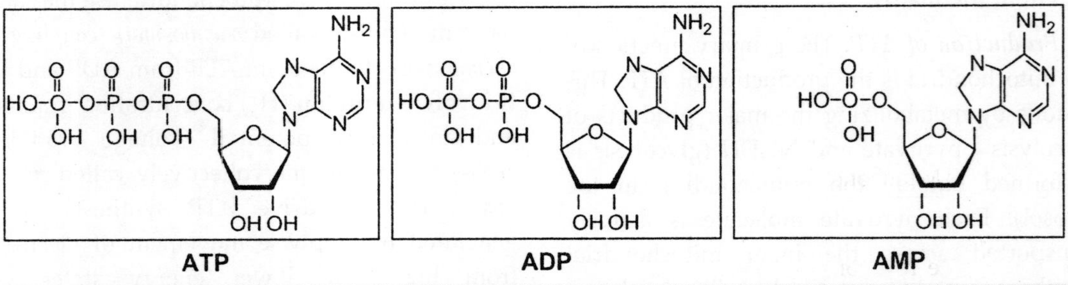

Fig. 11.3.14 : Formation of ADP and AMP from ATP to yield energy.

nucleotide binding site on the β subunits where ATP synthesis occurs are 9-10 nm away from the surface of the thylakoid membrane. As per ***binding change mechanism model*** (Boyer, 1999) protons from the cristae lumen pass through the proton half channels (I) and power the rotation of the *c* sub unit ring along with γ and ε subunits. It results in the cyclical changes in the conformation of the β subunits that differ in the binding affinities of ADP and Pi / ATP. [In fact each β subunit shows three states - an O (open) state that binds ADP and Pi very weakly, a L (low) state that binds ADP and Pi loosely and a T state that catalyzes the synthesis of ATP]. Initially ADP and Pi bind arbitrarily to O state of say β1 subunit. Proton flux causes 120° rotation of the β subunit that causes an increased affinity of β1 subunit for ADP and Pi (L state), next 120° rotation lead to their union while in next 120° rotation ATP formation takes place with its subsequent release. Thus each β subunit follows three conformational changes that collectively participate in the formation of ATP. After rotation of the *c* ring the proton is released into the matrix through the second half channel. The process of ATP synthesis by this process is called *chemiosmosis*. ATP is the source of energy for all vital activities. In ATP three phosphates remain attached by high-energy bonds and breakdown of one bond yields about 7.3 kcal energy. For these reasons mitochondria is considered as the ***power house of the cell.*** Under certain conditions, protons may be allowed to re-enter the mitochondial matrix without contributing to ATP synthesis. This process, known as *proton leak* or *mitochondrial uncoupling*, that results in the release as heat. This mechanism generates heat specially in the "brown fat" of newborn or hibernating mammals.

Major metabolic pathways converging on the TCA cycle: Sugars are converted anaerobically to pyruvate via glycolysis, which produce 2 ATP and 2 NADH per glucose molecule. Pyruvate then enters the mitochondria and gets enzymatically converted to Acetyl-CoA, which enters the Krebs cycle by giving 2 carbons to Oxaloacetate to regenerate Citrate. During the Krebs cycle, NAD+ gets converted to NADH by acquiring 2 electrons and a proton (FAD also gets converted to FADH2). NADH and $FADH_2$ will then give their electrons to the "electron transport chain" which will pump protons across the mitochondrial inner membrane and the resulting proton gradient will drive ATP synthase to convert ADP+Pi to ATP. Oxygen is consumed during this event and, by accepting 4 electrons and 4 protons, gets converted to water. Some intermediates of glycolysis pathway e.g. Fructose – 6 Phosphate may serve as substrate to amino sugars, glycolipids and glycoproteins. Dihydroxyacetone phosphate may also serve as substrate for lipid synthesis.

Several catabolic pathways converge on the TCA cycle. Most of these reactions add intermediates to the TCA cycle, and are therefore known as *anaplerotic reactions* while the processes that remove intermediates from the cycle are termed *cataplerotic reactions.* These increase the amount of acetyl-CoA so that the cycle ables to increase mitochondrion's capability to produce more ATP.

However, pyruvate may be carboxylated by pyruvate carboxylase to form *oxaloacetate* increasing the cycle's capacity to metabolize *acetyl-CoA* when the tissue needs energy (e.g. in muscle). In the TCA cycle all the intermediates (e.g. *citrate, iso-citrate, a -ketoglutarate, succinate, fumarate, malate* and *oxaloacetate*) are regenerated during each turn of the cycle. Addition of any one of them to the cycle has an anaplerotic effect, and its removal has a cataplerotic effect. These anaplerotic and cataplerotic reactions will, during the course of the cycle, increase or decrease the amount of *oxaloacetate* available to combine with *acetyl-CoA* to form *citric acid*. This in turn increases or decreases the rate of ATP production by the mitochondrion and thus the availability of ATP to the cell.

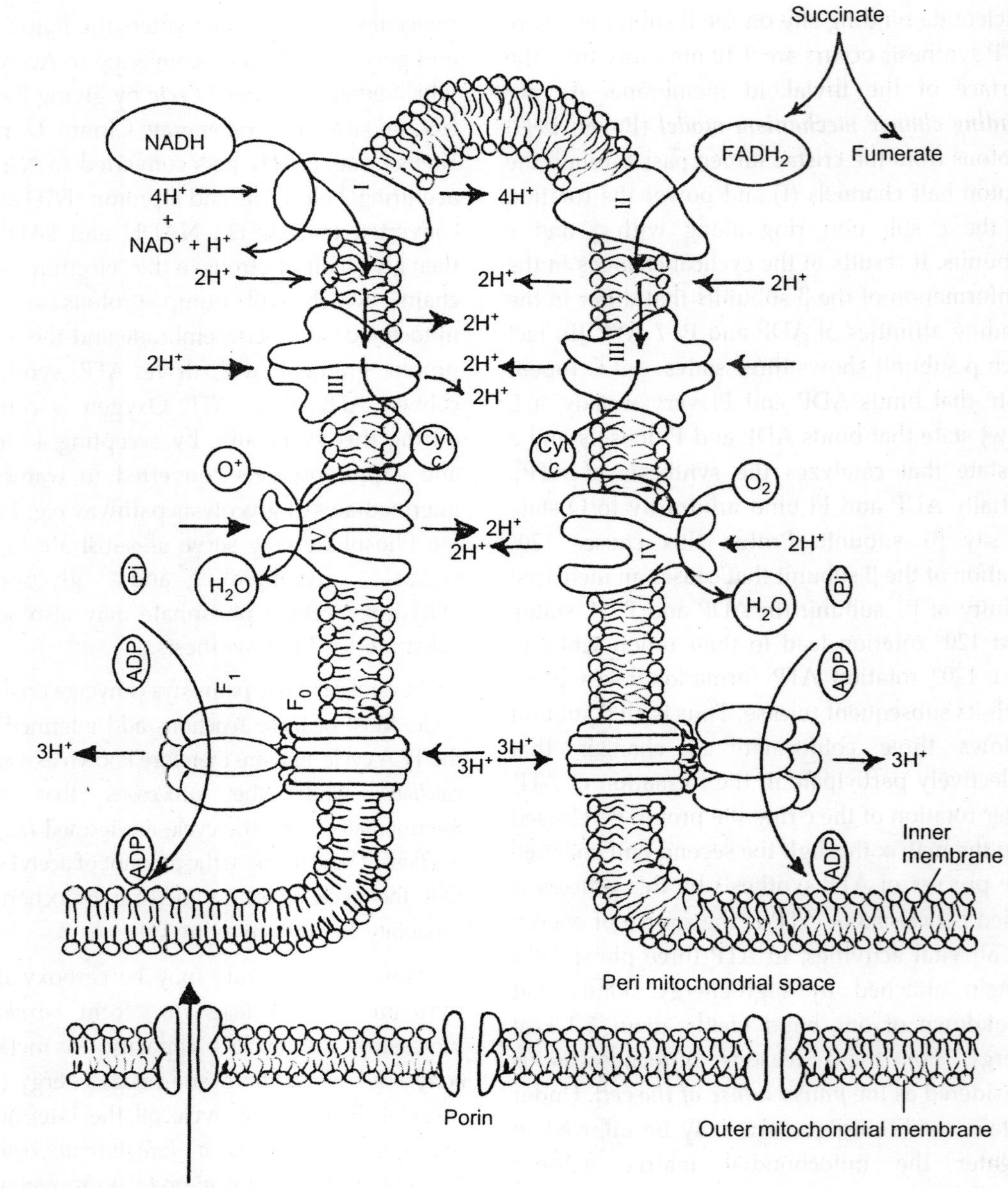

Fig. 11.3.15 : Molecular events in the crista and perimitochondrial space.

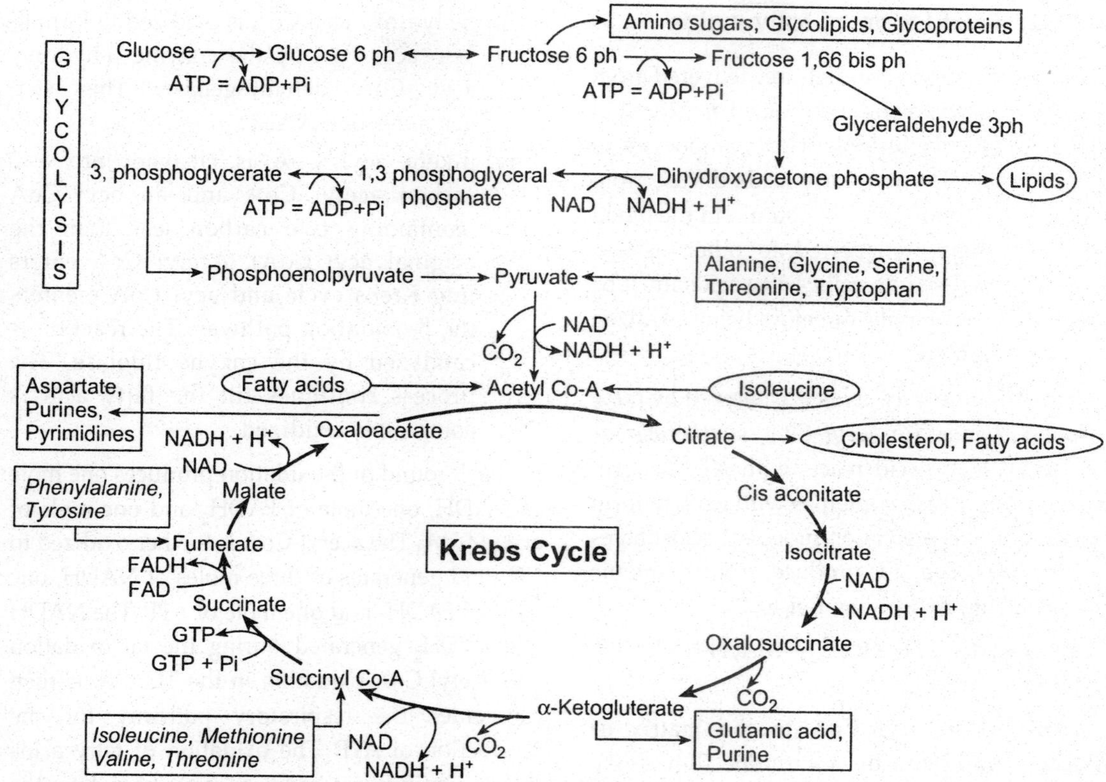

Fig. 11.3.16 : Interrelationship of glycolysis and Krebs cycle with fat and protein metabolism.

Acetyl-CoA, derived from pyruvate oxidation, can also be generated from the b-oxidation of fatty acids, or from the catabolism of isoleucine animo acid.

In protein catabolism Glutamic acid, Isoleucine, Methionine, Valine, Threonine, Phenylalanine, Tyrosine, Aspartic acid, etc. may either enter the TCA cycle as intermediates having an anaplerotic effect on the cycle, or Leucine, Isoleucine, Lysine, Phenylalanine, Tryptophan and Tyrosine are converted into *acetyl-CoA* which can produce CO_2 and water or used to form ketone bodies. These amino acids are therefore termed as "ketogenic" amino acids, whereas those that enter the TCA cycle as intermediates and can be converted into glucose are called "glucogenic" amino acids. De-aminated alanine, cysteine, glycine, serine, and threonine are converted to pyruvate and can consequently either enter the TCA cycle as *oxaloacetate* or as *acetyl-CoA* to form CO_2 and water.

In fat catabolism, triglycerides are converted into fatty acids and glycerol. In the liver the glycerol can be converted into glucose via dihydroxyacetone phosphate and glyceraldehyde-3-phosphate by way of gluconeogenesis. In heart and skeletal muscle, fatty acids are broken down through β-oxidation, producing *acetyl-CoA*, which can be used in the TCA cycle. β-oxidation of fatty acids with an odd number of methylene bridges produces propionyl-CoA, which is then converted into *succinyl-CoA* and fed into the TCA cycle as an anaplerotic intermediate.

11.3.5.2. β-oxidation of fatty acids

Fatty acids from the diet are delivered from the gut to cells via blood. Fatty acids are stored in the form of triacylglycerols within adipocytes of adipose tissue. The free fatty acids diffuse from adipose cells, combine with albumin in the blood and are thereby transported to other tissues, where they diffuse into cells. Fatty acids must be activated in the cytoplasm before being oxidized in the mitochondria.

Activation of fatty acids is catalyzed by ***fatty acyl-CoA ligase*** (or ***acyl-CoA synthetase*** or ***thiokinase***). Fatty acid reacts with ATP to form acyladenylate which combines with CoA to form acyl CoA. In this activation process 2 phosphates of ATP are used to produce Ppi. Inorganic pyrophosphate hydrolyses Ppi to Pi.

[Fatty acid + ATP + CoA $\longrightarrow$ Acyl-CoA + PP_i + AMP]

Transport of acyl CoA to the matrix of mitochondria occurs by a carnitine (β-hydroxy γ-trimethyl amino butyrate) carrier system. Acyl group of acyl CoA is transferred to carnitine present on the outer mitochondrial membrane by carnitine acyl transferase. Acyl carnitine is transported to the matrix by a carrier protein and is converted to acyl CoA by carnitine acyl transferase-II found on the IMM. Released carnitine returns to cytosol for reuse. Once inside the mitochondrion the fatty-CoA is a substrate for the β-oxidation machinery (Fig. 11.3.17). The process is termed β-oxidation since it occurs through the sequential removal of 2-carbon atoms by oxidation at the β-carbon position of the fatty acyl-CoA molecule. It involves following steps:

i) Acyl CoA undergoes dehydrogenase to form trans enonyl CoA showing double bond between α and β (2 and 3 carbons) and generating one molecule of $FADH_2$ by acyl CoA dehydrogenase.
ii) Transenonyl CoA forms β-Hydroxyacyl CoA by the hydration of the double bond by the enzyme enonyl CoA hydratase.
iii) β-Hydroxy acyl CoA is oxidized to form β-keto acyl CoA by the enzyme β-hydroxy acyl CoA dehydrogenase. This step generates one NADH.
iv) β-keto acyl CoA is cleaved into a 2 carbon acetyl CoA and an acyl CoA (containing two carbon less than the original acyl CoA). Acetyl CoA enters into Krebs cycle and acyl CoA reenters the β-oxidation pathway. The reaction is catalyzed by the enzyne thiolase. The process continues till the fatty acid is completely oxidized.

Each round of β-oxidation produces one mole of NADH, one mole of $FADH_2$ and one mole of acetyl-CoA. The acetyl-CoA is further oxidized to CO_2 and generates of three moles of NADH, one mole of $FADH_2$ and one mole of ATP. The NADH and $FADH_2$ generated during the fat oxidation and acetyl-CoA oxidation in the TCA cycle then can enter the respiratory pathway for the production of ATP. The oxidation of fatty acids yields significantly more energy than does the oxidation of glucose. The net result of the oxidation of one molecule of oleic acid (an 18-carbon fatty acid) and palmitic acid (16 C) will be 146 and 129 molecules of ATP (2 molecules of ATP are used during the activation of the fatty acid).

Other functions:

Liberation of energy by Glyoxylic acid pathway: During germination of lipid rich seeds in the initial stages, glyoxysomes appear in the cotyledons. At this stage by the cooperation of glyoxysomes and mitochondria fatty acids are utilized for the formation of energy by glyoxylic acid pathway.

Formation of ketone bodies: Acetyl CoA produced by β-oxidation is converted by the liver / kidney mitochondria to ketone bodies that are trans ported to muscles, brain and heart. Ketone bodies are the major source of fuel to be used after prolonged exercise or starvation.

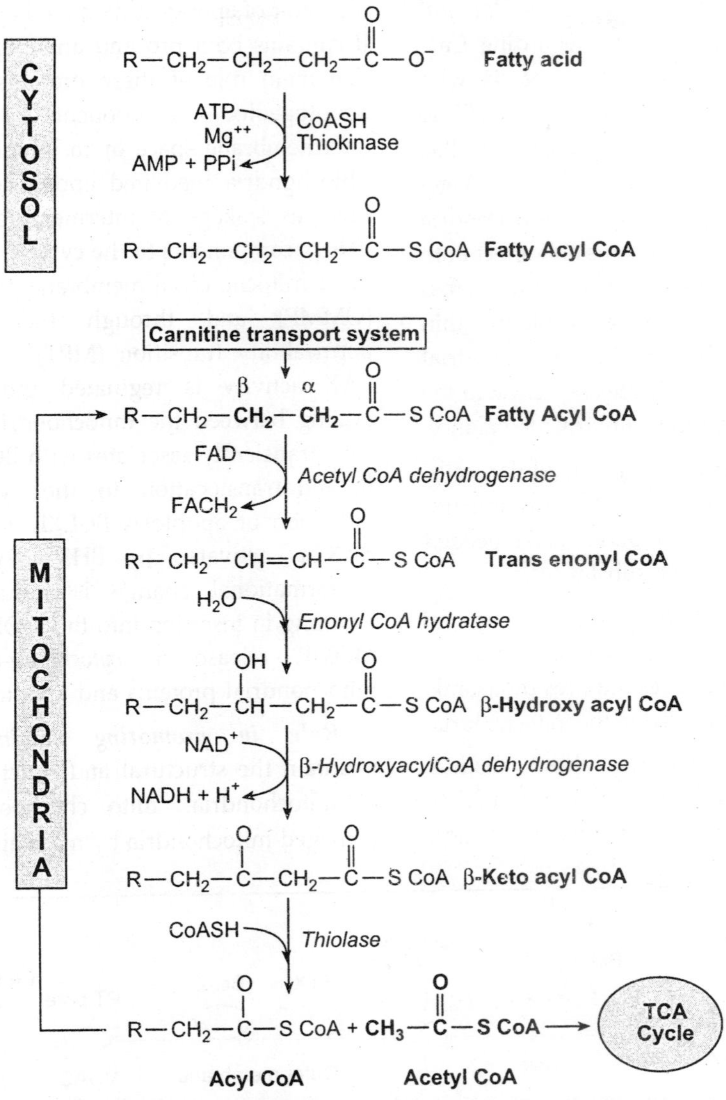

Fig. 11.3.17 : Reaction sequence of the β- oxidation of fatty acids in the mitochondria.

Urea cycle: Urea is the major catabolic end product of amino acids. In the hepatic cell mitochondria CO_2 and NH_3 are used for the production of urea. During its production, carbamyl phosphate acts as a carrier of nitrogen atom of NH_3 and carbon atom of CO_2. The whole reaction is catalyzed by the enzyme ornithine transcarbamylase. This enzyme occurs in large amounts in the matrix of mitochondria.

Synthesis of Heme: In the synthesis of heme, succinyl COA formed in the Krebs cycle in the matrix of mitochondria combines with glycine to form a pyrole molecule. Four pyroles combine to form a protoporphyrin IX that combines with Fe to form heme.

Ca^{++} homeostasis: They control not only their intra-organelle Ca^{++}, but also influence the entire network of cellular Ca^{++} signaling, including ER,

plasma membrane, and nucleus. As per ***Hot spot hypothesis*** (Rizzuto *et al*, 2004) regarding Ca^{++} homeostasis ER makes intimate contacts with mitochondria. Ca^{++} is transported from ER to mitochondria with the involvement of Bax mediated receptors and VDAC (voltage-dependent anion channel). Mitochondria accumulate Ca^{++} at microdomains and affect both Ca^{++} release from the ER and Ca^{++} entry across the plasma membrane. When Ca^{++} content within the ER falls, channels in the mitochondrial membrane open and Ca^{++} is released close to the ER thereby regulating intracellular Ca^{++} balance.

Glandular secretion: In the glandular cells, mitochondria are located near its base and use oxidative energy for the formation of ATP needed for the synthesis of organic substances.

Protein synthesis: Mitochondria play an indirect role in the protein synthesis because formation of peptide linkages requires thousands of ATP that are all synthesized in the mitochondria.

Apoptosis: Mitochondria has distinct proapoptotic and prosurvival roles (Fig.11.3.18). Proteins belonging to BCL-2 family are critical regulator of apoptosis mediated by mitochondria. It contains both pro and antiapoptotic proteins. The main role of these proteins is to liberate apoptogenic components from the intermembrane space or to inhibit their release. Mitochondria mediated apoptosis (Fig. 11.3.18) involves leakage of intermembrane as well as matrix components to the cytosol that is done by outer mitochondrial membrane permeabilization (OMMP) and through the mitochondrial permeability transition (MPT). In healthy cells BAX activity is regulated through constant cycling between the mitochondria and cytosol. BAX transiently associates with BCLXL,resulting in retrotranslocation to the cytosol. During induction of apoptosis BCLXL is inhibited and BAX is activated by BH3 only proteins. A conformational change is induced in BAX, resulting in insertion into the MOM, resulting in MOMP, release of *cytochrome-c* and other mitochondrial proteins and apoptosis.

Role in promoting carcinogenesis: To maintain the structural and functional integrity of mitochondria, 'mito checkpoints' monitor damaged mitochondria by maintaining a balance

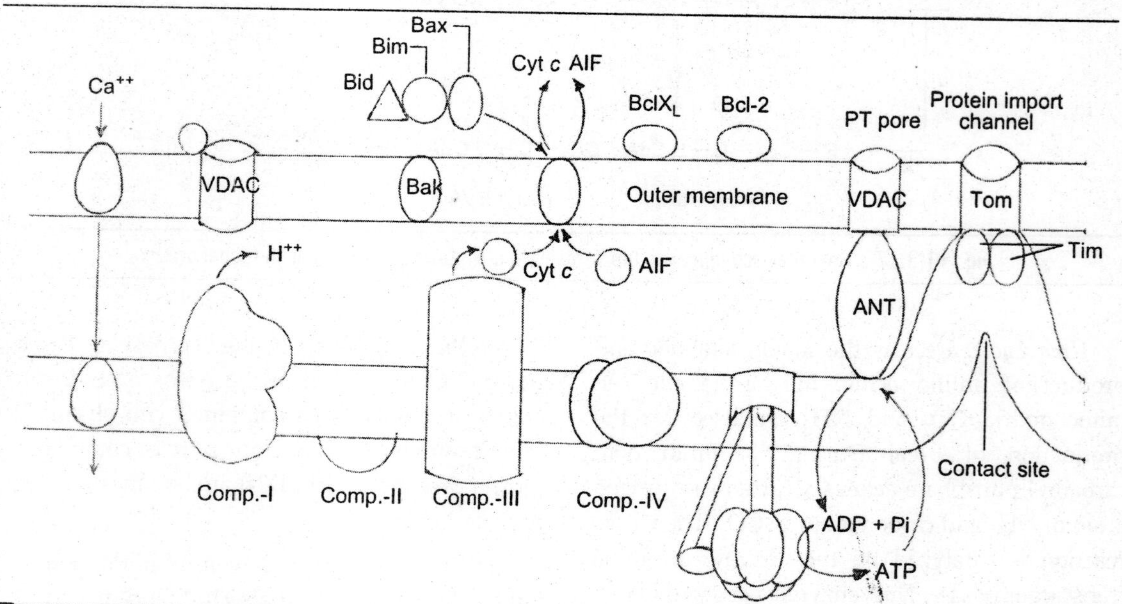

Fig. 11.3.18 : Mitochondrial architechture participating in apoptosis.

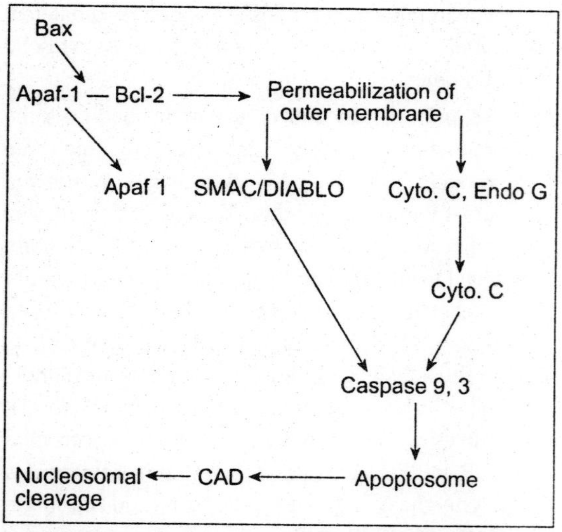

Fig. 11.3.19 : Mitochondria mediated apoptotic pathways.

between apoptotic and antiapoptotic signals. Severe mitochondrial damage or mutation in mtDNA may lead to genomic instability causing carcino genesis. In the pool of these cancerous cells a protein is expressed - surviving which prevents caspases activation and organizes a novel pathway to tumorigenesis.

Neurodegeneration: Neurodegeneration is strongly correlated with the formation of mitochondrial reactive oxygen and nitrogen species which in turn are correlated mitochondrial apoptotic pathway.

Managing trauma and shock: In response to trauma and shock, mitochondria destabilize the physiological integrity of the cell and many pores in the outer membrane of the mitochondria are opened. This causes leakage of the contents into the cytosol leading to enormous oxygen consumption and production of reactive oxygen species that damage genes and gene products.

11.3.6. ORIGIN OF MITOCHONDRIA

Mitochondria contain ribosomes and DNA, and are only formed by the division of other mitochondria. The mitochondrial fission apparatus is generally conserved. Higher plant mitochondria use dynamin-like and Fis-type proteins for division (Logan, 2006). It is generally accepted that they were originally derived from endosymbiotic prokaryotes. Studies of mtDNA employs a variant genetic code indicating their ancestor was a member of the Proteobacteria. A few groups of unicellular eukaryotes lack mitochondria: the symbiotic microsporidians, metamonads, and entamoebids, and the free-living pelobionts. On rRNA trees these groups appeared as the most primitive eukaryotes, suggesting their origin before the origin of these bacteria. They are speculated to retain genes or organelles derived from mitochondria.

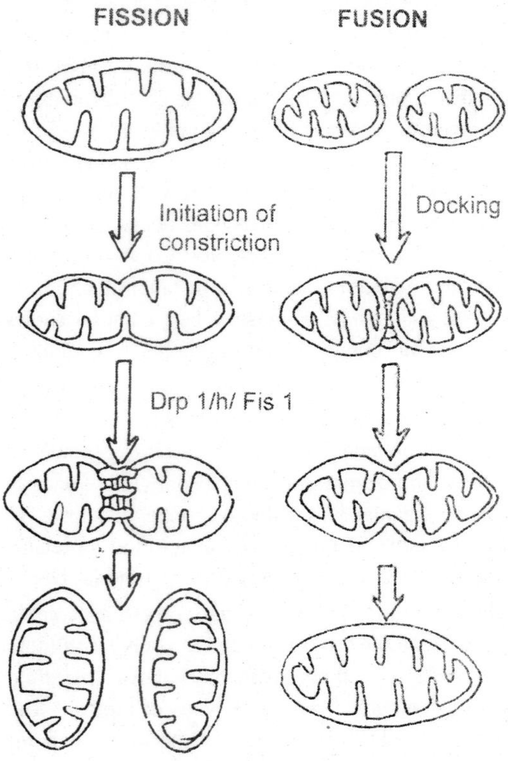

Fig. 11.3.20 : Diagramatic representation of fusion and fission in mitochondria.

11.3.7. MITOCHONDRIAL FUSION AND FISSION

A typical mammalian cell can have hundreds of mitochondria which are not fully autonomous. They exhibit fusion of neighboring mitochondria (*mitofusion*) and fission (*mitofission*) of existing mitochondrion. Mitofusins (Mfn1 and Mfn 2) and OPA1 are essential for mitochondrial fusion, whereas Fis1 and Drp1 are essential for mitochondrial fission. The mitochondrial outer membrane proteins Mff, MiD49 and MiD51 are involved in mitochondrial fission by recruiting Drp1 from the cytosol to the organelle surface. (Kristin *et al*, 2015).

SUMMARY

1. Mitochondria provide all the ATP needed for all the vital activities of the cell. In higher plants, all the mitochondria in a cell is called chondriome.
2. Mitochondria are invested by a double-layered envelope. Both membranes show numerous integral proteins. The inner membrane produces numerour cristae. Both membranes are firmly associated at many sites called contact sites. The protein, Mmm1, required for normal morphology and inheritance of mtDNA is associated with both membranes at the contact sites. Contact sites also control mitochondrial energy metabolism and Ca^{++} homeostasis. ATP synthase or F_1F_0 complex shows membrane bound F_0 and protruding F_1 complex.
3. In addition to their solitary existance frequent inter-mitochondrial fusion occurs enabling the mixing and recombination of mtDNA. The genome of human mitochondria contains 16,569 base pairs of DNA organized in a closed circle. It shows 37 genes. Only 13 proteins in humans and 8 proteins in yeast are encoded by the mitochondrial DNA.
4. Mitochondria show two types of proteins soluble and insoluble. Soluble proteins mainly constitute the enzymes. Inner membrane shows 14 while outer membrane shows 21 types of proteins. The outer membrane contains the characteristic porin protein that allows the passage of ions and small molecules through it. The inner membrane contains Cardiolipin that renders it virtually impermeable.
5. Human mitochondria possess about 1500 different proteins of which 99% are synthesized on cytosolic ribosomes.
6. The primary function of the mitochondria is the production of ATP. Each pyruvate molecule produced by glycolysis is actively transported across the inner mitochondrial membrane into the matrix where it is combined with CoA to form acetyl CoA.Mitochondria use the basic mechanism of chemiosmosis to generate ATP. It involves a proton (H^+) concentration gradient and an electric potential (voltage gradient) across the membrane collectively called proton motive force to drive ATP synthesis. The β-subunits of F_1 complex can bind ADP and Pi to catalyze ATP synthesis. Fatty acid is a long chain of 18 carbons and is oxidized in the matrix of mitochondria.
7. Mitochondria also stores Ca^{++} and affect both Ca^{++} release from the ER and Ca^{++} entry across the plasma membrane thereby regulating intracellular Ca^{++} signal. Mitochondria play an indirect role in the protein synthesis because formation of peptide linkages requires thousands of ATP which are all synthesized in the mitochondria. Mitochondrial outer membrane permeabilization is directly governed by the interaction between cytosol and outer mitochondrial membrane proteins like Bcl-2 family of proapoptotic proteins, which induce disruptions in the outer mitochondrial membrane and subsequent release of death-promoting proteins like cytochrome c. Perturbation of mitochondrial fusion results in defects in mitochondrial membrane potential and respiration; poor cell growth and increased susceptibility to cell death.
8. An apoptotic cell shows swelling of outer mitochondrial membrane and breakage of mitochondria releasing different mitochondrial enzymes into the cytosol that ultimately penetrate into the nucleus causing cleavage of inter nucleosomal DNA. Bcl-2 remains adhered to the outer mitochondrial membrane and bound with protein factor Apaf-1. Translocation of Bax from cytosol to mitochondria result in opening of pores causing cytochrome-c release into cytosol.

   To maintain the structural and functional integrity of mitochondria, 'mitocheckpoints' monitor damaged mitochondria by maintaining a balance between apoptotic and antiapoptotic signals.

Mitochondria are only formed by the division of other mitochondria. Higher plant mitochondria use dynamin-like and Fis-type proteins for division. TOM complex mediates the initial steps of the import of preproteins that are essential for mitochondrial biogenesis.

9. Only mother transmits mtDNA to the offsprings. Several human diseases are caused by mutations in mtDNA. Occurrence of DNA in mitochondria appears interesting. Mitofusins of adjacent mitochondria form homotypic and heterotypic complexes. In addition to normal mitochondria, slow sedimenting (ss)-mitochondria may represent a type of "nascent mitochondria" derived from a yet unidentified mitochondria-propagation mode operating during rapid seedling growth.

QUESTIONS

Long questions:

1. Describe the morphological and ultra structure of mitochondria.
2. Describe the process of ATP synthesis in mitochondria.
3. Describe the process of electron transport through the multiprotein complexes.

Short questions:

1. What do you mean by chondriome?
2. What are contact sites? Mention its different types and functions.
3. What do you mean by cristae junctions? What are their significances?
4. What do you mean by Oxysome? What is its relation with F_1 and F_0 complex?
5. What are different components of F_0 and F_1 complex?
6. What is mitoribosome? What are the S values of its subunits?
7. What is the nucleotide number in mt DNA? What are the basic differences between mt DNA and cytosolic DNA?
8. How many polypeptides are coded by mt DNA?
9. What do you mean by TOM and TIM complexes?
10. In which part of mitochondria Krebs cycle and electron transport system occur?
11. Which part of F_1 complex actively participates in ATP synthesis?
12. What are the additional processes that happens in mitochondria to form ATP other that Krebs cycle?
13. How mitochondria participate in Urea synthesis and apoptosis?
14. What are the positive evidences in support of the endosymbiotic origin of eukaryotes?

MCQs:

1. **Mitochondria are self replicating due to the possession of**
   (a) ribosome, (b) ATP (c) DN(A)
2. **Porin occurs in the**
   (a) wall of cristae,
   (b) outer membrane,
   (c) inner membrane.
3. **Cardiolipin, a phospholipid is present in**
   (a) outer membrane,
   (b) cristae wall and
   (c) inner membrane.
4. **The parts of F_0 that participate in proton movement are**
   (a) b, (b) c, (c) a.
5. **The part of F_1 that participate in ATP production is**
   (a) α, (b) β, (c) γ subunit.
6. **Chemiosmotic theory of ATP formation is based on**
   (a) Na^+, (b) K^+, (c) proton gradient.
7. **Mitochondria is considered as endosymbiont because** they
   (a) have membranes similar to that of bacteria,
   (b) they have nucleic acids,
   (c) both.
8. **Extranuclear DNA occurs in**
   (a) peroxisome,
   (b) glyoxysome,
   (c) chloroplast and mitochondria
9. **Cytochrome oxidase contains**
   (a) Mn, (b) Fe, (c) Zn.
10. **Cytochromes occur in the**
    (a) perimitochondrial space,
    (b) matrix of mitochondria,
    (c) cristae.

11. **Both respiration and photosynthesis require**
(a) cytochromes, (b) glucose, sunlight.
12. **The correct sequence of electron acceptor is**
(a) Cyt a, a_3, b, c; (b) Cyt b,c,a,a, (c) Cyt b_3, a, c, a
13. **The non proteinous part of ETC is**
(a) cytochrome, (b) cytochrome oxidase, (c) ubiquinone.
14. **During ATP synthesis electrons pass through**
(a) H_2O, (b) cytochromes, (c) CO_2.
15. **The mobile electron carrier in mitochondrial membrane is**
(a) Cyt a, (b) FeS, (c) CoQ.
16. **The final electron carrier is ETC is**
(a) NADP, (b) pyruvate, (c) cytochrome oxidase.
17. **In cytochromes electrons are picked up and released by**
(a) Fe, (b) Cu, (c) Zn.
18. **The cytochrome which hands over electrons to oxygen is**
(a) Cyt b, (b) Cyt c, (c) Cyt a_3.
19. **Final product of Krebs cycle is**
(a) ATP, (b) acetyl CoA, (c) CO_2.
20. **In the absence of carbohydrate,**
(a) fat, (b) protein, (c) amino acids will provide energy.
21. **Terminal acceptor of electron in respiratory chain is**
(a) H_2O, (b) *Cyt-b*, (c) O_2.
22. **Krebs cycle strats with the production of**
(a) acetyl CoA, (b) oxaloacetic acid, (c) malic acid.
23. **In the glycolytic pathway of glucose metabolism, which statement is FALSE?**
(a) Glucose-6-phosphate is formed from glucose plus ATP.
(b) Fructose-1, 6-Bisphosphate is a stimulator of pyruvate kinase,
(c) Lactate Dehydrogenase reaction is regulated by the [NAD^+] [NADH].[H^+] ratio
24. **In TCA cycle, which one of the following statements is false?**
(a) Oxaloacetate reacts directly with citrate.
(b) Citrate is converted to isocitrate.
(c) Citrate contains 6 carbon atoms.
25. **Final product of Krebs cycle is**
(a) ATP, (b) acetyl CoA, (c) CO_2.
26. **During oxidative phosphorylation in bacteria the electron transport chain is localized in the**
(a) Cytosol,
(b) the cell membrane,
(c) the inter-membrane space.

Answers to MCQs:

1. c, 2. b. 3. c, 4. b, 5. c, 6. c, 7. c, 8. c, 9. b, 10. c, 11. a, 12. b, 13. c, 14. b, 15. c, 16. c, 17. a, 18. c, 19. c, 20. a, 21. c, 22. b, 23. c, 24. c, 25. c, 26. b.

Chapter 11.4

Plastids

Plastids are major organelles found only in plants and algae. Plastids are responsible for photosynthesis, storage of products like starch and for the synthesis of many classes of molecules such as fatty acids and terpenes which are needed as cellular building blocks for the function of the plant.

Types of plastids: Plastids are commonly classified as *chloroplasts* (contains chlorophylls and other related pigments used in photosynthesis), *chromoplasts* (contains various coloured pigments meant for the colouration of leaves, fruits etc.), and *leucoplasts* that stores many food substances and are devoid of any pigment. All plastids are derived from proplastids, which are present in the meristematic regions of the plant.

Chromoplasts

Plastids that stores coloured pigments are called chromoplasts. These are double layered, round, filamentous or stellate shaped structures whose matrix is incompletely divided by numerous lamellae. Matrix, contain orange coloured carotene, lycopene and luteol pigments or yellow coloured xanthophylls. They impart the beautiful colours of petals and ripe fruits.

The chromoplasts that contain red pigments (phycoerythrin) are called *rhodoplasts* (found in the algae Rhodophyceae); yellowish brown pigments (phycoxanthine) are called *pheoplasts* (occur in Diatoms), yellow pigment – xanthine are called *xanthoplasts* (occur on the surface of the ripe fruits) or orange pigment – carotenoid are called *carotinoplasts* (occur in the carrots).

Leucoplasts: *Plastids that are devoid of any pigments and mainly store different food substances are called leucoplasts.* These are often called *amyloplasts* (for storing starch), *elaioplasts* (meant for storing fats), *aleuroneplasts* (mainly stores proteins). They also contain many lamellae that incompletely divide the matrix into many chambers.

Each plastid contains multiple copies of the circular 75-250 kb plastid genome whose number <100 in mature cells but > 1000 in rapidly dividing cells. The genome contains about 100 genes encoding rRNAs as well as proteins involved in photosynthesis. Nuclear genes encode the rest majority of plastid proteins.

Plastid DNA exists as large protein-DNA complexes associated with the inner envelope called *'plastid nucleoids'*. Each nucleoid particle contains more than 10 copies of plastid DNA. The developing plastid has many nucleoids, localized at the periphery of the plastid. Sometimes long thin protuberances called *stromules* extend from the main plastid body into the cytosol and several plastids remain interconnected.

Inheritance of plastids: Angiosperms generally inherit plastids from the mother while many gymnosperms inherit them from the father.

Origin of plastids: Plastids are thought to have originated from endosymbiotic cyanobacteria. Complex plastids originate by secondary endosymbiosis, when an eukaryote engulfed a red or green alga and retained the algal plastid, which is typically surrounded by more than two membranes. Apicomplexa, a phylum of obligate parasitic protozoa also harbors a complex plastid called *apicoplast*. Some dinoflagellates take up algae as food and keep the plastid of the digested alga to profit from the photosynthesis; after a while the plastids are also digested. These captured plastids are known as *kleptoplastids*.

11.4.1. CHLOROPLAST

Chloroplast is the principal pigment that participates in photosynthesis. In higher plants, chloroplasts remain mainly confined to the mesophyll cells. In different algae however their shape and distribution vary widely

11.4.1.1. Morphological features:

They may be spherical, oval, discoid, filamentous, saucer shaped, vesicular or club shaped. They have an ellipsoidal shape in the light but enlarged and round form during the dark period and the thylakoids appear swollen. They usually measure 2-3μ in thickness and 5-10μ in diameter. The chloroplasts of plants grown in shady areas and polyploid plants are larger and contain more chlorophyll. In algae their number is one but in higher plants it is usually 20 – 40 / cell.

11.4.1.2. Structure:

It shows following components:

i) It is invested by double layered lipo-proteinous membranes each of 9-10 nm thick separated by a space of 10-20 nm (Fig. 11.4.1).

ii) The outer membrane is permeable and contains porin while the inner membrane contains transport proteins for regulating the movement of metabolites into and out of the organelle.

iii) They contain a number of integral membrane proteins to which many pigments remain bound.

iv) The inner membrane encloses a jelly like *stroma* where calvin cycle occurs. It contains different enzymes (in C_3 plants, the most notable is RUBISCO – Ribulose 1, 5 bisphosphate carboxylase oxygenase. In C_4 plants, RUBISCO is absent in the stroma of mesophyll cell chloroplast but present in bundle sheath cell chloroplasts), tiny circular DNA (20-80 / chloroplast), RNA, ribosomes (55S), about 50% of the total proteins of the chloroplast, starch grains and fat droplets called *plastoglobuli.*

v) Within the stroma are stacks of flattened disks called *thylakoids* enclosing an empty space called *thylakoid space or lumen.* A stack of thylakoids is called *granum* (diam. varies from 0.5 – 2.0 m and number – 40-80/ chloroplast). No. of thylakoids varies from 1-50/ granum (Fig. 11.4.1). Number of grana are comparatively more in C_3 than C_4 plants. The grana are interconnected by

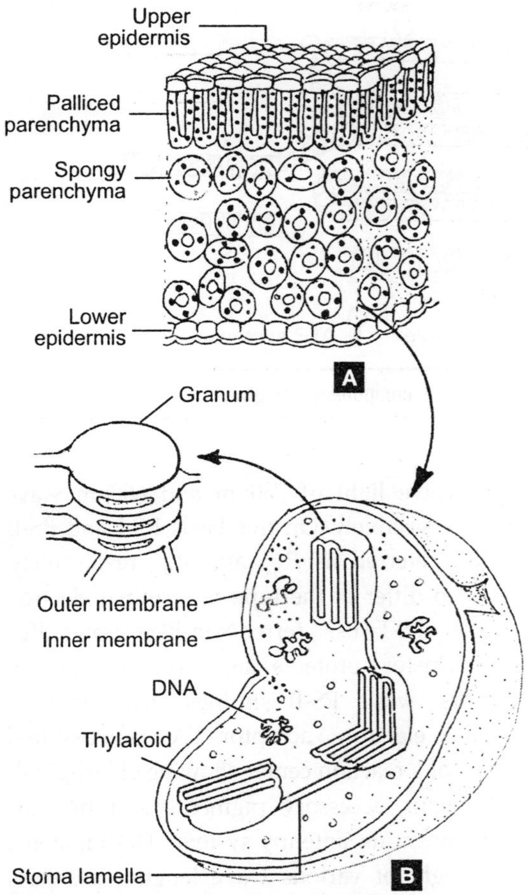

Fig. 11.4.1 : Position (A) and structure of chloroplast (B).

thread like *stroma lamellae.* Due to their structural continuity the grana are called *stacked thylakoids* and stroma lamellae are called *unstacked thylakoids.* Light reactions of photosynthesis takes place on the membrane of the thylakoids.

vi) The thylakoid membranes are noted to be connected with the inner chloroplast membrane through many small vesicles or tubules called *peripheral reticulum.*

Import of proteins to the chloroplast: Nuclear genes encode 95% of the chloroplast proteins that are escorted and translocated to the chloroplast by definite process. Proteins are targeted to the chloroplast by N-terminal sequences of 30-100 amino acids – *transit peptides.* A *guidance complex* recognizes these transit peptides and directs them to bind to the receptors – Toc159 and Toc34. Hsp70 molecules bind to the Toc complex and drive protein import. Hydrolysis of ATP and GTP provide energy for translocation. The preproteins are then transported to the translocase of the inner membrane – the Tic complex and transported to the stroma across the Tic complex with the help of Hsp100. In the stroma, the transit peptide is cleaved by a stromal processing peptidase (SPP) and the protein associates with stromal Hsp70 chaperones. Three pathways regulate protein import into the thylakoid membrane or lumen from the stroma.

i) *Sec pathway:* Sec A protein recognizes thylakoid signal sequence and target the protein to the Sec translocon and transfer the protein into the lumen where the signal sequence is cleaved by thylakoid processing protease (TPP).

ii) *Twin arginine translocation (TAT) pathway:* A cleavable thylakoid transfer sequence showing two arginines targets the fully folded proteins directly to the translocon which inserts them into the lumen. It mainly depends upon proton gradient across the thylakoid membrane.

iii) *SRP pathway:* The transmembrane, to be incorporated to the thylakoid membrane are recognized by the chloroplast signal recognition particle (SRP) and are inserted into the membrane.

11.4.1.3. Ultrastructure of the thylakoids:

Membranes of the thylakoids are called *fret membranes* made up of protein:unsaturated lipid in the ratio of 50:50. In addition to normal lipids and proteins, it consists of five special com ponents – pigment system-II (PS-II), cytochrome b6/f, ATP synthase, light harvesting complex and pigment system-I (PS-I). Lipids represent 50% of the membrane components. The structural lipids include glycolipids, sulpholipids, and a few phospholipids. Most of these lipids are

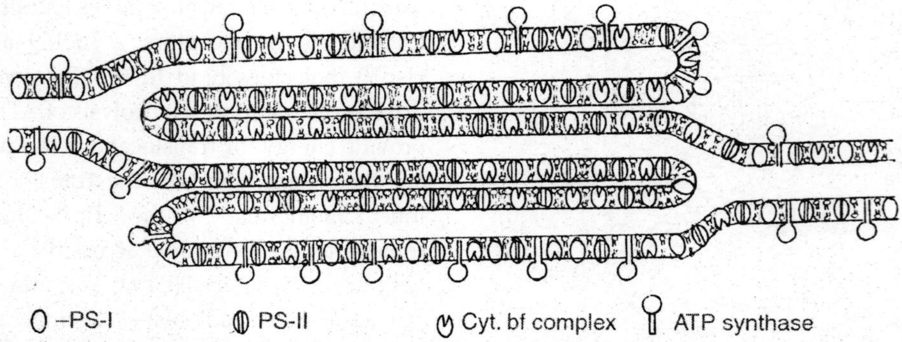

Fig. 11.4.2 : Diagramatic representation of different components of grana.

unsaturated which is mainly responsible for the fluidity of the membrane. PS-I shows binding sites for a number of lipids, either embedded in the protein interior or attached to the protein surface. The membrane proteins bind chlorophylls along with various accessory pigments. PS-I and ATP synthase are excluded from the stacked granal membranes; 85% of the PS-II particles are located in the granal stacks, they are called PS-IIα and remaining 15% are located in the stroma lamellae and are termed PSIIβ. Cytochrome-b_6-f complex is found in both stacked and unstacked regions.

Photosystems (PS): Each photosystem is differentiated by the wave length of light to which it is most reactive. Accordingly PS-I and PS-II receive light of 700nm and 680nm wave length respectively. On this basis PS-I and PS-II are also referred as P_{700} and P_{680} respectively. They also differ in the type of terminal electron acceptor. PS-I use ferredoxin-like iron-sulfur (Fe-S) cluster proteins as terminal electron acceptors, while PS-II shuffles electrons to a quinone electron acceptor. A photosystem consists of a reaction center showing chlorophyll-*a* and many accessory pigments and proteins which constitute antenna system. The pigments absorb light of various wave lengths and pass energy to the reaction center.

Photosystem - I: PS-I reaction center is a monomeric multiprotein complex (Fig. 11.4.3). At the enzymatic heart of the complex there are 3

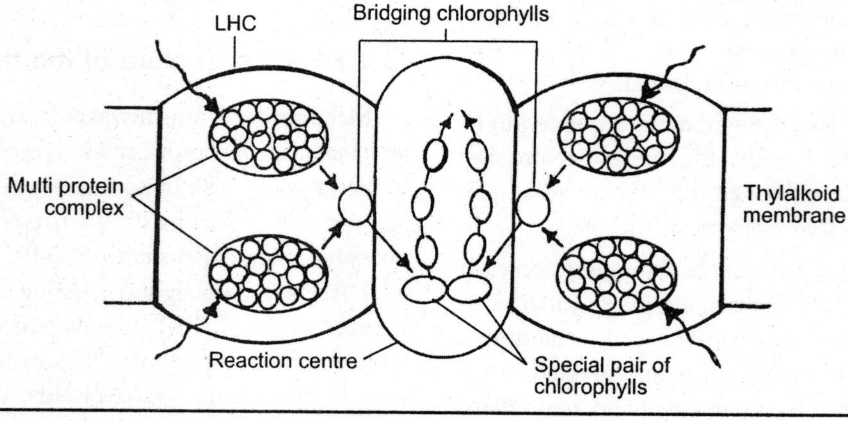

Fig. 11.4.3 : LHC and associated structure of PS I.

subunits known as PsaA, PsaB, and PsaC. At the core of PS-I there is a pair of chlorophyll-*a* molecules (***"the special pair"***) which are actual P_{700}. It contains additional about 200 - 400 core antenna chlorophylls (Chl-*b*) and 50 carotenoid molecules. The prosthetic groups, A_0, A_1, and F_X, are bound to PsaA and PsaB proteins. Electron acceptors include two membrane associated Fe-S proteins, Fe–S_A, and Fe–S_B. These are found in the PsaC protein. Additionally it is associated with plastocyanin, Cyt-f, Cyt-b_6 and feredoxin.

Photosystem - II: PS-II occupies distinct domains of the thylakoid membrane, the grana core, the grana margins, the stroma lamellae and the so called Y100 fraction. PS-II consists of about 25 polypeptides and is surrounded by a variety of chlorophyll-*a* and *b* binding proteins, known as *light harvesting complexes (LHC)*, which transmit light into the complex. At the enzymatic heart of the complex there are two proteins - D1 and D2 proteins and a special pair of chlorophyll-*a* molecules present in between D1 and D2. The pheophytin and Q_A are found on the D2 protein, the redox-active tyrosine Y_Z, and the Q_B plastoquinone are found on the D1 protein. In the D2 protein, the Q_A plastoquinone remains relatively fixed compared to the Q_B plastoquinone, which is free to move in and out of the D1 protein at the "Q_B site". The oxygen-evolving complex of PSII consists of three subunits, PsbO (33 kDa), PsbP (23 kDa), and PsbQ (17 kDa) attached towards the luminal surface. Ca^{++} is bound to the PsbO protein. PsbP and PsbQ are crucial for the stable architecture of the water-splitting system and optimize the efficiency of the oxygen evolution. Four Mn^{++}, one Ca^{++} and one Cl^- remain tightly bound to the PsbO subunit. Additionally, there is 10-kDa PsbR protein that is supposed to form a link between PsbP with the PSII core complex. It is also anticipated to play a role in water oxidation. The orientation of PS-II greatly differs in different parts of the thylakoids.

Dimeric PS-II (PS-II supercomplexes and PS-II core dimers) dominate in the stacked part of the thylakoid granum while the monomeric PS-II prevails in the stroma lamellae. Considerable amounts of PS-II monomers lacking the CP43 protein and PS-II reaction centers (D1/D2/Cyt-b559 complex) are found in the stroma lamellae. A large fraction of this PS-II do not evolve oxygen.

Light harvesting complexes: Both photosystems consists of two closely linked complexes – *a reaction centre* where the primary events of photosynthesis occurs and an *antenna complex* consisting of numerous proteins called *light harvesting complexes (LHCs)* which capture light and transmit it to the reaction centre. LHCs associated with respective pigment systems are

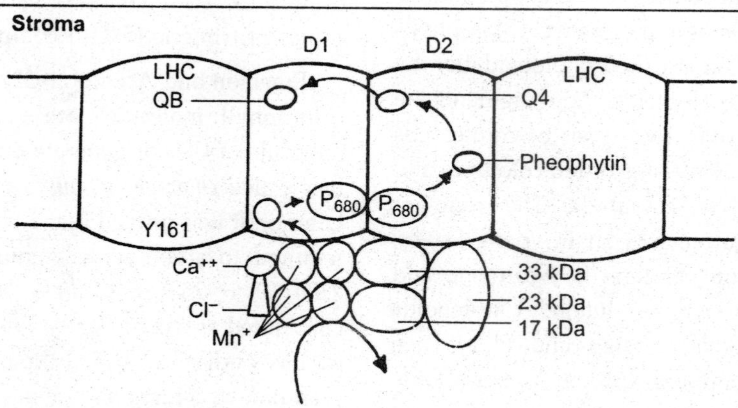

Fig. 11.4.4 : Diagramatic structure of the reacion centre of PS–II.

Table 11.4.1 : Differences between PSI and PSII

Pigment system I	Pigment system II
1. PSI shows 200-400 chlorophyll, 50 carotenoid, 2 cyt-b_6 and ferredoxin and 1 plastocyanin and cyt-f molecules.	1. PSII shows 200 chlorophylls, 50 carotenoid, plastocyanin and cyt-f molecules an unknown substance Z and inorganic molecules – Mn^{++}, Cl^-.
2. It has one molecule of reaction center P_{700}	2. It has one molecule of reaction center P_{680}
3. The number of chlorophyll-b is less	3. The number of chlorophyll-b is more.
4. Located on the outer surface of thylakoid.	4. Located on the inner surface of thylakoid.
5. Molecular oxygen is not evolved.	5. Photolysis results generation of oxygen.
6. NADPH + H^+ is formed in this system.	6. NADPH + H^+ is not formed in this system.
7. Participates in both cyclic and non-cyclic photophosphorylation.	7. Participates in only non-cyclic photophosphorylation.

referred as LHC-I and LHC-II respectively. Both reaction centres and LHCs contain tightly bound light absorbing pigments.

Chlorophyll-*a* is the principal pigment; in addition chlorophyll-*b*, *c*, *d*, *e*, *carotenoids* (yellowish or redish in color) and phycobillins act as light absorbing pigments. Chlorophylls have a tetrapyrrole porphyrin head and a long chain alcohol called phytol tail. A non-ionic Mg atom lies at the center of the porphyrin. The ratio of chlorophyll-*a* : *b* varies from 2.5-3.5 : 1.

Carotenoids are unsaturated polyhydrocarbons, composed of eight isoprene units. These are fat soluble, hence are also called *lipochromes.* Carotenoids protect the chlorophylls from a lethal condition called photooxidation by light of high intensity. Hence these pigments are called *shield pigments.* Two types of carotenoids are noted – *carotenes* (orange or red coloured) and *xanthophylls* (yellow coloured). The ratio of xanthophylls to carotene is 2:1 in young leaves. The most common carotene is β-carotene and common xanthophyll is lutein. Carotenoids mainly absorb violet, indigo and blue wave lengths but maximum absorption occurs at blue wave length.

Phycobillins are of three types phycoerythrins (red), phycocyanins and allophycocyanins (blue). They constitute red and blue pigments. special pairs, the photons are captured by the two bridging chlorophylls in each LHCs and thence to the special pairs. These pigments extend the range of absorbed light to be used for photosynthesis. The LHCs bind 90 chlorophyll molecules and 31 other small molecules, all held in a geometric fashion for optimal light absorption. In the reaction centre there are six chlorophyll molecules of which 2 are *"special pair"* that initiate photoelectron transport. Before energizing the onal LHC-I and / or LHC-II complexes. PS-II contains usually two to four copies of trimeric LHC-II complexes.

Emerson and Arnold (1932) proved that 2500 chlorophyll molecules are needed to fix one molecule of CO_2 or generate one molecule of O_2. It is called *chlorophyll unit* or *photosynthetic unit.* Later on it was proved that 10 flashes of light are required to attain same amount of yield. Thus 2500/10 = 250 molecules of chlorophyll are considered as a *physiological unit of function.* Later on EM studies confirmed the presence of certain granular structures in the chloroplast lamellae which Park and Biggins (1964) suggested to be the morphological manifestation of the physiological photosynthetic unit and referred them as ***quantasomes.*** They measure 180 × 160 × 100Å. They contain about 230 chlorophyll molecules.

Cytochrome b_6-f: PS-II and PS-I are connected by a trans membrane proton pump, *Cytochrome b_6 – f complex* (plastoquinol— plastocyanin reductase). It shows plastoquinone (PQ). Plastocyanin (PC) oxidoreductase activity. PQ acts as a shuttle between PS-II and Cyt b_6-*f* complex and PC in between Cyt b_6 – *f* complex and PS-I complexes. Electrons from PS-II are carried by a mobile, lipid- soluble electron carrier plastoquinone to b_6-*f*, where they are removed in a stepwise fashion and transferred to a water soluble electron carrier called *plastocyanin*. This process is coupled to the pumping of four protons across the membrane. The resulting proton gradient (together with the proton gradient produced by the water-splitting complex in PS- II) is used to produce ATP via ATP synthase.

11.4.1.4. Chloroplast Genome:

The size of chloroplast genome ranges from 1,10,000 bp to 160,000 bp. Complete chloroplast genome sequences have been reported for maize, rice, liverwort, *Chlamydomonas,* several other interesting plant species and *Euglena.* It shows circular DNA molecules present in multiple copies. It contains genes encoding the major rRNAs, tRNAs and proteins needed for gene expression as well as proteins required for photosynthesis. Chloroplast genome found in *Marchantia* contains 121,024 bp in a closed circle. These make up 4 genes encoding 23S, 16S, 4.5S, and 5S rRNAs; 37 genes encoding tRNA molecules; 21 genes encoding ribosomal proteins, 4 genes encoding some of the subunits of the RNA polymerase used for transcription within the chloroplast; one gene encoding the large subunit of the enzyme RUBISCO; 5 genes PSI and 12 genes for PSII; 6 genes encoding chloroplast ATP synthase, 4 genes for Cyt *bf* complex and genes for 19 of 60 proteins used to construct the chloroplast ribosome.

Chloroplast biogenesis: The fatty acids are assembled into glycerolipids at the thylakoid membranes or they can be assembled into lipids at the ER and return to the chloroplast where they are remodeled into the specific glycerolipids. Thus lipid biosynthetic enzymes of different organelles participate in the biogenesis of the thylakoid membrane system.

11.4.2. PHOTOSYNTHESIS

Photosynthesis is the complex biochemical anabolic process by which the chloroplasts synthesize carbohydrate type of food, liberate free oxygen and produce ATP. The principal end products of this process are sucrose (transported to the non photosynthesizing tissues like roots, seeds etc. and is metabolized for liberating energy) and starch (the primary storage carbohydrate in higher plants). Photosynthesis that prevails in photosynthetic protozoa (*Euglena*) and bacteria also generate oxygen. The overall reaction is – $6CO_2 + 6H_2O$ ‡ $6O_2 + C_6H_{12}O_6$ + ATP

11.4.2.1. Stages of Photosynthesis:

Photosynthetic is a oxidation-reduction process in which water is oxidized and CO_2 is reduced to form carbohydrates. The whole process is divided into four stages – (i) absorption of light; (ii) electron transport leading to the generation of oxygen from water and simultaneously reduction of $NADP^+$ to NADPH and generation of proton motive force; (iii) synthesis of ATP and (iv) conversion of CO_2 into carbohydrates. The first three reactions take place in presence of light, for which these are called ***light reactions*** while the fourth reactions is light independent reaction (but directly dependent on light reaction) called Calvin reaction or ***dark reaction.***

Light Reactions:

A. ***Absorption of Light:*** The initial step is the absorption of light by chlorophylls attached to the proteins in the thylakoid membrane and its conversion into chemical energy in the photosystems. Only 1-4% of the light is absorbed by chlorophylls [Sugarcane can absorb 0-12% of the solar energy]. The smallest group of pigment molecules to affect a photochemical act is *quantasome;* it contains chl-*a*, *b* and carotenoids. A particle of light (photon) containing a packet of energy is called *quantum (hv)*. The wave lengths of visible light ranges from 350-750 nm.The amount of absorption of light at different wave lengths by different pigments is called *absorption spectrum*. The effectiveness of different wave lengths of light for CO_2 fixation, O_2 evolution and NADP+ reduction is called *action spectrum.* Photons from 400nm – 700nm (far red) wave lengths of light can be absorbed by the pigment system of LHC. It is called *photosynthetically active radiation (PAR).*

Each PS contains a reaction centre and a light harvesting complex both of which contain chlorophyll-*a*, *b* and carotenoids. PS-I contains 200 - 400 molecules of chl-*a* and 50 carotenoids while PS-II contains 200 molecules of chl-*a*, 200 molecules of carotenoids, chl- *b, c, d* depending upon the species. At the initial stage of light acceptance, PS-II plays crucial role.

Chlorophyll-*a* molecules are energized by the energy transferred from LHC. The energy from the absorbed light is at first transferred to a bridging chlorophyll in each LHC and then rapidly transferred to one of the two *'special pair' chlorophyll-a* in the associated reaction centre.

B. ***Electron transport and generation of proton motive force:***

(i) *Transfer of electron to plastoquinone:* The absorption of photon by PS-II chl-*a* (P_{680}) increases its energy by 42 kcal/mol. This is the *first exited state* of the energized chl-*a*. The exited P_{680} has approximately 5×10^{-9} sec. life span, so it rapidly transfers an electron to an electron acceptor – plastoquinone (Q_B) via pheophytin and Q_A. It results Chl-a^+ towards the lumen and Q_B^- towards the stroma. Transfer of electron from P_{680} creates a gap which is filled up by the electron produced by the photooxidation of water.

(ii) *Photolysis and Evolution of Oxygen:* The bound ions (one Ca^{++}, one Cl^- and four Mn^{++}) in the PsbO and P_{680}+ splitt four molecules of H_2O to yield four electrons, four protons and one O_2. Photooxidation of water and evolution of oxygen requires the

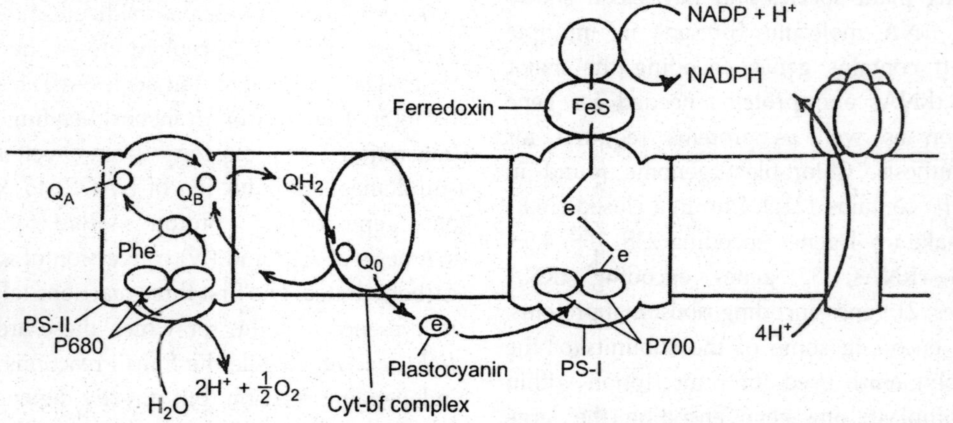

Fig. 11.4.5 : Photolysis and evolution of O_2.

presence of a water oxidizing enzyme complex that exists in five states S_0, S_1, S_2, S_3 and S_4. This complex is called *S state.* S cycle runs through these five states and each transition requires one quantum of light per electron transport. The evolution of O_2 becomes complete in the presence of some unknown compound called Z that accepts electrons from water and transfers it to the reaction center P_{680} . The released electrons are transferred via Mn^{++} and Y161 of D1 protein to $P_{680}+$ reducing it to the ground state $P_{680.}$ The OH^- molecules unite to form water and evolve molecular oxygen. This step requires the presence of Mn^{++}, Ca^{++} and Cl^-. The protons released, remain in the thylakoid lumen. The rate of yield of photosynthesis is measured in terms of *quantum yield* or oxygen evolution. It represents the number of oxygen molecules evolved per quantum of light absorbed. Eight quanta of light are required to evolve one molecule of oxygen.

During the formation of one molecule of O_2, 4 molecules of H_2O dissociate and 4 electrons are liberated. Therefore for the production of 6 molecules of O_2, $4 \times 6 = 24$ molecules of H_2O will dissociate and 24 electrons will be liberated. These 24 electrons and 24 protons (H^+) will combine with 12 molecules of NADP to form 12 NADPH + H^+

$$4H_2O \longrightarrow 4(OH^-) + 4H^+;$$

$$4(OH^-) \longrightarrow 2H_2O + O_2 + 4e^-;$$

$$2NADP + 4H^+ + 4e^- \longrightarrow 2NADPH + H^+$$

Quantum yield is slightly dropped in the region 440-520nm, remains constant and maximum in 600-680 nm but declines above 680 nm . This is called *'red drop'*. If combined beam of far red and red light is applied simultaneously then quantum yield increases significantly. This is called *Emerson effect or Photosynthetic enhancement effect*. It is due to the

i) presence of different types of harvesting molecules around a trap center in the photosynthetic unit and

ii) presence of two interconnected pigment systems with some common pigments.

C. ***ATP synthesis:*** *ATP synthase or F_1F_0 complex* shows membrane bound F_0 and protruding F_1 complex. F_0 part shows three types of components - *a, b, c* in the proportion $a_1b_2c_{14}$. All the components of *c* subunits are arranged as a ring in the plane of the membrane. Each *c* subunit participates in the proton movement. F_1 shows five polypeptides – $\alpha_3\beta_3\gamma\delta\varepsilon$. α and β subunits are arranged alternately as hexamer – $\alpha\beta\alpha\beta\alpha\beta$. δ subunit is linked with α and β subunits. γ Subunit fits rigidly in the centre of the *c* subunits. Two rod like β subunits is linked to one of the a subunits and δ subunit of F_1 and α subunit of F_0 complexes. ε subunit makes contact with both γ of F_1 and *c* sub- units of F_o complexes. β subunits form a stator that prevents the $(ab)_3$ from moving while it rests on the g subunit. Two proton half channels lie at the interface between *a* and *c* subunits of F_0 complex (Senior *et al*, 2002) (Fig. 11.4.3).

Chloroplasts use *chemiosmotic coupling or chemiosmosis* to generate ATP from ADP and Pi. It involves a proton (H^+) concentration gradient and an electric potential (voltage gradient) across the membrane collectively called *proton motive force* to drive ATP synthesis. It is generated by stepwise accumulation of proton concentration in the thylakoid lumen. Protons from the lumen are forced through the F_0 complex to drive proton motive force that rotates the β subunits of F_1 complex to produce ATP. Actually 2 protons are required to form one ATP molecule. The b subunits of F_1 complex can bind ADP and Pi to catalyze ATP synthesis. As per *binding change mechanism model* (Boyer, 1999) protons from the thylakoid lumen pass through the proton half channels and rotate the *c* subunit ring. It results in the cyclical changes in the conformation of the β subunits that differ in the binding affinities of ADP and Pi / ATP. [In fact each β subunit shows three states - an O (open) state that binds ADP and Pi very weakly, a L (low) state that binds

ADP and Pi loosely and a T state that catalyzes the synthesis of ATP]. Initially ADP and Pi bind arbitrarily to O state; proton flux causes 120° rotation of the β subunit that causes an increased affinity of β subunit for ADP and Pi (L state) causing their union while in next 120° rotation ATP formation takes place with its subse quent release. Thus each β subunit follows three conformational changes which collectively participate in the formation of ATP. After rotation of the *c* ring the proton is released into the stroma.

Transfer of electron to Cytochrome b_6 -f: As P_{680} absorbs a second photon and the semiquinone Q_B-accepts a second electron, it picks up two protons from the stroma generating QH_2 which diffuses into the intermembrane space and is bound to the Q_o site on a cytochrome b_6-*f* complex which accepts two electrons from QH_2 and releases two protons into the lumen to generate Q that shuttles to PSII exhibiting a *Q cycle*. Operation of Q cycle translo cates additional protons across the membrane to the thylakoid lumen. This type of electron transport generates increased proton motive force.

Cyclic photophosphorylation:

i) The chlorophylls and the accessory pigments of LHCI absorbs photons of specific wavelengths and transfer their energy to the reaction center – P700. It gets photoexited and emits an electron from its outer orbit (Fig. 11.4.8).

ii) The emitted electron is trapped by an intermediate electron acceptor A_1 (Chl-a). A_1 transfers the electron to A_2 (Fe-S protein) which again delivers the electron to A_3 (P_{430}). iii) A_3 then passes electron to ferredoxin (Fd) via some ferredoxin reducing substances.

The reduced Fd passes its electron to Cyt-b_6. This transport chain produces a proton-motive force, pumping protons (H^+) across the membrane; it produces a concentration gradient

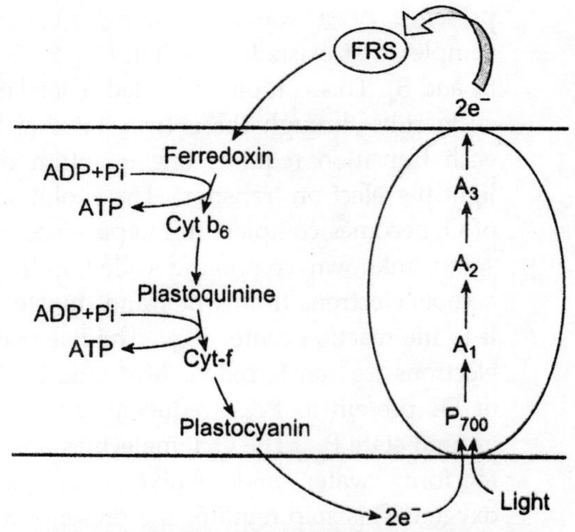

Fig. 11.4.6 : Biochemical pathways of cyclic photophosphorylation.

that can be used to power ATP synthase during chemiosmosis to produce one ATP. iv) The reduced Cyt-b_6 transfers its electron to hydrogen carrier plastoqionone (PQ) reducing it to plastohydroquinone (PQH_2). v) PQH2 passes its electron to Cyt f and $2H^+$ to the inner side of the thylakoid membrane. $2H^+$ generates proton motive force to produce one molecule of ATP. Cyt-f then transfers its electron to plastocyanin (PC). Reduced PC finally transfers its electron to photooxidized P_{700}.

The electrons return to the PSI but it does not involve PSII for which this type of electron flow is called *cyclic electron flow*. This pathway is known as *cyclic photophosphorylation*, and it produces neither O_2 nor NADPH. In bacterial photosynthesis, a single photosystem is used, and therefore is involved in cyclic photophosphorylation.

Noncyclic photophosphorylation: Noncyclic photophosphorylation is a two-stage process that involves two different photosystems - PSI and PSII (Fig. 11.4.7).

i) As P_{680} gets photoexited, losing two electrons from its outer orbit, the emitted electrons are accepted by pheophytin. Electrons from the photolysed water fill the gap and oxidize P_{680}.

ii) Pheophytin transfers electron to an unknown electron acceptor Q*a* (a quinone bound to thylakoid membrane) which in turn passes the electron to Q*b*.

iii) Q*b* transfers its electron to plastoquinone reducing it to plastohydroquinone (PQH_2).

iv) PQH_2 then transfers electron to Cyt-f or Cyt b_6 -*f* complex. In this step H^+ is released to the inner side of the thylakoid membrane. Therefore in this step one ATP is produced.

v) The reduced Cyt-*f* transfers its electron to plastocyanin(PC). It is a mobile carrier that shuttles between cyt-*f* and PSI complex.

vi) Simultaneously from the photoexited P_{700} two electrons are emitted so that the electrons passed from PC fills up the gap created by this emission and P_{700} becomes reduced.

vii) Electrons from P_{700} are accepted by A_1, then by A_2 and A_3. The reduced A_3 then passes electrons to ferredoxin (Fd) and ultimately to $NADP^+$ to form NADPH + H^+ in the presence of the enzyme ferredoxin-NADP – reductase.

The biochemical steps of synthesizing ATP and NADPH involving the transfer of high-energy electrons from one to another via the electron transport chain appears 'Z' like pattern, called ***"Z" scheme.*** These steps produce one ATP but the flow of electrons is linear/non-cyclic.This pathway is therefore called *non-cyclic photophosphorylation.*

Dark reactions or Calvin reaction:

CO_2 fixation (Calvin cycle): Synthesis of sugars from CO_2 is the most extensively studied biosynthetic pathway in plant cells. The first phase of this pathway is the fixation of CO_2.

(i) CO_2 is fixed in the stroma of the chloroplasts by *ribulose 1,5-bisphosphate carboxylase (RUBISCO).* It adds CO_2 to the 5C sugar ribulose 1,5- biphosphate (RuBP) to form two molecules of 3C 3-*phosphoglycerate (3PGA)* Due to the formation of this C_3 compound as the first formed stable compound in the CO_2 fixation these plants are called ***C_3 plants.***

(ii) Some of the formed 3 PGA is used in the synthesis of sucrose or starch while some are used in the regeneration of the enzyme. For every 12 molecules of PGA, 2

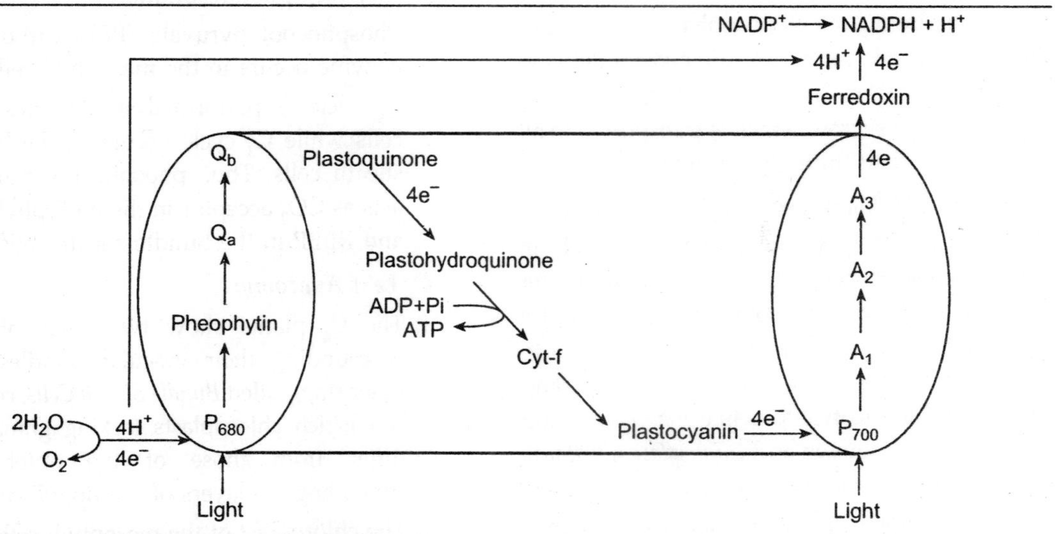

Fig. 11.4.7 : Biochemical pathways of non-cyclic photophosphorylation.

Table 11.4.2 : Difference between noncyclic and cyclic photophosphorylation	
Non cyclic photophosphorylation	**Cyclic photophosphorylation**
1. Both PSI and PSII participate in this system.	1. Only PSI participates in this system.
2. Electron emitted from the PSII, following a complex route is finally accepted by $NADP^+$ and never comes back to its original site.	2. Electron emitted from PSI, after following a complex route again comes back to its original site.
3. Photolysis causes oxygen evolution.	3. Oxygen is not evolved.
4. NADP+ is reduced to NADPH + H^+	4. $NADP^+$ is not reduced.
5. Transport of electrons in this system is PSII → Pheo → Qa → Qb → PQ → Cyt.f → PC → PSI → A_1 → A_2 → A_3 → Fd → $NADP^+$	5. Transport of electrons in this system is PSI → A_1 → A_2 → A_3 → FRS → Fd → Cyt b_6 → Cyt f → PC → PSI.
6. In this pathway 1 ATP is synthesized.	6. In this pathway 2 ATP are synthesized.

molecules are converted into glyceraldehyde 3 phosphate while 10 are converted into 6 molecules of ribulose bis phosphate. Fixation of 6 molecules of CO_2 generating 2 molecules of glyceraldehyde-3-phosphate requires the consumption of 18 ATPs and 12 NADPHs generated during light phase.

Synthesis of carbohydrates: After the formation of glyceraldehydes-3-phosphate in the stroma, it is transported into the cytosol where it forms dihydroxy acetone phosphate. It condenses with a second molecule of glyceraldehyde-phosphate to form fructose-1-6-bisphosphate. It is converted to sucrose which is transported to all parts of the plant through the phloem.

Hatch and Slack pathway or C_4 pathway: Hatch and Slack (1966) advocated that in sugarcane leaves the first stable compound after CO_2 fixation is not a C_3 compound (PGA) but a C_4 compound - oxaloacetic acid. It was also confirmed in *Zea mays, Sorghum, Panicum* and *Cyperus* and in dicotyledons like *Amaranthus* and *Atriplex* etc. It is known as *β-carboxylation pathway Co-operative photosynthesis* or *C_4 cycle* or *Hatch and Slack cycle*. Some C_4 plants — called CAM plants — separate their C_3 and C_4 cycles by time. Other C_4 plants have structural changes in their leaf anatomy so that their C_4 and C_3 pathways are separated in different parts of the leaf. C4 plants are

i) Found in tropical and subtropical regions. They grow fast in high temperature (optimum temp. varies from 30 – 45°C) and light intensities. They are also called *efficient plants.*

ii) Oxygen exerts no inhibitory effect since they lack photorespiration.

iii) Phosphoenol pyruvate (PEP) carboxylase enzyme occurs in the mesophyll cells.

iv) C_4 cycle is performed in the mesophyll cells while C_3 cycle occurs in the bundle sheath cells. Thus phosphoenol puryvate acts as CO_2 acceptor in the mesophyll cells and RuBP in the bundle sheath cells.

C_4 Leaf Anatomy:

i) The C_4 plants show two rings of cells surrounding their vascular bundles. The inner ring, called *Bundle Sheath Cells,* contain starch-rich chloroplasts lacking grana that differ from those of the outer ring containing 1-3 layers of mesophyll cells.

ii) The chloroplast of the mesophyll cells is of normal type but the chloroplasts of bundle

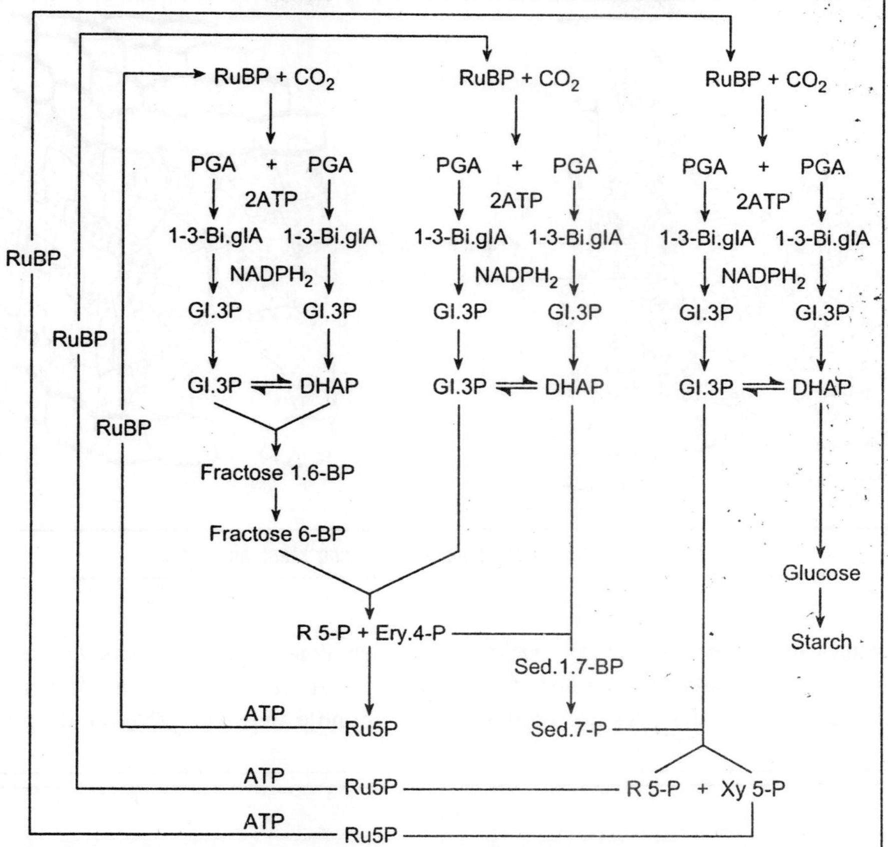

Gl3P = Glyceraldehyde 3—P; DHAP + Dihydroxyacetone phosphate; R 5-P = Ribose 5 phosphate; Er. 4-P = Erythrose 4-phosphate; Sec. 1.7-BP = Sedohepatulose 1.7 bisphosphate; R 5-P = Ribose 5 phosphate; Xy 5-P = Xylulose 5 phosphate.

Fig. 11.4.8 : Biochemical steps of Calvin cycle.

sheath cells are comparatively larger without grana or PSII and arranged centripetally. This peculiar anatomy is called ***Kranz anatomy*** (Fig. 11.4.9). Its primary function is to provide a site in which CO_2 can be concentrated around RUBISCO, thus reducing photorespiration.

Steps of the C_4 cycle:

A. **Reactions occurring in the chloroplasts of mesophyll cells:**

i) After entering through stomata, CO_2 diffuses into a mesophyll cell. CO_2 first combines with water to form bicarbonate ions (HCO_3^-) in the presence of the enzyme *carbonic anhydrase*. Being close to the leaf surface, these cells are exposed to high levels of O_2, but have no RUBISCO so they neither start photorespiration nor dark reactions of the Calvin cycle (Fig. 11.4.8).

ii) Phosphoenol pyruvic acid (PEP) reacts with HCO_3 with the help of the enzyme *PEP carboxylase* forming the first stable compound i.e. 4-carbon compound ***oxaloacetic acid*** (C_4).

iii) Oxaloacetic acid is very unstable and is converted into ***malic acid*** (C_4) by using light generated NADPH + H^+ under the

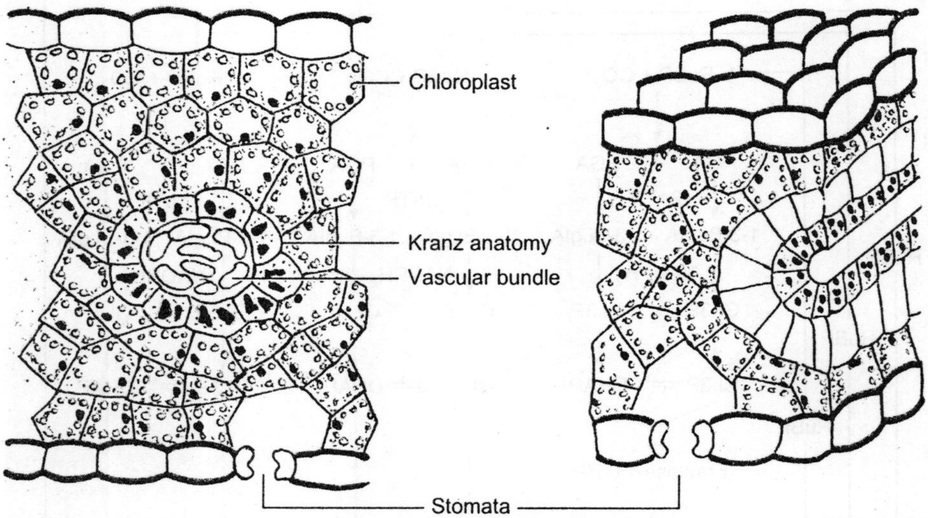

Fig. 11.4.9 : T.S. of sugar cane leaf showing kranz anatomy.

assistance of the enzyme *malic dehydrogenase*. It can also be converted into ***aspartic acid*** (C_4) in the presence of the enzyme *aspartate transaminase* that are transported by plasmodesmata into a bundle sheath cell that are deep in the leaf

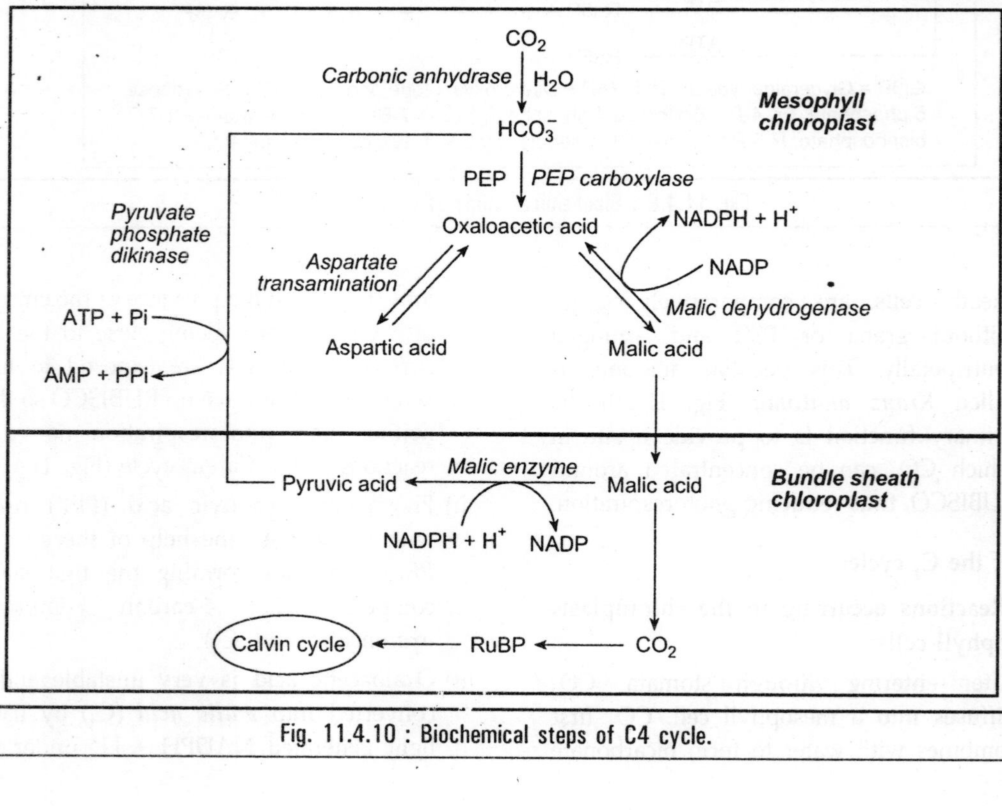

Fig. 11.4.10 : Biochemical steps of C4 cycle.

so atmospheric oxygen can not diffuse easily to them. They show a special anatomy – the Kranz anatomy and thylakoids with reduced PS II complexes (that produces O_2).

B. Reactions occurring in the bundle sheath cells:

i) In bundle sheath chloroplast the malic acid undergoes oxidative decarxylation to produce pyruvic acid and CO_2 and NADPH + H^+ in the presence of malic enzyme. In some cases, aspartic acid undergoes transamination to form oxaloacetic acid under the assistance of the enzyme *aspartate transaminase*. Oxaloacetic acid is decarboxylated to produce ***pyruvic acid***.

ii) CO_2 and NADPH + H^+ are directly delivered to RUBISCO and enters the Calvin cycle to form sugars and starch. iii) Pyruvic acid (C_3) is transported back to a mesophyll cell where it is phosphorylated to phosphoenol pyruvic acid (PEP) in the presence of the enzyme *pyruvate phosphate dikinase*. It splits ATP into AMP and PPi; PPi is again degraded into Pi. PEP again participates in the acceptance of CO_2.

Adaptive Value of C4 plants:

i) Photosynthesizes faster than C_3 plants under high light intensity and high tempe ratures because the CO_2 is delivered directly to RUBISCO, not allowing it to grab oxygen and undergo photorespiration.

ii) Has better water use efficiency because PEP carboxylase brings in CO_2 faster and so does not need to keep stomata open as much for the same amount of CO_2 gain for photosynthesis.

iii) In C_4 plants the photosynthetic yield is 2 or 3 times more than C_3 plants.

iv) C_4 plants can perform high rate of photosynthesis even when the stomata are closed. 5. It increases the rate of CO_2 fixation.

Many angiosperms (*e.g.* corn, maize, sugarcane, sorghum etc.) that thrive in hot environments keep their stomata closed in mid day to prevent loss of moisture. This causes the CO_2 level inside the leaf to fall below the Km value of RUBISCO. Therefore the rate of photosynthesis is slowed. To avoid this problem they utilize a two-step pathway to fix CO_2. It involves two types of cells, *mesophyll cells* that are adjacent to the air spaces and *chloroplast containing bundle sheath cells* surrounding the vascular bundle tissues (called Krantz anatomy). In the mesophyll cells *phosphoenol pyruvate (PEP) carboxylase* reacts with CO_2 to from *oxalo acetic acid* (OAA) that is converted to *malic acid or aspartic acid* (4C compounds). These compounds are then transferred to the cells surrounding the vascular bundle having enzymes of Calvin cycle. Here the malic acid is converved to pyruvic acid and CO_2. Pyruvic acid is transferred back to the mesophyll cells where it is converted to PEP and repeat the cycle.

11.4.2.2. CAM Photosynthesis:

CAM plants are also C_4 plants but instead of segregating the C_4 and C_3 pathways in different parts of the leaf, they separate them in time instead. (CAM stands for *Crassulacean Acid Metabolism* because it was first studied in *Bryophyllum calycinum* a member of the family Crassulaceae and because CO_2 is stored in the form of malic acid before use in photosynthesis). These include many succulents such as cactuses and agaves and also some orchids, pineapples, sedums, bromeliads and the "ice plant" that grows in sandy parts of the scrub forest biome.

Characteristic features of CAM plants:

i) They fix CO_2 in the dark because their stomata open at night (when evaporation rates are usually lower) and are usually closed during the day.

ii) They show diurnal pattern of malic acid

Table 11.4.3 : Difference between C_3 and C_4 cycles of photosynthesis	
C_3 cycle	**C_4 cycle**
1. The primary CO_2 acceptor is a C5 com- pound – Ribulose bisphosphate - RuBP.	1. The primary CO_2 acceptor is a C_3 com- pound – phosphoenol pyruvic acid – PEP.
2. The first stable compound is a C_3 compound - phosphoglyceric acid (PGA).	2. The first stable compound is a C_4 compound – oxaloacetic acid (OAA).
3. C_3 cycle is completed in only one type of chloroplast present in mesophyll cells.	3. C4 cycle is completed in two types of chloroplasts, one type occurring in mesophyll cells and other type in the bundle sheath cells.
4. Optimum temperature for this cycle is low to high.	4. Optimum temperature for this cycle is very high.
5. Photorespiration occurs.	5. Photorespiration usually is absent. 6.
6. Photosynthetically less efficient.	Photosynthetically more efficient.
7. CO_2 compensation point is very high.	7. CO_2 compensation point is low.
8. Carbonic anhydrase activity is higher.	8. Carbonic anhydrase activity is low.
9. Rate of translocation of the end product of photosynthesis is low.	9. Rate of translocation of the end product of photosynthesis is high.

formation and in fact they store this acid in the vacuoles of the leaf cells at night that decrease during the day.

iii) They exhibit sunken stomata, thick cuticle thorns and reduced leaves.

iv) They show decreased starch content in the photosynthetic cells during night.

v) They also show increased gaseous exchange during the night.

Biochemical steps of CAM photosynthesis: CAM cycle is completed in two phases acidification (night phase) and deacidification (day phase) (Fig. 11.4.11).

Acidification:

i) CO_2 joins with HCO_3^- produced from cellular water hydrolysis and PEP (formed from the hydrolysis of stored starch through glycolysis) to form the C4 ***oxaloacetic acid*** (OAA) in the presence of the enzyme *PEP carboxylase*.

ii) OAA is reduced to C4 ***malic acid*** in the presence of the enzyme malic dehydrogenase. It is facilitated by the presence of NADP formed during glycolysis. Malic acid accumulates during the night in the central vacuole of the cells.

Deacidification:

i) In the morning, the stomata close. The accumulated malic acid leaves the vacuole and is broken down to pyruvic acid or PEP and CO_2 that is taken up by RUBISCO into the Calvin (C_3) cycle for the formation of carbohydrate.

ii) Pyruvate is first converted into PEP in the presence of the enzyme pyruvate orthophosphate dikinase. PEP is converted into 3PGA (phosphoglyceric acid) by reverse reactions of glycolysis. 3PGA is utilized in Calvin cycle.

Adaptive Value:

i) Better water use efficiency than C_3 plants under arid conditions due to open stomata at night when transpiration rates are lower.

ii) When conditions are extremely arid, CAM

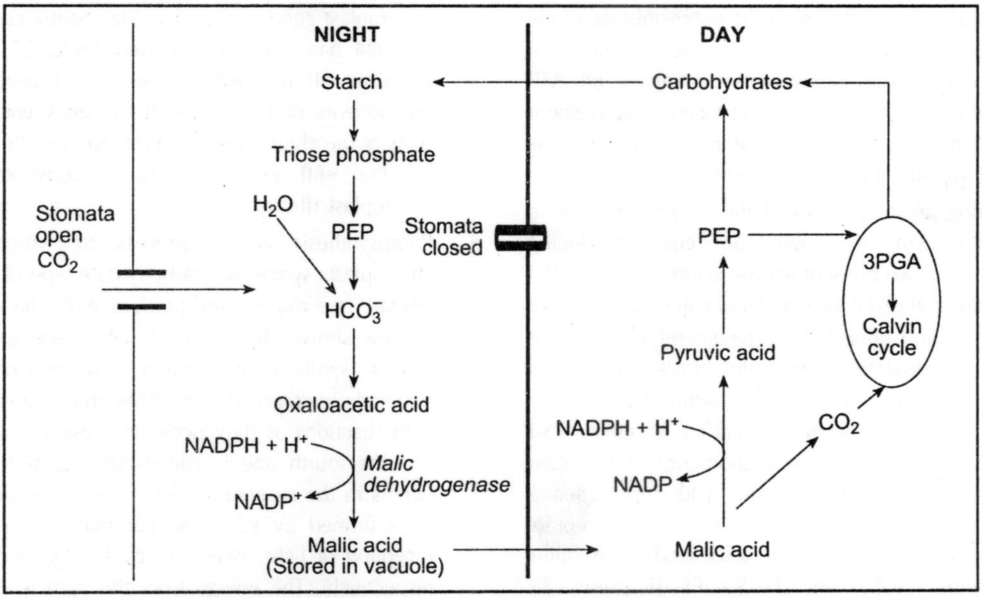

Fig. 11.4.11 : Biochemical steps of CAM photosynthesis.

plants leave their stomata closed night and day. Oxygen given off in photosynthesis is used for respiration and CO_2 given off in respiration is used for photosynthesis. This type of CAM-photosynthesis allows the plant to survive over dry spells, but they recover very quickly when water is available again.

iii) As CAM plants can fix CO_2 in the night, they can survive for longer periods in the light without CO_2 uptake.

iv) As CO_2 is taken during night, it limits photosynthesis. The stored organic acid causing slow growth rates as noted in CAM plants also augments it.

SUMMARY

1. Plastids are major organelles found only in plants and algae. The plastid genome contains about 100 genes encoding rRNAs and tRNAs as well as proteins involved in photosynthesis and plastid gene transcription and translation. However, these proteins only represent a small fraction of the total required proteins. Therefore nuclear genes encode the rest majority of plastid proteins. Plastid DNA exists as large protein-DNA complexes associated with the inner envelope membrane called 'plastid nucleoids'. Each nucleoid particle contains more than 10 copies of plastid DNA. In plant cells long thin protuberances called stromules sometimes form and extend from the main plastid body into the cytosol and interconnect several plastids. Proteins and presumably smaller molecules can move within stromules. Most plants inherit plastids from only one parent, usually mother. Algal plastids also differ from higher plant plastids in having pyrenoids. Plastids are thought to have originated from endosym biotic cyanobacteria.

2. **Chloroplast:** The shape and size of chloroplast vary widely. The chloroplasts of plants grown in shady areas and polyploid plants are comparatively larger and contain more chlorophyll. It is invested by a double layered membrane. The inner membrane encloses a jelly like substance called stroma. It contains about 50% of the total proteins of the chloroplast. It shows stacks of flattened disks called thylakoids enclosing an empty lumen. The stacks are called grana which are interconnected by thread stroma lamellae. Thus grana are called stacked thylakoids and stroma lamellae are called unstasked thylakoids.

3. Photosynthesis takes place on the membrane of the thylakoids. Thylakoid shows five special com ponents – pigment system II, cytochrome b/f, ATP synthase, light harvesting complex and pigment system I. The membrane proteins bind chlorophylls and various accessory pigments.

   During photosynthesis, light is absorbed by a photosystem to begin an energy-producing reaction. Two types of photosystems (PS) exist - PSI (P700) and PSII (P680). PSI and PSII receive light of 700nm and 680nm wave length respectively. PSI use ferredoxin-like iron-sulfur cluster proteins as terminal electron acceptors, while PSII shuffles electrons to a quinone electron acceptor. PS-I reaction center is a monomeric multiprotein complex. It contains about 100 core antenna chloro phylls. Additional electron acceptors include two membrane-associated iron–sulfur proteins, Fe–S_A, and Fe–S_B. PS II shows 25 polypeptides and is surrounded by a variety of chlorophyll a and b binding proteins, known as light harvesting complexes (LHC), which transmit light into the complex. Ca^{++} is bound to the PsbO protein of PSII. Dimeric PSII dominate in the stacked part of the thylakoid granum while the monomeric PSII prevails in the stroma lamellae. Considerable amounts of PSII monomers lacking the CP43 protein and PSII reaction centers (D1/D2/ Cyt b559 complex) were found in the stroma lamellae. PSII core complex shows two β-carotenes bound to the D2 protein that protect PSII from damage. Both photosystems consists of two closely linked complexes†– a reaction centre where the primary events of photosynthesis occurs and an antenna complex consisting of numerous proteins called light harvesting complexes (LHCs) which capture light and transmit it to the reaction centre. Both reaction centres and LHCs contain tightly bound light absorbing pigments. Chlorophyll a is the principal pigment; in addition chlorophyll b and carotenoids act as light absorbing pigments.

4. Chloroplast genome resembles large bacterial plasmids or small chromosomes. It contains genes encoding the major rRNAs and about half of the required ribosomal proteins. No major chloroplast function has been ascertained which can be fully accomplished by proteins encoded solely by the chloroplast. All chloroplast functions require imported proteins encoded by the nuclear genome. Chloroplast genome found in Marchantia contains 121,024 base pairs in a closed circle. 37 genes encoding all the tRNA molecules. 9 genes for components of PSI and PSII, 6 genes encoding parts of the chloroplast ATP synthase and 19 genes for the ~60 proteins used to construct the chloroplast ribosome.

5. Photosynthesis is the process by which the chloroplasts synthesize carbohydrate type of food, liberate free oxygen and produce ATP. The whole process shows absorption of light; generation of oxygen; synthesis of ATP and conversion of CO_2 into carbohydrates. The first three phases are called light reactions as they occur in presence of light and the fourth one is called dark reaction as it occurs in the absence of light. Absorption of light is performed by PS-II and PSI that as per their reception of light waves are called P_{680} and P_{700} respectively. The energy from the light is at first transferred to bridging chlorophyll and then to one of the two 'special pair' chl-a in the associated reaction centre. It rapidly transfers an electron to an electron acceptor – plastoquinone (Q_B) via pheophytin and Q_A. The protons released, remain in the thylakoid lumen. Operation of Q cycle translocates additional protons across the membrane to the thylakoid lumen. This type of electron transport generates increased proton motive force.

6. After the cytochrome bf complex accepts the electrons from QH_2, it transfers them to the Cu^{++} containing electron carrier plastocyanin reducing it to Cu^{+}· $NADP^{+}$ picks up one proton (H^+) from the stroma to form NADPH. This phase of electron flow furnishes linear electron flow. There it releases two electrons to the cytochrome bf complex and two protons to the thylakoid lumen. The electrons return to the PSI through plastocyanin but it does not involve PSII for which this type of electron flow is called cyclic electron flow. As a result, PSI and PSII equally function and exhibit linear electron flow.

7. ATP synthase or F_1F_0 complex shows membrane bound F_0 and protruding F_1 complex. 'c' subunit of F_0 participate in the proton movement. Chemiosmosis involves a proton (H^+) concentration gradient and voltage gradient across the membrane generating proton motive force to drive ATP synthesis. Protons from the thylakoid lumen rotate

its 'c' subunits and the b subunits of F_1 complex to produce ATP. Actually 2 protons are required to form one ATP molecule.

8. The first phase in the synthesis of sugars from CO_2 is the fixation of CO_2 in the stroma by Rubisco. It forms 3-phosphoglycerate. Due to the formation of this 3-carbon compound as the first formed compound these plants are called C-3 plants. Some of the phosphoglycerate are used in the synthesis of sucrose while some are used in the regeneration of the enzyme following Calvin cycle.
9. Many angiosperms (e.g. corn, maize, sugarcane, sorghum etc.) utilize a two-step pathway to fix CO_2. It involves two types of cells, mesophyll cells and chloroplast containing bundle sheath cells surrounding the vascular bundle tissues (called Krantz anatomy). In the mesophyll cells phosphoenolpyruvate (PEP) carboxylase reacts with CO_2 to from oxaloacetic acid (OAA) that is converted to 4 carbon compounds - malic acid or aspartic acid (These plants are thus called C-4 plants). These compounds are then transferred to the cells surrounding the vascular bundle having enzymes of Calvin cycle. Here the malic acid is converted to pyruvic acid and CO_2. Pyruvic acid is transferred back to the mesophyll cells where it is converted to PEP and repeat the cycle.

QUESTIONS

Long questions:

1. Describe the structure of chloroplast.
2. Describe the molecular steps in the light reactions of photosynthesis.
3. Describe the Dark reactions of photosynthesis.
4. Describe the Hatch and Slack pathway of photosynthesis. How C3 and C4 pathways differ.
5. What do you mean by CAM? Describe the pathways in CAM photosynthesis.

Short questions:

1. Classify plastids according to their functions.
2. Differentiate chloroplasts and etioplasts.
3. What do you mean by plastid nucleoid?
4. What are stromules?
5. Name the plastid that occurs in parasitic protozoa.
6. What are klep toplastids?
7. Which water conducting channeled proteins is noted in chloroplast?
8. What are stroma and grana?
9. What is the main enzyme found in the stroma of C_3 plants?
10. What do you mean by stacked and unstacked thylakoids?
11. Prove that chloroplasts are modified endosymbiotic bacteria?
12. What do you mean by antenna system?
13. What do you mean by 'special pair' of chlorophyll molecules?
14. What do you mean by 'light harvesting complex'?
15. What do you mean by 'oxygen evolving complex'?
16. Name the inorganic ions attached to the oxygen-evolving complex.
17. What are the criteria of the 'first exited state' of the energized chl-a?
18. Describe 'Q' cycle.
19. Differentiate cyclic and non-cyclic photophosphorylation.
20. How ATP is synthesized in the light phase of photosynthesis?
21. What do you mean by carbon fixation?
22. What is Calvin cycle?
23. What are the basic differences between C3 and C4 plants?
24. Define photon.
25. What happens to an electron when photon is absorbed?
26. Differentiate absorption spectrum and action spectrum.
27. Differenciate a pigment and photoreceptor.
28. Differentiate PS-I and PS-II.

MCQs:

1. **Granal membrane mainly contains**
(a) PSI, (b) PSII, (c) ATP synthase.
2. **Which of the components are found in both stroma lamellae and thylakoid membranes.**
(a) PSI, (b) PSII, (c) Cytochrome b6-f.
3. **PSI is also known as**
(a) P620, (b) P680, (c) P700.

4. **PSII is also called**
(a) P620, (b) P680, (c) P700. yield,
5. **The number of light quanta required for evolution of one molecule of O_2 is called**
(a) oxygen yield,
(b) photosynthetic
(c) quantum yield.
6. **Red drop is a decline in**
(a) O_2 yield,
(b) quantum yield,
(c) photosynthetic yield.
7. **In the light phase of photosynthesis**
(a) ATP,
(b) Glucose,
(c) both ATP and NADPH is formed.
8. **The photosynthetic unit is called**
(a) quantum, (b) quantasome, (c) oxysome.
9. **Photophos phorylation consists of**
(a) cyclic and non cyclic phosphorylation,
(b) oxidative phosphorylation,
(c) substrate phospho rylation.
10. **Rate of photosynthesis is maximum in**
(a) blue light, (b) green light, (c) red light.
11. **The excess energy of electron is used in the synthesis of ATP by the process**
(a) oxidative phosphorylation,
(b) photophosphorylation,
(c) phosphorylation.
12. **Photosynthetic pigments are located in the chloroplast which are found in**
(a) thylakoid lumen,
(b) thylakoid membrane,
(c) interthylakoid space.
13. **The source of proton within the chloroplast is**
(a) water, (b) exited chlorophyll, (c) CO_2.
14. **In cyclic photophosphorylation the electron released is accepted by**
(a) ferredoxin, (b) $NADPH^+$, (c) reaction center.
15. **In non cyclic photophosphorylation, the electron released by p_{680} is replaced by**
(a) NADP, (b) water, (c) CO_2.
16. **The wave lengtt effective in photosynthesis is**
(a) red, (b) violet, (c) green.
17. (a) PSI, (b) PSII, (c) plastocyanin **can transfer electron to $NADP^+$ during light reaction.**
18. **Sensitive to longer wave length of light is**
(a) PSII, (b) cytochrome b/f, (c) PSI.
19. **As compared to C_3 plants,**
(a) 6, (b) 10,
(c)12 more ATP molecules are needed for the formation of one molecule of hexose sugar.
20. **Emerson effect indicates the existence of two**
(a) pigment systems,
(b) photosynthetic units,
(c) photophosphorylations.
21. **In C_4 plants Calvin cycle enzymes are absent in the**
(a) mesophyll,
(b) bundle sheath,
(c) guard cell chloroplasts.
22. **The element essential for the photolysis of water is**
(a) N, (b) C, (c) Cl.
23. **Red drop occurs at a wave length**
(a) 600-650nm,
(b) 650-680nm,
(c) more than 680 nm.
24. **The specific function of light energy in the process of photosynthesis was**
(a) split water,
(b) activate chlorophyll,
(c) synthesize glucose.
25. **C_4 plants are adapted to**
(a) temperate dry,
(b) temperate humid,
(c) tropical dry climate.
26. **High CO_2 compensation point is found in**
(a) C_3, (b) C_4, (c) CAM plants.
27. **In C_4 pathway, CO_2 is accepted by**
(a) PGA, (b) PEP, (c) RuBP.
28. **In C_3 plants the first stable product is**
(a) PGA, (b) PEP, (c) RuBP.
29. **In PSII the active chlorophyll is**
(a) P_{680}, (b) P_{700}, (c) P_{660}.
30. **The quantum yield of photosynthesis is**
(a) 33%, (b) 12%, (c) 9%.
31. (a) 12, (b) 18, (c) 30 ATP **and $NADPH_2$ molecules are needed for the synthesis of one molecule glucose in C_4 plants.**
32. **The rate limiting step in photosynthesis is**
(a) O_2 evolution,
(b) CO_2 diffusion to photosynthetic site,
(c) dark reaction.
33. **Sugar cane is a**
(a) C3, (b) C4, (c) CAM plant.

34. **C_4 plants can start photosynthesis with a lower concentration of CO_2 because**
(a) respiration in C_4 plants is lower,
(b) C_4 plants do not have photorespiration,
(c) C_4 have photorespiration.

35. (a) O_2, (b) CO_2, (c) chlorophyll **is the limiting factor in photosynthesis.**

36. **If a plant is provided with 0.3% CO_2, the plant will**
(a) die,
(b) will show reduced photosynthesis,
(c) will show normal photosynthesis.

37. **The law of limiting factor was advocated by**
(a) Krebs, (b) Hill, (c) Blackman.

38. **The molecule used in electron transfer in photosynthesis is**
(a) FAD, (b) NADP, (c) cytochrome.

39. (a) Wheat, (b) rice (c) sugar cane **shows chloroplast dimorphism.**

40. **If day light hours is increased the rate of photosynthesis**
(a) increases, (b) decreases, (c) remains unchanged.

41. **Electron donor to PS-I is**
(a) *cyt-f*, (b) plastocyanin, (c) *cyt-b_6*.

42. **Cholorophyll absorbs**
(a) red light only,
(b) blue light only,
(c) blue and red light.

43. **In cyclic photophosphorylation,**
(a) PSI, (b) PSII, (c) both PSI and II operate.

44. **In cyclic photophosphorylation, ATP is synthesized when electrons pass from**
(a) ferredoxin to *cyt b6*,
(b) plastoquinone to *cyt-f*,
(c) ferredoxin to cyt-*b* and plastoquinone to *cyt-f*.

45. **In non cyclic photophosphorylation, ATP is formed when electron is passed from**
(a) plastoquinone to *cyt-f*,
(b) *cyt-f* to plastoquinone,
(c) *cyt-b* to *cyt-f*.

46. **CO_2 fixation in bundle sheath cells of**
(a) C_4 plants, (b) C_3 plants, (c) CAM plants.

47. **C_3 and C_4 plants differ with respect to**
(a) the number of ATP molecules consumed,
(b) substance that accepts CO_2,
(c) both.

48. **Photolysis is used in**
(a) oxidation of NADP,
(b) reduction of NADP, (c) oxidation of FAD.

49. **In C_4 plants the enzyme used in the fixation of CO_2 is**
(a) RuBP carboxylase,
(b) PEP carboxylase,
(c) PGA dehydrogenase.

50. **Carotenes protect the chlorophylls from**
(a) Photooxidation,
(b) photorespiration,
(c) desiccation.

51. **C_3 and C_4 differs in the**
(a) PSs,
(b) end products of photosynthesis,
(c) CO_2 acceptance.

52. **Krantz anatomy is found in**
(a) C_3, (b) C_2, (c) C_4 plants.

53. **CAM photosynthesis occurs in plants with** (a) thin leaves, (b) thick leaves, (c) coloured leaves.

54. **Calvin cycle takes**
(a) 4, (b) 6,
(c) 10 turns to generate one molecule of hexose sugar.

55. **Source of CO_2 in photosynthesis during day in CAM plants is**
(a) malic acid,
(b) PGA,
(c) phosphoenol pyruvic acid.

56. **Photorespiration involves**
(a) chloroplast,
(b) ER,
(c) chloroplast-peroxisome-mitochondria

57. **Which statement about oxygenic photosynthesis is FALSE?**
(a) Oxygenic photosynthesis is a type of photosynthesis in which CO_2 and water, in the presence of light and chlorophyll, is converted into glucose and oxygen.
(b) Oxygenic photosynthesis is responsible, at least in part, for the oxygen present in today's earth's atmosphere.
(c) Oxygenic photosynthesis is the only type of photosynthesis in existence today.

58. **The individual who demonstrated, using isolated chloroplasts, that splitting of water by light could**

reduce a dye solution as well as releasing oxygen is
(a) Robin Hill, (b) Calvin, (c) Krebs.

59. **All are true for non-cyclic electron transport EXCEPT:**
(a) The components of electron transport are arranged asymmetrically across the thylakoid membrane.
(b) Electron transport establishes a proton gradient across the thylakoid membrane.
(c) Electron transport between PS II and PSI produces ATP by substrate phosphorylation.

60. **Which one is correct? Photoexcited P680 and P700**
(a) have the same chemical properties of the ground state molecules.,
(b) Are strong oxidants—taking electrons from donor molecules.,
(c) Are strong reductants—giving electrons to acceptor molecules.

61. **Evolution of CO_2 and uptake of O_2 in photosynthetic tissues that occurs only in the light is called**
(a) dark reaction,
(b) photorespiration,
(c) Oxygenic photosynthesis.

62. **Pheophytin is a chlorophyll in which central Mg is replaced by**
(a) Fe, (b) 2H, (c) O_2.

63. **Photodynamic damage in chloroplast is prevented by**
(a) carotenoids, (b) phycobillin, (c) LH(C)

64. **Rieske center is**
(a) plastoquinone,
(b) Plastocyanin,
(c) Cyt b_6 –*f* complex.

65. **Non-cyclic electron transport is also described as**
(a) S, (b) Z, (c) Q scheme.

66. **Q_{10} for photosynthesis normally is**
(a) 1, (b) 2, (c)3.

67. **Kranz cells show**
(a) very minute
(b) significant amount
(c) no chloroplast.

Answers to MCQs:

1. b, 2. c, 3. c, 4. b, 5. c, 6. c, 7. c, 8. b, 9. a, 10. c, 11. b, 12.b, 13. a, 14. c, 15. b, 16. c, 17. a, 18. c, 19. c, 20. a, 21. b, 22. c, 23. c, 24. b, 25. c, 26. a, 27. b, 28. a, 29. b, 30. b. 31. c, 32. b, 33. c, 34. b, 35. b, 36. c, 37. c, 38.c, 39. c.40. c. 41. b, 42. c, 43. a, 44. c, 45. a, 46. a, 47. c, 48. b, 49. b, 50. a, 51. c, 52. c, 53. b, 54. b, 55. a, 56. c, 57. c, 58. a, 59. c, 60. c, 61. b, 62. b, 63. a, 64. c, 65. b, 66. b, 67. b.

Chapter 11.5

Ribosome

Ribosomes are the fundamental organelle needed for the synthesis of protein. Ribosomes are composed of rRNA and protein and play the most crucial role in the translation of the genetic code into proteins. Prokaryotic and eukaryotic ribosomes do not differ fundamentally because the genetic code is same in all organisms.

Ribosomes are naked cellular organelle composed of proteins and rRNAs, and are essential for protein synthesis.

11.5.1. TYPES OF RIBOSOMES

A. ***On the basis of sedimentation coefficient values:***

(i) On the basis of Sedimentation coefficient value (S) ribosomes are of three types – 64-72S (average 70S) type (mainly found in prokaryotes), 80S type (mainly found in eukaryotes) and 55S type (mainly occur in the mitochondria and chloroplast matrix).

(ii) They basically consist of one small and one large subunits.

(iii) Both subunits show separate S values. 70S ribosomes have 30S and 50S subunits, 80S ribosomes have 40S and 60S subunits and 55S ribosomes consist of 28S and 39S subunits. S values of the two subunits do not equal to the S values of the entire ribosome because the coefficients are not proportional to the particle weight.

(iv) In addition to these functional categories, in the nucleoli, during the biogenesis of ribosomes, the proteins and rRNA unite to form *pre ribosomes* (S value - 90S) that mainly gives rise to 80S ribosomes.

B. ***On the basis of occurrence in the cell:***

(i) Except in mature mammalian RBC and sperm, ribosomes are found in mainly two forms - *free ribosomes* (not attached to any organelle) and *bound ribosomes* (remain attached to the wall of ER or nucleus).

(ii) Free ribosomes occur in the cytosol and also inside the mitochondria (*mitoribosomes*) and chloroplast either singly or in groups. Mostly the subunits of the ribosomes remain separate but at the time of protein synthesis they unite with the involvement of Mg^{++}. The unified ribosomes get attached with the mRNA.

(iii) During protein synthesis many ribosomes assemble with one mRNA strand and are collectively known as *polyribosomes or polysomes* showing assemblage of up to 30 ribosomes on one strand of mRNA. Bound forms mainly remain actively engaged in the synthesis of secretory proteins. They insert the freshly produced polypeptide chains directly into the ER, from where they are transported to their destinations.

11.5.2. STRUCTURE OF RIBOSOMES

(i) Ribosomes are membrane less organelle, composed of mainly of two sub units - small and large. Diameter of the 70S ribosome is 260Å , it measures 170/230/250Å.

(ii) 30S subunit appears slightly bent or elongated. A cleft divides it into a globular *head* and *body*.

(iii)Head has two projections, nose and beak.

(iv)Body shows a shoulder, an arm like flexible platform and a compact foot with toe or spur like structures (Fig. 11.5.3).

(v) The upper part of the body and lower part of the head forms the *neck* that organizes mRNA – tRNA translocation and controls codonanticodon interaction. It creates a long curved channel. mRNA enters the channel through a pore, guarded by two proteins that from *nose.* These proteins are referred as *'latch'*.

(vi)Foot controls the orientation of the inter subunit interaction; platform contributes to the formation of *peptidyl or 'P' site* to which tRNA is transferred from *'A'* site.

(vii) P-site remains connected by a protein rich belt with shoulder. Thus the 30S subunit initiates mRNA engagement, decodes the message, governs mRNA and tRNA translocation and controls the fidelity of codon-anticodon interaction.

(viii) 30S subunit shows 16S rRNA (1542 nt long) and 21 types of proteins. The head part constitutes protein rich but the body part shows rRNA rich zones.

(ix)40S subunit shows three main parts - *head, body* and *platform.* Due to the presence of 20 -30

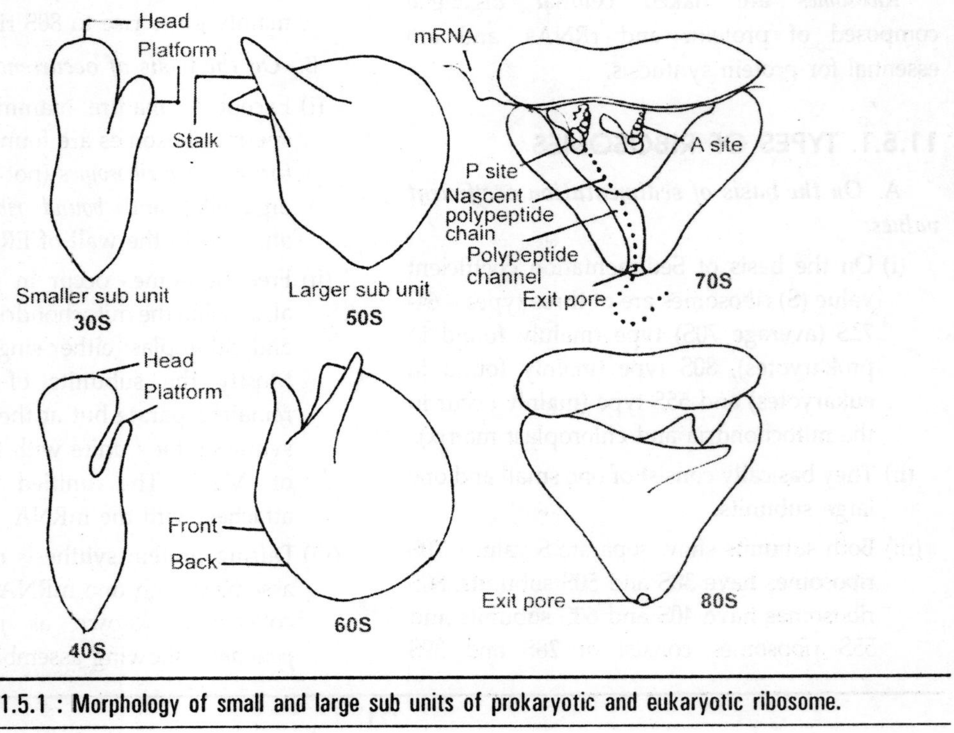

Fig. 11.5.1 : Morphology of small and large sub units of prokaryotic and eukaryotic ribosome.

additional proteins, it is longer than 30S subunit. Its extra density occurs at the bottom below the platform (forming the *foot* and *back lobe*) and at the head forming the *head lobe*. Foot shows two lobes called *'left foot'* and *'right foot'*.

(x) 40S subunit shows 18S rRNA [1798 nt long (256 nt longer than that of 30S unit) and 32 types of proteins (11 more than 30S type). From the head portion a beak like projection is given out.

(xi) 50S subunit shows a central projection, *a stalk* and *a ridge* like structure. It shows two types of rRNA - 23S (2904 nt) and 120 nt long 5S.

(xii) The 60S subunit shows 3 sub types of rRNAs 28S (3392 nt long), 5.8S (158 nt long) and 5S (121 nt long) and 49 types of proteins (Fig. 5.5.4). From its central core part some expansion segments project in different directions. 80S ribosome contains about 100 additional accessory proteins and a similar number of small nucleolar RNAs that are closely involved in ribosome biogenesis.

(xiii) 70S ribosome shows 6 while 80S ribosome shows 10 inter subunit bridges.

(xiv) In between two subunits, a tunnel is produced (10 nm long and 2.5 nm in diam.).

(xv) The unified ribosomes show slight movement of different parts. 16S rRNA that forms inter subunit bridge shows undulating movement.

(xvi) In the complete ribosomes there are three functional sites –acceptance site or *A site* where amino acyl tRNA are accepted, *peptidyl or P site*, where peptide bonds are formed between the amino acids of 'A' and 'P' sites and 'E' site where deacylated tRNAs stay temporarily before leaving the ribosome (Fig. 5.5.4). A and P sites are formed in both subunits but E sites occur only in large subunit. The unified ribosomes show several functional sites.

(xvii) The interior of the large subunit shows voids, inter connected channels that permeate the entire particle. Some of these interior voids may be cavities, with less than 10Å diameter. As most of the RNAs in the ribosome are helical and as they pack inefficiently, the gaps between them bind water molecules which makes the interior of ribosome wet.

(xviii) Newly synthesized polypeptide chain traverses a distance of about 100Å from the P site to the outer surface of the ribosome through an *exit tunnel*. The shapes and sizes of the molecules that pass through it determine the size and shape of the tunnel. Its wall shows 82% RNAs and 18% proteins. It suggests that nascent peptides leave the ribosome by two routes, one used by the

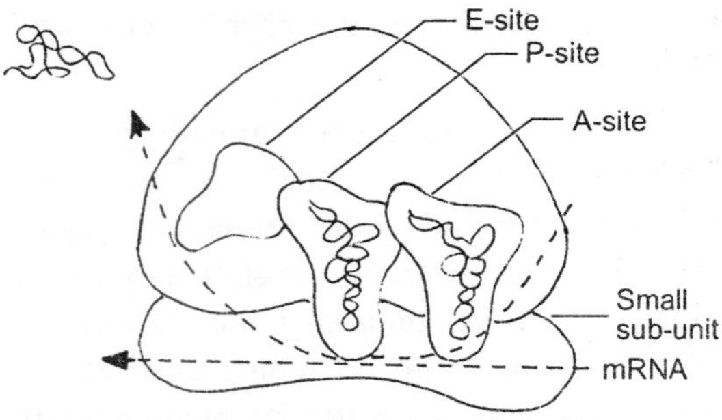

Fig. 11.5.2 : A, P and E sites of ribosome.

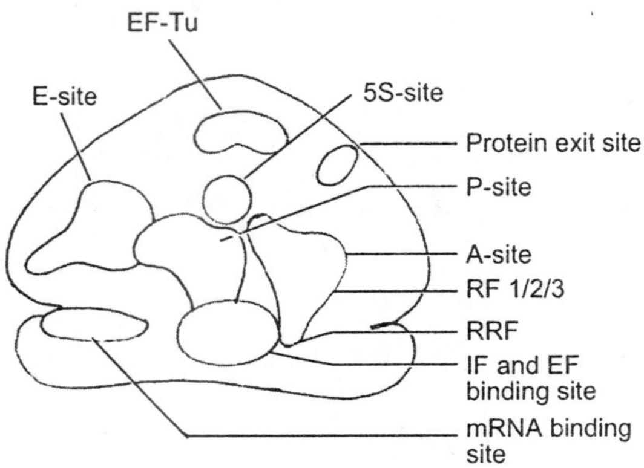

Fig. 11.5.3 : Important functional sites in ribosome. EF-Tu = Elongation factor Tu, RF = Release factor, RRF = Ribosome recycling factor, IF = Initiation factor, EF = elongation factor.

membrane proteins and the other by cytoplasmic proteins.

During the exponential growth phase of bacteria, after the "Recycling" stage, ribosomes can readily proceed to the "Initiation" stage of protein synthesis. During the stationary growth phase, *E. coli* 70S ribosomes are converted to 100S ribosomes, with no translational activity. This conversion is caused by binding of the ribosome modulation factor (RMF) to 70S ribosomes (Ortiz *et al*, 2010). 100S ribosome is formed by the dimerization of two tRNA-free 70S ribosomes 70S ribosomes mediated by face-to-face contacts between their constituent 30S subunits and binding of RMF near the ribosomal proteins S13, L13, and L2. The positions of these proteins indicate that the RMF binding site is near the peptidyl transferase center or the P site (Yoshida *et al*, 2004; Kato *et al*, 2010). Due to inactivation of ribosomes this phase is called "*Hibernation*" stage. Before the formation of 100S ribosome, hibernation promoting factor (HMF) binds and forms 90S ribosome to which YhbH protein binds and converts immature 90S ribosomes into mature 100S particles (Ueta *et al*, 2005, 2008). In fact prior to 100S ribosomes formation they change into forms which cannot bind with the dissociation factor - IF3 that aid in the formation of100S ribosomes by RMF and HPF (Yoshida *et al*, 2009). The binding site of RMF overlaps with that of the mRNA Shine-Dalgarno sequence, which prevents the inter action between the mRNA and the 16S rRNA. The binding sites. of HPF and a protein YfiA overlap with those of the mRNA, tRNA and initiation factors, which prevents translation initiation. The binding of RMF and HPF, but not YfiA, to the ribosome induces a conformational change of the 30S head 100S dimer formation (Polikanov *et al*, 2012). Thus two mechanisms of 100S ribosome formation are known: one involving RMF and HPF and the other involving only HPF in the majority of bacteria. The expression of RMF is regulated by ppGpp and cyclic AMP-cAMP receptor protein (cAMP-CRP) induced by amino acid starvation and glucose depletion, respectively. When stress conditions are removed, the 100S ribosome immediately dissociates into the active 70S ribosomes by releasing RMF (Yoshida and Wada, 2014). in addition to the mere presence of 100S, the correct timing of 100S formation and its break

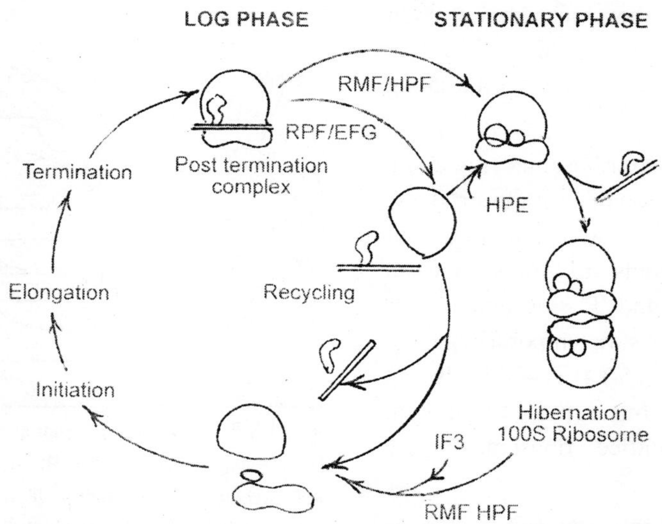

Fig. 11.5.4 : Diagramatic representation of 100S Ribosome formation.

down is required to maintain viability. Thus 100S acts as a tentative repository of ribosomes that are protected from degradation and provide a source of amino acids in later growth period (Shcherbakova *et al*, 2015). Similar to bacteria, rat cells contain stress-induced ribosomal dimers. Two 80S ribosomes are connected by a flexible structure to form 110S ribosomal complex during nutritional starvation (Krokowski *et al* 2011).

11.5.3. STRUCTURE OF MITORIBOSOME (after Kaushal *et. al.*, 2015)

(i) Mitochondria and chloroplasts are believed to have evolved from endosymbiotic bacteria; therefore their ribosomes exhibit closer structural and functional similarities with prokaryotic ribosomes. The overall 3D structure of the mammalian mitoribosome was found to be significantly porous as compared with its bacterial and eukaryotic cytoplasmic counterparts (Sharma *et al*, 2013).

(ii) Mitoribosomes consist of two subunits SSU - of 28S and LSU of 39S values. It shows a few large masses protruding from the relatively compact and globular core. The diameter of the unified ribosome is 320Å (60Å larger than prokaryotic ribosome).

(iii)The smaller subunit is 70Å longer and 15Å narrower than its bacterial counter part. Its head part is enlarged and displays a peculiar obliquely oriented channel. It lacks a beak like structure and its shoulder part is narrow. The most striking feature is the presence of a triangular gate like structure (*mRNA gate)* that covers the mRNA entry site; it may facilitate recruitment of leaderless mitochondrial mRNAs.

(iv)It shows a polypeptide exit tunnel that has an unusual nascent-polypeptide exit mechanism.

(v) Larger subunit is comparatively larger than 50S subunit. From the central protuberance a finger like projection is given out which interact with the 'P' site. Strong binding of tRNA to the 'P' site suggests that 'E' site is very weak. Mitochondrial tRNA has become an integral component of the central protuberance of the 39S subunit where it architecturally substitutes for the absence of the 5S ribosomal RNA, a ubiquitous component of all cytoplasmic ribosomes. The two subunits are held together by 15 inter subunit bridges, 5 of which are formed by RNA-RNA, 2 by RNA-protein and 8 by protein-protein interactions Kaushal *et al*, 2014). Both subunits can move relative to each other.

(vi)In between two subunits there is a tunnel like part having a wide opening before the 'E' site.

(vii) The evolutionary divergence of mitoribosomes has shown a reduction in their rRNA content and an increase in ribosomal proteins compared to their bacterial and cytoplasmic counterparts. The protein and RNA constituents of mitoribosomes differ widely from that of 70S ribosomes. The SSU shows 12S rRNA and 29 types of proteins while larger subunit shows 16S rRNA and 48 types of proteins. Protein: RNA ratio is 39:67 in respect of 69:31 in prokaryotes. 13 polypeptides encoded by the mitochondrial genome are synthesized by mitoribosomes (Richman *et al*, 2014). LSU is involved in catalyzing the peptidyl transferase reaction, and the SSU provides the platform for mRNA binding and decoding (Greber *et al*, 2015).

11.5.4. CHEMICAL COMPOSITION OF RIBOSOMES

Chemically ribosomes show protein and rRNA (Table 11.5.1). The small ribosomal subunit (SSU) of *E. coli*, consists of 16S rRNA (1540 nts) and 24 proteins. The large ribosomal subunit (LSU) consists of 23S rRNA (2913 nts) and 34 proteins. The protein distribution is *E. coli* ribosomes is customarily written as (24/34) meaning 24 proteins in the SSU and 34 proteins in the LSU. Thus bacterial ribosome has 58

Table 11.5.1 : Comparison of Prokaryotic, Eukaryotic, and Mitochondrial Ribosome

	Prokaryotic (70S)	*Eukaryotic (80C)*	*Mitochondrial (55S)*
Small Subunit	**30S**	**40S**	**28S**
rRNAs	16S (1542 nts)	18S (1798 nts)	12S (950nts)
Proteins	21	33	29
Large Subunit	**50S**	**60S**	**39S**
rRNAs	5S (120 nts), 23S (2904 nts)	5S (121 nts), 5.8S, (158 nts) 28S (3392 nts)	16S (1560 nts)
Proteins	33	49	48

proteins, archaeal ribosomes have 68 proteins (28/40) and Eukaryotic ribosomes have 78 proteins (32/46). In fact rRNA constitutes 37 - 62% of the ribosome. unlike other RNAs it becomes physiologically mature in the nucleolus. All the major functional features of ribosome mainly depend upon rRNAs and proteins.

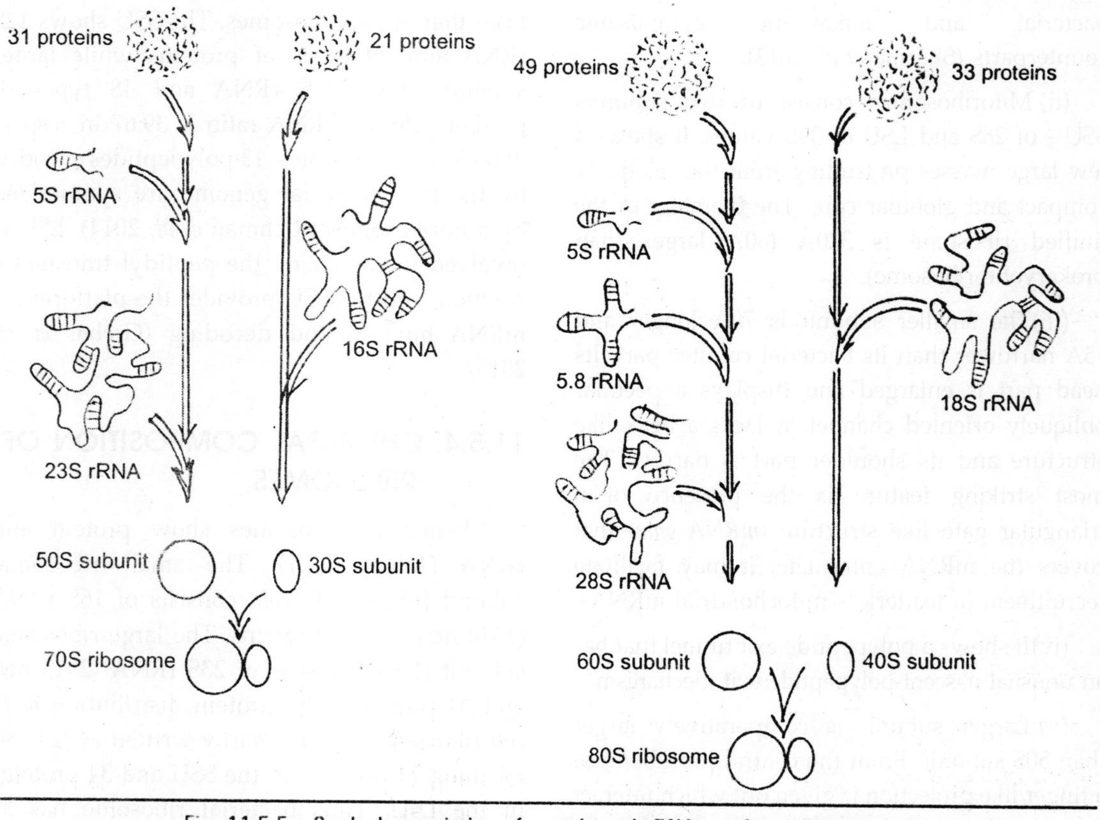

Fig. 11.5.5 : Gradual association of protein and rRNAs to form ribosomes.

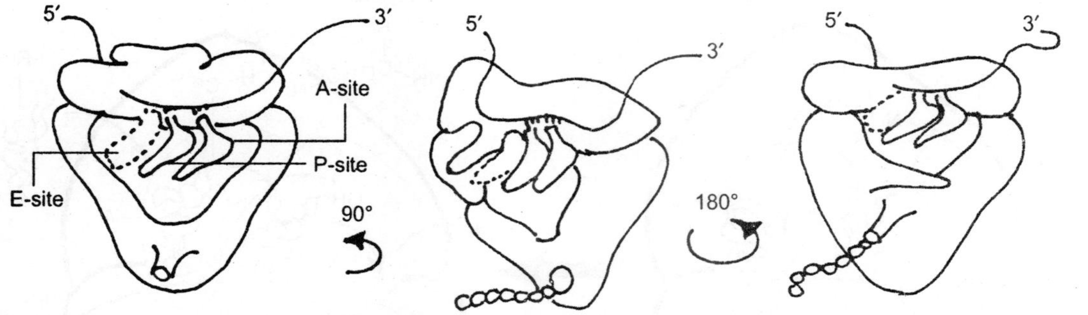

Fig. 11.5.6 : Role of three sites of ribosomes during protein synthesis.

11.5.5. FUNCTIONS

Protein synthesis: Ribosomes are the sole machinery, which play critical roles in the synthesis of protein. Each subunit consists of one or two very large rRNA molecules and many protein molecules. rRNAs are the crucial components in protein synthesis; while protein components act as a scaffold that may enhance the ability of rRNA to synthesise protein. Ribosome displays many active sites, each dedicated to one of the tasks required for translation of mRNA into protein. Two kinds of ribosomes (free and bound) vary based on the nature of the synthesized protein. Free ribosomes synthesize proteins that are used within the cytosol (e.g, enzymes of glycolysis) while bound ribosomes synthesize proteins, destined for secretion (by exocytosis), or plasma membrane receptors and for lysosomes. Prior to the beginning of protein synthesis, several ribosomes attach to a single mRNA molecule, giving rise to *polyribosome or polysome*. The major involvements of ribosome in the protein synthesis are as follows:

i) The process begins with the capture of the tRNA carrying an amino acid. This binds to the 'P' site of the small ribosomal subunit.

ii) Initiation complex binds to the 5' end of an mRNA molecule.

iii) The large subunit of the ribosome now joins the complex.

iv) A second tRNA is now brought to the 'A' site of the ribosome by the elongation factor.

v) Peptidyl transferase, which is part of the large ribosomal subunit, mediates the separation of the first amino acid from its tRNA and a peptide bond is formed between the initial methionine and the second amino acid.

vi) The ribosome now moves down the mRNA by one codon, a process called *translocation*. This movement shifts the second AA-tRNA to the 'P' position, as a result a new charged tRNA enters to the empty 'A' site and the uncharged tRNA that previously occupied the 'P' site is translocated to the 'E' site and ultimately leaves the ribosome Translocation of peptidyltRNA and mRNA within the ribosome during protein synthesis is promoted by EF-G and the energy needed for this process is supplied by the hydrolysis of guanosine triphosphate (GTP). After about 30 codons of mRNA get translated a new ribosome joins at the 5′ end of mRNA. As many as 20 ribosomes may join the same mRNA depending upon its length. The process continues until the first stop codon is encountered.

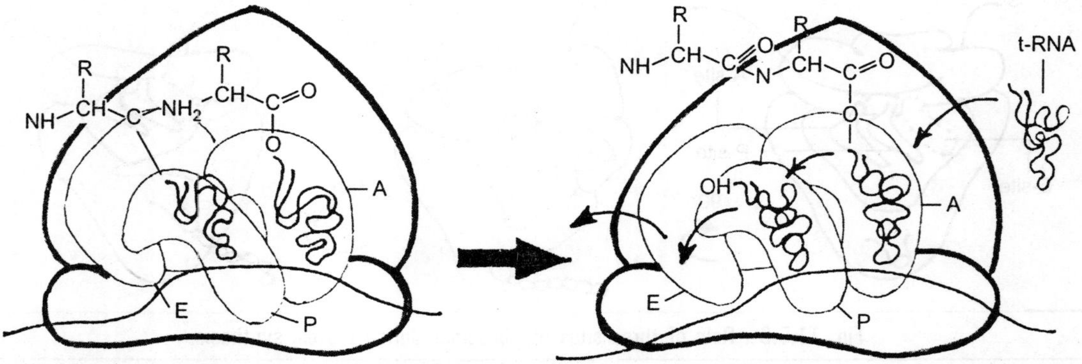

Fig. 11.5.7 : Biochemical steps performed by the three sites of ribosomes during protein synthesis.

A termination factor releases the completed protein from the last tRNA and the ribosome dissociates into its component parts. It is speculated that the nascent peptide chain grows through a channel or groove in the large ribosomal subunit since experimentally ribosomes protect a segment of 30-40 amino acids from degradation. After protein synthesis, 70S ribosomes are dissociated by cooperative action of ribosome recycling factor (RRF), elongation factor G (EF-G), and initiation factor-3 (IF-3). RRF and EF-G not only split ribosomes into subunits but also participate directly in the release of deacylated tRNA and mRNA for the next round of translation. This is called *ribosome recycling*. In case of protein synthesis by mitoribosomes the mRNA occupies an RNA-rich groove that encircles the neck of the SSU The interior of the channel is lined with basic and aromatic amino acids that may directly interact with mRNA.

Protein transport and folding: Eukaryotic ribosome has to play other additional biological processes like protein transport and folding. For the cotranslational translocation of proteins they bind to the Sec61p channel on the ER membrane which allows direct transfer of the polypeptides into the channel. Translocation is mediated by a ribonucleoprotein complex, the signal recognition particle-54 (SRP-54), and its membrane-associated receptor (SR). SRP-54 binds to signal sequences of nascent proteins as they emerge from the exit tunnel of the ribosome. The resulting targeting complex, composed of the SRP and the ribosome-nascent chain complex (RNC), joins with the receptor in a GTP-dependent manner. SRP-54 and SR deliver the RNC to the translocon, which in turn mediates protein translocation across the ER mem brane. Thus it involves three steps - the assembly and interaction of the SRP ribonucleoprotein core, the GTP-dependent SRP-SR association, and the interaction between SRP-54 and the ribosome.

11.5.4. RIBOSOME BIOGENESIS

The synthesis of ribosomes in eukaryotes involves synthesis and processing of pre-rRNA in the nucleolus and their sequential assembly with a large number of ribosomal proteins. In prokaryo the genes coding 16S, 23S and 5S rRNAs are arranged in a series. rRNA transcribed from DNA is a 30S unit which is called p30S. 16S component occurs at the 5′ end while 5S component occurs at the 3′ end. 23S component occurs in the middle. All the components are separated by spacer units. During processing, p30S unit is cleaved into 25S and 18S units by

RNaseIII. These secondary elements are reduced into 23S and 16S units by the enzyme RNase M or Maturase. Several ribosome assembly intermediates and a large number of novel ribosome assembly factors have been identified that are involved in the formation of ribosomes. In higher eukaryotic cells at the beginning of mitosis when rDNA transcription is repressed, the nucleoli disassemble and are no longer observed throughout mitosis. Conversely at the exit from mitosis concomitantly with restoration of rDNA transcription nucleoli assemble and are functionally active throughout interphase.

SUMMARY

1. Ribosomes are naked cellular organelle composed of proteins and rRNAs, and are essential for protein synthesis. Ribosomes basically consist of one small and one large subunits. On the basis of S values these are of three types. 70S ribosomes (prokaryotic) have 30S and 50S subunits, 80S ribosomes (eukaryotic) have 40S and 60S sub units and 55S ribosomes (mitoribosomes) consist of 28S and 39S subunits. Mostly the subunits of the ribosomes remain separate but at the time of protein synthesis the unified ribosomes may remain attached with an mRNA. During protein synthesis many ribosomes assemble with one mRNA strand and are collectively known as polyribosomes or polysomes showing assemblage of up to 30 ribosomes on one strand of mRNA.
2. The diameter of the unified ribosome is 320 A^0 (60A^0 larger than prokaryotic ribosome). Larger subunit is comparatively larger than 50S subunit. Its head part constitutes protein rich but the body part shows rRNA rich zones. 40S subunit shows three main parts - head, body and platform. 70S ribosome shows 6 while 80S ribosome shows 15 inter subunit bridges. The unified ribosomes show slight movement of different parts.
3. Ribosomes are composed of mainly protein and RNAs. 30S subunit shows 16S rRNA (1542 nt long) and 21 types of proteins. The 60S sub unit shows 3 sub types of rRNAs 28S (4718 nt long), 5.8S (160 nt long) and 5S (120 nt long) and 49 types of proteins. 16S rRNA that forms inter subunit bridge shows undulating movement.
4. In the complete ribosomes there are three functional sites – acceptance or A site, where amino acyl tRNA are accepted, peptidyl or P site, where peptide bonds are formed between the amino acids of 'A' and 'P' sites and 'E' site where deacylated amino acids stay temporarily before leaving the ribosomes.
5. The two subunits are held together by 15 inter subunit bridges, 5 of which are formed by RNA-RNA, 2 by RNA-protein and 8 by protein-protein interactions. The protein and RNA constituents of mitoribosomes differ widely from that of 70S ribosomes. The small subunit shows 12S rRNA and 29 types of proteins while larger subunit shows 16S rRNA and 48 types of proteins.

The inner mitochondrial membrane protein Mba1 functions as a ribosome receptor. Chemically ribosomes show protein and rRNA. In fact rRNA constitutes 37 - 62% of the ribosome. Ribosomes are the sole machinery, which play critical roles in the synthesis of protein. Each subunit consists of one or two very large rRNA molecules and multiple smaller protein molecules. rRNAs are the crucial components in protein synthesis; while protein components act as a scaffold that may enhance the ability of rRNA to synthesise protein. Free ribosomes synthesize proteins that are used within the cytosol (e.g., enzymes of glycolysis) while bound ribosomes synthesize proteins, destined for secretion (by exocytosis), or plasma membrane receptors, and for lysosomes. Prior to the beginning of protein synthesis, several ribosomes attach to a single mRNA molecule, giving rise to polyribosome or polysome. The major involvements of ribosome in the protein synthesis are as follows:

This binds to the 'P' site of the small ribosomal subunit. iii) The large subunit of the ribosome now joins the complex. Hirokawa et al (2005) reported that after protein synthesis, 70S ribosomes are dissociated by cooperative action of ribosome recycling factor (RRF), elongation factor G (EF-G), and initiation factor 3 (IF-3). Eukaryotic ribosome has to play other additional biological processes like protein transport and folding. Regulation of ribosomal activities is directly related with the synthesis of ribosomal proteins. Prokaryotes avoid overproduction of ribosomal proteins. The synthesis of ribosomes in eukaryotes involves synthesis and processing of pre-ribosomal RNA (pre-rRNA) in the nucleolus and their sequential assembly with a large number of ribosomal proteins.

QUESTIONS

Long questions:

1. Describe the molecular structure of ribosome and their role in protein synthesis.
2. Describe the structure and mitoribosome and compare it with prokaryotic ribosome.

Short questions:

1. What do you mean by S values? On the basis of this value classify ribosomes.
2. What is preribosome? What is its significance?
3. Name some cells where ribosomes are not found.
4. What is polysome?
5. What are the three sites of ribosomes? How they are formed? What are their roles?
6. What are the basic differences between pro and eukaryotic ribosomes?
7. What are the basic differences between prokaryotic ribosome and mitoribo some? 8
8. What are the functional differences between proteins and rRNAs?
9. What are the functional differences between free and bound ribosomes?

MCQs:

1. **Both subunits of ribosomes unite with the help of**
   (a) Ca^{++}; (b) Na^{+}; (c) Mg^{++}.
2. **Polysomes are the assem blage of**
   (a) 30S; (b) both; (c) 50S subunits.
3. **30S subunit shows**
   (a) 16S, (b) 12S, (c) 18S rRNAs.
4. **50S subunit shows**
   (a) 16S, (b) 23S, (c) 5.8S rRNAs.
5. **40S subunit shows**
   (a) 16S, (b) 18S, (c) 12S rRNAs.
6. **60S subunit shows**
   (a) 16S, (b) 18S, (c) 5.8S rRNAs.
7. **30S subunit shows**
   (a) 33, (b) 29, (c) 21 types of proteins.
8. **40S subunit shows**
   (a) 33, (b) 29, (c) 49 types of proteins.
9. **28S subunit of mitoribosome shows**
   (a) 12S, (b) 18S, 5S rRNAs.
10. **28S subunit of mitoribosome shows**
    (a) 21, (b) 33, (c) 29 types of proteins.
11. **39S subunit of mitoribosome shows**
    (a) 12S, (b) 16S (c) 18S rRNAs.
12. **39S subunit of mitoribosome shows**
    (a) 33, (b) 48, (c) 49 types of proteins.
13. **Chloramphenicol inhibits**
    (a) binding of AA-tRNA,
    (b) binding of small and large subunits,
    (c) peptide bond formation between two amino acids.
14. **Tetracycline interferes with the**
    (a) binding of ribosomal subunits,
    (b) binding of AA-tRNA with the small subunit,
    (c) translocation of tRNA from A to P site.

Answers to MCQs:

1. c; 2. b; 3. a; 4. b; 5. b; 6. c; 7. c; 8. a; 9. a; 10. c. 11.b; 12. b. 13. c, 14. b.

Chapter 11.6

Lysosomes

Lysosomes are cells' garbage disposal organelle. It was discovered in 1950 by Rene de Duve. It is a tiny, roughly spherical organelle found in all eukaryotic cells (both in plants and animals). Lysosomal morphology and number varies with the physiological state of the cell.

11.6.1. STRUCTURE

Lysosome is bounded by a single membrane having unusual phospholipid and cholesterol content. Lysosomal membrane contains several N-glycosylated proteins, including the lysosome-associated membrane proteins - LAMP-1 and LAMP-2 and lysosomal integral membrane protein-2 / lysosomal membrane glycoprotein-85 (LIMP- 2 / LGP85). LIMP II. Transport proteins in the membrane allow the final product of digestion of macromolecules like amino acids, sugars and nucleotides to be transported to the cytosol. In prolonged starvation, lysosomes appear long, thread like called *nematolysosomes*. They contain vacuoles, granules, hydrolytic enzymes that digest macromolecules and break down damaged or old cell parts as well as bacteria. These enzymes are only active within the acidic membrane of the lysosome so that in case of leakage or rupture, the enzymes will not cause any harm for the cell untill the cytosol becomes acidified.

Lysosomal enzymes: Lysosomes contain about 60 different kinds of hydrolytic enzymes - lipase, phospholipases, polysaccharidases, proteases and nucleases. These are produced in the ER and processed during passage through the Golgi apparatus. As they need the acidic environment to function correctly, a H^+-ATPase, found in the membrane of lysosomes acidifies the environment (Fig. 11.6.2). They degrade nucleotides, proteins, lipids, phospholipids and also remove carbohydrate, sulfate or phosphate groups from molecules. All these enzy mes are active at pH 4-5, therefore if they leak out of the lysosome, they can not damage the cell since the nor mal pH of cytosol is around 7.2 but if accidentally pH of the cytosol becomes acidic and lysosomal membrane is ruptured then its

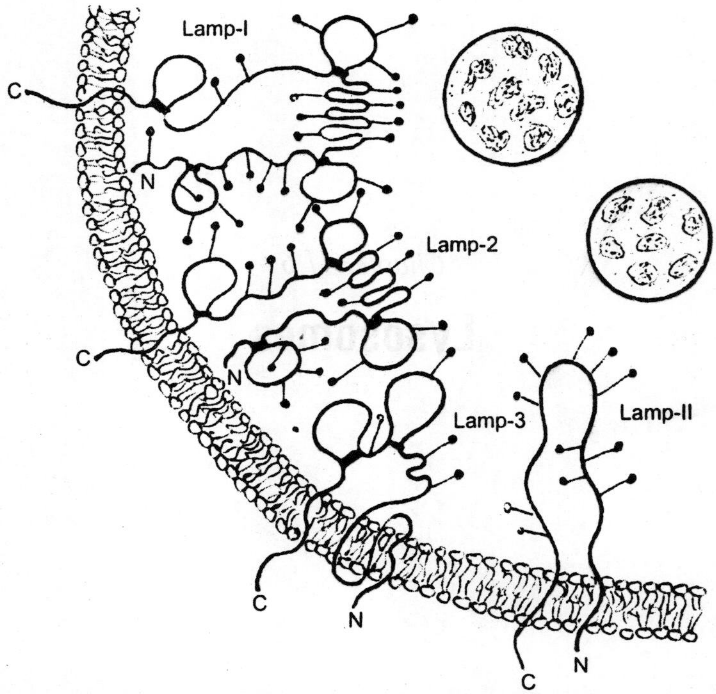

Fig. 11.6.1 : A part of lysosome membrane.

enzymes digest the cell itself. For this reason it is allegorically called *'suicide bag of cell'*.

Lysosome, a meeting place of multiple pathways: Several streams of intracellular traffic converge to lysosome (Fig. 11.6.2). Digestive enzymes are synthesized in the ER and are delivered to the lysosome via Golgi. The endocytosed macromolecules are initially taken up by small irregularly shaped organelle known as *early endosome* from where some of the materials are recycled back to the plasma membrane via *sorting endosomes*. The rest of the materials that are deposited in the early endosome are delivered to *multivesicular bodies* and ultimately to *late endosomes* where the materials first come in contact with the acid hydrolases. Though the interior of the late endosome is slightly acidic (pH 6.0) even then intracellular digestion first begins in this compartment. Mature lysosomes are formed from this endosome by further addition of hydrolase containing vesicles that originate from the TGN and with the accumulation of more H^+ in the interior.

The second pathway that merge with the lysosome is involved with autophagy since the aged and non functional organelles are encircled by the membrane of unknown origin to create autophagosome that fuses with the lysosome or late endosome for intracellular digestion. The third pathway that involves lysosome for degradation is heterophagy or phagocytosis of larger particles and microorganisms.

11.6.2. FUNCTIONS

Autophagy: Lysosome mediated degradation of the non-functional organelles is called autophagy. Initially a membrane cisterna called *isolation membrane* or *phagophore* curves to form a double layered vesicle, called *autophagosome which* are translocated through the microtubules to the lysosome for degradation (Fig.11.6.3). One or

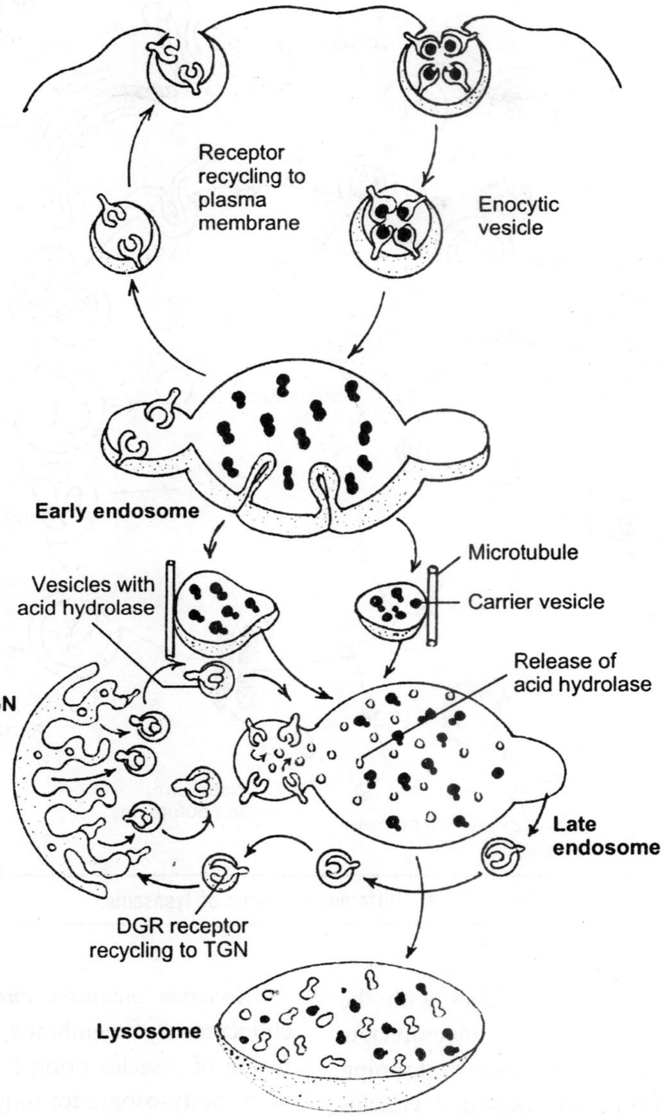

Fig. 11.6.2 : Biogenesis of lysosome, a meeting place of many pathways.

more primary lysosomes fuse with the phagosome to form *secondary lysosome.* These gradually degrade the dying organelle. Autophagosomes retain some post degradative undigested materials called *residual bodies.* It is a cell survival mechanism and can be treated as an alternative form of apoptosis. In response to starvation, when constant flow of macromolecules is stopped, it maintains the physiological needs of the cell supplying amino acids for the synthesis of proteins. It also helps to replace non-functional receptors of the plasma membrane.

Heterophagy: Lysosome mediated degradation of any microbe or macromolecules is called *heterophagy.* It proceeds through four

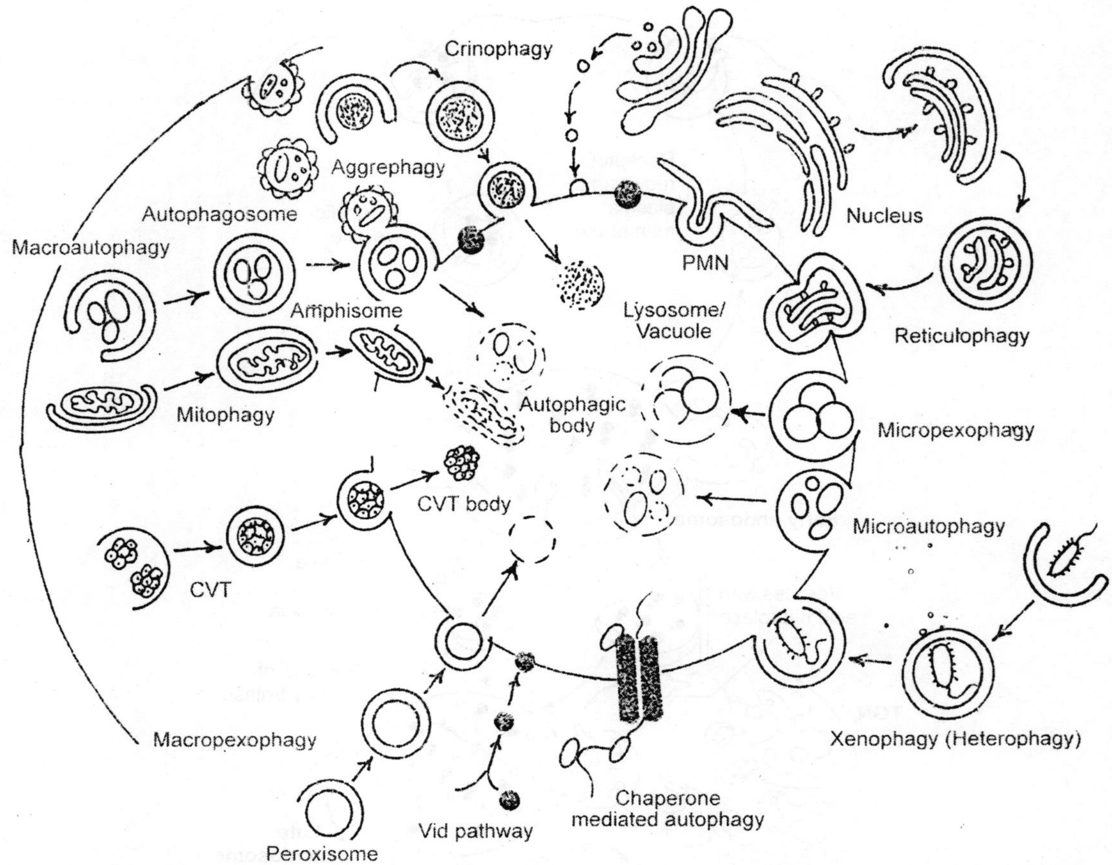

Fig. 11.6.3 : Different functions of lysosome.

phases – attachment of the microbes with the plasma membrane of the macrophage, engulfment of the organism forming heterophagosome, fusion of lysosomal vesicles with the heterophagosome and degradation of the microorganism (Fig. 11.6.3).

Specific antibodies of the macrophage - opsonin recognizes specific antigens on the surface of the bacteria which triggers its engulfment by the heterophagosome. Fusion of heterophagosome with lysosome produces a hybrid vacuole called a *phagolysosome*. These products of digestion are transported across the mem brane of the phagolysosome and are used for the synthesis of the macromolecules.

Receptor mediated endocytosis: In involves endocytosis of membrane receptors and ligands; fusion of vesicles bring the receptor and ligand with the lysosome forming *endosome*. Lysosomal enzymes lower pH cause release and degradation of the contents that are used for recycling of the receptors and other membrane components.

Role in malignancy: Lysosomal hydrolases are reported to be related with malignancy including loss of cell growth regulation, altered regulation of apoptosis, acquisition of chemoresistance and metastasis.

Induction of apoptosis: In addition to activating caspases, tumor necrosis factor receptor 1 (TNF-R1) causes the release of cathepsins B, from the lysosome which induces apoptosis. Accumulation of lipofuscin and non-functional mitochondria in lysosomes seems to reduce their autophagic capacity and induce apoptosis. Accidental release of lysosomal proteases, cysteine cathepsins and the aspartic protease cathepsin D into the cytosol induces apoptosis.

Plasma membrane repair: Elevation in intracellular Ca^{++} triggers fusion of lysosomes with the plasma membrane which is required for the repair of disrupted plasma membrane.

Secretory Role: Some secretory cells store their secretory proteins in a dual-function organelle, called *secretory lysosome*. These store additional secretory products and fuse with the plasma membrane in response to extra cellular stimuli releasing their contents by exocytosis. This is called *lysosomal secretion or defecation*. Functionally they serve both as a degradative and as a secretory compartment.

Biogenesis of Lysosome: Lysosome biogenesis involves vesicular delivery of newly synthesized lysosomal proteins from the TGN, fusion of lysosomes with late endosomes and sorting of membrane proteins into luminal vesicles. Lysosomal membrane proteins LAMP-2 and LIMP-2/LGP85 play specific role in lysosome biogenesis.

Lysosomal Storage Diseases: There are a number of diseases that are caused by the malfunction of the lysosomes or failure to manufacture an enzyme needed for the breakdown of endocytosed materials. Lysosomal storage diseases (LSDs) are manifested as neurodegenerative disorders. They severely affect eye, motor and cognitive functions and in most cases, abbreviate the lifespan. *Hurler's Syndrome* results in severe physical deformity and damage. *Tay-Sachs disease* and *Gaucher's disease* - both caused by a failure to produce an enzyme needed to break down sphingolipids. *I-cell disease ("Inclusion-cell disease")*, is caused by a failure to "tag" all the hydrolytic enzymes that are transported from TGN to the lysosomes.

Defects in the proper targeting of glycoproteins to the lysosomes can also lead to clinical complications. Two disorders related to deficiencies in the targeting of lysosomal enzymes are *I-cell disease or mucolipidosis II* and *pseudo-Hurler polydystrophy or (mucolipidosis III / mucolipidosis-HI)*. Mucolipidosis-II is characterized by severe psychomotor retardation, skeletal abnormalities, coarse facial features, painful restricted joint movement, and early mortality. Pseudo-Hurler polydystrophy is less severe; it progresses more slowly, and afflicted individuals live to adulthood.

SUMMARY

1. Lysosomes are garbage disposal organelle that is involved to maintain the hygienic condition of the cell. They are globular structure but in prolonged starvation, they appear threadlike called nematolysosomes.
2. Lysosome membrane shows pumps and transportters. Lysosomal membrane contains several proteins - lysosome-associated membrane proteins LAMP-1 and LAMP-2 and lysosomal integral membrane protein-2/lysosomal membrane glycol protein-85 (LIMP-2/LGP85).
3. A H^+-ATPase is found in the membrane of lysosomes to acidify the environment. Acid hydrolase is targeted to lysosome by phosphorylation. Golgi complex receives lysosomal enzymes from the RER in the cis region. Hydrolase-M6P- receptor complex then buds clathrin coated vesicles that fuse with late endosomes at pH 6.0 and finally with the lysosome. Cellular organelle can be senile or damaged rendering putrefactive damage to the cell. Lysosome mediated degradation of the non-functional cellular organelles is called autophagy.
5. During autophagy, membrane-bound organelles are delivered for destruction to lysosomes. Upon induction of autophagy, a membrane cisterna known as the isolation membrane (IM or phagophore) appears and curves around part of the cytoplasm. Sealing of the edges of the IM forms a unique double membrane vesicle, the

autophagosome that fuse with the lysosome. Macroautophagy is the major lysosomal catabolic pathway for macromolecules and organelles in eukaryotic cells. Gradually the membrane fusion results in the formation of autophagosome with which lysosomes merge delivering acid hydrolases into the lumen forming an autophagosome. For normal Schwann cell function, the levels of PMP22 are tightly regulated by autophagy.

6. Lysosome mediated degradation of the extra cellular invaders (bacterium) or macromolecules are called heterophagy. The interaction between the membrane receptors of the cell (macrophage) with specific proteins on the microorganism, triggers engulfment. Fusion of heterophagosome with lysosome produces a hybrid vacuole called a phagolysosome. Introduction of lysosomal enzymes and lower pH causes release, and degradation of the contents.
7. Lysosomal Storage Diseases are caused by the accumulation of improper macromolecules (proteins, polysaccharides, lipids) in the lysosomes. Resulting in impaired cell metabolism. The lysosomal storage disease, I-cell disease ("inclusion-cell disease"), is caused by a failure to "tag" (by phosphorylation) all the hydrolytic enzymes that are supposed to be transported from the Golgi to the lysosomes.
8. Plasma membrane wounds are repaired by a mechanism involving Ca^{++} -regulated exocytosis. Elevation in intracellular Ca^{++} triggers fusion of lysosomes with the plasma membrane. Ca^{++}-regulated exocytosis of lysosomes is required for the repair of plasma membrane disruptions. It proves a novel role of lysosomes as Ca^{++} -regulated exocytic compartments responsible for plasma membrane repair.

QUESTIONS

Long questions:

1. Describe the structure and functions of lysosome.
2. What is autophagy? Describe the types and functional divesrsities of autophagy.

Short questions:

1. What do you mean by nematolysosomes?
2. What are the constitutional peculiarities of lysosome membrane?
3. Name two characteristic lysosomal membrane proteins.
4. How proteins are translocated into the lyso some?
5. Which chemical acts as a sorting signal in TGN during processing of enzymes?
6. How lysosomal enzymes are processed in the Golgi?
7. What are autophagosomes and residual bodies?
8. What is pexophagy?
9. What are the major roles played by autophagy in plants?
10. What is the specialty of endocytotic pathway mediated autophagy?
11. What do you men by crinophagy?
12. Differentiate micro and macroautophagy.
13. What do you mean by phago lyso some?
14. What are endosomes? What is the basic factor for their differentiation?
15. Why lysosomal enzymes cannot cause harm to the cytosol?
16. Name some important lysosomal storage diseases.
17. What do you mean by secretory lysosome?
18. What do you mean by aggresome?

MCQS:

1. **Lysosomes are**
   (a) membrane less,
   (b) single membrane,
   (c) double membrane organelle.
2. **Danon disease is the result of**
   (a) LAMPI, (b) LAMPII, (c) LIMP-II deficiency.
3. **Cytosol is slightly**
   (a) acidic, (b) basic, and (c) neutral.
4. **Autophagosome formation is stimulated by**
   (a) insulin, (b) thyroxine, (c) glucagon.
5. **Heterophagosome formation is assisted by**
   (a) actin, (b) MT, (c) intermediate filaments.
6. **Organelle involved in the transformation of cell membrane is**
   (a) ER, (b) lysosome, (c) Golgi.
7. **Hydrolytic enzyme containing organelle is**
   (a) RER, (b) SER, (c) lysosome.

8. **Fusion of lysosome and autophagosome forms**
(a) primary lysosome,
(b) secondary lysosome,
(c) autophagosome.

9. **Autolysis is related with**
(a) glyoxysome, (b) peroxisome, (c) lysosome.

10. **Direct fusion of lysosome with secretory vesicles containing non-functional protein is called**
(a) crinophagy, (b) microautophagy, (c) heterophagy.

11. **Phagophores are related with**
(a) autophagy, (b) heterophagy, (c) pexophagy.

Answers to the MCQs:

1. b; 2. b; 3. b; 4. c; 5. a; 6. b, 7. c, 8. c, 9. c, 10. a, 11.a.

Chapter 11.7

Paroxisome

Of the organelles of eukaryotic cells, peroxisome was the last to be discovered by *Christian de Duve* (*1965*). These are also called *microbodies*. These are almost round (0.5–1.5 µm in diam.) encased in a lipid bilayer and resemble lysosomes in being filled with enzymes but show a crystalline core structure (concentrated catalase, thiolase or urate oxidase) (*Hu et al. 2012*). Their main function is to detoxify the cells. They are numerous in the liver where toxic byproducts accumulate.

11.7.1. PEROXISOMAL MEMBRANE

They are bounded by a single layered *membrane* that contains specific membrane proteins which act as pumps and transporters used in importing proteins into the organelles and in proliferation (Fig. 11.7.1).

11.7.2. PROTEIN IMPORT INTO THE PEROXISOME

Peroxisomal enzymes and other proteins, including the integral membrane proteins are synthesized on free cytoplasmic ribosomes and are selectively imported into peroxisomes. There are about 32 known peroxisomal proteins, called *peroxins*, which participate in the process of importing proteins. Each contains a peroxisomal targeting signal (*PTS*) that binds to a receptor molecule that takes the protein into the peroxisome and then returns for another load. Two peroxisomal targeting signals have been identified -a 9-amino acid sequence at the N-terminal and a 3 amino acid sequence at the C-terminal end. Each has its own receptor to take it to the peroxisome. The protein receptors, *Pex5* and *Pex7* accompany their cargos (containing PTS1 or PTS2 respectively) into the peroxisome where they release the cargo and then return to the cytosol. This is called "*recycling*" of receptors. Proteins do not have to unfold to be imported into the peroxisome. Overall, the import cycle is referred to as the "*extended shuttle mechanism*" (Fig. 11.7.2). PTS1 binds with the cytosolic receptor Pex5 which carries the bound protein and interacts with Pex14 receptor of the membrane. Pex5 complex is then translocated into the peroxisomal matrix via Pex 2,

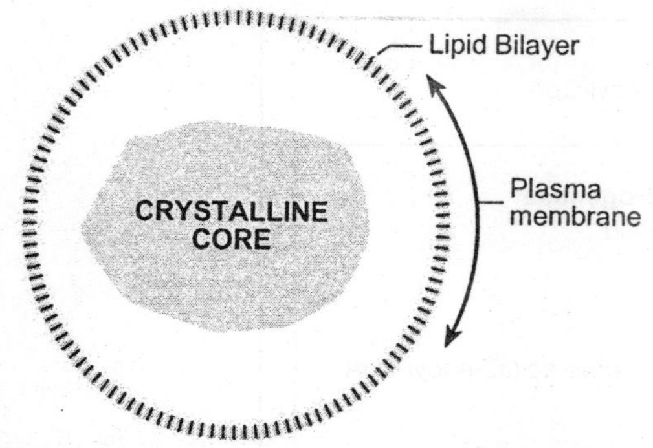

11.7.1 : Diagramatic structure of peroxisome.

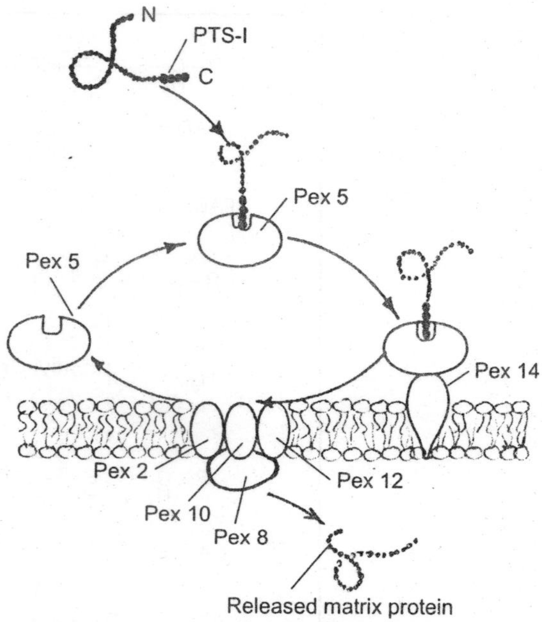

Fig. 11.7.2 : Import of peroxisomal matrix protein.

Pex8, Pex10 and Pex12. Inside the matrix, Pex5 is released and is recycled back to the cytosol. Protein translocation requires ATP hydrolysis. Other proteins like thiolase are synthesized as precursor with an N – terminal targeting sequence - PTS2 that bind with Pex7 and is translocated through another multiprotein channel complex. In higher eukaryotes Pex5 and Pex7 form a complex and may function as a single import receptor for all peroxisomal matrix protein. Import of peroxisomal membrane protein is less characterized because the targeting sequences are ill defined though it is sure that they are unrelated to PTS1/PTS2. Moreover membrane protein transport occurs in the absence of Pex5 and Pex7 and also in Pex mutants.

But Pex3, Pex16 and Pex19 play crucial role in the membrane protein import (Fig.11.7.3). An evolutionary analysis of the peroxisomal proteome found homologies between the peroxisomal import machinery and the ERAD pathway in the ER which indicates that peroxisome does not have an endosymbiotic origin, instead it originates from the ER and its proteins were recruited from the cytosolic pools.

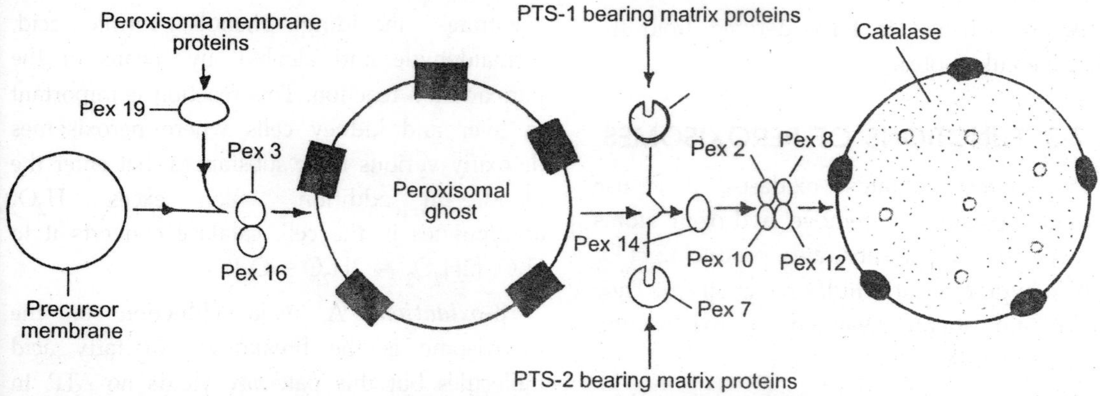

Fig. 11.7.3 : Import of peroxisomal membrane and matrix proteins.

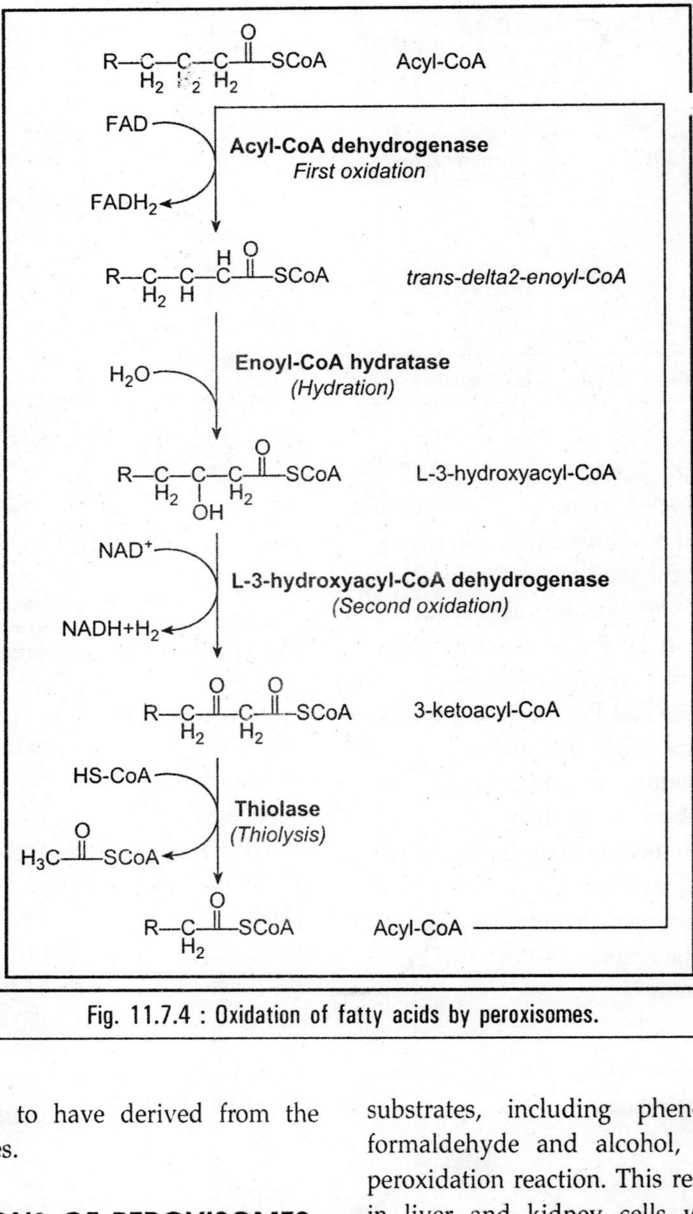

Fig. 11.7.4 : Oxidation of fatty acids by peroxisomes.

These are believed to have derived from the primitive eukaryotes.

11.7.3. FUNCTIONS OF PEROXISOMES

Peroxisome contains oxidases that use molecular oxygen, to remove hydrogen atoms from specific organic substrates, producing H_2O_2 (a toxic byproduct of cellular metabolism). They also contain catalase, D-amino acid oxidase and uric acid oxidase.

Role of Catalase: It uses H_2O_2 generated by other enzymes in the peroxisome to oxidize other substrates, including phenols, formic acid, formaldehyde and alcohol, by means of the peroxidation reaction. This reaction is important in liver and kidney cells where peroxisomes detoxifiy various toxic substances that enter the blood. In addition, when excess H_2O_2 accumulates in the cell, catalase converts it to H_2O. [$2H_2O_2 \rightarrow 2H_2O + O_2$]

β-oxidation: A major function of the peroxisome is the breakdown of fatty acid molecules but this pathway yields no ATP. In yeast and plant cells it is exclusive for the peroxisome. Branched chain fatty acid, as well as very long chain fatty acid, β-oxidation occurs only in peroxisomes. Peroxisomes degrade fatty acids acids with more than 12 carbon atoms by a series of reactions similar to those used by liver mitochondria. This is called α-oxidation. However, the electrons and protons transferred to FAD and NAD^+ during the oxidation reactions are subsequently transferred to oxygen, forming H_2O_2. In this pathway the carboxyl carbon of the fatty acid is lost as CO_2 and the carbon is oxidized to an aldehyde group at the expense of H_2O_2. This reaction is catalyzed by fatty acid peroxidase. The fatty aldehyde, formed is oxidized to the corresponding carboxylic acid. This two-enzyme reaction is repeated on the shortened free fatty acid. Since fatty acid peroxidase attacks only fatty acids having 13 – 18 carbon atom, this pathway cannot lead to complete oxidation of long chain fatty acids. The aldehydes produced by β–oxidation may alternatively undergo reduction to yield long chain fatty alcohols.

Plasmalogen formation: The first reaction in the formation of plasmalogen in animal cells occurs in peroxisome. It is the most abundant phospholipid in myelin. Deficiency of plasma logens causes profound abnormalities in the myelination of nerve cells, which is one of the reasons that many peroxisomal disorders lead to neurological disease.

Formation of bile acids: Peroxisomes play an active role in the formation of bile acids.

Cholesterol synthesis: Peroxisomes participates in the synthesis of cholesterol.

Uric acid formation: Peroxisomes play a major role in the break down of excess purines (AMP, GMP) to uric acid.

Nitrogen fixation and photorespiration: In plant cells where they participate in symbiotic nitrogen fixation and photorespiration.

Role in Apoptosis: Peroxisome proliferators (PPs) activate the PP-activated receptor alpha (PPARα). PPs could suppress both spontaneous and TGF-β1-induced apoptosis that ultimately lead to tumours.

11.7.4. BIOGENESIS

Unlike lysosomes, which are formed in the secretory pathway, peroxisomes usually self-replicate by enlarging followed by division, although there is some indication of their de novo formation (Fig. 11.7.5). The ability of peroxisomes to divide and import proteins post-translationally, just like mitochondria and chloroplasts, led workers to suggest their endosymbiotic origin. But this view has been challenged by recent workers. Brown and Baker (2003) advocated three main steps of their biogenesis: (a) import of peroxisomal membrane proteins; (b) import of peroxisomal matrix proteins and (c) peroxisome proliferation. Combined genetic and biochemical approaches led to the identification of 32 PEX genes-encoding proteins so-called peroxins required for their biogenesis.

The proteins required for peroxisome are translated on free ribosomes in the cytosol. Peroxisome membrane is formed firstly by Pex3, followed by Pex19, in association with the ER (Kunau, 2005) or any endomembrane (Rucktaschel *et. al.* 2010). The matrix protein catalase and thiolase were detected in the specialized sub domains of the ER that were continuous with a peroxisomal reticulum from which mature peroxisomes are formed. Peroxisomes are not autonomously multiplying organism rather are formed through self-assembly. The existing peroxisomes can take in more proteins then divide via fission to produce new peroxisomes (Smith and Aitchison, 2013). Proteins translocate to the interior of the peroxisome using one of two pathways, which are highly conserved across species. They can have a Ser-Lys-Leu motif at the carboxy terminus, known as Peroxisome Targeting Signal 1 (PTS1), or a 9 residue sequence (PTS2). A major protein

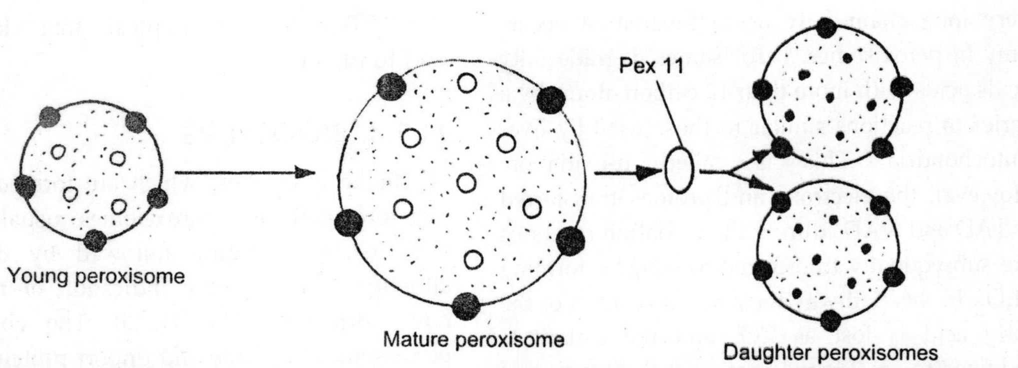

Fig. 11.7.5 : Formation of new peroxisome from the old one.

that is imported into the peroxisome during its biogenesis is Catalase. The organelle also requires phospholipids, which are transported to the peroxisome by phospholipid transfer proteins.

Peroxisomal biogenesis disorders (PBDs): Biogenesis of peroxisome involves a dozen of different genes, and mutations of any of them can cause either complete failure or the incorrect formation of peroxisomes. Without correctly formed and functioning peroxisomes, a number of important chemical processes within the cell do not occur leading to a group of severe and often fatal diseases, the PBDs (Peroxisomal Biogenesis Disorders).

There are four PBDs - Zellweger syndrome (ZS), Neonatal adrenoleukodystrophy (NALD), infantile Refsum disease (IRD) and Rhizomelic chondrodysplasia punctata *(RCDP)*. ZS is the most severe form, IRD the least, with NALD in the middle. These three diseases are known as the *Zellweger spectrum,* and affected children share a common set of abnormalities and disabilities. RCDP is the fourth PBD, and affected children have some things in common with those in the Zellweger spectrum. Protein import to peroxisome is an important event in maintaining the physiological balance of the organelle. A deficiency in the protein import can lead to empty peroxisomes, leading to abnormalities in the brain, called *Zellweger syndrome.* This disorder results from the inheritance of two mutant genes for one of the receptors (PXR1) needed to import proteins into the peroxisome. A deficiency in the *Pex2* has shown to be responsible for one form of the syndrome. A milder inherited disease is caused by a defective receptor for the N-terminal import signal.

Deficiency of the formation of plasmalogens can also cause severe brain disorders, leading to neurological disease. Mutant versions of one or another of the enzymes found within peroxisomes may lead to ***X-linked adrenoleukodystrophy (X-ALD)***. It results from a failure to metabolize fatty acids properly. One result is deterioration of the myelin sheaths of neurons. The disorder occurs in young boys. This is peroxisomal "single-enzyme" disorders. Another one of the single enzyme disorders is ***Refsum disease,*** from which infantile Refsum disease has taken its name.

SUMMARY

1. Peroxisomes are filled with enzymes showing a central core surrounded by a membrane. They help to get rid the cell of toxic or other harmful metabolites. Peroxisomal enzymes and other integral membrane proteins are synthesized on free cytoplasmic ribosomes and are selectively imported into peroxisomes.
2. Protein import to peroxisome is an important event in maintaining the physiological balance of the

organelle. There are at least 32 known peroxisomal proteins, called peroxins, which partici pate in the process of importing proteins. Each such protein has a targeting sequence PTS1 that binds with the cytosolic receptor Pex5 that carries the bound protein and interacts with Pex14 receptor of the membrane. Pex5 complex is then translocated into the peroxisomal matrix via Pex 2, Pex8, Pex10 and Pex12. Protein translocation requires ATP hydrolysis. In higher eukaryotes Pex5 and Pex7 form a complex and may function as a single import receptor for all peroxisomal matrix protein. Moreover membrane protein transport occurs in the absence of Pex5 and Pex7 and also in Pex mutants. But Pex3, Pex16 and Pex19 play crucial role in the membrane protein import. A deficiency in the protein import can lead to empty peroxisomes, leading to abnorma lities in the brain, called Zellweger syndrome.

3. Peroxisomes play a major role in the break down of excess purines (AMP, GMP) to uric acid. Peroxisome proliferators (PPs) activate the PP-activated receptor α (PPARα). The first reaction in the formation of plasmalogen in animal cells occurs in peroxisomes. Peroxisomes participate in the synthesis of cholesterol.
4. Import of peroxisomal membrane proteins and matrix proteins are important step of peroxisome proliferation.

QUESTIONS

Long questions:

1. Describe the structure and functions of peroxisome.

Short questions:

1. What are the basic differences between lysosome and peroxisome?
2. How proteins are imported into peroxisome?
3. What is the basic difference in fatty acid metabolism in mitochondria and peroxisome?
4. What are the main functions of peroxisome?
5. What are the effects of improper protein import and other roles of peroxisome.

MCQs:

1. **Peroxisomes are usually found in higher organisms and**
   (a) Monera, (b) Algae, (c) Yeast, (d) Fungi.
2. **In higher animals Peroxisomes are abundauntly found in**
   (a) Lungs, (b) Liver, (c) Kidney (d) Liver and kidney.
3. **The organelle which deals with the formation of decomposition of H_2O_2 is**
   (a) Peroxisomes, (b) Ribosome, (c) Mitochondria, (d) ER.
4. **The peroxisomal core contains concentrated**
   (a) Catalase, (b) Thiolase, (c) Urate oxidase, (d) All.
5. **Peroxisomal matrix proteins are synthesized in**
   (a) RER, (b) SER,
   (c) Free ribosomes in the cytosol, (d) ER.
6. **The peroxins which play vital role in biogenesis are**
   (a) Pex 3, (b) Pex 7,
   (c) Pex 9, (d) Pex 3 and Pex 19.
7. **Deficiency in the protein import leading to abnormalities in the brain is called**
   (a) Neonatal adrenoleukodystrophy,
   (b) infantile Refsum disease,
   (c) Zellweger syndrome,
   (d) Rhizomelic chondrodysplasia punctata.

Answers to the MCQs:

1. c, 2. d, 3. a, 4. d, 5. c, 6. d, 7. c.

Chapter 11.8

Centrosome

Centrosome is a unique cytoplasmic domain, localized in the vicinity of nucleus in the animal interphase cells. It is one of the few intracellular organelles that distinguish animal and higher plant cells. Usually plant cells lack centrosome though it occurs in lower plants (except in blue green and red algae). In unicellular eukaryotes and fungi a large range of centrosomal structures are found. Mostly they show one or two central granules called *centrioles* that are homologous with the basal bodies found at the base of the cilia and flagella. In fact, in primitive flagellates, they are mainly concerned with the locomotion and their role in cell division is secondarily acquired. In vertebrates centrioles occur in male gametes while female gametes lack it. Moreover muscle cells, neurons and mature RBC are devoid of centrosomes. Some cells (megakaryocytes and osteoblasts) however contain multiple centrosomes. Due to its unique role in the nucleation of microtubules (MT) during cell division it is called *microtubule organizing centers (MTOC)*. It regulates the polarity, number, and spatial arrangement of microtubules. It plays crucial roles in the segregation of chromosomes during cell division, the organization and transport of cytoplasmic organelles, morphogenesis, and determination of cell shape. Through its well-established role in the assembly of MTs and impact on other cytoskeletal structures, it is crucial for cell motility, proliferation and differentiation.

Flemming and Van Beneden (1875) discovered centrosome. Boveri (1888) coined the term 'centrosome' and established its role in cell division. He regarded it as the dynamic center of the cell.

In most cells, centrosome and nucleus are closely apposed to each other. It is proved that the motor protein, dynein mediates attachment of centrosome with nuclear membrane. Dynein-mediated attachment of centrosome with the nucleus is mediated by an intranuclear protein -

Zyg-12 with the help of another protein Sun-1 (Kemp *et al*, 2007).

Centrosome appears as a cylindrical, membrane less organelle having 0.16 to 2.5μ diameter and 0.16μ to 8μ length. In some cases the length of the centrosome increases before transformation into basal bodies. Some somatic cells show smaller centrosomes but oocytes and spermatocytes show giant sized centrosomes.

11.8.1. STRUCTURE

(i) Each centrosome consists of a pair of orthogonally connected *centrioles;* collectively called *diplosomes* surrounded by an electron dense matrix of pericentriolar materials (PCM) called *centrospheres*. These are morphologically identical from the basal bodies of the cilia or flagella.

(ii) Each centriole shows a central tube measuring 0.3-0.5μ long /0.15-0.2μ in diameter. The wall of each centriole consists of 9 triplet MTs each of which exhibits 3 sub tubules often marked as A, B and C. These sub tubules are tilted at an angle of 40° to the center of the centriole Each sub tubule is 250A° in diam. and the A sub tubule appears round while B and C appear cup shaped in T.S. The wall of the sub tubules contains filamentous subunits whose number is 13 in A while 10 in B and C sub tubules because B sub tubule shares A and C sub tubule shares B sub tubule (Fig. 11.8.1).

(iii)The distal end of each centriole contains many distal and sub distal appendage like structures attached with the peripheral triplets by specific proteins.

(iv)PCM contains a matrix with 12-15 nm fibers termed the centromatrix, that directs the recruit ment of other centrosomal components.

(v) PCM proteins include pericentrin / kendrin, C-NAP-1 and GMAP210. These in association with γ-tubulin assist in MT nucleation and anchorage. Pericentrin binds calmodulin, dynein, γ-tubulin ring complex and several other

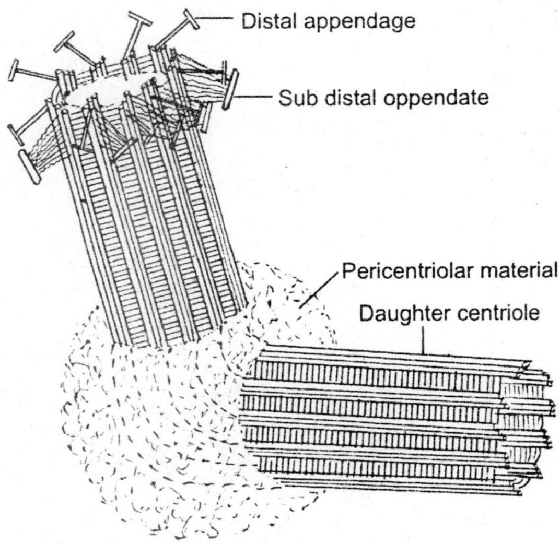

Fig. 11.8.1 : Structure of centrioles.

enzymes. The kinase Aurora-A localized at PCM is required for chromosome segregation.

(vi)Centrioles are composed of many different proteins like α, β, δ and ε tubulins. They also contain centrin and Ca^{++} binding proteins that are essential for centriole biogenesis and for spindle pole biogenesis. Centrin acts as links between nucleus and centrioles. Centrin-2 participates in centriole segregation and cytokinesis while centrin-3 is essential for centriole duplication. Ninein help in anchoring the minus (-) ends of the MTs to the centrosomes. Centriolin anchors cell cycle controlling proteins and regulate MT stability (Xia *et al*, 2007).

11.7.2. CENTROSOME DUPLICATION

Maintenance of single centrosome per cell cycle is monitored by several factors. Separase, which is the key regulator of anaphase also restricts centrosome duplication to a single per cell cycle. Plk4/SaK a member of the Polo kinase family, also regulates centriole formation (Nigg, 2007) (Fig. 11.8.2). In G1, centrosome shows a mother and a daughter centriole linked together by protein fibers. The daughter centriole moves

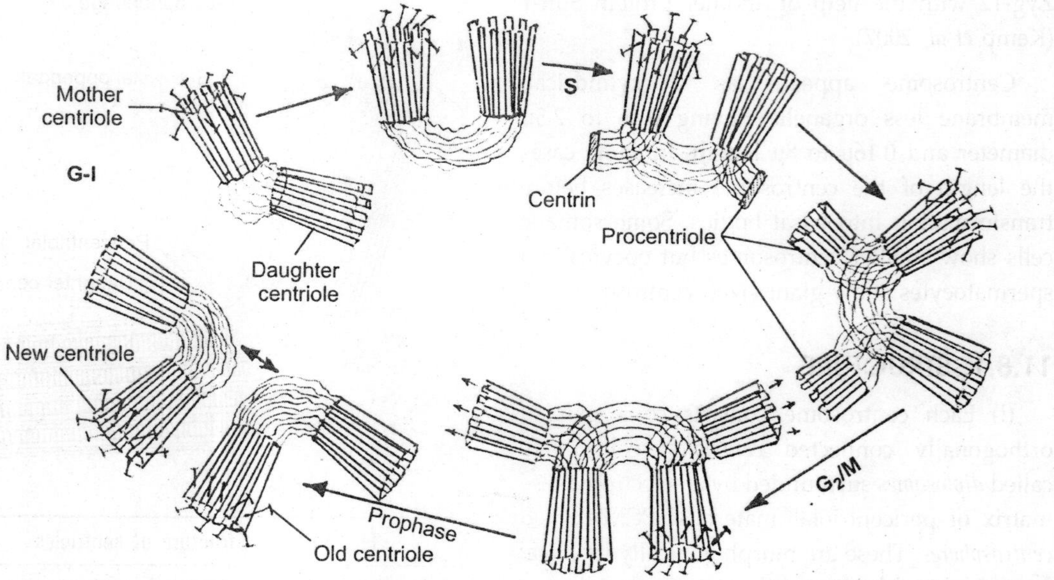

Fig. 11.8.2 : Schematic representation of centrosome duplication.

slightly away from the mother and initiates the formation of new daughters at the onset of 'S' phase mediated by Cdk2. At first a disc like structure is formed at right angles to the proximal end of each mother centriole called *generative disc* that acts as the precursor of the daughter centriole. Daughter centrioles elongate as the cell cycle progresses attaining full length by G2/M during which the centrioles become functionally fully mature accumulating γ-tubulin and other PCM materials.

Newly formed daughter centrioles are oriented at right angles to the proximal end of the mother centrioles. The old centriole as usual shows distal and sub distal appendages but the new centriole does not acquire these appendages until G2/M. Cep135 and dynamitin, p50 dynactin subunits play an important role in the assembly and maintenance of functional MTOC in mammalian centrosome. PCM that surround both old and new centrioles nucleate MTs but these are mainly anchored near old centriole. In the early S phase, at the proximal end of both old and new centrioles, a disc like structure is formed that begins to nucleate a ring of 9 MTs called *'procentrioles'* It is gradually converted into an array of 9 triplets mediated by δ and ε tubulins. The daughters gradually lengthen and become fully mature by late G2/M.

Cohesion of two newly formed centrioles with the mother centrioles is maintained by C-NAP1 and Nek2, concentrated at the proximal ends of the centrioles. At the onset of prophase, phosphorylations of these proteins trigger their separation to set up two poles of the mitotic spindle. Polo and Polo-like kinases (Plks) are essential for the activity of the MTOC and localize to centrosomes or spindle pole bodies. Following mitosis, each daughter cell retains a single centrosome from its mother. Thus, centrosome duplication is semi-conservative in that one centriole is new and the other is inherited from the parental centrosome. The kinase ZYG-1 is required for centrosome duplication. *Szy* (suppressor of zyg-1) gene has been identified that interact with ZYG-1 (Kemp *et al*, 2007).

Failure in centrosome duplication can cause the formation of monopolar spindles that are incapable of segregating chromosomes blocking the cell division process. Excessive rounds of centrosome duplication, on the other hand, can result in the formation of multipolar spindles and ultimately, chromosome missegregation which are characteristic of many malignant tumors. Abnormally high numbers of centrosomes have been found in many kinds of tumor cells.

11.8.3. FUNCTIONS

Microtubule nucleation: Centrioles play crucial role in the nucleation of MTs that are essential for cell division. Two centrioles are positioned at right angles to each other. They are not directly related with the nucleation of MTs rather, they form a scaffold for the assembly of pericentriolar material. The γ-tubulin ring complex of PCM provides a direct template to nucleate MTs. The centrosomal levels of γ-tubulin and centrosome size appear maximum at metaphase and anaphase respectively while MT nucleation remained high in telophase. Ran GTPase is associated with the centrosome and is involved in MT assembly. Centrosome is also linked with the production of MTs required for intracellular transport.

Controls cell division: It has a crucial role in cell division since it determines the spindle poles that segregate duplicated chromosomes between dividing cells. Su48 a centrosomal protein plays a vital role in chromosome segregation and normal cytokinesis. Su48 abnormalities may cause malignancy (Marshall, 2007).

Formation and positioning of cytokinesis furrow: In animal cells, the position of the cytokinesis furrow is specified by the position of the mitotic spindle that is generated by the centrioles. The furrow bisects the spindle producing two daughter cells. Centrin an important PCM protein plays a crucial role in the localization of cytokinesis furrow. The furrow is first positioned by a signal determined by microtubule asters and then by a second signal that is derived from the spindle midzone.

Prevention of different diseases: Alsin, a PCM protein and a kinase anchoring protein (AKAP-450) are linked with the auto prevention of different diseases since its over expression caused accumulalation of early endosomes, impairment of mitochondria trafficking and fragmentation of the Golgi apparatus.

Control of cell cycle: Centrosome acts as a scaffold for anchoring many cell cycle regulatory proteins that are required for several cell-cycle transitions, like G-1/S, G-2 /M, metaphase to anaphase and general cell cycle progression. The kinase Aurora-A accelerates cyclin B1-Cdc2 activation and mitotic entry.

Aggresome formation: Protein degradation in the cell is catalyzed by the ubiquitin-proteasome system (UPS) that monitors and eliminates potentially toxic misfolded or damaged proteins. When the capacity of this system is exceeded, these proteins aggregate and are assembled to form *aggresome*.

Ontogenic viability: In *Drosophila,* a centrosomal protein CP190 bind directly to MTs and is essential for com pletion of the development since flies homozygous mutant for Cp 190 gene die at late pupal stages of development.

Ciliogenesis: Throughout the animal kingdom and in the lower plants, centrioles prior to the formation of of cilia and flagella divide and form basal granules that generate one cilium or flagellum (Fig. 11.8.3)). In the development of the sperm also it generates the flagellum.

Regulation of cell shape, polarity and motility: Centrosome is the major MT nucleation center of animal cells and through its influence on the cytoskeleton it is involved in maintaining the cell shape, polarity and motility.

Nucleoplasmic transport: Ran GTPase, importins/ karyopherins the key regulators of

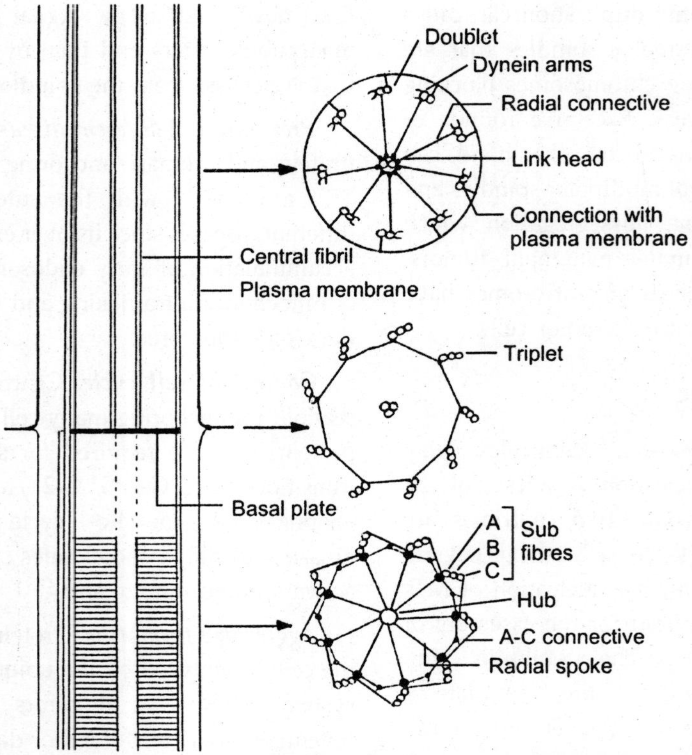

Fig. 11.8.3 : Transformation of centriole to basal body and cilia.

nucleo-cytoplasmic transport are associated with some centrosomal proteins.

Neurogenesis: Centrosome acts as a key marker for cell polarity. The position of centrosome determined by extracellular factors, directs multiple aspects of neuronal development (Higginbotham and Gleeson, 2007).

SUMMARY

1. Centrosome is a unique cytoplasmic domain, localized in the vicinity of nucleus in the animal cells. The motor protein, dynein mediates attachment of centrosome with nuclear membrane. Some somatic cells show typical smaller centrosomes but oocytes and spermatocytes show giant sized centrosomes.
2. Each centrosome consists of a pair of centrioles surrounded by a pericentriolar materials centro sphere. Each centriole shows a central double and nine peripheral triplet tubules. Its walls show 13 in the first and 10 in other sub tubules.

   PCM proteins includeta, b, d and e tubulins, SAS-4, EVI5, pericentrin/kendrin, C-NAP-1 and GMAP210 is association with tubulins regulating centrosomal stability and dynamics. Centrioles also show centrin-2 that participates in centriole segregation and cytokinesis while centrin-3 is essential for centriole duplication. The sub distal appendages of centriole contain a number of proteins like ninein, centriolin and many unidentified proteins. Centriolin anchors cell cycle controlling proteins and regulate MT stability.
3. Centrosome undergoes a cycle of maturation and division during the cell cycle. In G1, single centrosome shows a mother and a daughter centriole linked together by protein fibers. Newly formed daughter centrioles are oriented at right angles to the proximal end of the mother centrioles. PCM that surround both old and new centrioles nucleate MTs but these are mainly anchored near

old centriole. Following mitosis, each daughter cell retains a single centrosome from its mother. Centrioles play crucial role in the nucleation of MTs that are essential for cell division. Due to this fact, it is called microtubule-organizing centers (MTOC). The centrosomal levels of γ-tubulin and centrosome size appear maximum at metaphase and anaphase respectively while MT nucleation remained high in telophase. †Centrosome is also linked with the production of MTs required for intracellular transport.

4. Even in centrosome-containing cells, non-centrosome-associated MTs may contribute to spindle formation. QN1/ KIAA1009 a MT-associated ATPase, located at the spindle poles and centrosomes seems to be related with chromosome segregation and centrosome separation. Su48 a centrosomal protein plays a vital role in chromosome segregation and normal cytokinesis. Centrin an important PCM protein plays a crucial role in the localization of cytokinesis furrow. It serves as a scaffold for anchoring cell cycle regulatory proteins that are required for several cell-cycle transitions, like G-1/S, G-2 /M, metaphase to anaphase and general cell cycle progression.

QUESTIONS

Long questions:

1. Describe the structure and functions of centrosome.

Short questions:

1. In algae, centrosome occurs except —— and —— algae.
2. In flagellates and ciliates, they are mainly concer ned with ——.
3. In vertebrates, centrosome is absent in —— gamete.
4. Why pericentriolar materials are collectively called ——.
5. Name a protein that regulates centrosome size.
6. Name a protein that regulates centro some stability.
7. Name some important proteins of PCM.
8. Name some important Centriolar proteins.
9. Name some proteins of sub distal appendages of centrioles.
10. Differentiate centriole and procentriole.

Fill up the blanks:

1. Neurons and mature mammalian RBC are —— of centriole.
2. Megakaryocytes and osteoblasts contain —— centrioles.
3. —— coined the term centrosome.
4. The relation between centrosome and nucleus is mediated by ——.
5. Each centrosome consists of a pair of centrioles that are collectively called ——.
6. Centrioles are surrounded by pericentriolar materials called ——.
7. The number of the filamentous subunits in the sub tubules is —— in A while —— in B and C.
8. The fibrous elements of PCM is called ——.
9. Centriolar division starts at —— phase of the cell cycle under the regulation of ——.
10. During duplication, the disc, formed near the mother centriole is called ——.
11. Centriolar division becomes complete by ——.
12. —— a PCM protein plays a crucial role in positioning the cytokinesis furrow.

MCQs:

1. **Megakaryocyte contains**
(a) 1, (b) 2, (c) many centrioles.
2. **Pair of centrioles is called**
(a) centrospheres, (b) diplosomes, (c) dikaryons.
3. **Pericentrin is a protein of**
(a) PCM, (b) centriole, (c) sub distal appendages.
4. **Centrin is a protein of**
(a) PCM, (b) centriole, (c) sub distal appendages.
5. **Centriolin is a protein of**
(a) PCM, (b) centriole, (c) sub distal appendages.
6. (a) α, (b) β, (c) γ **tubulin pays crucial role in the nucleation of microtubules.**

Answers to MCQs:

1. c; 2. b; 3. a; 4. b; 5. c; 6. c.

Chapter 11.9

Interphase Nucleus

In eukaryotes, the the genetic material is highly organized into compact chromosomal structures which remain separated from rest of the cytoplasm by a membrane bound organelle, called **nucleus.** Almost all cells of higher organisms contain nucleus (except mature mammalian RBC and sieve tube cells in the phloem tissue of plants). However their number may be one (*mononucleate*), two (*binucleate*) or many (*polynucleate*). Polynucleate conditions in animals, is called *syncytium* and in plants, is called *coenocyte.*

11.9.1. MORPHOLOGICAL FEATURES

It may be disc shaped (in squamous epithelial cells), spherical, cuboidal, polyhedral, elliptical, lanceolate, spindle shaped, ovoid, elongated, C-shaped (in *Vorticella*), pyriform (in *Paramoecium*), moniliform (in *Spirostomum*), branched (in spinning gland of *Platyphylax*), bilobed, multilobed etc. In *Paramoecium caudatum*, out of two nuclei, the smaller micronucleus is round while the larger macro nucleus is kidney shaped. Usually in embryonic condition, it occupies central position but may be displaced later on. In adipose cells, it occupies peripheral while in glandular cells basal positions. However the position of the nucleus is maintained by an array of cytoskeletons. It is basically anchored in the cytoplasm by intermediate filaments and ER.

LINC complexes providing structural support to the nucleus: The position of nucleus is maintained by some skeletal protein. SUN and KASH domain proteins (mainly Nesprins) provide the essential linkage between the three main cytoskeletons and the nucleokeleton especially the nuclear lamina. Plectin acts as an intermediary linker connecting the intermediate filaments while a motor protein (Kinesin) is generally involved in linking microtubules.

11.9.2. STRUCTURE OF NUCLEUS

Irrespective of the positions and number, nucleus remains in its position by some fibrillar structures. Structure of nucleus varies widely in

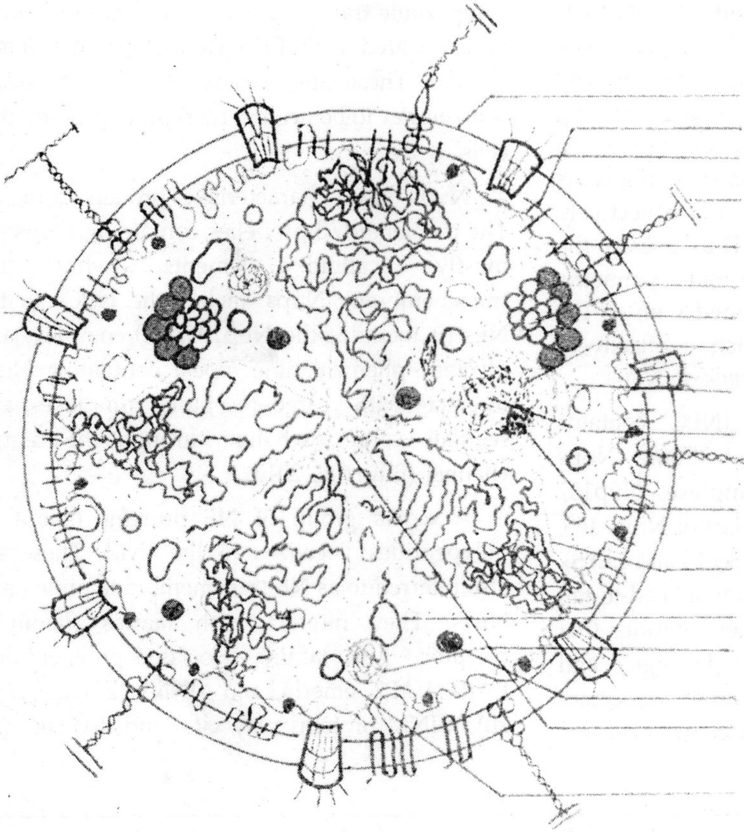

Fig. 11.9.1 : Different parts of a Nucleus.

different phases of the cell cycle. However the structure of a typical nucleus can be illustrated by the interphase nucleus whose different parts are as follows (Fig. 11.9.1).

Nuclear envelope / Nuclear membrane / Karyotheca : Nucleus is enveloped by a double layered envelopes (NE) or karyotheca, enclosing a very narrow 50 nm wide - *peri nuclear space (PNS).* The outer membrane (ONM) remains directly connected with the RER. Ribosomes also remain attached with the outer membrane. Despite the continuity between the NE and ER, both ONM and INM are comprised of diverse groups of proteins.

Structure of Nuclear pore (After Kabachinski and Schwartz, 2015): The most distinguishing feature of NE is that about 10% of its surface area is occupied by pores.

(i) Each pore consists of 3 concentric *annuli* (average diam.-130 nm) (Fig. 11.9.2) – a barrel like ***luminal ring*** flanked by an outer ***cytoplasmic*** and an inner ***nuclear ring***. The nuclear ring is anchored in the nuclear lamina.

(ii) Three rings are joined by 8 *radial spoke* like structures which anchor the rings with the nuclear envelopes. These show central buldging that fuse with each other to form a *central ring* having a lumen (diam. 55 nm). In some cases a central plug or granule guards the apical part of the central ring.

(iii) The intervening parts of the radial spoke like structures create 8 channels of 10 nm diam.

(iv) 8 filaments project out both from cytoplasmic and nuclear rings. The cytoplasmic filaments are 30-50 nm long while nuclear filaments are 100 nm long and are joined at their outer ends to form a *terminal ring*. The nuclear ring, filaments and the terminal rings are collectively known as *nuclear basket*. Thus usually the nuclear pore presents 8-fold symmetry. Each pore remains associated with 60-100 different proteins which collectively constitute *Nuclear Pore Complex (NPC)*.

The nuclear pore complex (NPC) contains about 30 nucleoporins (Nups). Some Nups assemble together into stable complexes. Nup107 is one such complex and it consists of Nups 160, 133, 107, 96, 85, 43 and 37, Shel Sec13 and ELYS). These protein complexes interact in a head-to-tail fashion to form ring-like octameric entities that are integral part of the scaffold (D'Angelo and Hetzer, 2008). Next to Nups other important protein is ALADIN which largely controls nucleo-cytoplasmic transport. The NPC is also found to be associated with ER cisternae called *annulate lamellae*. These are considered as *pre assembled pore complex* to be needed by rapidly proliferating cells.

Nups have characteristic protein domains. The α-solenoid/β-propeller containing Nups are mostly scaffold elements, while the transmembrane Nups anchor the NPCs to the NEs (it includes the Nup107/160 complex)and glycine –phenylalanine repeat containing Nups interact with cargo–receptor complexes and regulate cargo exchange through the central channel (Bui *et al*, 2013).

A *second* group of NE proteins (about 60 polypeptides) localizes to the INM. These are also referred to as NE transmembrane proteins or NETs. They interact with lamins (lamin B receptor - LBR, lamina-associated polypeptide - LAP1, LAP2, emerin, and MAN1 (Eibauer *et al*, 2015). INM proteins play vital and diverse roles

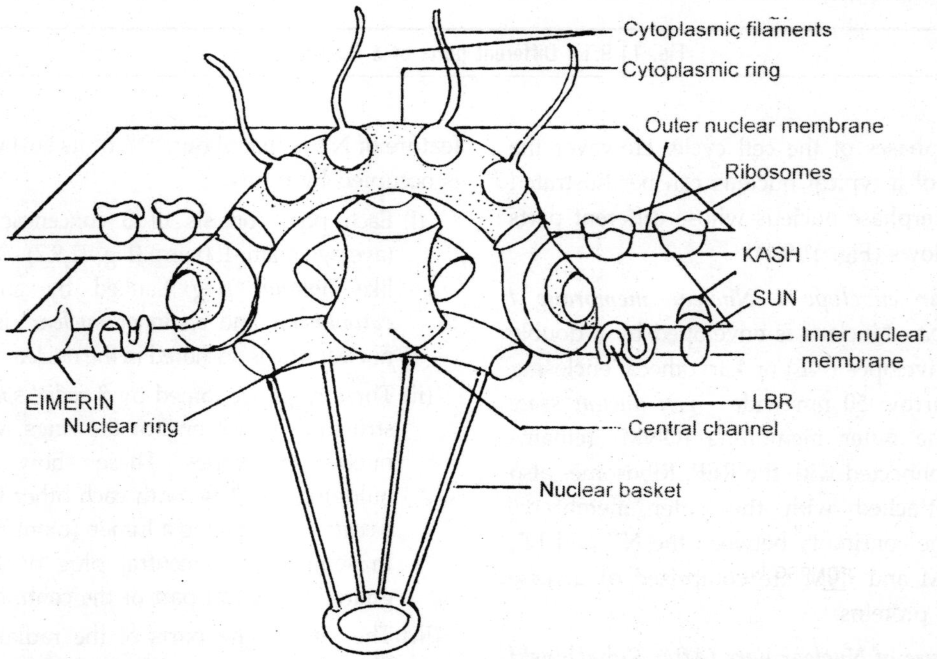

Fig. 11.9.2 : Structure of a nuclear pore.

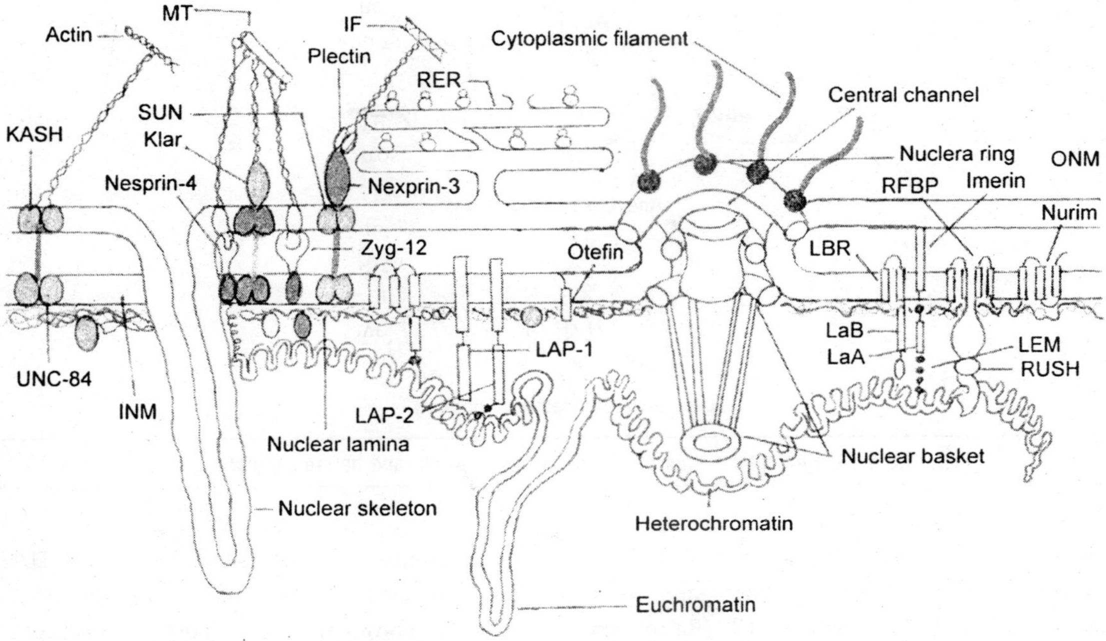

Fig. 11.9.3 : Structure of nuclear lamina.

in nuclear function such as chromatin organization, gene expression, and DNA metabolism (Reddy *et al.* 2008).

ONM and INM proteins form "bridges" across the perinuclear space. These proteinaceous bridges establish physical connections between the cytoskeleton and chromatin, which might be relevant for transcription, replication, and DNA repair mechanisms (Stewart *et al.* 2007). Thus,NE fulfills a critical role in shielding the genome from cytoplasmic components, but also represents a highly specialized membrane that provides anchoring sites for chromatin and the cytoskeleton.

Nuclear lamina: INM is followed by a filamentous scaffold structure (30-100 nm thick), called nuclear cortex or nuclear lamina. Over 2/3rds of lamins are located in the nuclear interior. Lamins are nuclear-specific intermediate filament (IF) proteins showing a N-terminal "head," long coiled-coil "rod," and globular C-terminal "tail" domains. They interact to form lamin dimmers which form polymers, which then associate with each other. Lamin proteins include nesprin, emerin, lamina-associated proteins 1 and 2 (LAP1 and LAP2b), the lamin B receptor (LBR) MAN1, Nurim and Dpy19L1 to L4. The transcription factors that bind to the lamina include the retinoblastoma transcriptional regulator (RB), germ cell-less (GCL), sterol response element binding protein (SREBP1), FOS and MOK2. Integral proteins of the INM are synthesized in the RER and bind to cytoskeletal elements - lamins A, B and C.

It maintains the shape of nucleus, spacing of nuclear protein complex, organization of heterochromatins, bind NPCs, many NE proteins and transcription factors. It conveys signals to the chromosomes that show differential structural organisation during cell division, development, differentiation and normal cell function. Some signals trigger chromosome condensation and nuclear breakdown during mitosis, DNA replication and nuclear integrity. There are

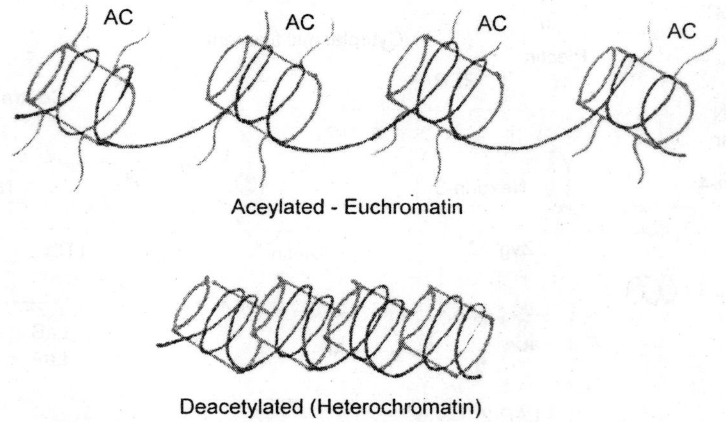

Fig. 11.9.4 : Diagramatic status of the nucleosomes in eu- and heterochromatin states.

anionic and cationic channels on the inner nuclear membrane which are permeable to K^+ and Na^+, but impermeable to Cl^- (Raices and D'Angelo, 2012). Nuclear lamina depolymerizes during cell division due to phosphorylation.

Nucleoplasm: Nucleoplasm is not an open space with haphazard arrangement of chromatins and othe sub nuclear structures. It is a jelly like structure showing enzymes, different RNAs, ions and chromatins which occupy distinct territories (CT) and positions in the nucleoplasm. Individual chromatins are folded into active and silent configurations are technically known as *euchromatins and heterochromatins.* Heterochromatin is usually localized to the periphery of the nucleus. Heterochromatins are of two types - *constitutive heterochromatins* (which remain permanently condensed) and *facultative heterochromatins* (which remain condensed in some stages but in decondensed state during differentiation or morphogenesis). In dividing stages due to the appearance of condensin and cohesin proteins, the chromatins become highly condensed to form chromosome. Euchromatin comprises the most active portion of the genome within the cell nucleus. 92% of the human genome is euchromatic. The remainder is heterochromatin. From molecular viewpoint, in the presence of acetylated histones (HAT mediated) and absence of methylase (HMT) activity, chromatin is loosely packaged. Additional nucleosome repositioning by chromatin remodeler complex, SWI/SNF opens up DNA region where transcription machinery proteins, like RNA Pol II, transcription factors and co-activators bind to turn on gene transcription. In the absence of SWI/SNF, nucleosomes can not move farther and remain tightly aligned to one another. Additional methylation by HMT and deacetylation by HDAC proteins condenses DNA around histones and thus, make DNA unavailable for binding by RNA Pol II and other activators, leading to gene silencing.

Condensed heterochromatin and silent genes are also bound to the nuclear lamina and constitute LAD. These constitute constitutive LADs (cLADs) that remain associated with the lamina while LADs become detached as the genes that they contain become active and form facultative LADs (fLADs). After mitosis, some LADs relocate to the periphery of the nucleolus and constitute nucleolus-associated chromatin domains; NADs. NADs are enriched for pericentric satellite repeats, A- and T-rich sequences and gene-poor regions

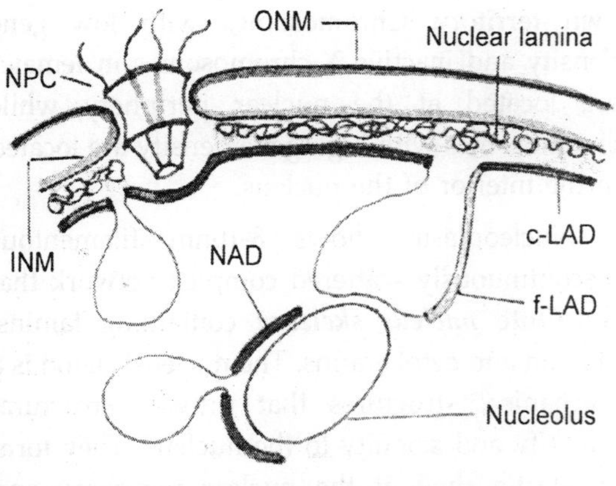

Fig. 11.9.5 : Diagramatic positions of different LADs).

Within compartments, the chromatin are packaged in the form of topologically associated domains (TADs). Chromatins are further folded into sub-TADs. Ultimately, DNA is wrapped around nucleosomes, which represents the first level of genome folding (Lieberman-Aiden *et al*, 2009). The CT position tends to be stable within a given cell though a significant amount of intermingling between different CTs occur (Misteli, 2010). CT positions and NE bonds are major players in genome organization because they largely affect the structural organisation of the chromatin fibres. Lamin polypeptides bind with specific DNA sequences called *matrix*

Table. 11.9.1 : Comparison between Euchromatin and Heterochromatin

Euchromatin	Heterochromatin
1. It is relaxed.	1. It is highly condensed with compact nucleosomes.
2. Transcriptionally active.	2. It is transcriptionally inactive.
3. It replicates early.	3. Heterochromatin DNA is late replicating.
4. Histone3 methylated at H3K4 (in active regions).	4. Heterochromatin DNA is methylated.
5. Histone is hyper acetylated.	5. Histones are hypo-acetylated.
6. Histone 3 is methylated at K4.	6. Histones are methylated on H3 lysine 9.
7. Involved in genetic recombination.	7. Does not participate in genetic recombination.
8. Not involved in epigenetic phenomenon.	8. Involved epigenetic and genetic imprinting phenomena.

Table 11.9.2 : Comparison between Constitutive heterochromatin and Facultative heterochromatin

Constitutive HC	Facultative HC
1. It is stable and conserves its heterochromatic properties during all stages of development in all tissues until not changed through mutation, hence also called obligate heterochromatin.	1. It is reversible, it can change from the heterochromatic state to an euchromatic state in different issues upon requirement.
2. Deeply stained by DAPI	2. Less deeply stained
3. HC is strongly stained by the C-band technique. This staining could be the result of the very rapid renaturation of the satellite DNA following denaturation.	3. It is not stained by the C-band technique.
4. Exhibits C-bands	4. No C-bands
5. Rich in satellite DNA, is highly polymorphic *Examples:* Centromeric and Telomeric heterochromatins	5. Not enriched in satellite DNA-non polymorphic *Example:* Barr body in Mammalian females.

attachment regions (MAR) spanning 300–1000 bp having high A/T content (Jerabek and Heermann, 2014).

There are two models regarding the arrangement of CTs. As per ***Inter Chromatin Domain (ICD) model,*** chromatins are separated by an ICD compartment with little or no chromatin. Active genes are in direct contact with the ICD compartment, as they lie at the surface of CTs. Rare chromatin loops extending from CTs may invade the ICD space. As per ***Intra chromosomal network model – (ICN) model,*** chromatins from different chromosomes are not separated by compartments but expand into the surrounding territories. According to ***Chromosome territory-interchromatin compartment model*** (Cremer and Cremer, 2001) the transcription complexes are restricted to inter chromatin compartments. Active genes are positioned on the surface of the chromatin domains that line the interchromatin compartment. Each chromosome occupies its own territory. Chromosomes with low gene density and inactive X chromosomes in females are located at the nuclear periphery while chromosomes with high gene density are located in the interior of the nucleus.

Nucleoplasm shows 8-10nm filamentous discontinuously scattered complex network that constitute ***nuclear skeleton*** containing lamins, vimetin and cytokeratins. The nucleoskeleton is a mechanical structures that provide structural integrity and stability to the nucleus. They form an elastic shell at the nuclear periphery and scaffolds for localized stiffness in the viscoelastic interior; both of which are thought to serve as platforms for different genome functions.

In the inter-chromosomal spaces ***interchromatin granules*** (200-250Å in diam.) are found that contain aggregates of snRNPs and other proteins involved in RNA processing. RNA transcription and processing occur at about 10,000 discrete sites on the surface of ***perichromatin fibrils*** containing hnRNAs.

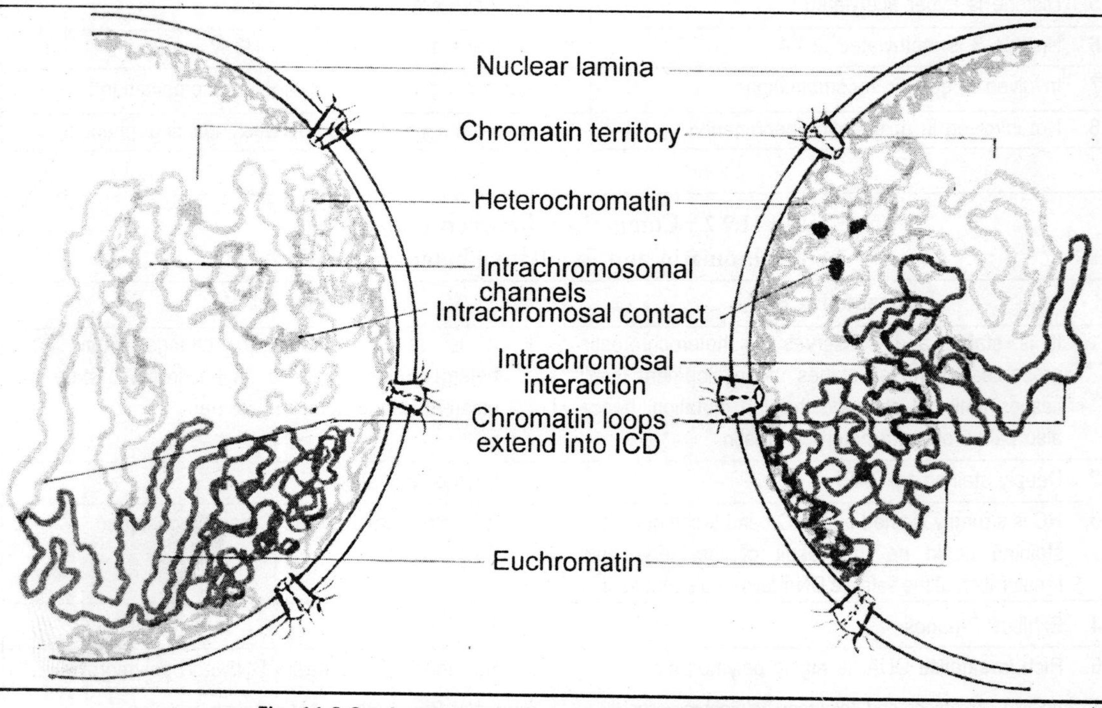

Fig. 11.9.6 : Arrangement of chromatins within nucleoplasm.

In plant and human nuclei, in the interchromosomal domains, following subnuclear bodies are found:

Cajal bodies (= coiled bodies) are spherical sub-organelles of 0.3-1.0 μm in diameter. CB are found 1-10 in the nucleus of proliferative cells like embryonic cells and tumor cells, or metabolically active cells like neurons. A major component of CBs is the protein p80 coilin. CBs resemble balls of tangled thread and densely packed coilin. They were first reported by Santiagó Cajal in 1903, who called them *nucleolar accessory bodies* due to their association with the nucleoli in neuronal cells. They were rediscovered with the use of the EM and named *coiled bodies,* according to their appearance as coiled threads on EM images. Their maximum number occurs in mid G_1 and towards G_2 they become larger and their number decreases. CBs disassemble during the M phase and reappear in G_1 phase. CBs are possibly sites of assembly of the transcription machinery of the nucleus. CBs are involved in mRNA processing, snRNP biogenesis and in the trafficking of snoRNPs, histone mRNA modification and snRNPs, which move through the CBs to nucleoli or splicing speckles. Cajal bodies in plants have been implicated as sites of siRNA and miRNA biogenesis.

Gemini of Cajal bodies, or Gems, whose name is derived from the Gemini constellation in reference to their close "twin" relationship with CBs. Gems are similar in size and shape to CBs, and in fact are virtually indistinguishable. Unlike CBs, gems do not contain snRNPs, but do contain a protein called survival of motor neuron (SMN) whose function relates to snRNP biogenesis. CBs are SMN positive and coilin positive but gems are SMN positive and coilin negative.

Promyelocytic leukaemia bodies/Nuclear domain 10 / Kremer bodies / PML oncogenic domains often seen in association with Cajal bodies and cleavage bodies and appear as spherical bodies found scattered throughout the nucleoplasm, measuring around 0.1–1.0 μm. PML bodies regulate many nuclear functions like DNA replication, transcription or epigenetic silencing.

Speckles (= Nuclear speckles, Splicing factor compartments or Interchromatin granule clusters) are enriched in pre-mRNA splicing factors and are located in the interchromatin regions of the nucleoplasm of mammalian cells. They appear as clusters of interchromatin granules. Speckles are often found close to actively transcribed genes and are thought to act as a reservoir for the splicing of nascent pre-mRNA at nearby genes. Many of the larger speckles correspond to ***interchromatin granule clusters (IGCs)*** - 0.8-1.8Å in diameter and are composed of 20-25-nm diameter particles that appear connected in places. IGCs have been proposed to be involved in the assembly and/or modification of pre-mRNA splicing factors. Because of a cell's changing requirements, the composition and location of these bodies changes according to mRNA transcription and regulation via phosphorylation of specific proteins.

Paraspeckles are irregularly shaped compartments in the interchromatin space. They are known to exist in all human primary cells, transformed cell lines and tissue sections. They are dynamic structures that are altered in response to changes in cellular metabolic activity. They are transcription dependent and in the absence of RNA Pol II transcription, they disappear and all of its associated protein components form a crescent shaped perinucleolar cap in the nucleolus. In the cell cycle, they occur during interphase and all stages except for telophase when they form a perinucleolar cap.

Perichromatin fibrils are located next to the transcriptionally active chromatin and are hypothesized to be the sites of active pre-mRNA processing.

Polycomb bodies - are a family of proteins that can cause epigenetic silencing of genes. These are well known for silencing Hox genes through modulation of chromatin structure during embryonic development

RAFA domains or polymorphic interphase karyosomal associations (PIKA) were found to be associated with transcription of small nuclear RNA (snRNA) (Saunders *et al*, 1991).

Clastosomes: These nuclear bodies are of irregular shape and size that contain both catalytic and regulatory subunits of the proteasome. These are normally absent in most cell types but when proteins targeted for proteolysis are rapidly encountered, recognized and degraded by the diffusely distributed proteasome containing clastosomes in the nucleoplasm. Clastosomes become prominent when cells are exposed to stimuli that suddenly increase the levels of proteins targeted for degradation.

Transcription sites / OPT domains: These are observed on the periphery of IGCs. Several transcription factors have been shown to be concentrated in 1-3 compartments termed OPT (Oct1/PTF/transcription) **domains**. These domains are 1.0-1.5 °m in diameter and contain nascent transcripts as well as other transcription factors, but they contain few factors involved in RNA processing. OPT domain appears in G1, next to nucleoli but it disappears during S phase (Mannen *et al*, 2016).

Chemically nucleoplasm shows 2% DNA, 30 - 60% RNA (hnRNA, snRNAs etc.), nascent ribosomal subunits and 10 -15% proteins (40 – 50 major polypeptides) of the nucleus. Nuclear matrix shows 6% of total cell phospholipid that regulates lipid-mediated signaling events; Na^{++}, K^+, Cl^- etc.; actin and nuclear myosin I (NMI) that assist in transcription. The chemical composition of the nucleoplasm presents a dynamic medium which changes during development, ageing and diseases. These components exhibit cell and tissue-specific differences which are mainly noted in the nuclear peripheral regions. It may act as a "hub" in integrating cellular functions including chromatin organization, transcriptional regulation, mechanosignaling, cytoskeletal organization, and signaling pathways (Burke and Stewart, 2014).

11.9.3. STRUCTURE OF NUCLEOLUS

Nucleolus is a multifunctional "sub organelle" of the eukaryotic nucleus. Most mammalian cell nuclei have one to five nucleoli, which are specialized regions 0.5 to 5.0Å in diameter surrounding transcriptionally active ribosomal RNA (rRNA) gene clusters. It is surrounded by a layer of condensed chromatin. In fact nucleolus is a knot of chromatin, organized from the "nucleolar organizing regions" NOR) of different chromosomes. EM observation reveals that nucleolus shows three distinguishable regions: the innermost *fibrillar centers* (FCs), surrounded by the *dense fibrillar component* (DFC), which in turn is bordered by the *granular component* (GC).

Fibrillar centres (FC): It is made up of a network of 4-5 nm thick fibrils (diam. 50 nm - 1Å). Number and size of FCs / nucleolus varies with the cellular activity and the need for ribosome production. Generally cells with a high rate of ribosome biogenesis possess numerous small FCs. On the contrary, cells with greatly reduced metabolic and transcription activities, present small nucleoli with one large-sized FC such as in lymphocytes and in inactive mammalian neurons. FC is likely the site where the primary rRNA transcripts are generated. The nucleoli of *Drosophila* and other insects lack FCs.

Dense fibrillar component (DFC): It surrounds FCs and is made up of very fine (3-5 nm) and densely packed fibrils forming a mesh work. Its amount is directly proportional with the ribosome biogene sis. Sometimes this meshwork occupies large areas of the nucleolus. The FCs and DFC constitute a nucleolar substructure called the *nucleolonema*

Granular component (GC): After the formation of rRNA it is linked to proteins to form ribonucleoprotein particles (RNP) (15 – 20 nm in

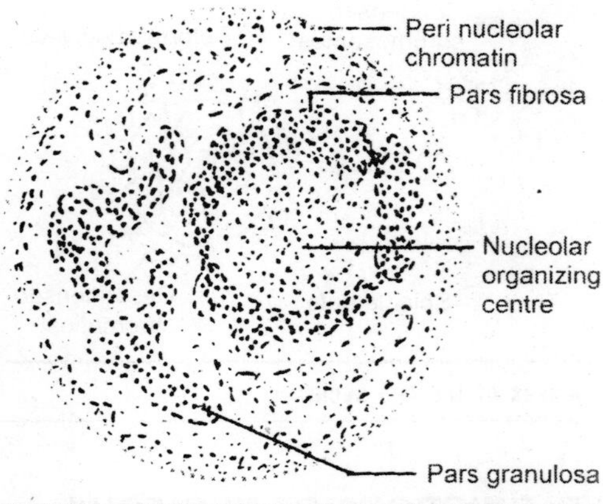

Fig. 11.9.7 : Structure of nucleolus.

diam) that constitute pars granulosa (PG) or granular components. They typically form a mass surrounding the FC. Fibrillarin and plant nucleolin homologue NopA64 are two important nucleolar proteins involved in pre-rRNA processing were found in the transition zone between FC and DFC. Although nucleolus is devoid of any membrane dense peripheral layer of GC marks the border between nucleoplasm and nucleolus.

The most outstanding feature is the high concentrations of nucleic acids surrounding the nucleolus and also in the form of network within the nucleolus (Sirri *et al*, 2008).

Specific functions are assigned to specific compartments of the nucleolus. Nascent transcripts appear at the junction between the FCs and DFC and accumulate in the DFC. Processing of the 47S pre-rRNA starts at the site of transcription in the DFC. The nucleolar proteins such as fibrillarin and nucleolin that participate in the early stages of rRNA processing, localize in the DFC, whereas proteins - nucleophosmin and PM-Scl 100 that are involved in intermediate or later stages of processing have been localized to the GC.Thus the distribution of the machineries involved in ribosome biogenesis correlates with the different processing steps of the biogenesis of the ribosome subunits.

In higher eukaryotic cells at the beginning of mitosis when rDNA transcription is repressed, the nucleoli disassemble and are no longer observed throughout mitosis. Conversely at the exit from mitosis concomitantly with restoration of rDNA transcription nucleoli assemble and are functionally active throughout interphase.

NE remodeling in different phases of cell division: The NE of metazoan cells completely disintegrates during cell division to allow the mitotic spindle to access chromosomes. As a consequence, every dividing cell has to reform the NE and re-establish the identity of the nuclear compartment (Guttinger *et al.* 2009). By the end of interphase in G2, the nuclei duplicate their genome, double the number of NPCs, and increase the surface area of the NE. The surrounding ER network remains continuous with the NE, but not enriched in NE proteins. The prophase, is marked by NE breakdown (NEBD) and the loss of the nucleo-cytoplasmic compartmentalization. Between NEBD and early anaphase, the majority of soluble NE proteins are distributed throughout the cytoplasm and NET proteins reside in the ER. In anaphase, ER membranes begin to reassociate and rapidly enclose the chromatin mass; along with association of Nups, decondensation of chromatin and the assembly of new NPCs concomitantly with NE formation (Anderson and Hetzer 2008c). At the end of cell division, the NE has reformed as a closed membrane barrier and re-establishes the nuclear compartment by enabling selective nucleo-cytoplasmic transport (Dultz et al. 2008).

The disassembled NE components serve critical functions in mitosis. E.g. Nup107-160 and Nup358, complex are detected with kinetochores; ELYS/MEL-28 and Nup107-160 are also found with spindle poles (Rasala *et al.* 2006). Moreever, in mammals, the absence of the Nup107-160

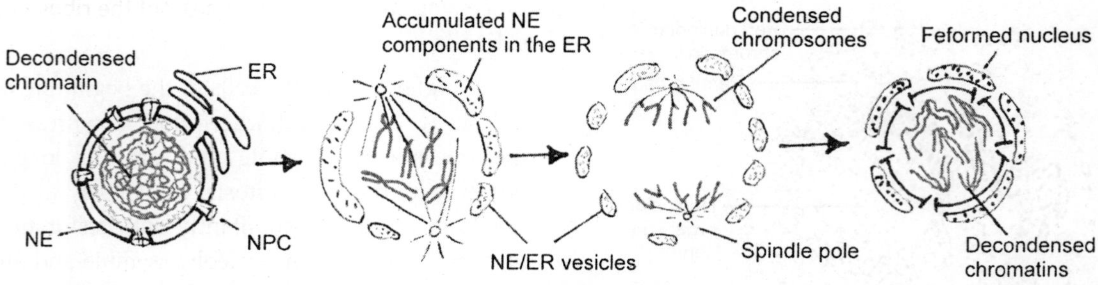

Fig 11.9..8 : Existance of NE in different phases of the cell cycle.

perturbs bipolar spindle formation (Cho *et al.* 2009). A group of nucleoporins Nup107/160 complex, and NETs - Ndc1 and POM121 have been shown to be essential for pore assembly (Anderson *et al.* 2009). Thus NE reformation is linked with the reformation of NPCs.

11.9.4. FUNCTIONS OF NUCLEUS

(i) It controls the heredity characteristics of an organism and stores hereditary materials in the form of DNA.

(ii) It regulates protein synthesis, cell division, growth and differentiation.

(iii)It stores proteins and RNA in the nucleolus.

(iv)It is a site for transcription.

(v) In the nucleus, pre-mRNA is associated with a variety of proteins to form heterogeneous ribonucleoprotein particles (hnRNPs). They must undergo post-transcriptional modification like capping, polyadenylation and splicing in the nucleus before being exported to the cytoplasm;

(vi)Aids in exchange of DNA and RNA (heredity materials) between the nucleus and cytoplasm.

(vii) Nucleolus produces ribosomes that are involved in protein synthesis.

(viii) It regulates the integrity of genes and gene expression.

(ix)It mediates the replication of DNA during the cell cycle.

11.9.5. FUNCTIONS OF NUCLEOLUS

(i) The main role of nucleolus is RNA biogenesis. All the proteins, necessary for the biogenesis of ribosome, are found in the nucleolus. It controls biogenesis and export of mRNAs. Moreover 5S, 8S, 18S and 28S rRNAs are processed and assembled with proteins in the nucleolus. snRNAs are formed and transported to their final desti nation by the nucleolus. It modifies small RNAs.

(ii) It controls aging and modulate telomerase function.

(iii)It regulates apoptosis.

(iv)It also controls gene silencing.

(v) The size of the nucleolus is generally enlarged in cancer cells, and this has been correlated with cell proliferation.

The non-traditional functions of nucleolus are:

i) It interacts with viral components.

ii) It regulates tumor suppression and oncogene activities.

iii) It helps in signal recognition particle assembly.

iv) It modifies small RNAs.

v) It controls aging and modulate telomerase function.

vi) It is reported to accumulate specific proteins that play vital roles in apoptosis.

vii) It plays crucial role in the processing and

assembly of a variety of RNPs and is involved in cell cycle control and senescence and as a sensor of stress in plants (Brown and Shaw, 2008). In humans, reduction of DNA helicase levels (which are localized in the nucleoli) helps to the development of those diseases associated with premature senescence. Furthermore, the presence of some telomerase components in the nucleolus, suggests role of nucleolus in the synthesis of the telomeres which is linked within the regulation of cell senescence.

viii) The striking correlation between perturbation of nucleolar function, elevated levels of p53, and induction of apoptosis suggest that nucleolus is a stress sensor that regulates p53 activity.

11.8.6. ORIGIN OF NUCLEUS

Various hypotheses have been proposed on the evolutionary origin of eukaryotic nucleus. The formation of nuclear membrane should have been a key event in eukaryotic evolution. Because one of the major cargoes in the nucleocytoplasmic export in the eukaryotic cell is the ribosome, its stimulating proteins called Ribosome Export Factors (REFs) might be attributed to be a crucial factor in the origin of eukaryotic nucleus (Ohyañagi *et al*, 2008). REFs are of two types – non membranous and membranous. The non-membranous REFs (non-mREFs) are known to be originated exclusively from eubacterial proteins, whereas the membranous REFs (mREFs) are from both archaeal and eubacterial proteins. Non-mREFs work inside the nucleus while mREFs shuttle between the nucleus and the cytoplasm. It suggests that the extant REFs working inside the nucleus have derived exclusively from eubacterial proteins. It implies that the nucleus arose in a cell that contained chromosomes showing a substantial fraction of eubacterial genes. It supports several models advocating endosymbiosis as a crucial factor under lying eukaryote origins. Evolutionary appearance of mREFs in the later periods may be one of the crucial factors for ensuring the transport from the nucleus to the cytoplasm when the nucleus was formed in the process of eukaryotic evolution.

SUMMARY

1. In eukaryotes, DNA remains separated from the cytoplasm by a membrane bound organelle, called nucleus. It is basically anchored in the cytoplasm by intermediate filaments and ER. In addition nesprins, a nuclear envelope-anchored spectrin protein is directly associated with the inner nuclear membrane protein Sun1 that in turn is connected with the actin cytoskeleton.
2. Nuclear membrane (NE) is a double-layered membrane, enclosing a very narrow peri nuclear space (PNS). A fibrous nuclear lamina supports the inner nuclear membrane. Nuclear envelope (NE) performs this role of sequestration of imamture RNAs. NE contains numerous pores.
3. The nuclear ring is anchored in the nuclear lamina. Each nuclear pore consists of a barrel like luminal ring embedded in the NE. It is flanked by an outer cytoplasmic and an inner nuclear ring. The nuclear ring, filaments and the terminal ring are collectively known as nuclear basket. Each pore is associated with 60-100 proteins constituting Nuclear Pore Complex (NPC). The dominant protein is nucleoporins (nups) that are essential for nucleocytoplasmic transport. Nuclear pore is the main channel through which cytoplasmic proteins (mainly histone, non histone and ribosomal proteins) are imported into the nucleus and ribosomal subunits and mRNA are exported from the nucleus but NPC is not simply a tunnel between the nucleus and cytoplasm but appears as flexible structure actively involved in the transport process by a series of structural modifications.
4. The importable proteins bear a nuclear localization signal or sequence - NLS (cargo) that is recognized by specific carrier proteins – adapters. The exportable proteins also bear a nuclear export sequence – NES. Another signal nuclear retention signal (NRS) is present on a number of proteins that

need not transport beyond the nucleus e.g. immature RNAs. Inside the nucleus this complex encounters Ran GTP that binds with importin b leaving importina-cargo complex. Ran GTP-importin β shuttles back into the cytoplasm through the pore.

5. During cell division phosphorylation of two NE proteins Nup200 and Nup97 promote disassem bly of the NE and NPC components which remain scattered in the whole cytoplasm in vesicles. At telophase, nuclear reassembly requires a timely coordinated and dephosphorylation dependent association of lamin binding proteins. Nup153 plays pivotal roles in NE disassembly, NPC regeneration and nucleo-cytoplasm ic export and import. Nup155 is required for the formation of a continuous nuclear membrane. A protein complex Nup 107-160 regulates NPC assembly. Moreover, NE contains some specific RNA species that are distinct from nuclear and cytoplasmic RNAs.
6. Inner nuclear membrane is followed by a filamentous scaffold structure called nuclear cortex or nuclear lamina; it maintains the shape of the nucleus, spacing of nuclear protein complex, organization of heterochromatins, DNA replication and regulation of transcription. It shows three distinct proteins, lamin A, B and C that are the predominant DNA binding proteins. Lamin A is expressed in all differentiated cells; lamin B in all somatic cells. Lamin A and C cross link with each other to form an orthogonal lattice, lamin B maintains the link between inner nuclear membrane and nuclear lamina. Various inner nuclear membrane proteins like Nurim, REBP, Olefin, Emerin, MANI and Lamina associated proteins 1 and 2 maintain the integrity of nuclear lamina and activity of nucleus.
7. Throughout the nucleoplasm, anastomosing filaments remain scattered discontinuously show ing large voids that constitute nuclear skeleton. In the interchromosomal spaces bright speckles are noted called interchromatin granules which contain aggregates of snRNPs and other protein factors involved in RNA processing.
8. Mammalian nuclei show 10-20 highly condensed bodies called pro myelocytic leukemia (PML) bodies or PML oncogenic domains (PODs) or nuclear domain 10.
9. Chemically nucleoplasm shows 2% DNA, 30 -60% RNA and 10 -15% proteins. Nuclear matrix contains actin and nuclear myosin I (NMI) which are localized in condensed chromatin.
10. Nucleolus is a "sub organelle" of the nucleus surrounded by a layer of condensed chromatin. It shows fibrillar centres (FC), dense fibrillar component (DFC), granular component (GC) and rDNA zones. The FCs and DFC constitute nucleolonema. The FCs alone remains associated with the mitotic chromosomes forming nucleolar organizer regions (NORs). The nucleolar protein nucleolin and RNA pol I remain bound at NORs. Dense peripheral layer of GC marks the border between nucleoplasm and nucleolus. Nucleolus is mainly involved in rRNA biogenesis.

QUESTIONS

Long questions:

1. Describe the structure of interphase nucleus.
2. Describe the structure of nuclear pore with a note on its disassembly and reassembly.
3. What is the basic difference between prokaryotic and eukaryotic genome?
4. What do you mean by tetrasome?
5. What do you mean chromatosome?
6. What is linker DNA?
7. Describe the finer details of different histone proteins.
8. Which factor determines the spacing between two nucleosomes?
9. Describe 'multiscale model' of nucleosome.
10. What do you mean by cold sites?
11. What is the significance of Cse4?
12. Describe 'Semi conser vative model' of nucleosome assembly.
13. What do you mean by 'nuclease hypersensitive sites'?
14. Describe general roles of nucleo somes.
15. Point out the roles of N terminal tails in the regulation of the structures of nucleosome.
16. Des cribe the molecular structure of octamer with special emphasis of the inter amino acid interactions and thermodynamic nature.
17. Describe different models regarding the structural organization of nucleosomes.

Short questions:

1. What do you mean by coenocyte and syncytium?
2. Where do you find binucleate conditions in animals?
3. Diferentiate nucleoid and nucleus.
4. Describe the structure of nuclear pore.
5. Name important proteins associated with NPC.
6. What do you mean by annulate lamellae?
7. Which proteins regulate disassembly of nuclear envelope during cell division?
8. Which proteins are required for the reassembly of NPC and NE after cell division?
9. Name the important nuclear lamina proteins.
10. What do you mean by NLS, NES and NRS?
11. What role does Ran-GTPse play in the nucleo-cytoplasmic transport?
12. Nucleo-cytoplasmic transport is not universally dependent upon Ran-GTPase. Explain.
13. What is the basic cause of Emery-Dreifuss Muscular Dystrophy?

MCQs:

1. **Peri nuclear space is continuous with**
   (a) Peri mitochondrial space, (b) ER, (c) Golgi.
2. **Nuclear pore usually presents**
   (a) 8, (b) 10, (c) 12 fold symmetry.
3. **The diameter of the transporter of the NPC is**
   (a) 10 nm, (b) 11 nm, (c) 9 nm.
4. **Karyophorins recognize**
   (a) NLS, (b) NES, (c) both NLS and NES.
5. **The cytoskeletal material that play important role in nucleocytoplasmic transport is**
   (a) Actin, (b) Intermediate filament, (c) Microtubule.

Answers to MCQs:

1. b; 2. a; 3. c; 4. c; 5. a.

Chapter 11.10

Chromosome

In eukaryotes, the DNA molecules remain separated from rest of the cytoplasm in a membrane bound organelle, called nucleus. Among other constituents, the most widely studied and most important constituent of the nucleus is the enormously coiled DNA molecules that are organized into discrete structures called ***chromosome*** (*chroma* = color; *soma* = body) having intense affinity for basic dyes. In eukaryotic cells among the notable alterations of the cellular organelles in interphase and M phase of the cell cycle the structural and functional changes of chromosomes appear to be most noteworthy. These marvelous tiny thread like intra nuclear structures regulate all the enzymatic activities of the cell through protein synthesis and play the most crucial role related with inheritance of characters from generations after generations.

Definition: *Chromosomes are tiny, thread like; coloured highly condensed intranuclear DNA fibres that play the most crucial role in association with histone and non-histone proteins in the maintenance of biochemical reactions by the formation of enzymes, inheritance of characters and evolution through mutations.*

11.10.1. NUMBER, SIZE AND SHAPE OF CHROMOSOMES

Every species has a definite number of chromosomes in each cell. Usually in a sexually reproducing organism each set of chromosomes exhibit two identical copies; father and mother contribute one of the copies. Such genetically identical paired chromosomes are called ***homologous chromosomes.*** Each set of homologous chromosome is called ***haploid set (n)*** of chromosome while two haploid sets constitute ***diploid set (2n)*** of chromosome. Usually somatic cells contain diploid and germ cells contain haploid set of chromosomes. Haploid sets of chromosomes represent the ***genome*** of the organism. Sometimes somatic cells exhibit chromosome numbers in multiplication of the haploid set, such cells are called ***polyploid*** (may be 3n/4n/5n/6n and so on).

Size of the chromosome may be as short as 0.25μ (μ = 0.001mm) in some fungi and birds to as large as 32μ in a plant *Trillium*. Usually most metaphase chromosomes fall within the size range 2 – 50m / 0.2-2m [3m (in *Drosophila*), 6μ (in Man) and 8-12m (in Maize)]. Usually the number and size of the chromosomes are inversely proportional. The sizes of the chromosomes of monocotyledons are usually larger than dicotyledones. The chromosomes of the salivary glands of many dipteran insects and oocytes of amphibians appear to be exceptionally large (about 2mm in length) and are referred as polytene and lampbrush chromosomes respectively. Those chromosomes that regulate the vital activities of an organism are called ***autosomes*** while those that regulate sex determination are called ***sex chromosomes.*** Other than autosomes, all chromosomes are called ***allosome.***

Some other chromosomes:

B-chromosome: These are heterochromatic supernumerary chromosomes that are smaller than normal chromosomes. They result loss of vigour. In *Tradescantia* there are five supernumerary chromosomes in addition to 12 somatic chromosomes.

m-chrmosomes: These are minute but functional chromosomes, found in some bryophytes and insects.

L-chromosomes: Found in the germ line cells that are eliminated during the formation of somatic cells.

11.10.2. KARYOTYPE

Every species has a definite number, size and shape of chromosomes. This species specific characteris tics (number, type, shape etc) of the chromosomes is called ***karyotype.*** It gives an idea about primitiveness or advanced feature of that species. The karyotype showing remarkable differences between smallest and largest chromosomes and fewer metacentric and mostly acrocentric chromosomes are called ***asymmetric karyotype*** while less differences between smallest and largest chromosomes and presence of mostly metacentric chromosomes is called ***symmetric karyotype.*** Karyotypes are used to study chromosomal aberrations, and may be used to determine other macroscopically visible aspects of an individual's genotype, such as sex. The chronological arrangement of the photographs of the chromosomes exhibiting a definite diagram is called ***idiogram.***

In order to be able to see the chromosomes and determine their size and internal pattern, they are chemically labeled with a dye ("stained"). The pattern of individual chromosomes is called ***chromosome banding,*** whereas the study of whole sets of chromosomes is known as ***karyology.***

11.10.3. STRUCTURE OF A TYPICAL CHROMOSOME

The nucleoplasm in interphase exhibits thin nucleoproteinous thread like structures, called chromatin fibers that follow gradual condensation to form thick distinct structure, called chromosome in divisional or M phase of

Table 11.10.1 : Comparative accounts of autosome and allosome

Autosome	Allosome
1. The chromosomes that are identical in morphology in both sexes of a species.	1. The chromosomes that are specific for either sex in a species.
2. They regulate the expression of all somatic features.	2. They regulate the expression of mainly sex specifc features.

the cell cycle. The physical structure of chromosome can best be studied in metaphase stage whose main structural components are as follows:

Chromonema / Chromatid: In interphase and early periods of prophase, in the nucleoplasm the visible, tiny, thread like indistinct structures are called *chromonemata.* These appear paired, identical structures, joined at one point. Both chromonemata remain coiled with each other in parallel or spiral fashion; these patterns of arrangement are respectively called *paranemic* and *plectonemic coiling.* Irrespective of the nature of coilings, each chromonema shows minute coilings throughout the axis, called *minor coils*. The coiled thread again forms large coils at right angle to the longitudinal axis of the chromonema; these coils are called *major coils*. Towards the end of prophase, the chromonemata show gradual condensation to form *chromatids*. Thus chromonema and chromatids are same structure in different phases of cell cycle.

Polytene chromosomes and leptotene chromosomes, due to non-uniform condensation, show numerous bead like structures throughout the longitudinal axis of the chromosomes. These artifact like structures are called *chromomeres.*

Primary constriction: Throughout the longitudinal axis, one or more constricted, faintly stained parts are noted, called *primary constriction.* Its central part, where the DNA fibers remain united, is called *centromere.* In fact centromere is a chromosomal locus that ensures binding of the chromosomes with the kinetochore microtubules leading to the delivery of chromatids to each daughter cell during mitotic cell division.

On the basis of the positions of centromeres, chromosomes may be classified as *metacentric, sub metacentric, acrocentric* and *telocentric*, (Fig. 11.10.1) when centromere is located in the center, near the center, near the telomere and at the telomere respectively. Telocentric chromosomes are very rare. True telocentric chromosomes were first reported in some plants, protozoans and mammals by Marks (1957). In human, however, there is no telocentric chromosome. On the basis of the number of centromere, chromosomes may

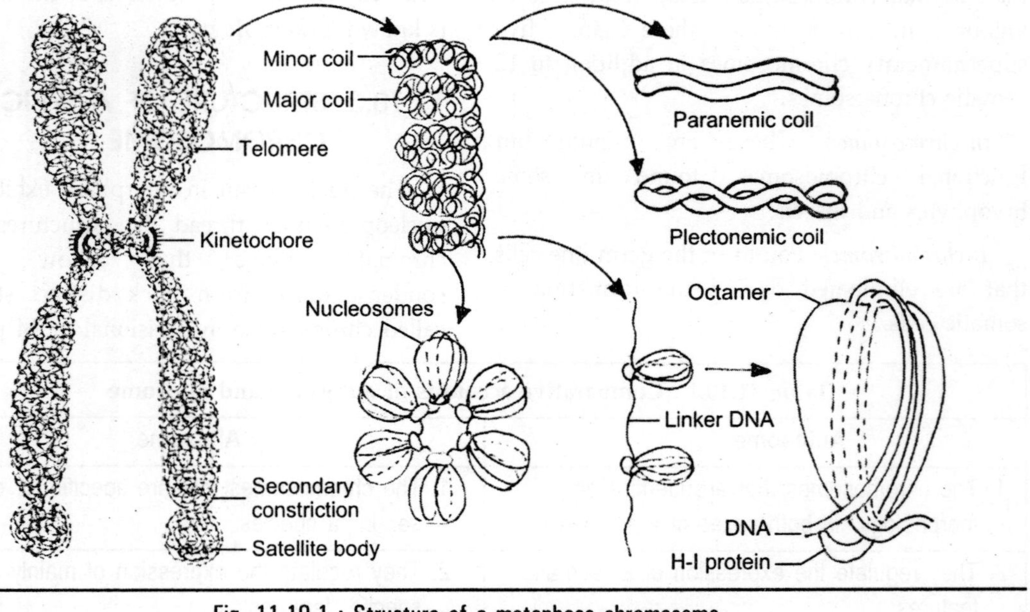

Fig. 11.10.1 : Structure of a metaphase chromosome.

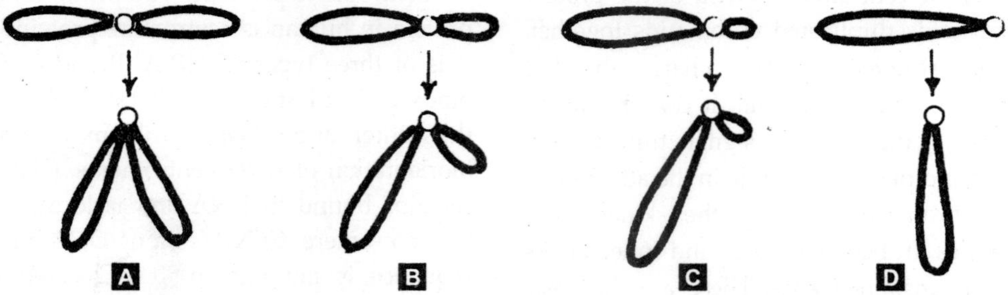

Fig. 11.10.2 : Types of chromosomes on the basis of the positions of centromeres in metaphase and anaphase. A-Metacentric, B-Submetacentric, C- Acrocentric, D- Telocentric chromosomes.

be ***monocentric, dicentric, holocentric*** and ***acentric*** having the number of centromeres - one, two, many and no centromere - at all (due to accidental breakage of chromosomes).

The parts of the chromosome on each side of the centromere are called *arms.* The long and short arms are designated as **'p'** and **'q'** respectively. The ratio of the length of the longer and shorter arms is called ***arm ratio.*** As the position of centromere is fixed, the arm ratio is also fixed for each chromosome. In case of metacentric chromosomes the arms are equal while in other cases the arms are unequal. Such chromosomes are called ***isobrachial*** and ***heterobrachial chromosomes*** respectively. During cell division, on each side of the centromere, a proteinaceous structure (0.2 - 0.25 nm diam.) assembles called ***kinetochore*** that act as special site for spindle microtubule attachment. It shows four layers during cell division. The ***innermost layer*** is continuous with the centromeric heterochromatin domain that links two sister kinetochores. It is a locus of cohesins and regulatory proteins like Aurora B and Kin I.

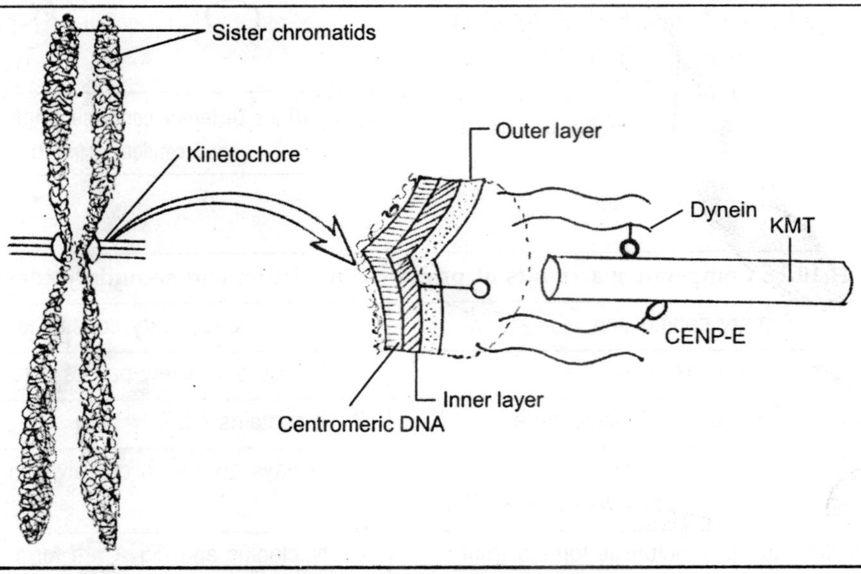

Fig. 11.10.3 : Structure of kinetochore.

Cohesin is the central component of the protein glue that holds duplicated chromatids together. The ***outermost layer*** appears as a narrow band of dense non-chromatin material. It contains many microtubule binding and signal transduction molecules and plays a vital role in the attachment of the chromosome with the kinetochore microtubules. In between outer and inner layers there is an unstained area. These are the basic layers of kinetochores showing attached microtubules. The kinetochores that are free of microtubules show a fourth layer that radiates from the outer layer called ***corona***. It mainly shows motor proteins that move chromosomes during mitosis (Fig. 11.10.3).

Molecular structure of Centromere:

Centromeric DNA: The chromatin of centromere is entirely composed of constitutive heterochromatin. Each centromere has a particular DNA sequence called CEN sequence that has specific protein binding sites required for the assembly of kinetochore proteins. In humans it spans 0.3 – 5 Mbp of DNA and shows 1,500 – 30,000 copies of repeated 171 bp sequence termed α *satellite*. In all eukaryotic kinetochores, a unique histone Cse4 occur between microtubules and outer kinetochore layer. Cse4, together with centromere-specific DNA-binding factors, regulate assembly of the kinetochore. Centromeric DNA shows two major α satellite DNAs – α-I and α–II.

Centromeric proteins: The prime centromeric protein in human is ***cen****tromere* ***p****rotein* – CENP. It is of three types CENP-A, B and C. CENP-A binds with α-I at the peripheral sides and α-II at the center and plays a role in cohesion and biorientation of sister centromeres. CENP-B also remains bound to DNA and is located beneath the centromere. CENP-C binds DNA but specific sequence is not known. It is located in close association with the inner kinetochore.

Secondary constrictions: Near the telomeric end, there may be one colored constricted part, called secondary constriction. Usually in one genome their number is one but it may be more than one. However in those cases where more than two secondary constrictions per diploid set are present, at least two are involved in the synthesis of rRNA that ultimately forms

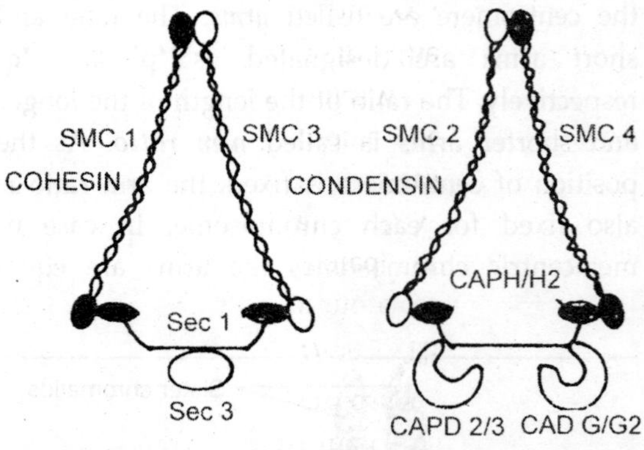

Fig. 11.10.6 : Different components of cohesion (left) and condensin (right)

Table 11.10.2 : Comparative accounts of primary constriction and secondary constriction	
Primary constriction	Secondary constriction
1. It occurs in all chromosomes.	1. It occurs in a few pair of chromosomes.
2. It contains centromere and kinetochore.	2. It contains NOR.
3. It plays a crucial role in cell division by attaching the chromosomes with spindle fibres.	3. It plays no role in cell division.
4. No specific nuclear structure is formed from this part.	4. Nucleolus and rRNA are formed from this part.

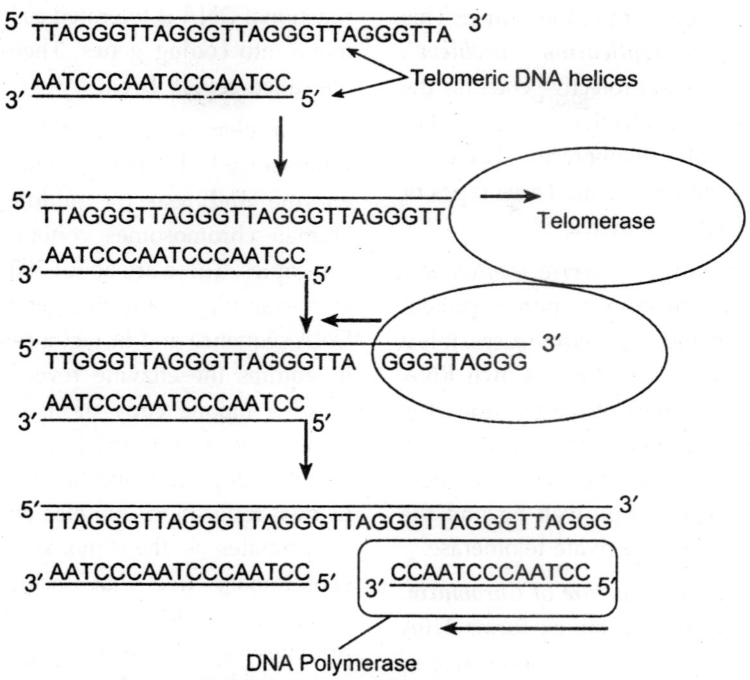

Fig. 11.10.5 : Role of telomerase in telomere extension.

nucleolus. This part is thus called ***nucleolar organizer region (NOR)*** (in humans, 13, 14, 15, 21, 22 and Y chromosomes contain NORs and chromosomes 1, 10 and 16 contain simply secondary constrictions). ***Thus all NORs are secondary constrictions but all secondary constrictions are not NOR.*** To distinguish these two definite regions NORs are called *secondary constriction I* and simply secondary constriction are called *secondary constriction II* or *pseudo NORs.* In fact, NOR contain special DNA region called rDNAs that represents 0.3% of the total nuclear DNAs which is destined to synthesize 18S and 28S rRNA that are transcribed by RNA polymerase I.

Satellite body: The tiny part, which lies beyond the secondary constrictions are called satellite body. Such chromosomes are called *SAT - chromosomes.* SAT stands for *Sine Acid Thymonucleinico* i.e. without thymine containing nucleic acid. The DNA of this part is inactive and represents repetitive DNA.

Telomeres: Telomeres are the special nucleoprotein complexes found at the end of all linear chromosomes that cap the ends of chromosomes (Fig.11.10.1). They keep the identity of the chromosomal ends and protect them from acci dentally becoming attached to each other chromosome. In all organisms (except some flies), it shows many repeats of G rich bases at 3′ end along with an array of associated proteins. Each time a cell divides, telomeres get shorter until finally they are so short that damage begins to happen to the "interior" DNA and chromosomes start to stick to one another on the ends. The number of times a cell divides before going into this decline is called the "*Hayflick limit.*" Synthesis of the lagging strand of normal DNA double helix requires a short primer that will be removed in due course. At the extreme end of a chromosome, there is no way to synthesize this region when the last primer is removed. Therefore, the lagging strand of telomeric DNA is always shorter than its template by at least the length of the primer. This is the so-called"***end-replication problem***". Telomeres provide special protective ends for the DNA molecule with protective proteins. For normal function of telomerase, RNA is complexed with specific proteins. Three types of proteins associate with telomeres.

Telomerase is an unusual *reverse transcriptase* because unlike the conventional purely protein reverse transcriptase, found in retroviruses, it is a ribonucleoprotein which contains its own RNA template as an integral part of the enzyme. It is present in high concentrations in the fast-growing embryo's stem cells, and declines with age. Most cancer cells are able to divide well beyond the Hayflick limit and they re-activate telomerase.

Electron microscopic structure of Chromatin: DNA and histone protein complex forms 110Å fibre or chromatin. The haploid genome size in human is 3.3×10^9 bp but the predicted number of protein coding genes are 35,000. Non-coding DNAs shows large amounts of repetitive DNA sequences (that occur in many copies in contrast with the coding sequences that are represented by single copies per genome) while the coding DNAs are called ***unique-sequence DNA***. Repetitive DNAs show two patterens of distribution, some are clustered in discrete areas (e.g. around centromere) and are called ***satellite DNA*** while others are dispersed throughout the genome which are composed of ***short interspersed repeated sequences (SINES)*** and ***long interspersed repeated sequences (LINES)***. Both SINES and LINES along with other similar DNA sequences are found in almost all eukaryotes (except yeast) and together account for at least 45% of the human genome. They show similarities to mobile genetic elements – retroposons that move from one place to another through the production of an RNA intermediate. Most of the SINES and LINES elements appear to be non functional. Their physiological roles are uncertain, they do not play any advantageous roles rather it is presumed that such non-functional DNA is tolerated as long as it does not move into coding genes. Therefore such DNAs are called ***selfish DNA***.

Alu elements are classified as SINEs (Short **IN**terspersed Elements) which contains the enzyme *Alu*I near the middle of the element. Human chromosomes contain about 1,000,000 *Alu* copies. These are about 300 bp in length. *Alu* is an example of "jumping gene" – a transposable DNA sequence and in fact a retroposon, because it requires the enzyme reverse transcriptase to make a mobile copy of itself. Some scientists regard *Alu* as an example of "selfish DNA" – it encodes no protein and appears to exist only for its own replication. *Alu* elements are found only in primates – the "monkey" branch of the evolutionary tree that includes humans. So, thousands of *Alu* copies have accumulated in primates since their separation from other vertebrate groups about 65 million years ago. Once an *Alu* integrates into a new site, it accumulates new mutations. *Alu* elements can be sorted into distinct lineages, or families, according to inherited patterns of new mutations. Once an *Alu* inserts at a chromosome locus, it can copy itself for transposition, but there is no evidence that it is ever excised or lost from a chromosome locus. So, each *Alu* insertion is stable through evolutionary time. Like genes, *Alu* insertions are inherited in a Mendelian fashion from parents to children. Thus, all primates showing an *Alu* insertion at a particular locus have inherited it from a common ancestor.

Chromatin proteins: Chromatin shows two types of proteins, non-histone and histone proteins.

Non-histone protein: It is the central fabric with which DNA remain attached. Two types of non-histone proteins are noted; some bind relatively non-specifically to the DNA, while some bind to specific short DNA sequence. These are respectively called *non-specific non-histone proteins* and *specific non-histone proteins*. The first discovered sequence specific non-histone protein

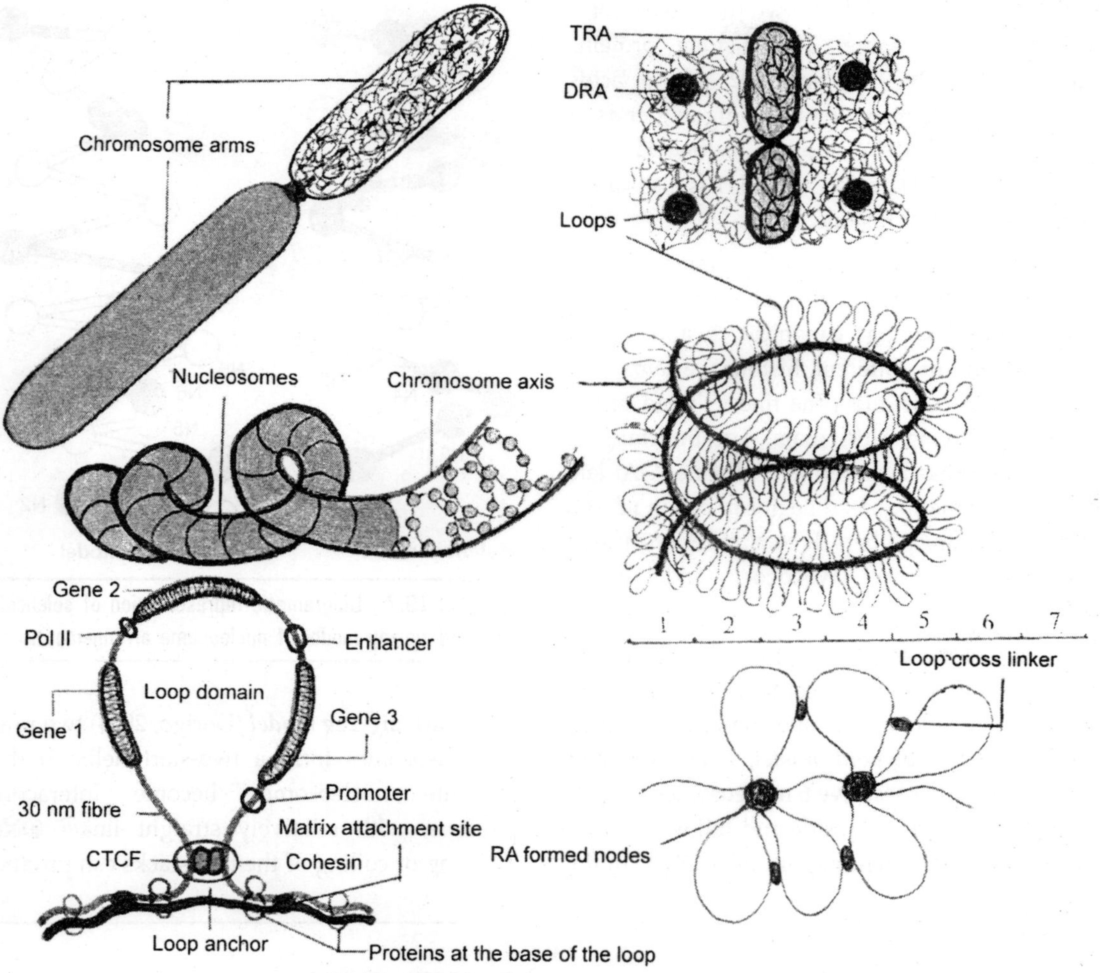

Fig. 11.10.6 : Diagramatic representation of chromatin loop formation.

is transcription factor III A (TF IIIA). It forms 9 folds and each fold shows a central Zn atom. Such folds with central Zn atom is called ***Zinc finger***. Another such protein is ***Cro protein***. It binds to a cluster of specific DNA sequences. In all chromosomes the central fabric with whom the loops of DNA fibers remain attached is called chromosome scaffold containing DNA topoisomerase IIα. Another scaffold protein is structural maintenance of chromosomes-2 protein (SMC proteins). Condensin (essential for the condensation of the chromosome) and cohesin (essential for sister chromatid pairing) are the important SMC protein (Fig. 11.10.4.). According to unified model of chromosome structure an axial distribution of topoisomerase-IIα, condensin and structural maintenance of chromosomes 2 (SMC2) occurs throughout 50–200-nm - diameter core. Chromatin loops are anchored to this central scaffold (Woodcock and Ghosh, 2010).

Histone protein: Histones are responsible for the formation of nucleosomes (110Å × 110Å × 57Å dimension) situated at 56 bp (190.4Å) apart that constitute linker DNA. Four histones H2A,

H2B, H3 and H4 (each with two molecules) form octamer or tetrasome. The histone proteins contain 129, 125, 135 and 102 amino acids respectively. The incoming and outgoing ends of DNA are clamped by another histone H1 that contains about 224 amino acids. H1 is also called extra nucleosomal histone.

Levels of Chromosome Organization: The primary chromatin structure showing arrays of nucleosomes having canonical histones and combinations of different histone variants and linker DNAs constitutes the finest thread of the chromatin having 110Å diam. The bunching of nucleosomes has been termed "clutches". At late G-2, the chromatin thread begins to coil in its axis to form 300Å thick fibers. It constitutes secondary chromatin structure (Woodcock and Ghosh, 2010).

So far as the arrangement of nucleosomes to form 300Å diam. is concerned, there are two models –solenoid model and zig-zag model. In the 'one-start' ***solenoid model,*** (Robinson and Rhodes, 2006) consecutive 6 nucleosomes interact with each other side by side and follow a helical arrangement with bending of linker DNA. In the 'two-start' ***zig-zag model*** (Dorigo, 2004) two rows of nucleosomes form a two-start helix so that alternate nucleosomes become interacting partners, with relatively straight linker DNA Twisting or coiling of the two stacks can produce

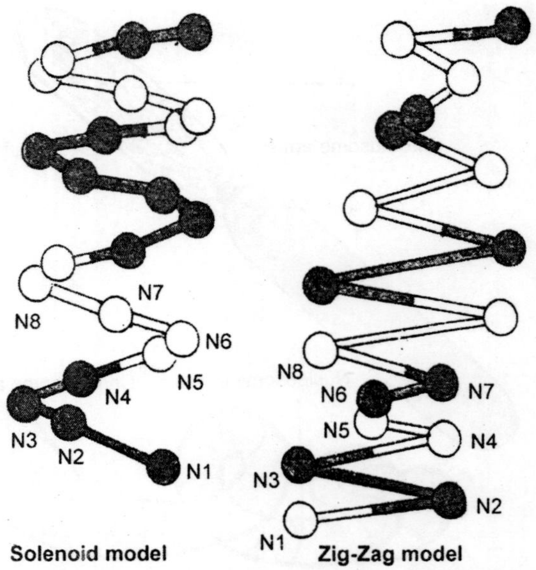

Fig. 11.10.7 : Diagramatic representation of solenoid and zig-zag model of nucleosome arrangement.

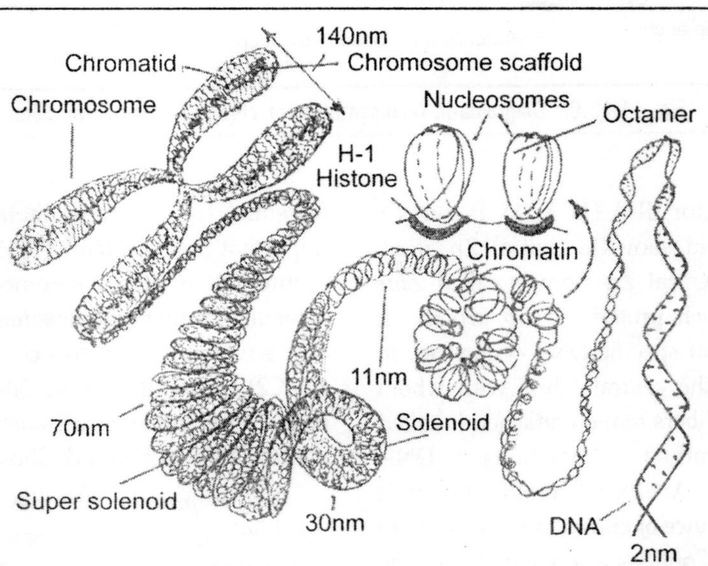

Fig. 11.10.8 : A hypothetical organizational model of chromonema.

different forms of the zig-zag model (Shogren-Knaak *et al*, 2006). Stem cells have been found to have less specific chromosome structures with less densely packed clutches of nucleosomes. In differentiated cells, the packaging of the nucleosomes is more dense.

EM studies however favour zig-zag model suggesting that nucleosomes are arranged in a zig-zag fashion forming a ribbon like structure which coil up to form a helical 300Å filament with 12 nucleosomes / turn. During coiling, the flat surfaces of the octamers are arranged parallel to the long axis of the filament. Lugar *et al* (2012) reported that there is no uniformity of nucleosome arrangement rather conformational heterogeneity of nucleosome interactions is noted. Hence chromatin fibre is referred as ***'heteromorphic' fibre***. It showed predominantly zig-zag organization, interspersed with partially bent linker DNA where interactions between consecutive nucleosomes exhibited one-start solenoid arrangement. This model with predominant zig-zag type interspersed with solenoidal conformations is more favourable than uniform zig-zag or solenoid arrangement that promoted the most compact folding in the presence of linker histone and Mg^{++}. It is also confirmed that linker length influences the chromatin fibre organization: short to medium nucleosome repeat length - NRLs (173–209 bp) favoured zig-zag structures, whereas longer NRLs (218–226 bp) favoured solenoid features (Perisic *et al*, 2010).

SUMMARY

1. In eukaryotes, the DNA molecules remain confined in a membrane bound organelle, called nucleus. The nucleoplasm in interphase exhibits nucleoproteinous threads called chromatin fibers that are gradually condensed to form thick chromosome in M phase of the cell cycle. Chromosomes exhibit a species-specific number. Each chromosome shows a replica that are called homologous chromosome. Each set of homologous chromosome is called a haploid set (n) or genome; two haploid sets constitute diploid set (2n) of chromosome. Usually somatic cells contain diploid and germ cells contain haploid set of chromosomes. The chromosomes that regulate vital activities are called autosomes and all other chromosomes are called allosomes.
2. Polytene chromosomes and leptotene chromosomes, due to non-uniform condensation, show numerous beads like structures throughout the longitudinal axis of the chromosomes called chromomeres.
3. Metaphase chromosome shows distinct terminal ends called telomeres, one or more constricted part called centromere which contains plate like two lateral structure called kinetochore that act as anchoring structure for microtubules and chromosomes. It mainly shows motor proteins that move chromosomes during mitosis.
4. Centromeric chromatin is entirely composed of constitutive heterochromatin. It shows a particular DNA sequence - CEN sequence showing specific DNA elements like CDE I, II and III. that has specific protein binding sites. In all eukaryotic kinetochores, there is a unique histone variant, called CenH3 (centromere histone H3) that together with centromere-specific DNA-binding factors directs assembly of the kinetochore. Centromere function depends critically on sequence specific DNA-protein interactions.
5. Near the telomeres there may be one to many constricted parts, called secondary constriction. In one genome at least one is involved in the synthesis of rRNA that gives rise to nucleolus. Such constriction is called nucleolar organizer region (NOR). The tiny part beyond the secondary constriction is called satellite body. Such chromosomes are called SAT - chromosomes.
6. Telomeres provide special protective ends. In all organisms, it shows many repeats of G rich bases at 3′ end along with an array of proteins. With each cell division, telomeres get shorter; finally they become so short that they stick one another. The number of cell division needed for this decline is called the "Hayflick limit". Telomeric end shows a "quartet" of G residues. The 3′ repeat units unite with the 5′complementary bases displacing the other strand to form a displacement loop or 'D' loop.

7. Three types of proteins associate with telomeres. The first type regulates the length of telomere by regulating action of telomerase. Second type regulates telomere elongation and third type protects against breakage repair machinery that recognizes telomeric ends.
8. Telomere length is a marker of aging. The senescent cells after a certain critical points suffer from chromosomal instability followed by apoptosis.

   Telomerase is a reverse transcriptase that regulates telomere length. It acts as template for the synthesis and subsequent addition of the repeat bases at the end of 3′ end of the DNA helix.

   It shows two molecules an enzyme moiety and RNA.
9. The metaphase chromosome (14,000Å diam) after repeated decondensation and uncoiling yields chromatin (110Å diam) fibers that are mainly composed of DNA and proteins. Each chromatin shows a continuous DNA thread having 3 different sequences - telomeric, centromeric and replication sequences that serve specific functions. Telomeric sequence maintains the identity of the chromosome and hinders the chromosomes to join with another chromosome. Centromeric sequence regulated kinetochore activities and Replication sequences participate in the DNA replication
10. Chromatin shows two types of proteins, non-histone and histone proteins. Non-histone proteins are of two types; some bind relatively non-specifically to the DNA, while some bind to specific short DNA sequence. Non-specific non-histone proteins remain attached with DNA haphazardly without following definite arrangements. In all chromosomes the central fabric with whom the DNA loops remain attached is called chromosome scaffold. Histone proteins are responsible for the formation of nucleosomes while non-histones form the fabric of the chromosomes. Nucleosome contains a core of eight histones H2A, H2B, H3 and H4, each represented by two molecules. It is called octamer; at one end of the octamer the incoming and outgoing ends of DNA remain attached with H1 histone protein. 147 bp DNA wraps around the octamer to form nucleosome that remain separated by 56 naked DNA called linker DNA.
11. At late G-2, the chromatin thread begins to coil in its axis to form 30 nm thick fibers called solenoid. It is again condensed to 700Å thick super solenoid and ultimately 14000Å thick chromatids. Condensin occurs throughout the central axis of the chromosome plays a crucial role in condensation. Cohesin binds to double stranded DNA and induces the formation of large protein-DNA aggregate which in the presence of Topoisomerase II stimulates inter molecular attraction of DNA molecules.

QUESTIONS

Long questions:

1. Describe the structure of a metaphase chromosome.
2. Describe the ultrastructure of chromosome.
3. Classify chromosome on the basis of the position of centromere.

Short questions:

1. What is karyotype?
2. What is ideogram?
3. What is chromonema?
4. What is chromatids?
5. What is kinetochore?
6. What is telomere and satellite body?
7. What do you mean by euchromatin and heterochromatin?
8. Which stains are used to visualize chromosome and DNA?
9. Differentiate secondary constriction and NOR.
10. Define plectonemic and paranemic coiling.
11. Differentiate autosome and allosome.
12. Differentiate prokaryotic chromosome and eukaryotic chromosome.

MCQs:

1. **Each species has a specific chromosome number. It is called**

   (a) Species specificity;

   (b) Idiogram;

   (c) Karyotype;

   (d) Hay Flick number.

2. **In interphase nucleus chromosomes appear faintly stained because**
(a) DNA is not tightly coiled;
(b) Amount of nuclear sap is more;
(c) The stain cannot adhere with the chromatins properly; (d) All.

3. **In one chromosome, the number of DNA thread is**
(a) 2; (b) 1; (c) 4; (d) 6.

4. **The chromatids remain attached at**
(a) Kinetochore, (b) Centromere,
(c) Cen- trosome, (d) NOR.

5. **In human beings the no. of allosome is**
(a) 44; (b) 46; (c) 2; (d) 23.

6. **In human beings the no. of autosome is**
(a) 44; (b) 46; (c) 2; (d) 23.

7. **X chromosome is**
(a) Telocentric; (b) Metacentric;
(c) Acrocentric, (d) Submetacentric.

8. **Y chromosome is**
(a) Telocentric; (b) Metacentric;
(c) Acrocentric, (d) Submetacentric.

9. **The chemical used to study metaphase chromosome is**
(a) Carmine, (b) Haematoxylene,
(c) Colchicine, (d) Fuelgen.

10. **During cell division, the astral fibres get attached with**
(a) Centromere, (b) Secondary constriction,
(c) Telomere, (d) Kinetochore.

11. **Telomeres at 3′ end shows many repeats of**
(a) A; (b) T; (c) G; (d) C **rich bases.**

12. **The diameter of a chromatin is**
(a) 120Å; (b) 110Å; (c) 150Å; d. 200Å.

13. **During metaphase, the condensation of chromosome occurs due to the action of**
(a) Cohesin, (b) Condensin,
(c) Colchicine, (d) Reverse transcriptase.

14. **The central axis of chromosome shows**
(a) Basic proteins, (b) Neutral proteins;
(c) Acidic proteins; d. No protein at all.

Answers to the MCQs:

1. c; 2. d; 3. b; 4. b; 5. c; 6. a; 7. b; 8. c; 9. c; 10. d; 11. c; 12. b; 13. b; 14. c.

Chapter 11.11

Nucleosome

One of the basic differences between prokaryotic and eukaryotic genome is the presence of histone protein along with DNA in eukaryotic chromosomes. *E. coli* genome shows 1.5 mm long DNA while haploid genome of man shows1000 mm long DNA. This enormously large DNA is packaged in chromosomes with definite orientation of histone proteins and DNA forming the basic structural unit called *Nucleosome.* Human chromosomes with 6.4×10^9 bp contain about 30 million nucleosomes.

Nucleosome was discovered in 1974 by Roger Kornberg. He digested the chromatin fibre by nuclease to obtain bead like structure containing both histone protein and wrapped DNA that was named nucleosome.

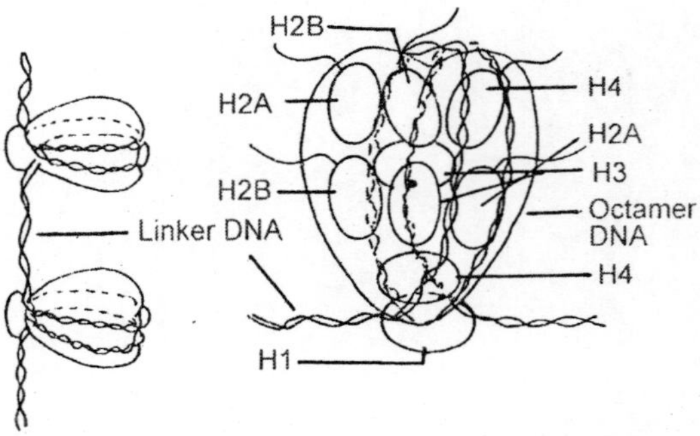

Fig. 11.11.1 : Structure of nucleosome.

11.11.1. GROSS MORPHOLOGY OF NUCLEOSOME

(i) Bead like structures (110Å × 110Å × 57Å dimension), occurs usually at 56 bp apart and basically shows 8 histone protein molecules - H2A, H2B H3 and H4 (each represented by 2 molecules) forming ***nucleosome core particle*** (Fig. 11.11.1) *(NCP) or Octamer.* The core histones share a averaging about 70 amino acids and histone fold extensions (***histone tails***) responsible for structurally conserved motif, the ***histone fold,*** protein–protein interactions within the histone

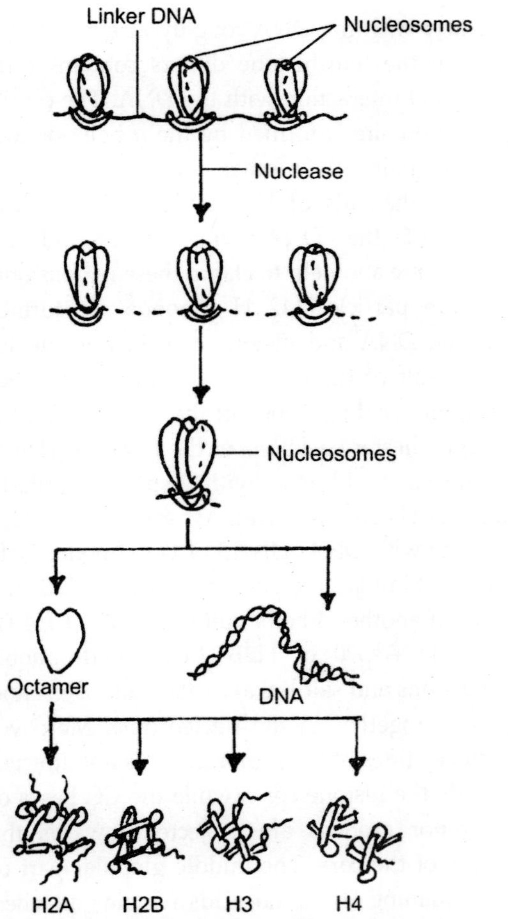

Fig. 11.11.2 : Isolation and Characterization of different histone proteins in Octamer.

octamer and thus is important for packaging DNA into chromatin.

(ii) The shape of the octamer is roughly flat, for which it is also called ***platysome.*** The octamer is mobile and can swing around its own axis.

(iii)The length of DNA between NCPs is called linker DNA; it varies not only between species, but also between tissues of the same organism and within a single nucleus. Mean linker lengths range from 20 bp in budding yeast to 75 bp in echinoderm sperm. For a typical vertebrate nucleus it is 35 bp. (Jiang and Pugh 2009). Linker DNA is associated with non-histone protein. ATP-dependent chromatin-assembly factor (ACF) determines equal spacing between nucleosomes.

(iv)147 bp long DNA is wrapped around octamer in left handed direction showing 1.67 turns.

(v) At one end of the octamer the incoming and outgoing ends of DNA are attached with one histone protein molecule, called ***H1 protein or Linker histone or Extra nucleosomal histone protein.*** Nucleosome with H1 and associated DNA showing a total length of 166 bp constitute *chromatosome.*

6. The path of DNA around the histone core is not smooth rather shows four kinks which is the cause of the non uniform surface of the core.

11.11.2. MOLECULAR ASPECT OF THE CORE PARTICLE

DNA wrapped around the octamer is predominantly of B-form. Curvature into the superhelix occurs when the minor or major grooves face the octamer. Major groove bending occurs smoothly; minor groove bending is facilitated by arginine side chains inserted into the groove and occurs smoothly around H3-H4 tetramer, but is kinked around the H2A/H2B dimer regions. The interspaces between the histone molecules remain filled up with water molecules. Nucleosomal DNA consists of two distinct structures, the terminal DNA domains are less tightly bound to the histone than the middle domain DNA. All four histones have central globular domains of 70 – 100 amino acids and a flexible peripheral N-terminal tails of approximately 30 amino acids. They project outward from the nucleosomal core as well as through minor groove 'tunnels' of the DNA. As per ***'multiscale model'*** of Nucleosome histone tails bind to nucleosomal DNA via strong salt-bridge interactions, suggesting that histone tails regulate inter- and intranucleosomal assemblies via binding with nucleosomal DNA and they have direct functional role in stabilizing higher order chromatin structure, mediated by salt-bridge interactions with adjacent DNA. Specific regions of the histone core H2A/H2B-H4/H3-H3/H4-H2B/H2A the *"cold sites"* are clustered around H3-H3, H2A-H4 and H4-H2A interhistone interfaces regulating nucleosome stability. Wrapping of ~80 bp DNA around $(H3–H4)_2$ tetramer forms ***tetrasome,*** due to a strong interaction between two H3 proteins. Nucleosomes lacking one H2A–H2B heterodimer forms an intermediate state called ***hexasomes.*** Histone H2A exhibits non-allelic variants that vary by only a few amino acids. Notable variants include H2A.1, H2A.2, H2A.X, and H2A.Z (Bargaje *et al*, 2012).

Positions of nucleosomes: Except 15 nucleosomes surrounding the centromeric DNA, majority of nucleo somes are not precisely positioned. Many specific sites of DNA throughout its length lack nucleosomes that are called *nuclease hypersensitive sites.*

At late G-2, the chromatin thread begins to coil in its axis to form 300Å thick fibers. During coiling the flat surfaces of the octamers remain arranged parallel to the long axis of the solenoid filament. It gradually forms *super solenoid* and ultimately *chromonema* (100 – 300 nm diam). H-I proteins are supposed to play vital roles in the coiling of chromatin.

Stability of Nucleosomes: The nucleosome is stabilized by a multitude of protein–protein interactions within the histone octamer and by numerous electrostatic and hydrogen bonds between protein and DNA over its entire length (Killian *et al*, 2012). The vast majority of DNA–histone interactions are between structured regions of the histones and DNA whereas the flexible histone tails extend away from nucleosomal DNA and are mainly involved in interaction with neighbouring nucleosomes (Tan and Davey, 2011).

Interaction of DNA with histones and interaction among histones: The histone octamer interacts with the DNA roughly at every 10 bp. Each of the four-histone dimers contains three regions of interaction with the DNA. The central interaction site is formed by the α-helix of each histone pair. Two other interacttions occur through the tails of H3 at the entry and exit points of the DNA wrapped around the nucleosome and help to clamp these regions onto the core particle. 142 H - bonds are formed between DNA and histone core in each nucleo some. Half of these bonds are formed between the amino acid backbone of the histones and the phosphodiester backbone of the DNA. Arg116 of H3 forms 2 H-bonds with Asp123. Similarly, Arg78 of H4 H-bonds with Asp 85. Arg83 of H2B interacts with DNA. His113 of H3 H-bonds with Asp123. Similar interactions between H2B and H4 form another 4-helix bundle. His75 of H4 H-bonds to Asp90 of H2B. Many hydrophobic interactions and salt linkages also hold DNA and proteins together in the nucleosome. Moreover AT nucleotides of the minor grooves are directed towards the histone cores while the CG bases of the minor grooves are projected towards the outside of the core. The middle globular part of H1 containing 90 amino acids remains attached with DNA as well as with octamer.

Assembly of nucleosomes: After DNA replication, new core particles are assembled in a definite manner. As per ***Semi conservative model of nucleosome assembly,*** the nucleosides are disrupted at the replication fork; the parental H3-H4 tetramers are segregated equally into daughter DNA strands as H3-H4 dimer with which the newly synthesized H3-H4 dimers are assembled by the action of H1. Initially H2A - H2B and H3-H4 dimers are formed. H3-H4 dimers associate to form H3-H4 tetramer onto which two H2A-H2B dimers associate to form the octamer. Chromatin assembly factor 1 (CAF1), has been shown to have the ability to wrap newly synthesized DNA around histones forming the first level of higher order chromatin structure beyond naked DNA.

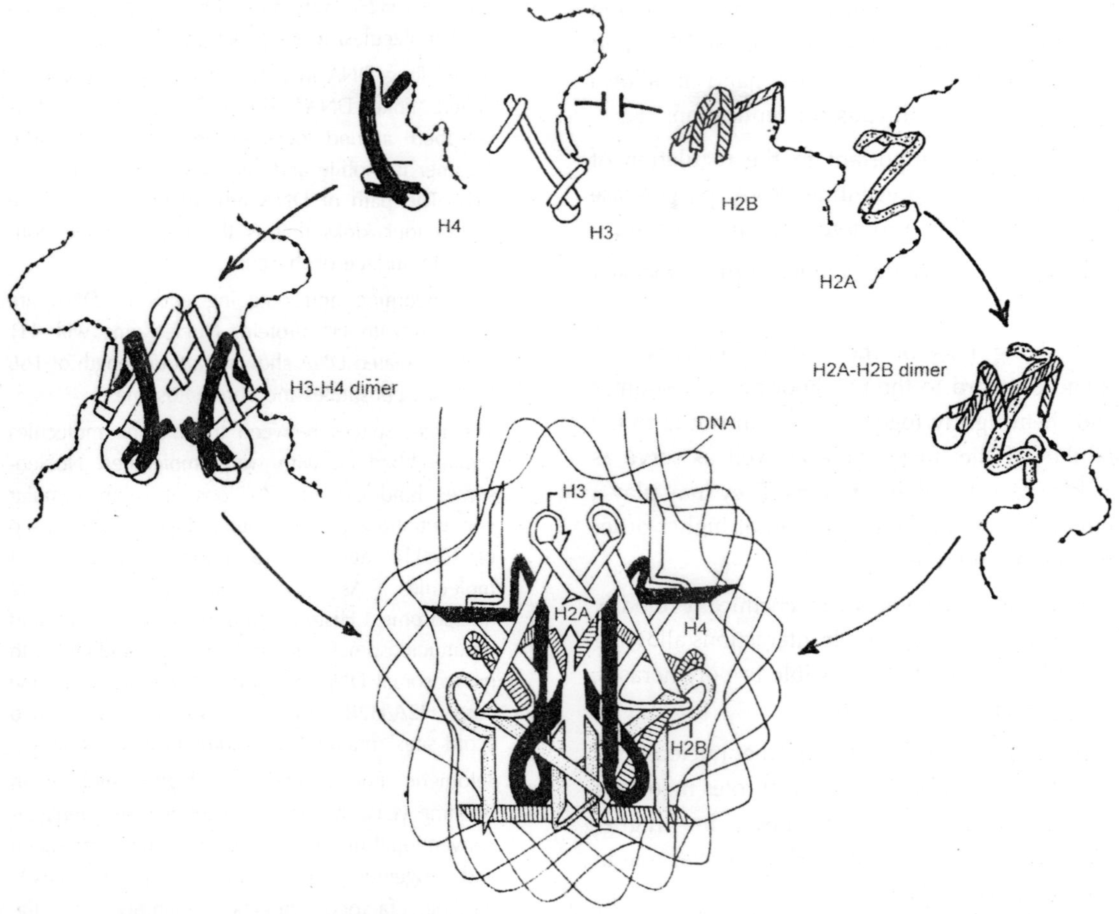

Fig. 11.11.3 : Assembly of different histone proteins,

Alternative structure of Nucleosome: Acetylation of histone H3 at Lys56 and H2A and H3 in the terminal DNA segments lead to spontaneous unwrapping and rewrapping, as a result, DNA partially dissociates from the histone octamer (Li *et al*, 2005). Moreover, the histone variants H2A.Bbd and CenH3 do not tightly organize the terminal DNA segments and thus exist in an 'open state' (Poirier *et al*, 2008). The human histone H2A. Bbd variant lacks three of the six H2A acidic amino acids and consequently, H2A.Bbd nucleosome arrays cannot fold (Zhou *et al*, 2007). DNA not only fluctuates at the entry and exit sites but also that nucleosomes can undergo spontaneous sliding, 'splitting' and complete dissociation without the involvement of additional proteins such as ATP-dependent chromatin-remodelling factors (Miyagi *et al*, 2011). The structural transitions and non-canonical nucleosomes exist *in vivo*. This might be the intermediates of assembly and disassembly processes or they might occur during transcription (Torigoe *et al*, 2011).

11.11.3. ROLES OF NUCLEOSOME

1. They provide the lowest level of compaction that is required to fit DNA into the nucleus. The H1 histones are involved in the folding the DNA into higher-order chromatin

structures. On one hand, it stabilizes chromatin structure forming tight association of DNA and histone protein; on the other hand it allows uncoiling of DNA threads for replication.

2. They are important in the regulation of transcription by preventing RNA polymerase from accessing the promoter regions of genes.

3. They modify the activity of the associated genes.

4. Long tails of the histone proteins that extend outward to the neighboring nucleosomes bind them tightly together and contribute to the stability of the nucleosome as well as serve as docking sites for other proteins. They may play a significant role in the generation of higher order chromatin structure.

5. Nuclear enzymes can chemically modify histone tails to weaken their interactions allowing particular genes more accessible to polymerases, causing their transcription.

6. In addition to uniting the nucleosomal DNA fibers, H1 also plays crucial roles in joining the neighboring nucleosomes. In higher eukaryotes, H1 additionally plays crucial role in embryonic development.

7. Phosphorylation of histone H2A at Serine129 plays a crucial role in double-strand break repair.

SUMMARY

1. One of the basic differences between prokaryotic and eukaryotic genome is the presence of histone protein along with DNA in eukaryotic chromosomes. Chromatin, which is basic structure of chromosome, shows highly organized DNA and histone protein structures called nucleosomes.
2. All histones have a central domain of 70 – 100 amino acids and a peripheral N-terminal tails of 30 amino acids. They project out from the nucleosomal core as well as through minor groove 'tunnels' of the DNA to form 30 nm fibers.
3. These measure110 A° X 110A° X 57 A° dimension and occurs at 56 bp apart. They show 8 histone proteins - H2A, H2B H3 and H4 (each represented by 2 molecules) forming octamer.

   56 bp long DNA in between two nucleosomes is called "linker DNA". The 147 bp long B-DNA is wrapped around octamer in 1.67 turns. The octamer is mobile and can swing around its own axis. The path of DNA around the histone core shows four kinks that is the cause of the non-uniform surface of the core.
4. The incoming and outgoing ends of DNA are attached with H1 protein. Nucleosome with H1 and associated DNA showing a total length of 166 bp DNA constitutes the chromatosome.
5. The inter spaces between the histone molecules remain filled up with water molecules. Nucleosomes bind Co and Zn ions at submicromolar concentrations at a zinc finger-like site, composed of Cys110 and His113 residues of both H3 molecules. As per 'multiscale model' of nucleosome. Histone tails regulate inter- and intranucleosomal assemblies via binding with nucleosomal DNA. Specific regions of the histone core H2A/H2B-H4/H3-H3/H4-H2B/H2A, termed "cold sites" maintain inter protein association.
6. Telomeric nucleosomes are highly mobile; in budding yeast, histone H3 variant, Cse4, replaces conventional histone H3 in centromeric chromatin and, together with centromere-specific DNA-binding factors, directs assembly of the kinetochore; H3 has three isotypes -H3.1, H3.2 and H3.3.

   Mammals show three sub types of H1 histone. In S. cerevisiae, H1 is absent .In nucleated RBC and sperms; unlike H1 quite different extra nucleosomal histone protein is noted called H5 that is rich in arginine. The octamer interacts with at every 10 bp. 142 H- bonds are formed between the amino acids of the histones and the phosphodiester backbone of the DNA. Many hydrophobic interactions and salt linkages also hold DNA and proteins together in the nucleosome. Moreover AT nucleotides of the minor grooves are directed towards the cores while the CG bases are projected towards the outside of the core.
7. After DNA replication, new core particles are assembled in a definite manner; the parental H3-H4 tetramers are segregated equally in to daughter DNA strands as H3-H4 dimer with which the

newly synthesized H3-H4 dimers are assembled by the action of H1.

8. Except 15 nucleosomes surrounding the centromeric DNA, majorities are not positioned properly. Many sites of DNA lack nucleosomes that are called nuclease hypersensitive sites. At late G-2, the chromatin coils in its axis to form 30 nm thick fibers. So far as the arrangement of nucleosomes is concerned, there are three models – solenoid model, zigzag model and two start model.

   In solenoid model, 6 nucleosomes form a whorl of 30-nm diam. In zig-zag model nucleosomes are arranged in a zigzag fashion forming a ribbon like structure with 12 nucleoso- mes / turn.

   As per two-start model compacted nucleosomes are stabilized by the disulfide cross-links and chromatin fibers show two stacks of nucleosomes.

QUESTIONS

1. What is the basic difference between prokaryotic and eukaryotic genome?
2. What do you mean by tetrasome?
3. What do you mean chromatosome?
4. What is linker DNA?
5. Describe the finer details of different histone proteins.
6. Which factor determines the spacing between two nucleosomes?
7. Describe 'multiscale model' of nucleosome.
8. What do you mean by cold sites?
9. What is the significance of Cse4?
10. Describe 'Semi conservative model' of nucleo some assembly.
11. What do you mean by 'nuclease hypersensitive sites'?
12. Describe general roles of nucleosomes.xiii) Point out the roles of N terminal tails in the regulation of the structures of nucleosome.
13. Describe the molecular structure of octamer with special emphasis of the inter amino acid interactions and thermodynamic nature.
14. Describe different models regarding the structural organization of nucleosomes.

Chapter 12

Cell Division

The types of Cell Division described in this section are :

Chapter 12.1

Mitosis Cell DIvision

Cell division is the prime feature of all living organisms in which a parent cell forms its replica and thereby transmits its genetic elements to the next generation resulting in the formation of many daughter cells and indirectly helping in the perpetuation of species as well as evolution.

Cell division is the prime event of the M phase of cell cycle. Except for the division of primordial germ cells for the formation of gametes, all cells divide into two identical cells. This type of cell division is called *mitotic cell division* or *equational cell division.*

Definition: *Mitosis is the process of cell division in which a cell accurately replicates its cytoplasm and nuclear contents and distributes identical copies of the replicated genome into two daughter cells, maintaining the exact replica of the parental cell.*

12.1.1. Phases of mitosis:

Mitosis is conventionally divided in to two phases - karyokinesis and cytokinesis. Karyokinesis involves the division of nucleus or chromosomes and cytokinesis involves the division of cytoplasm. Karyokinesis is divided in to five phases (prophase, prometaphase, metaphase and telophase) having distinct features. The main events of each phase are as follows (Fig. 12.1).

Prophase: It is the initial phase and is characterized by gradual condensation of the replicated chromosomes, centrosomal duplication and separation regulating microtubular reorganization culminating in the complete separation of exactly the half of the cellular contents producing two daughter cells.

Microtubular reorganization forming spindle fibres: Each daughter centriole is the focal point of the origin of microtubules (MTs) having one attached '–' end at MTOC and one free '+' end. Formations of MT occur at the expense of gradual binding of α and β monomers of the tubulin proteins. The cytoskeletal reorganization into MT and ultimately into spindle fibers is carried out by two proteins, caldesmon and vimetin being synthesized by mitotic maturation factor (MPF) at late G2 phase.

Formation of Spindle fibers: As the nuclear membrane breaks down, the MTs (now these are called *spindle fibers)* get attached with the condensed chromosomes and the daughter centrioles in each pole - *spindle pole*. Splitting up of the centrioles requires the activity of Cdk2-Cyclin-A aided by nucleoplasmin Mps1p, CP110 and a kinase – Zyg1. Two motor proteins- dynein and kinesin play critical roles in the separation of centrioles and spindle pole formation. Four different categories of spindle fibers are noted - the small fibers that occur around each pole and remain connected with the cortex of the cell are called *astral fibers*, these; long fibers that connect two poles are called *inter polar fibers*, in the mid line, these fibers overlap with each other; the fibers which remain connected with the kinetochores are called *kinetochore fibers* and the fibers which connect the chromosomal arms are called *chromosomal fibers* (Fig. 12.1).

Two pathways can achieve spindle formation - the *centrosome directed pathway* in which MT assembly is nucleated by MTOC and *chromosome directed pathway* in which chromosomes play an active role in the assembly of spindle fibers. In centrosome directed pathway, MT nucleation requires g-tubulin, which exists as two main complexes: the γ-tubulin small complex, and the γ-tubulin ring complex. These complexes accumulate at the centrosome to support spindle formation. The MTOC plays the most vital role in proper assembly of bipolar spindle.

Chromosome directed pathway is regulated by some nuclear proteins. During interphase, in the nucleoplasm there occur three important proteins - aster promoting activator (APA), importin α and β all of which unite to form a multiprotein complex that hinders the formation of spindle fibers. In prometaphase, guanine nucleotide exchange factor, Ran GTPase is generated around each chromosome that causes disassembly of APA and importins. It causes microtubule assembly.

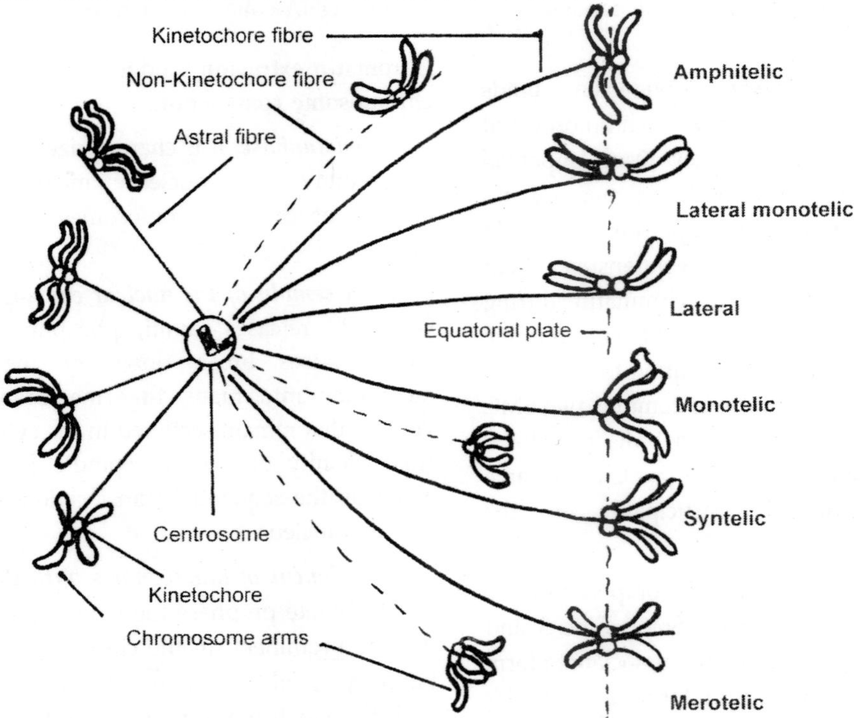

Fig. 12.1 : Kinds of spindle fibres.

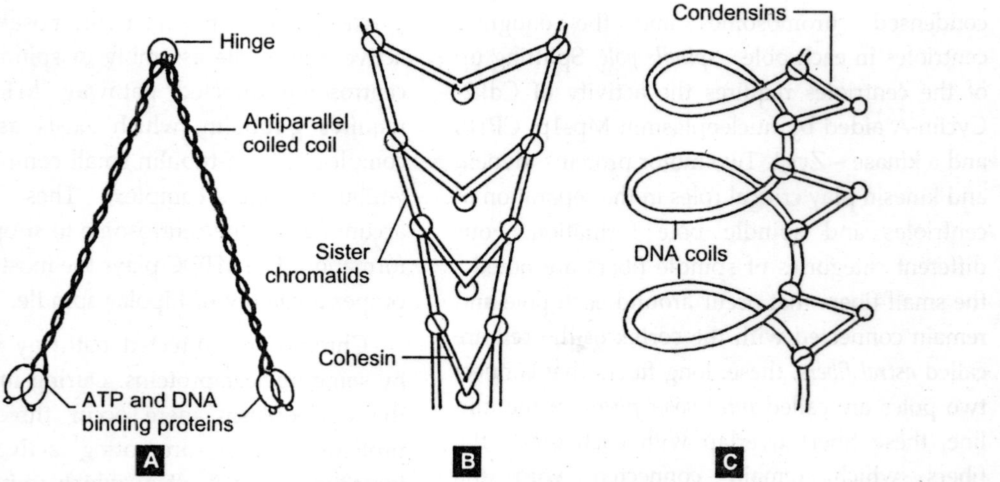

Fig. 12.2 : Diagrammatic structure of cohesin and condensin and their mode of action.

Condensation of chromosomes: As prophase continues the water content of the nuclear sap becomes greatly reduced and the chromatin material gradually condenses into chromosomes (1400Å diam.). Cohesin and condensin are the principal proteins bringing about condensation (Fig. 12.2.).

Role of condensin: Condensin binds throughout the central axis of the chromosome at every 5-10 kb of DNA. They result supercoil formation in the presence of topoisomerase-I and knots in the presence of topoisomerase-II. In mammals two types of condensins occur *condensin I* (interacts with chromatin during prophase) while *condensin II* bind to chromosome throughout the mitosis. In addition to condensin I and II, a five subunit protein complex two SMC (SMC2 and 4) and three non SMC proteins - CAP-D2, G and H [CAP = Chromosome Associated Proteins] participate in the condensation process.

Role of cohesins: Cohesin is composed of five subunits - Scc1/Mcd1, Scc3, Smc1, Smc 3 and Pds1p proteins. Smc1 and Smc3 associate to form a 'V' shaped structure. Scc1 forms a triangular ring that holds the sister DNA molecules. Following DNA replication, cohesin complex are attached to several AT rich regions at S phase. Cohesins bind along the chromosomal arms and around the centromere. As per *ring entrapment model,* cohesin rings concatenate individual sister minichromosome DNA molecules forming a dimeric DNA-cohesin structures.

Chromatin extrusion model of chromosome compaction:

Prometaphase: It is characterized by the total disassembly of nuclear membrane and attachment of the chromosomes with spindle fibers.

Disassembly of the nuclear envelope: At late G2, MPF releases lamin; phosphorylation of lamin causes break down of the nuclear membrane into many tiny membrane bound vesicles that remain scattered in the cytoplasm in close vicinity of the ER. Spindle formation is aided by the sequential translocation of tubulin into the nucleus.

Attachment of kinetochores with the spindle fibers: By late prophase the kinetochore proteins become prominent. In the chromosome directed pathway, spindle formation and there attachment with the kinetochore occurs simultaneously but in centrosome directed pathway the microtubules

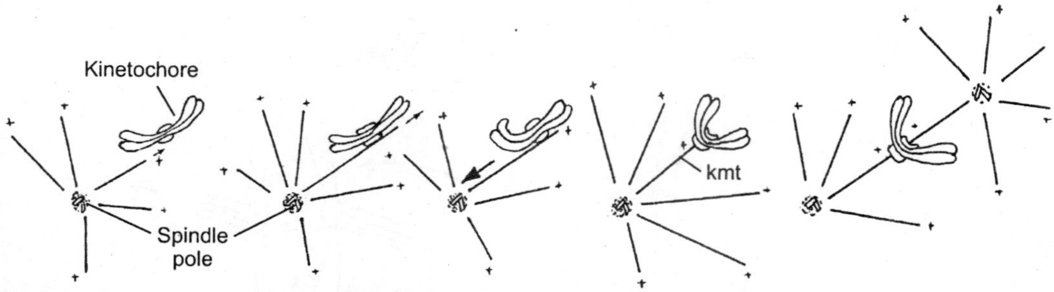

Fig. 12.3 : Capture of kinetochores with the kMTs.

grow and shrink conti nuously and capture kinetochores. According to *"search-and-capture"* model, microtubule (+) ends are associated with many proteins which search and capture kinetochores (Fig. 12.3.). The two kinetochores of chromatids get attached with the microtubules emerged from opposite poles so that they can segregate properly. Chromatid arms are attached with additional chromosomal MTs in metaphase.

Mitotic spindle checkpoint: Kinetochores perform four major roles:

i) attach with the '+' ends of the kMTs.

ii) regulate spindle assembly;

iii) generate force for chromosome movement and

iv) prevent anaphase until all the kinetochores attach to the kMTs (Fig. 12.4.) In pro-metaphase, some proteins (MAD1,2 and BUB1, 2, 3) assemble to form mitotic checkpoint complex that monitors proper attachment of the kinetochores with the kMTs. If any irregularity is noted mitotic progression is arrested.

Metaphase: It is characterized by proper alignment of the chromosomes to the equatorial *metaphase plate.* Once the kinetochore becomes attached with the spindle fibers, the chromosomes

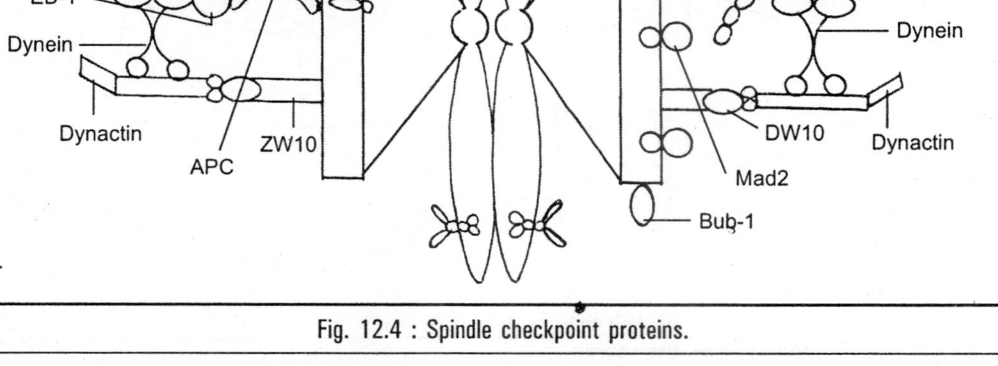

Fig. 12.4 : Spindle checkpoint proteins.

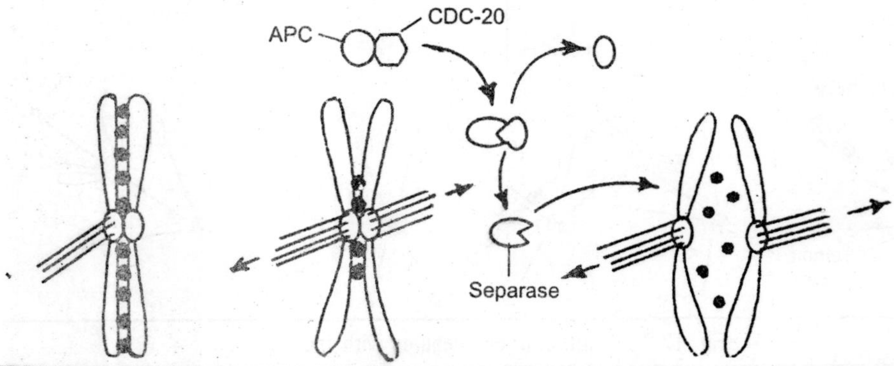

Fig. 12.5 : Role of separase in anaphase A.

show back and forth movement to adjust their position on the metaphase plate by poleward (P) and anti poleward (AP) forces and ultimately these are arranged exactly half way between two poles. This type of chromosomal movement is called *congregation.*

Anaphase: At the onset of anaphase loss of cohesion between two chromatids triggers separation of the chromatids towards the opposite poles. The whole process is regulated by many factors. It is divided into two sub phases - anaphase-A and anaphase-B.

Anaphase-A: It is characterized by the poleward separation of the sister chromatids. Up to metaphase the attraction between the sister chromatids is maintained by cohesin. At M-A transition APC releases the enzyme – *separase* which degrade the protein causing poleward movement of the chromatids towards opposite poles (Fig. 12.6).

Anaphase B: It involves slight movement of the spindle poles towards the cortex. As the chromatids reach the poles, the motor molecules of the cortical cytoplasm - dynein pulls the spindle poles towards the cortex. Keeping pace with this pulling tension the kinesins - BimC at the mid zone of the inter polar fibers slowly glides, as a result both poles move towards the cortex slightly.

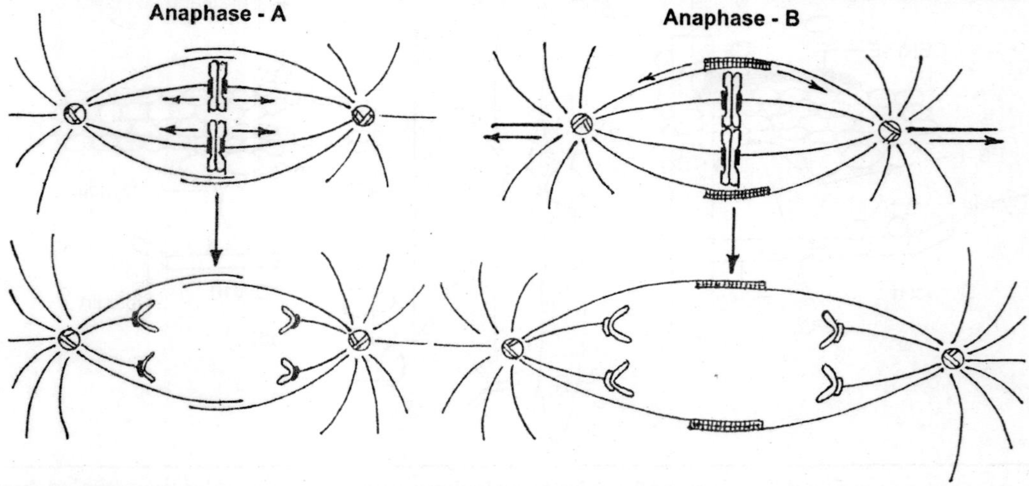

Fig. 12.6 : Comparative representation of Anaphase-A and B.

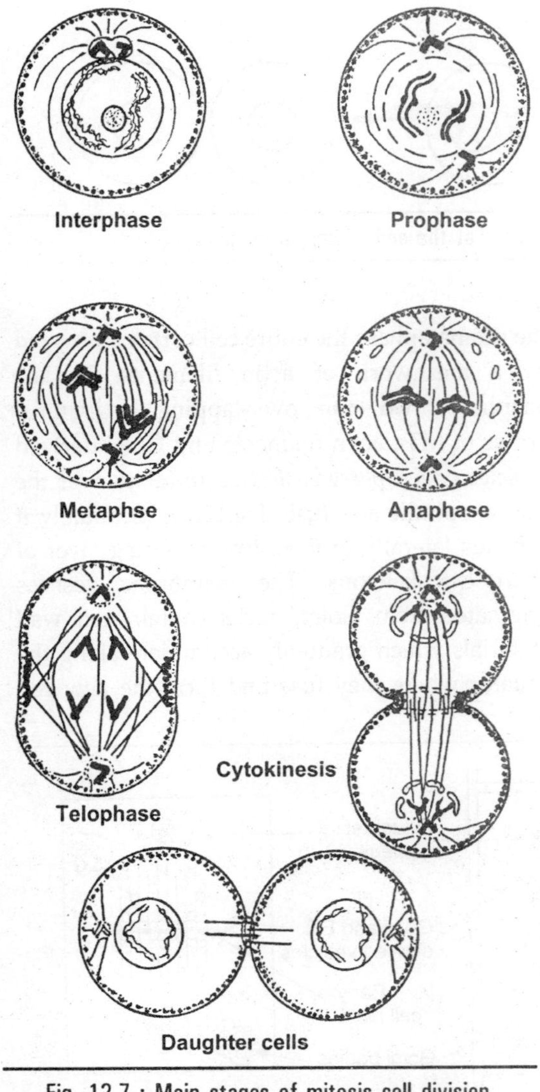

Fig. 12.7 : Main stages of mitosis cell division.

Telophase: After segregation of the chromatids, almost all the organelles are segregated equally to each pole. The disassembled nuclear material that adhered with the ER. bind to the decondensing chromatins and fuse with each other to form double membrane around each chromatid.

Assembly of nuclear membrane is followed by the assembly of Nuclear Pore Complex traversing the inner and outer nuclear

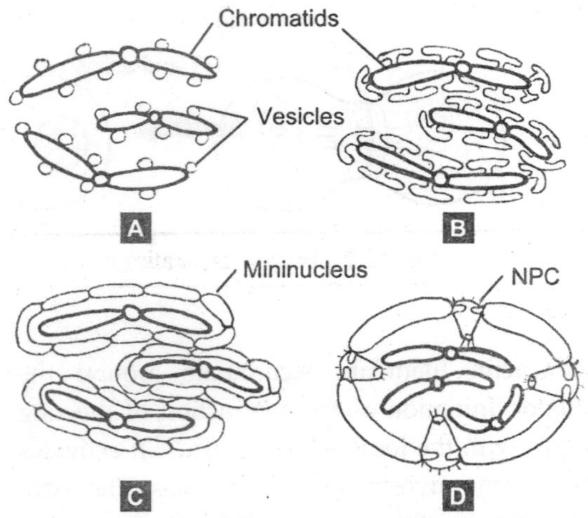

Fig. 12.8 : Formation of mininucleus.

membranes. Reassembly of other structures form a *'mini nucleus'* or *'karyomere'* (Fig.12.8). Subsequent fusion of these mini nuclei generates daughter nucleus.

12.1.2. Cytokinesis

It involves separation of cytoplasmic organelles equally into two halves. In animal cells it starts a few minutes after the onset of anaphase with the appearance of a *contractile ring* at the mid periphery of the cytoplasm at right angle to the longitudinal spindle axis (Fig. 12.9). The chromosomes play a vital role in determining cleavage furrow because *'chromosome passenger proteins'* (INCEP, Aurora B kinase and survivin) translocate from chromosomes to the spindle microtubules before the onset of cytokinesis. The vesicles, are carried by kinesin, followed by myosin driven motility through cortical actin that specifically concentrate around the contractile ring. In addition to actin and myosin II, cell cycle regulated kinase CD43 largely contribute to the initiation of the furrow that begins in the mid anaphase.

During the formation of contractile ring, INCEP, anillin, radixin proteins interdigitated

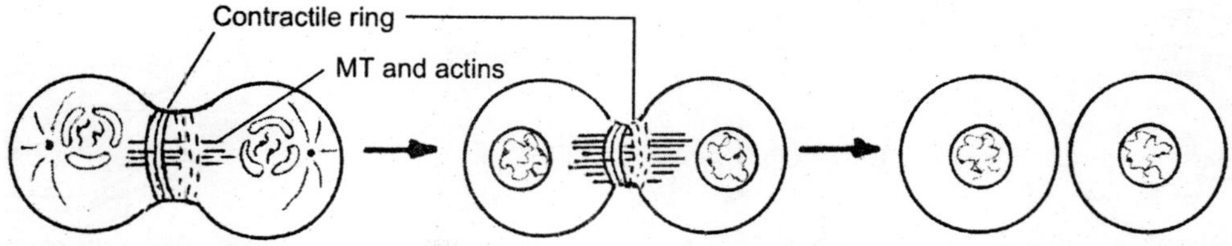

Fig. 12.9 : Gradual separation of two daughter cells at the end of animal cytokinesis.

with actin filaments accumulate around the equator. Initiation of the contraction of the ring begins with the local release of Ca^{++}. It activates the enzyme myosin light chain kinase.The extra membrane synthesized, completes the ingression of the furrow. This step is also called *'abscission'* characterized by the cutting of the mid body channel between the dividing cells. In plant cell, early in mitosis, a band of microtubules and actin filaments is formed around the equator above the nucleus called *preprophase band*. As the cell enters the prometaphase the entire cell cortex is covered by a meshwork of actin filaments. In late anaphase, two non over-lapping antiparallel arrays of MTs, actin filaments and Golgi derived vesicles called *phragmoplast* accumulate over the central spindle and fuse (Fig.12.10). Gradually it expands laterally to form two apposing layer of short cytoskeletons. The membrane vesicles originated from Golgi stacks contain cell wall materials which gradually accumulate along the equator where they fuse and form the new cell

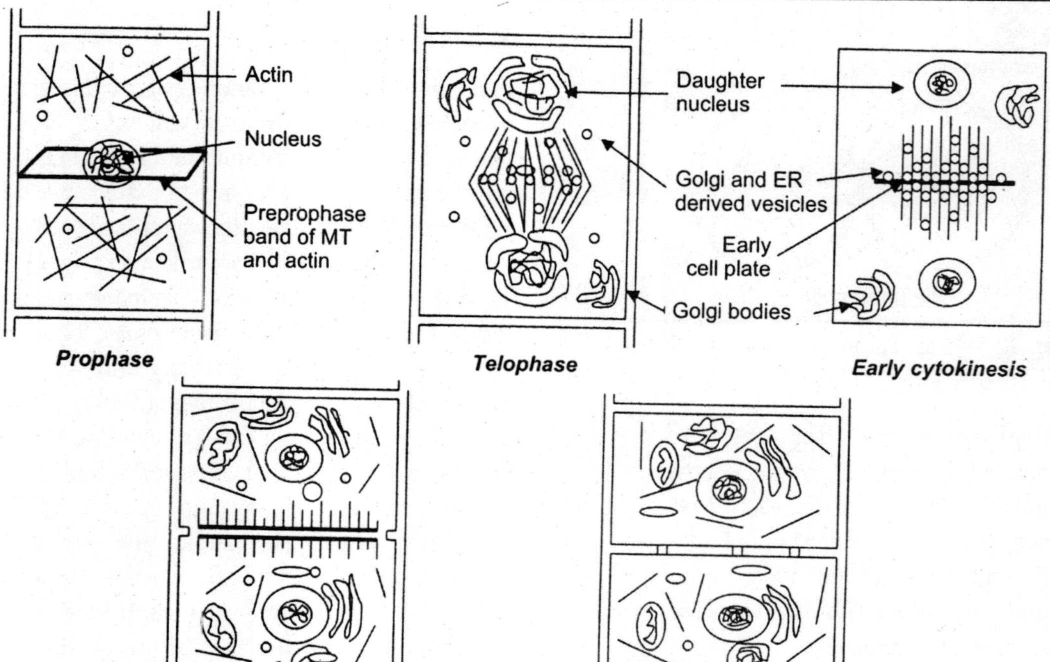

Fig. 12.10 : Differerent phases of plant cytokinesis.

wall called *cell plate.* Actin microfilaments accumulate near the daughter nuclei towards the emerging cell plate, and play significant roles in cell plate formation and expansion. It forms the primary cell wall and gradually by the deposition of secondary cell wall materials the cytokinesis is completed.

SUMMARY

1. Cell division is the result of the involvement of so many proteins meant for the formation of microtubules, compaction and proper alignment of the chromosomes with the spindle fibers and flawless segregation of the sister chromatids. It is the prime event of the M phase of cell cycle. Except for the formation of gametes, all cells divide into two identical cells. This type of cell division is called mitotic cell division. It is divided into prophase, pro-metaphase, metaphase, anaphase, telophase and cytokinesis.
2. The main objective of prophase is the chromosomal reorganization that makes them ready to attach them with the spindle fibres. For the formation of spindle fibres centrosomes are divided into two centrioles. Daughter centrioles help in the assembly of spindle fibres. A protein kinase, Zyg-1 helps in the centrosomal duplication.
3. Formations of MT occur at the expense of binding of α and β monomers of the tubulin proteins. The non-microtubular elements - Nuclear Mitotic Apparatus protein (NuMAP) occur in the nucleoplasm and concentrate near the centriole and play crucial role in the stabilization of spindle apparatus. Those cells that lack centrosome involve their chromosomes to generate spindle fibers.
4. As the nuclear membrane breaks down in prometaphase, the MTs get attached with the condensed chromosomes, now these are called spindle fibers and the daughter centrioles in each pole is called spindle pole. Two distinct motor proteins- dynein and kinesin help in the separation of centrioles and spindle pole formation. Four types of spindle fibers are noted, the fibers that occur around each pole, are called astral fibers, these remain connected with the cortical fibers of the cell; long fibers that connect two poles are called inter polar fibers, in the mid line, these fibers overlap with each other; those fibers which remain connected with the kinetochores are called kinetochore fibers and the fibers which connect the chromosomal arms are called chromosomal fibers.

   Two pathways can achieve spindle formation - the centrosome directed pathway and chromosome directed pathway. In centrosome directed pathway, MT nucleation requires γ-tubulin, which exists as two main complexes: the γ-tubulin small complex, and the γ-tubulin ring complex. During mitosis, these accumulate at the centrosome to support spindle formation. Some nuclear proteins like Ran GEF, RCCI proteins on the chromosomes also promote microtubule assembly. In chromosome directed pathway the MTs haphazardly organized around chromosomes become cross-linked with the help of kinesins.
5. As prophase proceeds the chromosomes condense. Condensing and cohesin carry out chromosome condensation. Condensin binds directly through out the central axis of the chromosome. In mammals two types of condensin occur condensin I (interacts with chromatin after prophase) while condensin II bind to chromosome throughout the mitosis. Following DNA replication, cohesins assemble to several AT rich regions at S phase. Establishment of cohesin with DNA threads requires Ctf7 protein in S phase and dissociates at M-A transition. Cohesin association is closely related with DNA polymerase because just before cohesin association DNA polymerase dissociate. Cohesin and condensin at first remain bound to the kinetochore and after full condensation of the chromosomes they dissociate. Cohesin induces the formation protein-DNA aggregate and increases the intermolecular ligation of linear DNA molecule while condensin causes intermolecular knotting.
6. Prometaphase is characterized by the total disassembly of nuclear membrane and attachment of the chromosomes with spindle fibers mediated by kinetochores. According to "search-and-capture" model, MT plus ends [associated with many plus-end-binding proteins like (+TIPs), APC, CLIPs, CLASPs, cytoplasmic dynein/ dynactin, and EB1 search and capture kinetochores.

   Chromatid arms get attached to additional chromosomal MTs in metaphase. Attachment of the spindle fibers with the kinetochores is a key factor that regulates proper arrangement of the chromosome to the metaphase plate. In prometaphase, some proteins assemble to form mitotic checkpoint complex that monitors proper attachment of the kinetochores with the kMTs. The centromeric proteins MAD and BUB regulate mitotic spindle checkpoint. In mammals, mitosis checkpoint complex shows Mad1, Mad2, BubR1, Bub1, Bub3, Mps1, Cdc20 and Cdc27. Movement of the chromosomes along the spindle fibers is controlled by motor proteins kinesin dynein, chromokinesins and CENP-E which accumulate at the '+' end of the spindle fibers.

7. Cohesion between the condensed sister chromatids resists the pulling forces exerted by the kMTs emanating from the opposite spindle poles which is essential for the congregation and alignment of chromosomes from prometaphase to metaphase. At the onset of anaphase loss of cohesion between the two chromatids triggers separation of the chromatids towards the opposite poles. Anaphase-A is characterized by the cleavage of the Sccl subunits of cohesin by separase leading to its destruction and allowing the spindles to pull the chromatids towards opposite poles. According to one model, in between kinetochore and spindle fibers there are some motor molecules that guide the movement of kinetochore and ultimately the chromatid.
8. After telophase, the dephosphorylated nuclear membrane proteins bind to the decondensing chromatins and fuse with each other to form double membrane around each chromatid.
9. In animal cells cytokinesis starts a few minutes after the onset of anaphase with the appearance of a contractile ring at the mid periphery of the cytoplasm at right angle to the longitudinal spindle axis. The proteins – 'chromosome passenger proteins' (INCEP, Aurora B kinase and survivin translocate from chromosomes to the spindle microtubules before the onset of cytokinesis. Actin, myosin II, Rho and related GTPase Rac and CD43 largely contribute to the initiation of the furrow. INCEP and anillin bind to the actin filaments during furrow formation. Initiation of the contraction of the ring begins with the local release of Ca^{++}. Due to the interaction of actin and myosin II, the contractile ring pulls the cell membrane inward.

   In plant cell, a preprophase band of MTs and actin filaments is formed around the equator. In prometaphase it disassembles. In late anaphase, two antiparallel arrays of MTs, actin filaments, Golgi and ER derived vesicles containing cell wall materials called phragmoplast accumulate over the central spindle. The vesicles gradually accumulate along the equator where they fuse and form cell plate. It forms the primary cell wall, gradually by the deposition of secondary cell wall materials, cytokinesis is completed.

QUESTIONS

Long questions:

1. Describe the steps of centrosome duplication.
2. Describe the process of chromosome condensation during cell division.
3. Describe the process of mitosis in brief.

Short questions:

1. Why mitosis is called equational cell division?
2. Why centrosome is the crucial point of spindle fibre formation?
3. Why centrosomal duplication is considered as semi conservative type?
4. Name the components that maintain the cohesion of nascent centrioles up to G2.
5. Name two MPF derived factors that assist in the maturation of MTs into spindle fibres.
6. Name a non-microtubular element that helps in the stabilization of spindle fibres.
7. Describe the process of centriolar separation.
8. Which motor proteins help in the separation of centrioles?
9. How does centriolar separation differs before and after nuclear membrane breakdown?
10. Which motor proteins participate in the chromo some directed pathway of spindle assembly?
11. What is basic difference in the action of condensing and cohesin?
12. How condensin carries out compaction of chromosomes?
13. Explain the role of cohesions in the compaction of chromosomes.

14. Name the components of MPF that carries out nuclear breakdown.
15. Name the motor proteins that mediate attachment of kinetochores with kMTs.
16. What are the roles of kinetochore?
17. Describe the mitotic spindle checkpoint mechanism.
18. What is congregation?
19. What is APC? How it regulates anaphase?
20. Describe the molecular processes of anaphase-A and –B.
21. Describe the process of the formation of mininucleus.
22. Describe cytokinesis in animal and plant cells.
23. What is phragmoplast? How it takes part in cell plate formation in plants?

MCQs:

1. **Centrosome divides in**
(a) G1, (b) S, (c) G2.
2. **The change of attachement position of kinetochores from sides to the tip of the kMTs is regulated by**
(a) Cdc42, (b) TIP, (c) EBI.
3. **APC is inactivated up to M-A transition by**
(a) Mad1, (b) Mad2, (c) Bub1.
4. **Chromatin àxis is maintained by condensin and**
(a) cohesin, (b) SMC proteins, (c) topoisomerase.
5. **Cut2 and Pds1 are collectively called**
(a) APC, (b) Separase, (c) Securin.
6. **The disassembled NE up to anaphase remains adhered to**
(a) Golgi, (b) ER, (c) actins.
7. **The dominant chromosome passenger protein is**
(a) syntaxin, b, survivin, (c) radixin.
8. **Contraction of cortical ring of cytokinetic furrow is facilitated by**
(a) K^+, (b) Na^+ (c) Ca^{++}.
9. **Cell plate forms**
(a) primary, (b) secondary (c) tertiary cell wall.
10. **Equal distribution of cytoplasmic organelles is done by**
(a) cytokinesis, (b) karyokinesis, (c) amitosis.
11. **Colchicine inter-feres**
(a) DNA replication,
(b) spindle formation,
(c) congregation.
12. **Phragmoplast is the precursor of**
(a) Primary cell wall. (b) Middle lamella, (c) Cell plate.
13. **DNA replication and protein synthesis occurs in**
(a) Prophase, (b) Interphase, (c) Anaphase.
14. **The cell doubles in size in**
(a) G1, (b) S, (c) G2.
15. **Centromere does mot divide up to the end of metaphase because**
(a) Produces spindle fibres,
(b) Holds the replicated DNA together,
(c) Contains genes that control prophase.
16. **Cytokinesis begins at**
(a) Anaphase, (b) Telophase, (c) Metaphase.
17. **Histone protein is synthesized in**
(a)G1, (b) S, (c)G2.
18. **Congregation occurs during**
(a) Early prophase, (b) Interphase, (c) Early metaphase.
19. **Post mitotic phase is**
(a) G1, (b) M. (c) G2.
20. **Colchicine is**
(a) Prophase poison,
(b) Cytokinesis poison,
(c) Mitotic poison.
21. **Spindle fibres are made up of**
(a) Flagellin, (b) Tubulin, (c) Intermediate filament.
22. **In which stage of cell division chromosomes are most condensed?**
(a) Metaphase, (b) Anaphase, (c) Telophase.
23. **Chromosomes exhibit minimum coiling at**
(a) Prophase, (b) Anaphase, (c) Interphase.
24. **Karyokinesis is the division of**
(a) Cytoplasm, (b) Nucleus, (c) Both.
25. **Spindle fibres arise from**
(a) Centromere, (b) Nucleolus (c) Centriole.

Answers to MCQs:

1. b, 2. a, 3. b, 4. c, 5. c, 6. b, 7. b, 8. c, 9. a, 10. a, 11.b, 12. c, 13. b, 14. a, 15. b, 16. b, 17. b, 18. c, 19. a, 20. c, 21. b, 22. a, 23. c, 24. b, 25. c.

Chapter 12.2

Meiosis Cell DIvision

Maintenance of the species specific chromosome number from generations after generations is the fundamental feature of all advanced species. In sexually reproducing organisms haploid gametes fuse to form zygote.

Meiosis thus is a specialized cell division process that allows one diploid cell to divide in a special way to generate haploid cells (gametes) in eukaryotes.

Meiosis cell division is divided into two phases - ***meiosis I*** and ***meiosis II***. Meiosis I shows complex remo- deling of chromosomes leading to the formation of two haploid cells. This phase is thus called *reductional or heterotypic division.* Meiosis II is basically homologous with mitosis cell division producing two identical haploid cells i.e. the replica of the meiosis I daughter cells. It is therefore called *equational or homotypic cell division.* In the whole process nucleus or chromosome divides once while cytoplasm divides twice.

Biochemically, meiosis uses some of the same mechanisms employed during mitosis to accomplish the redistribution of chromosomes but there are several features unique to meiosis, most importantly the pairing and recombination between homologous chromosomes, which enable them to separate from each other. During meiosis, the genome of a diploid germ cell, undergoes one round of DNA replication followed by two rounds of chromosome segregation with no intervening round of DNA replication resulting in the formation of four haploid gametes. Each gamete contains one complete set of genome. As the chromosomes of each parent undergo genetic recombination during meiosis-I it exhibits a reciprocal process of joining the two genomes of the opposing genders.

12.2.1. Types of Meiosis cell division:

Three types of meiosis cell division are noted.

A. *Zygotic meiosis:* In lower plants meiosis cell division follows fertilization producing haploid cells that form the adult gametophytic generation.

B. *Sporic meiosis:* In some thallophytes, meiosis takes place at some indeterminate point producing haploid mega and microspores which undergo repeated mitosis forming male and female gametophytes respectively. Certain cells of the gametophytes differentiate to form gametes.

C. *Gametic meiosis:* In higher plants and animals, the germ cells divide to form spermatogonia (in male) and oogonia (in female) that undergo meiosis cwll division to form sperms and ova respectively.

Meiosis - I: It is divided into four phases - prophase, metaphase, anaphase and telophase.

Prophase – I : Prophase I is the longest stage of meiosis I and shows several highly complicated chromosomal modifications. For the understanding of different substages it is divided into five substages leptonema, zygonema, pachynema, diplonema and diakinesis:

Leptonema (Leptotene; *Leptos = thin, Nema = thread)***:** Its distinguishing features are as follows:

i) Nuclear envelope and nucleolus remain intact.

ii) The nucleoplasm appears so hydrated that chromosomes appear faintly stained.

iii) Individual chromosomes appear as long thread like structure. The two sister chromatids are still so tightly bound that they are indistinguishable from one another.

iv) Under light microscope the chromosomes appear beaded. These granular structures are called *chromomeres.*

v) The chromosomes may be irregularly arranged or may be polarized showing a striking telomeric movement in which all the telomeres are clustered on the inner nuclear envelope (INE) and are attached with an electron-dense plate. This meiosis-specific telomere clustering is defined as the *'Bouquet arrangement' or 'Polarized arrangement'* due to its resemblance to a bouquet of cut flowers. Telomere proteins, SCP2, and the meiosis-specific cohesin STAG3 seem to regulate this arrangement. Actin polymerization in the nucleoplasm facilitates the clustering of telomeres of the homologous chromosomes together.

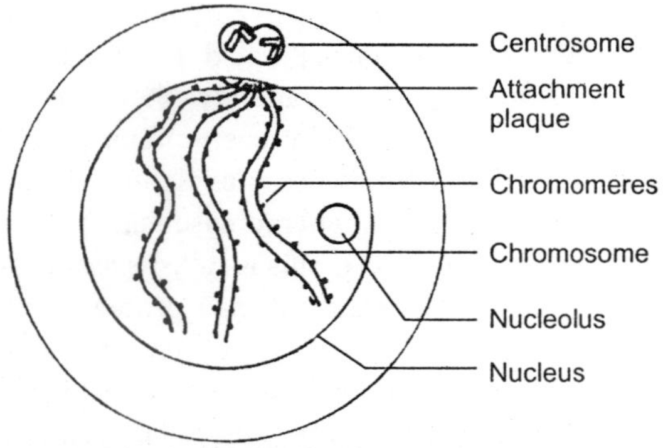

Fig. 12.11 : Leptonema sub stage.

vi) Along the bases of the chromatids special nucleoprotein structures assemble called axial elements (AE) whose ends attach to the INE. AE is required for chromosome condensation, rapid exit from the bouquet stage and fine-tuning of homologue pairing. Chromatin fibres are attached as a series of lateral loops on the sides of the AE.

vii) The centrosome separates into two centrioles which begin to separate.

Zygonema *(Zygotene; Zygos = pairing):* Its distinguishing features are as follows:

i) The nuclear envelope and nucleolus remain intact; the chromosomes appear comparatively thick and deeply stained.

ii) The homologous chromosomes are arranged side by side. This process of arrangement is called synapsis or *pairing.* The paired homologous chromosomes are called *bivalent*. Pairing is not simply lateral arrangement of the homologous chromosomes but it is the initial phase of recombination since throughout the longitudinal axis of the chromosomes there are several contact points between two homologous chromosomes which closely correspond with the recombination points.

iii) At the initial stage, two chromosomes of a bivalent remain separated by a gap of 3000Å gap but gradually they are so intimately paired that every minor part of the chromosomes becomes tightly wound up and the interchromosomal gap is reduced to 1000Å. This is called *synapsis* or *syndesis* or *synizesis*. It may proceed from the telomere towards the centromere or from the centromere towards the telomere which are called *proterminal* and *procentric synapsis* respectively. Synapsis results due to the appearance of a glue like structure called *synaptinemal complex* in between two chromosomes. Sometimes synapsis occurs in between non homologous chromosomes, it is called *pseudosynapsis*. Sometimes pairing of the homologous chromosomes occurs in leptotene substage also, which is called *premeiotic pairing*.

iv) The distance between two centrioles increases.

Molecular mechanism of pairing: As per one model, each chromosome has a unique array of transcription units showing numerous loops tethered to specialized sites along its length. Only homologues share similar strings of clouds and promoters. Interactions between promoters and RNA polymerases mediate pairing.

Pairing can be divided into three distinct steps - recognition, presynaptic alignment and *synapsis* with the help of synaptonemal complex (SC). Recognition of DNA/ DNA homology plays an important role in the pairing. The Mnd1 and Hop2 proteins localize to chromatin promote pairing.

Synaptinemal complex (SC) (Fig. 12.13): The assembly of SC starts from zygotene but in leptotene, prior to pairing, AE is formed at the base of each chromatid that ultimately fuses to form lateral elements. Small ubiquitin-like protein modifier (SUMO) localizes specifically to discrete foci on chromosomes which coincide with the axial elements where proteins involved

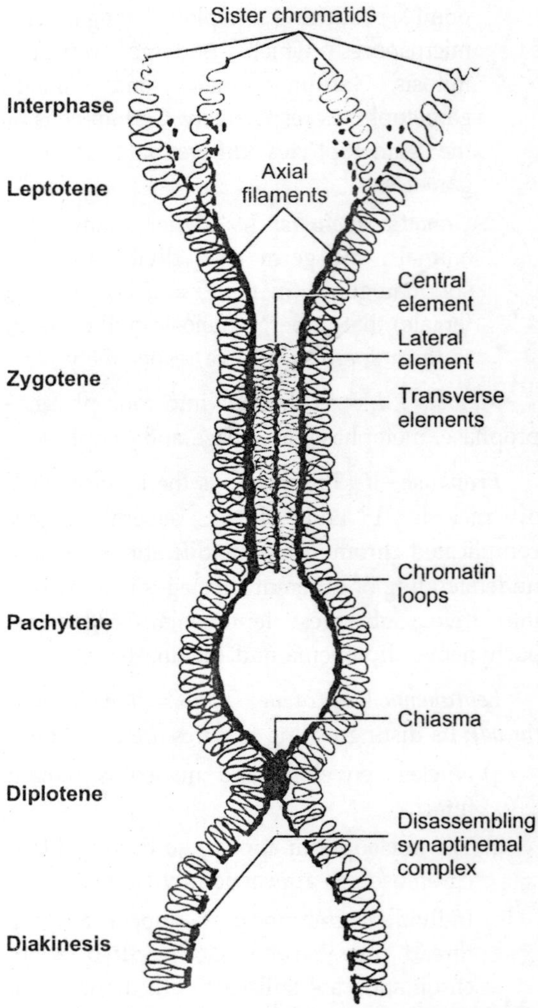

Fig. 12.12 : Structure of synaptinemal complex.

in the initiation of synapsis are located. Synaptinemal complex is a proteina- ceous structure formed along the lengths of the bivalents. Its overall structure shows two *lateral elements* and a *central element* in between two lateral elements. Many *transverse filaments* are arranged perpendicular to the lateral filaments connecting the lateral and central filaments. Most of the chromosomal DNA extends laterally from the lateral elements as *loops*. SC formation follows intimate association of both lateral elements by the formation of nodule like structures. These are

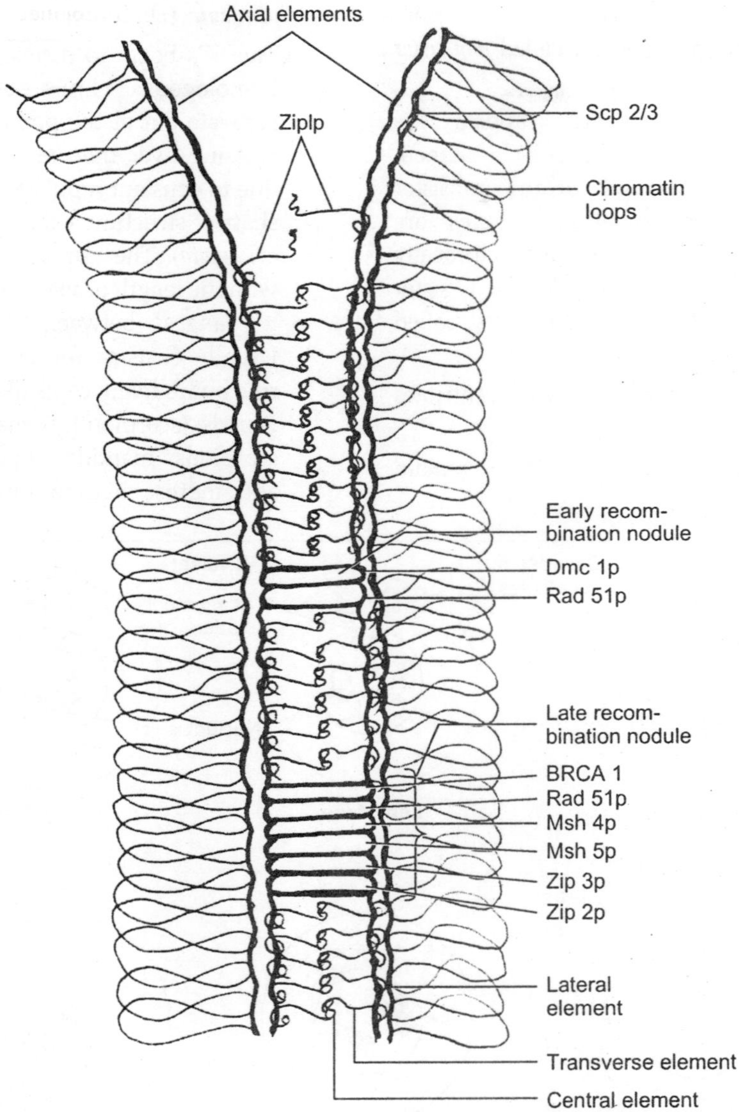

Fig. 12.13 : Formation of recombination nodule.

divided into early and late nodules. Early nodules appear in the early zygotene and are mediated by Rad51 and Dmc1. These act as sites for double strand break. In pachytene, these form late recombine tion nodules and sites of chiasma formation with the involvement of Rad51, BRCA1, Msh4, Msh5, Mlh1, Zip2 and Zip3 proteins.

Pachynema/Pachytene: *(Pachytene; Pachy = thick):* Its main features are as follows:

i) It is the longest substage of prophase-I. During the early periods of pachytene, each chromosome of a bivalent forms two chromatids; thus four chromatids form *tetrad*. The chromatids produced from one chromosome are called *sister chromatids* while those that are produced from different chromosomes are called *non sister chromatids.*

ii) Pachytene chromosomes appear as condensed, thick, deeply stained. Throughout the length of the nonsister chromatids recombination nodules are formed where they break and reciprocally unite with that of another non sister chromatid. This phenomenon is called ***crossing over.***

iii) The distance between two centrioles increases.

iv) Nuclear membrane and nucleolus remain intact.

Diplonema (Diplotene):

i) The SCs begin to disintegrade so that the homologous chromosomes begin to separate but at the point of contact of the crossing over they remain united so that due to constant repulsion it looks like a 'X' shaped structure called ***chiasma*** (plural, *chiasmata*). The number of chiasma varies with the length of the chromosome, usually it ranges between 1-14/chromosome. Usually chiasma formation is the normal rule but in many cases like male *Drosophila*, *Panorpa* (scorpion fly), many grasshoppers, scorpions, mantids, copepods, butterflies and moths chiasma formation does not

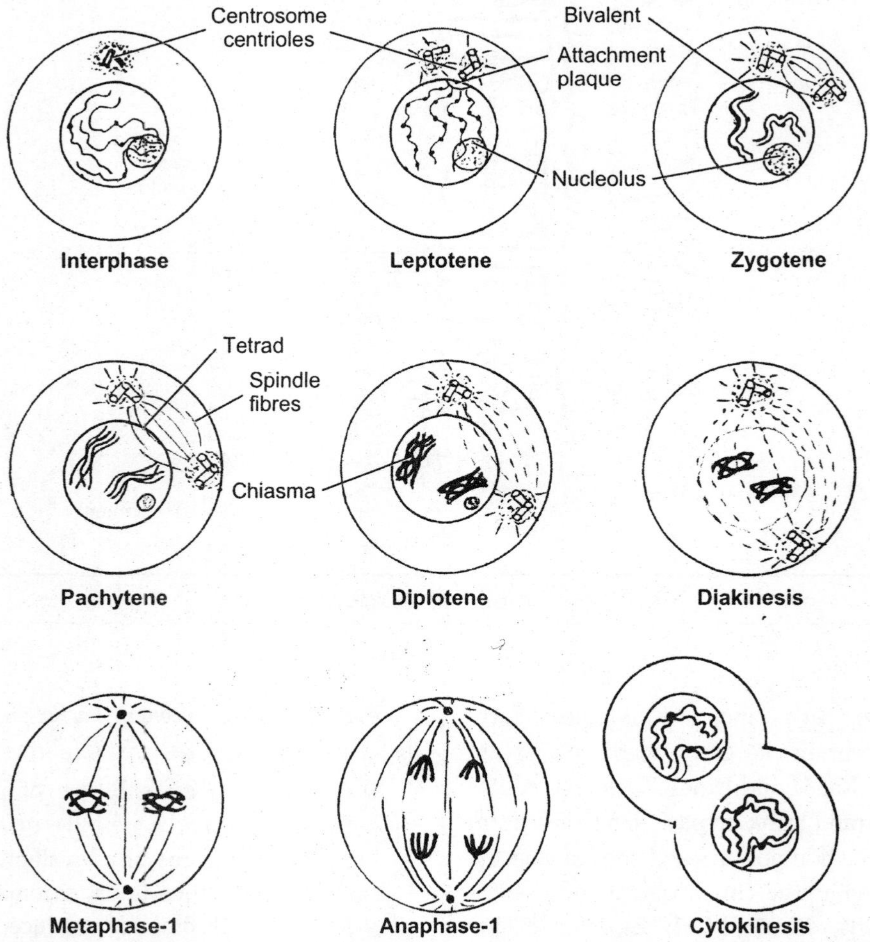

Fig. 12.14 : Main stages of meiosis cell division.

take place. This is called *achiasmatic meiosis.*

ii) As the repulsion continues the bivalents rotate 90° to 180° on its axis, it is called *rotation.*

iii) Due to continued repulsion, the chiasma gradually move towards the telomere. It is called *terminalization.*

iv) At this substage the nucleolus is reduced in size.

v) Distance between the centrioles increases.

Diakinesis:

i) The bivalents appear highly condensed and deeply stained.

ii) Due to continuous repulsion these appear round with terminal chiasma.

iii) The nuclear membrane and nucleolus begin to disintegrate.

iv) Centrioles occupy almost opposite positions and spindle fibres are formed between them.

Cohesion of meiotic chromosomes: Apparently from leptotene onwards up to diakinesis the bivalents show gradual condensation exhibiting more and more thick configuration showing deeper stains. This phenomenon is called *cohesion* which is fundamentally similar with that of mitotic chromosomal condensation involving condensin and cohesins but meiotic cohesin differs from its mitotic counterpart because partly it has to participate in the recombination process and partly its centromeric part has to be protected from separase at the onset of anaphase I holding sister chromatids together so that the whole chromosome can continue up to meiosis II. Up to metaphase-I, the whole bivalent is treated as a unified entity since the arms of the sister chromatids are firmly bound to each other by cohesin sub units but at the onset of Anaphase-I all the cohesin subunits except from the vicinity of centromere are lost. There is difference between the regulation of sister chromatid cohesion at centromeres and along chromosome arms because centromeric cohesion, but not arm cohesion, persists throughout anaphase of the

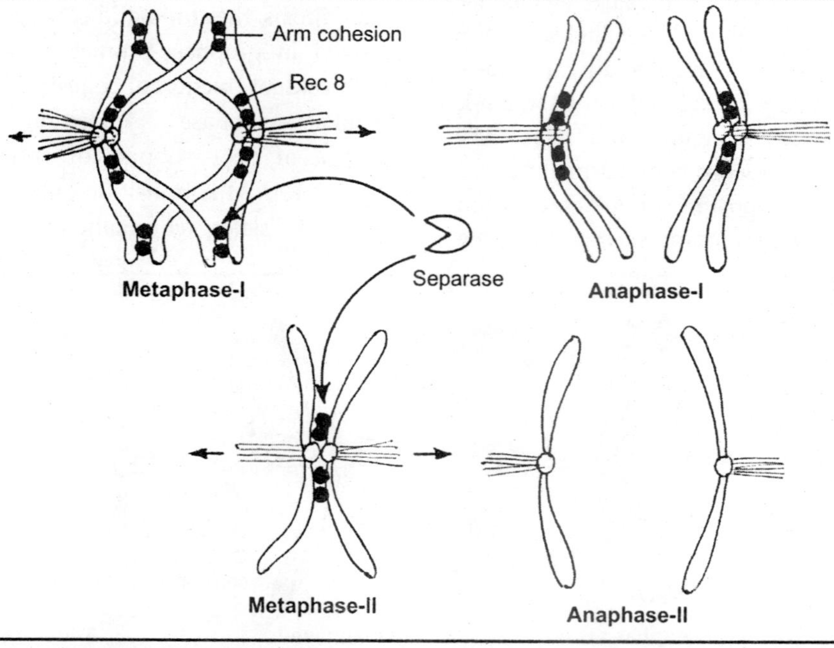

Fig. 12.15 : Differential cohesin disintegration from chromosome arms and centromere during anaphase I and II.

first meiotic division. Thus sister chromatid cohesion is much more acute at the centromere at metaphase, where it antagonizes the pulling force of spindle microtubules that attach to the kinetochores from opposite poles.

Checkpoint: Cell cycle checkpoints ensures faithful chromosome segregation by linking the onset of anaphase to the establishment of bipolar kMT attachment. It also does not activate anaphase promoting complex/ cyclosome (APC/C) until all the chromosomes correctly align on the spindle apparatus via their kinetochore. Mad2, Mad3 are the spindle checkpoint that plays a role in reorienting chromosomes. From molecular view point, the checkpoint exhibits two arms, one dependent on Bub1, the other on Aurora B. Both of these arms converge on the meiotic checkpoint complex (MCC), They promote formation of stable bipolar kMT attachments.

Metaphase-I:

i) The most notable feature of metaphase –I is the attachment of kinetochore microtubules (kMTs) with the sister kinetochores of each chromosome of the a bivalent emanating from the same pole. This is called *monopolar attachment* which is regulated by a meiosis specific *monopolin* subunit called Mam-I. Kinetochores contain multiple MT attachment sites but cells normally avoid MTs with apposing orientations from attaching to the binding sites on the same chromatid. Thus all the MTs generated from the same pole do not attach with the fused kinetochore. This type of unusual attachment is called *merotelic attachment*.

ii) The chromosomes are aligned throughout the equatorial plate of the cell. This type of movement is called *congregation*.

Anaphase-I: At the onset of anaphase, separase becomes active and cleaves all the cohesins but centromeric cohesin represented by Rec8. It is protected from separase by the protein shugoshin (or Sgo1). to maintain the continuity of the tight association of the sister centromeres (Fig. 12.16).

12.2.2. Differences between mitotic and meiotic chromosome behavior:

During meiosis, two rounds of chromosome segregation follow a single round of DNA replication (no DNA replication occurs in between meiosis I and II). In mitosis single round of chromosome segregation follows each round of DNA replication.

In mitosis, the chromatid cohesins are mostly removed in prophase-prometaphase transition and a residual fraction of cohesin is removed in metaphase-anaphase transition following cleavage of Scc1 subunit of cohesin but in meiosis - I, in M-A transition only arm cohesins from the sister chromatids are lost but

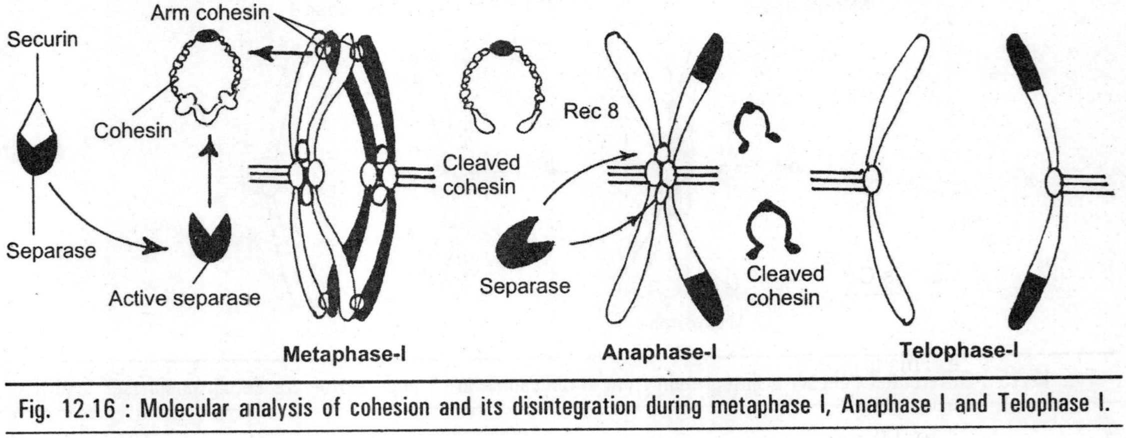

Fig. 12.16 : Molecular analysis of cohesion and its disintegration during metaphase I, Anaphase I and Telophase I.

Table 12.1 : Comparative account of mitosis and meiosis cell divisions

MITOSIS	MEIOSIS
1. This cell division occurs in all body cells.	1. In advanced organisms, it occurs in the primary spermatocytes and oocytes of testis and ovary respectively. In primitive unicellular (n) animals it occurs after the zygote formation.
2. It produces two daughter cells of same quality.	2. It produces two haploid cells.
3. Nucleus and chromosomes divide once.	3. It involves division of nucleus twice but chromosomes once.
4. Daughter cells show same type of genic characters as the mother cell.	4. Daughter cells exhibit completely new patterns of genic recombinations.
5. In S phase, DNA replication becomes complete and single round of chromosome segregation follows each round of DNA replication.	5. In S phase, DNA replication remains incomplete which becomes complete in Zygotene/Pachytene but no DNA replication occurs in between meiosis I and II.
6. Except a few cases mutation does not accompany this division.	6. Mutation is a regular feature of this division.
Prophase :	
1. Occurs once in the whole division.	1. Occurs twice in the whole division.
2. Short lasting stage and is not divided into different substages.	2. Long lasting stage and is divided into different substages.
3. Pairing, synapsis, crossing over and recombination are not found.	3. Pairing, synapsis, crossing over and recombination are found.
4. Volume of nucleus does not increase.	4. Volume of nucleus increases.
Metaphase :	
1. Occurs once in the whole division.	1. Occurs twice in the whole division.
2. Separate chromosomes lie in the equatorial plane.	2. Homologous chromosomes bound by chiasma lie in the equatorial plane.
3. Centromeres perfectly align in the equatorial plane and arms extend towards the poles.	3. Centromeres are not perfectly aligned in the equatorial plane rather slightly inclined towards the poles. However the arms extend towards the center. In Meiosis-II the alignment of chromosomes perfectly resemble mitosis.
4. Centromere divides at the end of this stage.	4. Centromeres do not divide in Meiosis – I, they divide in Meiosis-II.
5. Centromeres of one chromosome binds to the kMTs generated from the opposite poles.	5. Centromeres of one chromosome binds to the kMTs generated from the same pole.

Table contd. next page

MITOSIS	MEIOSIS
Anaphase :	
1. This stage occurs once in the whole division.	1. It occurs twice in the whole division.
2. Chromatids move towards the poles.	2. In Meiosis-I chromosomes and in Meiosis-II chromatids move towards the poles.
3. Chromatids appear long and thread like.	3. Chromosomes appear short, thick and highly condensed in Anaphase-I. In Anaphase-III chromatids are long thread like.
4. The arrangement of genes remain unchanged.	4. The arrangements of genes are changed due to crossing over in Meiosis-I.
5. Centromeric cohesion is disrupted to allow separation of chromatids.	5. Centromeric cohesion is protected in Anaphase-I but disrupted in Anaphase-II.
Telophase :	
1. It occurs once and is noted always.	1. It occurs twice and somes Telophase-I does not commence and the cell directly leads to Prophase-II.
2. The chromosome number in daughter cells resembles mother cell.	2. The chromosomes number in daughter cells is always halp in respect of the mother cell.
Cytokinesis :	
1. It always occurs after each division.	1. It may not occur after telophase-I.

centromeric cohesin only disintegrates at M-A transition of meiosis II.

Significance of meiosis:

i) Advanced organisms, who practice sexual reproduction, maintain species specific chromosome number through the production of haploid gametes which are produced by meiosis cell division.

ii) Crossing over and recombination cause an admixture of paternal and maternal characters which produces genetic variety in gametes that propagate to offspring.

iii) Random crossing over leads to the random recombination and production of numerous new genetic combinations that may help to maintain heterozygocity or genetic load or homozygocity of the mutated genes leading to genetic death.

iv) Recombination and independent assortment allow for a greater diversity of genotypes in the population and it allows a species to maintain stability under environmental changes.

SUMMARY

1. To solve this problem of maintaining the constancy of species-specific chromosome number, in sexually reproducing organisms a special type of cell division called meiosis cell division is noted whereby the gametes show half the chromosome number in respect of the parental cells. Meiosis cell division is divided into two phases - meiosis I and meiosis II. Meiosis I shows complex remodeling of chromosomes leading to the formation of two haploid cells. Meiosis II is basically homologous with mitosis cell division producing two identical haploid cells i.e. the replica of the meiosis I daughter cells. It is therefore called equational or homotypic cell division. In the whole process nucleus or chromosome divides once while cytoplasm divides twice. Three types of meiosis cell division are noted. Zygotic meiosis (in lower plants it follows fertilization producing haploid adult gametophytic generation), Sporic meiosis (in some thallo phytes, it occurs at some

indeterminate point producing haploid mega and microspores forming gametophytes) Gametic meiosis (in higher plants and animals, the primordial germ cells divide to form sperms and ova).

2. Prophase-I is divided into five substages leptotene, zygotene, pachytene, diplotene and diakine sis. Telomere proteins, SCP2, and the meiosis-specific cohesin STAG3 seem to regulate bouquet arrangement of leptotene.. Actin polymerization in the nucleoplasm also facilitates the clustering of telomeres of the homologous chromosomes together. Along the bases of the chromatids special nucleoprotein structure assemble called axial elements that is required for chromosome condensa tion. Chromatin fibres are attached as a series of lateral loops on the sides of the axial element. The homologous chromosomes are arranged side by side to ensure genomic and physical contacts between them. This process of lateral arrangement is called pairing. The paired homologous chro mosomes are called bivalent. Two chromosomes of a bivalent become so intimately paired that every minor part of the chromosomes becomes tightly paired. This is called synapsis. It results due to the appearance of glue like structure called synaptinemal complex in between two chro mosomes. Synaptinemal complex (SC) is a proteinaceous structure formed along the lengths of paired homologues that apparently mediates synapsis between homologous chromosomes. The overall architecture of SC resembles a railroad having two outer lateral elements and a central element. SC formation follows intimate association of both lateral elements by the formation of recombination nodules like structures. These may have a role in the formation of chiasma.

3. During early pachytene, each chromosome of a bivalent is called tetrad. The chromatids produced from one chromosomes are called sister chromatids while those that are produced from different chromosomes are called non-sister chromatids.

4. Pachytene chromosomes are comparatively condensed. Throughout the length of the non sister chromatids recombination nodules break and reciprocally unite exchanging segments of chroma tids. This phenomenon is called crossing over

5. The cohesin complex establishes sister-chromatid cohesion. There is a difference between the regulation of sister chromatid cohesion at centromeres and along chromosome arms during meiosis, because centromeric cohesion persists throughout anaphase-1.

6. Mad2 reorients chromosomes that are incorrectly attached to the spindle as well as delaying the cell cycle, whereas Mad3 is needed for the cell cycle delay, but not for chromosome reorientation. Bub1 monitors kMT attachment while the aurora B monitors biorientation

7. kMTs originating from the same pole attach to both kinetochores of a chromosome. It is called monopolar attachment which is regulated by a meiosis specific monopolin subunit called Mam1. Mam1 enforces monopolar MT attachment of sister kinetochores either by promoting fusion or copolarization of sister kinetochores.

8. The bulk of the cohesions are dissociated from the arms but still they are held together due to the occurrence of chiasmata. At the onset of anaphase, separase cleaves all the cohesins but cen tromeric cohesin represented by Rec8 is protected from separase to maintain the continuity of the tight association of the sister centromeres.

9. The essence of meiosis is the breakage and reunion of the chromatids so that both paternal and maternal chromosomes exchange parts of chromosomes. The resultant cells show admixture of characters of both parents.

QUESTIONS

Long questions:

1. Describe the molecular steps of prophase I.
2. Describe the process of crossing over with the help of Holliday model.
3. Describe the process of crossing over involving double strand break.

Short questions:

1. Differentiate gametic and sporic meiosis.
2. What do you mean by Bouquet arrangement? How it is regulated?
3. Differentiate pairing and synapsis.
4. Differentiate procentric and proterminal synapsis.
5. What do you mean by pre meiotic pairing?

6. What do you mean by pseudosynapsis?
7. What are recombination nodules? How they mediate chiasma formation?
8. Name the proteins involved in the formation of late recombination nodules.
9. What do you mean by achiasmatic meiosis?
10. What are rotation and terminalization?
11. Describe chromosome compaction during prophasè-I. How it differs from mitotic condensation process?
12. Differentiate the action of separase in mitosis and meiosis-I.
13. What do you mean by 'Holliday Junction'?
14. What is called branch migration?
15. What do you mean by 'heteroduplex DNA'?
16. What do you mean by 'planar molecule'?
17. What do you mean by 'first end capture' and 'second end capture'?
18. What do you mean by 'first Holliday junction' and 'second Holliday junction'?
19. What do you mean by 'D-loop'?

MCQs:

1. **Synapsis initiation complex shows**
(a) Zip1/2, (b)2/3, 3/4.
2. **The central element of SC is formed by**
(a) Zip1, (b) Zip2, (c) Zip3.
3. **One of the proteins mediating early recombination nodules is**
(a) BRCA, (b) Msh5, (c) Rad51.
4. **The centro- meric cohesin**
(a) Rec8, (b) Scc1, (c) Scc3 protects centromere from separase.
5. **The single strand nick during crossing over is created by**
(a) Rec8, (b) RecBCD, (c) Rad51.
6. **Double strand break is initiated by**
(a) Rad51, (b) Spo 11, (c) Rec8.
7. **Telomere proteins, SCP2, and the meiosis-specific cohesin STAG3 seem to regulate**
(a) synapsis,
(b) terminalization,
(c) bouquet arrange ment.
8. **The Mnd1 and Hop2 proteins of chromatin promote meiotic chromosome**
(a) pairing, (b) recombination, (c) double strand break.
9. **In mammals Scp2 and Scp3 proteins are required for the assembly of**
(a) axial elements,
(b) bouquet arrangement,
(c) crossing over.
10. (a) Scp2 and Scp3, (b) Mnd1 and Hop2, (c) Rad51 and Dmc1 **mediate early recombination nodule formation.**
11. **A protein** (a) Wap1, (b) Scp2, (c) Hop2 **controls the interaction between cohesin and chromatin.**
12. **The protein required to protect centromeric cohesin Rec8 in anaphase-I is**
(a) Sgo1, (b) Hop2, (c) Rad51.
13. **During recombination, the enzyme**
(a) recombinase –Rec BCD,
(b) separase,
(c) Aurora kinase creates a nick in one strand of a DNA double helix.
14. (a) Scp2 and Scp3, (b) RuvA and RuvB, (c) Mnd1 and Hop2 **catalyze branch migration.**
15. **Crossing over usually contributes to genetic variation by exchanging chromosomal segments between**
(a) chromatids of non homologues,
(b) sister chromatids of a chromosome,
(c) nonsister chromatids of homologues.
16. **Meiosis II is similar to mitosis in that**
(a) homologous chromosomes synapse,
(b) the daughter cells are diploid,
(c) sister chromatids separate during anaphase.
17. **A human cell containing 22 autosomes and a Y chromosome is**
(a) zygote,
(b) a somatic cell of a male,
(c) a sperm cell.
18. **Homologous chromosomes move to opposite poles of a dividing cell during**
(a) meiosis-I, (b) mitosis, (c) meiosis II.
19. **The DNA content of a diploid cell in the G1 phase is *x*, then the DNA content of the same cell at metaphase-I would be**
(a) 0.25*x*; (b) *x*, (c) 2*x*.
20. **In meiosis chromosomes are replicated in**
(a) Zygotene, (b) Interkinesis, (c) Interphase.
21. **At which phase of meiosis separated sister chromatids are noted?**
(a) Anaphase-I, (b) Anaphase-II, (c) Interphase.

22. **In which stage of prophase I crossing over occurs?**
(a) Diplotene, (b) Diakinesis, (c) Pachytene.

23. **The point at which crossing over occurs is called**
(a) Synaptinemal complex,
(b) Branch migration,
(c) Chiasma.

24. **Terminalization is related with**
(a) Diplotene, (b) Diakinesis, (c) Anaphase-I.

25. **Disjunction refers to (a) Chromosome separation in**
(a) Anaphase-I (b) Anaphase-II, (c) Telophase-II.

26. **The no. of bivalents is 8 in Zygotene; what would be the no. of chromosomes in anaphase-II?**
(a) 16, (b) 32, (c) 8.

27. **Name the stage where sister chromatids are equatorially aligned in two cells.**
(a) Anaphas-I, (b) Metaphase-II, (c) Metaphase-I.

28. **The enzyme that splits the chromosomes in pachytene is**
(a) Ligase, (b) Lipase, (c) Endonuclease.

29. **Chromosomes are highly condensed in**
(a) Metaphase, (b) Anaphase, (c) Telophase.

30. **Longest phase of meiosis is**
(a) Prophase-I, (b) Prophase-II, (c) Metaphase-I.

Answers to MCQs:

1. b, 2. a, 3. c, 4. a,5. b, 6. b, 7. c, 8. a, 9. a, 10. c, 11. a, 12. a, 13. a, 14. b. 15. c, 16. c, 17. c, 18. a, 19. b, 20. c, 21. b, 22. c, 23. c, 24. b, 25. a, 26. c, 27. b, 28. c, 29. a, 30. a.

Chapter 13

Cell Cycle

Cell cycle is the repetition of G_1, S, G_2 and M phases in the life cycle of a cell. The main objective of cell division is to replicate the genome and transmit its counterparts into daughter cells through the segregation of chromatids. The main divisional phase i.e. mitosis or meiosis (*M phase*) remain separated from the DNA replication phase (*S or Synthesis phase*) by two preparatory or gap periods, one preceding the S phase - *Gap-1 or G-1* and other succeeding the S phase - *Gap-2 or G-2* phase. S, G1 and G2 phases are collectively called *interphase* (Fig. 13.1). G1 is the most crucial phase and if appears unfavourable it enters into a specialized resting phase called G0 in which they remain for days, weeks, years or even permanently. If extracellular signals appear favourable for the cell to grow and divide, the cell then resumes division and reenters the G1 phase. In mammalian cell, the total duration of cell cycle is 24 hours of which interphase lasts for 23 hr. and M phase for only one hr; S phase occupies10-12 hr. It is the most most vital phase in which the total amount of DNA i.e. 2n is replicated into 4n. Thus in G2 the cell has two complete 2n set of chromosomes.

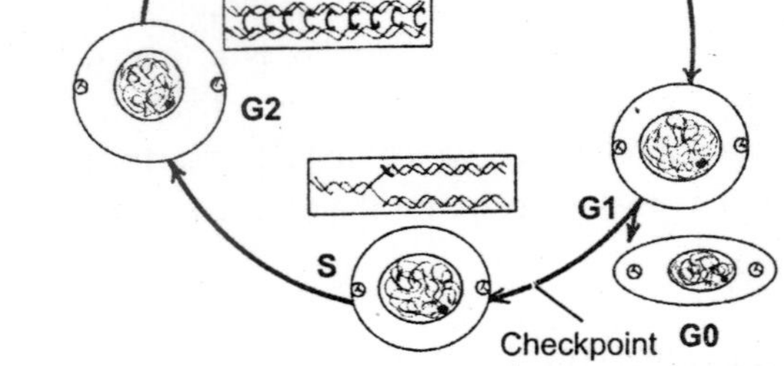

Fig. 13.1 : Main phases of cell cycle.

In interphase, in the course of metabolism the volume of both nucleus and cytoplasm change. In addition to duplicating their genome, cells duplicate other organelles and macromolecules. When nuclear volume becomes approximately double in respect to the volume of cytoplasm then cell tends to divide.

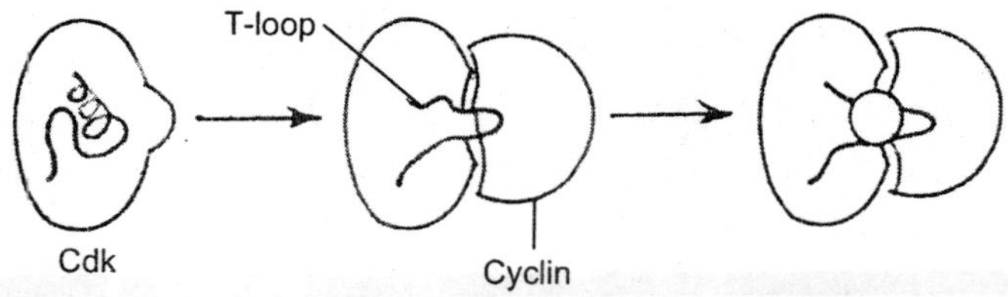

Fig. 13.2 : Structural basis of Cyclin-Cdk activation.

Cell cycle is basically regulated by a complex, consists of a structural protein - *Cyclin* and a kinase, called *Cyclin dependent kinases (Cdks)*. Cdks are the principal protein that drives the molecular events of cell cycle but unless they are tightly bound to cyclin they have no kinase activity). However their ability to trigger various events completely depends upon the structural protein - Cyclin. Cyclins activate their partner Cdks to form a complex which induces cell cycle progression.

Three kinds of Cdks are noted -Cdk 2, 4 and 6. The timing of expression of cyclins varies in different parts of cell cycle e.g. Cyclin D occurs in G1; A and E occur in G1 and S while A and B1 occur in S, G2 and M. If there is any defect, Cyclin-Cdk complex formation is inhibited and if there is no defect, the complex is activated by phosphorylation that leads to the activation of a transcription factor which then transactivate a specific gene necessary for the next cell cycle step

13.1. CONTROL OF CELL CYCLE EVENTS IN G_1 PHASE

G1 is the most vital phase of cell cycle in which it responds to extracellular signals by either advancing towards another division or withdrawing from the cycle into a resting or quiescent phase (G0). As the newly formed cell enters the G1 phase, pCdc-25C regulate the activity of cyclins by inactivating the M phase cyclins and inducing specific Cdks. In G1, Cdk-2/4/6 combines with specific Cyclin D to carry out kinase activity. In response to the growth factors, Cdk 4 and 6 combine with Cyclin D which is the major complex for the progression of cell cycle through G1. The activation of Cyclin-Cdk complex requires phosphorylation of Cdk at threonine by Cyclin activating kinase (CAK). The catalytic activity of the assembled holoenzyme are first manifested by the mid G1 and reaches to maximum concentration at G1-S transition. Cyclin D over expression occurs in most cancers and members of this family are prognostic markers in tumors of pancreas, endometrium, breast, head and neck. Four INK proteins p16, p15, p18 and p19 can inhibit the binding of Cdk 4/6 with cyclin D. This prevention often blocks the cell to cross the restriction point.

Restriction point: Progression through G1 is controlled by many factors. To regulate the phase flawlessly, in mid G1, there is a point after which cells become independent of growth factors. This special point is called *restriction point*. Disruption of restriction point is a common biological feature in human cancer. It is coordinated by two families of cyclins – Cyclin D and E. Cyclin Ds activate Cdk4/6 and Es activate Cdk2. Activities of both cyclins are needed for regulated progression through the restriction point.

Normal cell cycle ensures duplication of undamaged DNA. There are many censoring proteins which ensure this flawless duplication. This is performed prior to the entrance to S and M phases and failure of these screening phases are the main causes of cancer. In fact, checkpoints are the assemblage of several proteins which

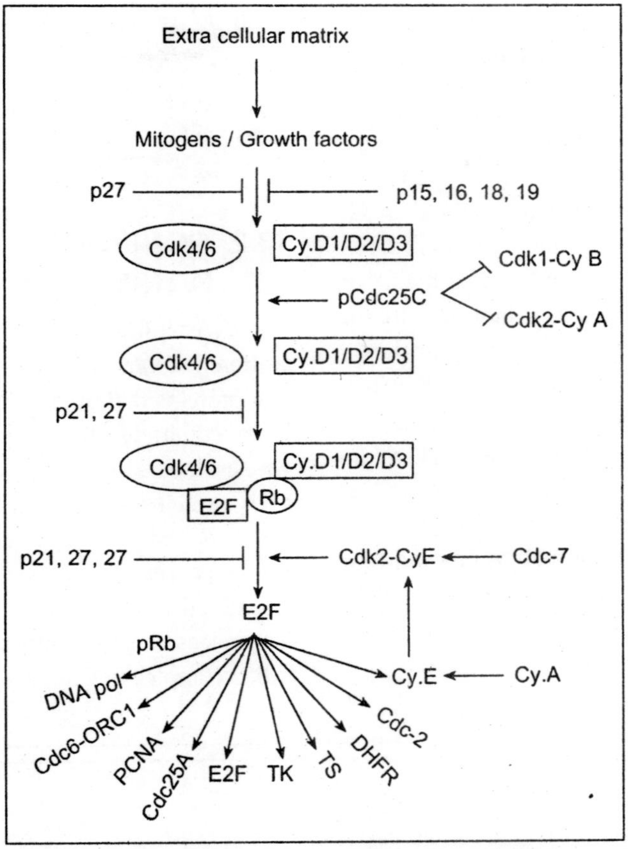

Fig. 13.3 : Role of Rb and other factors in controlling G1.

mainly censor DNA damage and dysfunction repair of damaged DNA. The role of different censoring proteins is as follows:

Role of Retinoblastoma (Rb) protein: Retinoblastoma (Rb) protein is a tumor suppressor protein and is one of the mammalian regulators to cross the G1-S restriction point. It is considered as cell cycle master switch. Loss of Rb occurs in retinoblastoma, osteosarcoma, small cell lung carcinoma and bladder carcinoma. In G1, Rb binds with Cdk 4/6-Cyclin-D and also with a transcriptional factor E2F. Cdk 4/6 and CyclinE-Cdk2 phosphorylate RB for G1 exit. RB has multifaceted job, it is implicated in DNA replication, differentiation, DNA repair, cell cycle checkpoints and apoptosis. Binding of E2F with Rb inhibits the ability of E2F to transactivate the target genes so that RB-E2F complex represses some target genes whose expression depends upon their dissociation (Fig. 13.3). In G1/S transition, released E2F inturn triggers the expression of Dihydrofolate reductase (DHFR), thymidine kinase (TK), Thymidine synthase (TS), DNA poly merase (POL), Cdc-2/Cdk1, Cyclin E, Cyclin-A and possibly E2F-1. Cyclin E expression peaks at the time of entrance to S phase. Rb is a target of several pathways that can inhibit growth in G1/G0. The growth inhibitory factor, TGFβ acts through inhibitors of Cdk-Cyclin kinases (CKIs) and prevent phosphorylation of Rb making it impossible to release E2F so that the cell gets arrested in G1. The normal progression of the cell cycle requires increased levels of Cdc-Cyclin dimer to overcome this inhibitory threshold.

Role of p53 protein: p53 is another potent regulator of G1 checkpoint. In response to genotoxic damage, cells become arrested in G1 and always try to repair the damaged DNA before passing to S phase. p53 senses the DNA damage and if the damage is minor, it halts the cell cycle in G1 but if the damage is major then it triggers apoptosis. p53 binds with E2F which prevents E2F from binding to the promoters of protooncogenes like *c-myc, c-fos* etc. Therefore p53 indirectly inhibits the expression of protooncogenes. Thus cell cycle arrest and apoptosis appear to be alternative p53 induced outcomes. Apoptosis is arguably the most potent natural defense against cancer since it eliminates the pre malignant cells that enter S phase.

Role of ATM: ATM (Ataxia Telangiectasia Mutated), ATR and Rad 3 related (ATR) proteins are involved in detecting DNA damage and can arrest cell cycle progression at G1/S, S and even G2/M.

13.2. CONTROL OF CELL CYCLE EVENTS IN S PHASE

DNA replication begins at *origin of replication* scattered at various locations in the chromosome. The proteins that bind to the origin of replication is called *origin replication complex* (ORC) with which binds several regulatory proteins. One of such proteins is *Cdc6;* it increases abruptly at early G1 and attracts a protein *Mcm* to form *pre replicative complex or pre-RC.* Once pre-RC is assembled in late G1, the cell enters S phase. In fact, released E2F, synthesized Cyclin-A and E-Cdk-2 complex (S-Cdk) acts as ***S phase promoting***

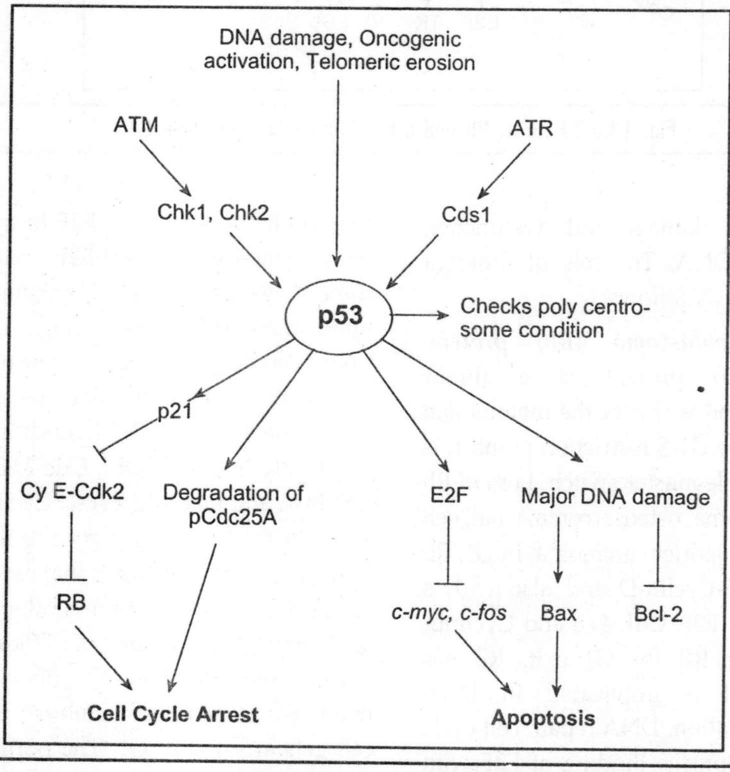

Fig. 13.4 : Role of p53 in controlling cell cycle in G1.

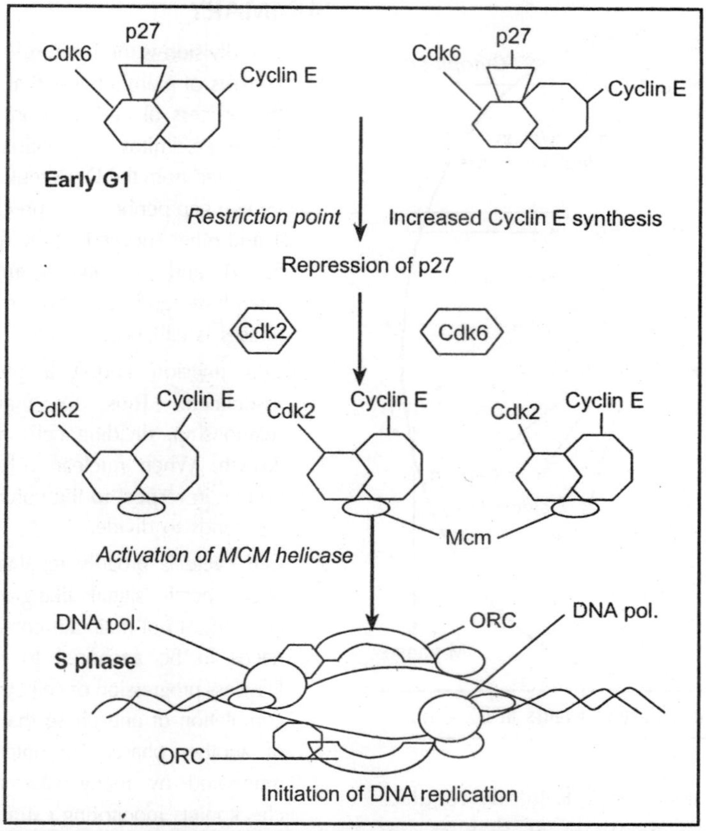

Fig. 13.5 : Molecular events at G1-S.

factor (SPF). As replication begins, E2F and Cyclin E are inactivated.

13.3. CONTROL OF CELL CYCLE EVENTS IN G2 PHASE

After the completion of DNA replication, the cell enters into G_2 phase with the accumulation of sufficient amount of Cyclin B1. In early G2, Cdk-2 unites with Cyclin B1 to form ***maturation promoting factor (MPF).*** It remains in an inactive state by phosphorylation at threonine 14 (by Wee1), tyrosine 15 (by Myt1) and threonine 161 (Fig. 13.6). MPF mainly synthesizes lamins (that causes degradation of nuclear envelope), vimetin and caldesmon (reorganizes cytoskeleton leading to the formation of spindle fibers). Moreover phosphorylation of caldesmon helps in the forma- tion of contractile ring during cytokinesis. In addition, MPF acts on H1 that causes chromosome condensation. MPF is also known to activate microtubule associated protein kinase (MAPK) – responsible for the reorganization of the microtubules at the onset of M phase. In G2, a protein - geminin assembles on freshly synthesized DNA which inhibit further replication of DNA. Initiation of M phase is activated by rapid activation of MPF by its dephosphorylation mediated by Cdc25B followed by Cdc-25C is favored by the inactivation of Wee1, Myt 1 as well as hyperphosphorylation of Cdc25C.

G2 - M checkpoint: In response to DNA damage, if cell cycle proceeds, an array of factors participate to arrest the cell cycle. In response to

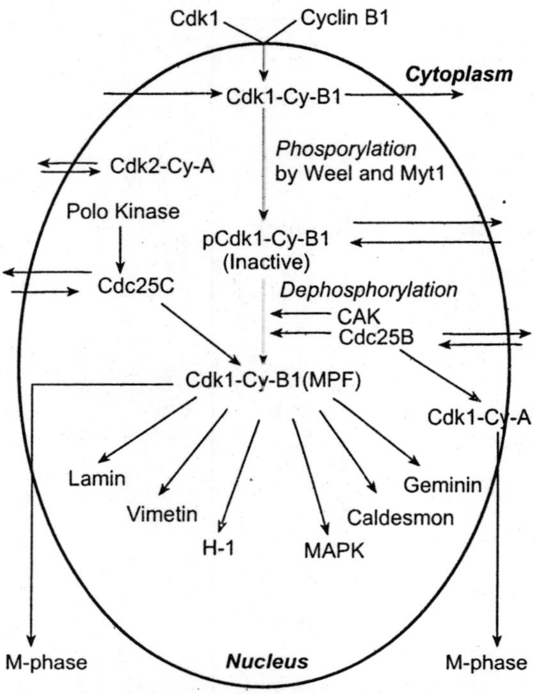

Fig. 13.6 : Different molecular events in G2 – M.

DNA damage, members of Pl3K family regulate DNA repair. Several members of Pl3k family phosphorylate p53, ATM and ATR. ATM down regulates CHK1 and CHK2 kinases which in turn phosphorylate p53 for stabilization of protein after DNA damage. CHK1 and CHK2 also phosphorylate Cdc25C leading to export of Cdc25C from nucleus to cytoplasm that eventually lead to the inhibition of Cdc2-Cyclin B1 activity and G2-M arrest. In addition to p53, p21 is necessary for G2 arrest. Moreover, pCdc14B translocates from the nucleolus to the nucleoplasm and causes degradation of Plk1, a prominent mitotic kinase. This process induces the stabilization of Claspin, an activator of the DNA-damage checkpoint, and Weel, an inhibitor of cell-cycle progression, and allows an efficient G2 checkpoint. Thus the main biochemical event in G2 progression is the activation of Cyclin B-Cdk1 complex by Cdc-25C. Therefore inhibition of the formation of this complex is the key event involved in G2-M progression.

SUMMARY

1. Cell division is the basic rule of all living beings. It consists of many phases that collectively carry out the process of cell division. The main divisional phase i.e. mitosis or meiosis (M phase) remain separated from the DNA replication phase (S phase) by two gap periods, one preceding the S phase -G-1 and other succeeding the S phase or G-2 phase. S, G1 and G2 phases are collectively called interphase and repetition of G_1, S, G_2 and M phases is called cell cycle.
2. Cell division and cell growth are intimately associated. Thus to maintain proper size relationship, dividing cells must coordinate their growth. When nuclear volume becomes nearly double in respect to the volume of cytoplasm then cell tends to divide.
3. Cell cycle is broadly regulated by 3 aspects – It needs specific signals like growth factors, cytokines etc; it must not maintain contact with other cells; it needs to be anchored to extra cellular matrix. Flawless progression of cell cycle involves successful completion of one phase that triggers the initiation of another phase. The intermediate periods are monitored by many factors that constitute the checkpoints monitoring extra cellular signals, DNA damage and faithful DNA replication etc.
4. Cell cycle is basically regulated by a complex, consists of a structural protein - Cyclin and a kinase, called Cyclin dependent kinases (Cdks) that drives the molecular events of cell cycle. Each cyclin-Cdk complex phosphorylates different set of substrate proteins regulating a particular event. Three kinds of Cdks are noted -Cdk 2, 4 and 6. The activity of cyclin-Cdk complex can be inhibited by a kinase -Wee1 and by the binding CKIs in G1 and S phases. Cdk-2 combines with cyclin A and E, Cdk 4/6 with cyclin D and Cdk-7 with cyclin H. The activation of Cyclin-Cdk complex requires phosphorylation of Cdk at threonine by Cyclin activating kinase (CAK). However Cdk inhibitory subunits can bind some substrates and inactivate Cdk-cyclin complex. Four INK proteins p16INK4a, p15INK4b, p18INK 4C and p19INK 4d can inhibit the binding of Cdk 4/6 with cyclin D.
5. For the progression through G1 flawlessly, in mid G1, many factors monitors physical features of

DNA, growth, anchorage and intercellular adhesion of the cell. This special point is called restriction point after which cells become independent of growth factors.

6. Retinoblastoma (Rb) is a tumor suppressor protein and is one of the mammalian regulators to cross the restriction point. It is considered as cell cycle master switch. RB monitors any damage in DNA, if it is minor it delays the cell cycle but if it is major it induces apoptosis. Binding of E2F with Rb inhibits the ability of E2F to transactivate the target genes so that RB-E2F complex represses some target genes whose expression depends upon their dissociation. Cdk 4/6 and CyclinE-Cdk2 phosphorylate RB for G1 exit. The inhibitors of Cdk-Cyclin kinases, called Ckis prevent phosphorylation of Rb making it impossible to release E2F so that the cell gets arrested in G1. Cyclin E, which is essential for the initiation of DNA replication, peaks at the time of entrance to S phase.

7. ATM-ATR can activate cell cycle arrest at G1/S, S and even G2/M. ATR propagates the signal via Cds1 and ATM via Chk1 and Chk2 that in turn phosphorylate Cdc25A. Chk1 mainly censors double stranded break of DNA.

8. PTEN-Mdm_2-p53 interaction is another tumor supperssor pathway recruits the protein like serine/threonine kinase- Akt. Phosphorylated Mdm_2 enters into the nucleus where it degrades p53. PTEN blocks Akt signaling pathway and thereby inhibits the entrance of Mdm_2 into the nucleus, protecting p53 and thus allowing the cells to respond to DNA damage by apoptosis. In PTEN mutation, p53 is not protected so that DNA damage cannot be rectified which lead to genetic instability.

9. DNA replication begins at origin of replication scattered at various locations in the chromosome. Released E2F, synthesized cyclinA and E-Cdk-2 complex (S-Cdk) acts as S phase promoting factor (SPF). As replication begins, E2F and cyclin E are inactivated. Initially Cdc6 increases abruptly at early G1 and attracts a protein Mcm to form pre-RC.

10. After the completion of DNA replication accumulation of sufficient amount of cyclin B allows cell to enter into G2 phase. In early G2, Cdc-2 unites with Cyclin B1 to form maturation promoting factor (MPF). MPF synthesizes lamins, vimetin and caldesmon. It also acts on H1 that causes chromosome condensation. It also activates microtubule associated protein kinase (MAPK). In G2 geminin accumulates that prevents rereplication of synthesized DNA. Initially it remains in an inactive state by phosphorylation at threonine 14, tyrosine 15 and threonine 167 but at late G2 its dephosphorylation mediated by Cdc25B followed by Cdc-25C led MPF to enter into the nucleus to initiate M phase.

11. In response to any DNA damage or flaw in replication, activation of Cyclin B-Cdk1 complex by Cdc-25C. In response to DNA damage, a member of PI-3K family becomes activated and initiate signal transudation pathways that regulate DNA repair. Several members of PI-3k family can phosphorylate p53, ATM and ATR. ATM down regulates Chk1 and Chk2 kinases that in turn phosphorylate p53 for stabilization of protein after DNA damage. Chk1 and Chk2 also phosphorylate Cdc25C on Serine-216 leading to export of Cdc25C from nucleus to cytoplasm that eventually lead to the inhibition of Cdc2-Cyclin B1 activity and G2-M arrest. In addition to p53, p21 is necessary for G2 arrest.

QUESTIONS

Long Questions:

1. Describe the molecular events of G1 phase of the cell cycle.
2. What do you mean by checkpoints? How checkpoint proteins in G1 monitor cell cycle progression?
3. Discuss the role of p53 in the regulation of G1 phase of the cell cycle.

Short questions:

1. What do you mean by interphase? Name its different phases.
2. Cell division is related with neigh bouring cells – explain.
3. What is called checkpoints? How checkpoints of G1 differ from G2?
4. The appearance of Cyclin-Cdk facilitates complexes is specific to cell cycle phases – explain.
5. Why Rb is considered as 'master switch' for G1?

6. Which enzyme E2F to participate in transcription of target genes?
7. How p53 saves the cell from genotoxic damage?
8. How PTEN protects p53 and saves the cell from malignancy?
9. Which cyclin is prevalent in S phase?
10. What do you mean by SPF?
11. What do you mean by MPF?
12. How MPF becomes inactivated in early G2 and reactivated in late G2?
13. Name the components synthesized by MPF.

MCQs:

1. **DNA replication occurs in**
(a) M phase, (b) G1, (c) S phase.
2. **The ploidy of the genome in G2 is**
(a) 2n, (b) n, (c) 4n.
3. **Cyclin D is predominantly noted in**
(a) G1, (b) S, (c) G2.
4. **NK proteins inhibit the binding of**
(a) cyclinD/Cdk4/6, (b) CyclinB/Cdk2, (c) Cyclin A/ Cdk2.
5. **Inactivation of MPF at threonine14 is done by**
(a) Myt1, (b) Wee1, (c) Chk1.
6. **Inactivation of MPF at tyro- sine15 is done by**
(a) Myt1, (b) Wee1, (c) Chk1.
7. **Rereplication of synthesized DNA is prevented by**
(a) vimentin, (b) caldesmon, (c) geminin.
8. **Contractile ring formation during cytokinesis is mediated by**
(a) lamin, (b) caldesmon, (c) Chk1.
9. **During G1**
(a) RNA and proteins,
(b) DNA and proteins,
(c) DNA are synthesized.
10. **The cell doubles in size during**
(a) G1, (b) G2, (c) S
11. **Histone protein is synthesized in**
(a) G1, (b) G2, (c) S.
12. **In the cell cycle DNA replicates**
(a) twice, (b) Once, (c) Many times.
13. **Which of these proteins are found in G2?**
(a) Histone, (b) DNA polymerase, (c) Tubulin.
14. **Post mitotic phase is**
(a) G0, (b) G1, (c) G2.
15. **Cebtrosome replication occurs in**
(a) G1, (b) S, (c) G2.
16. **Which one is stored in G1?**
(a) Tubulin, (b) Histone, (c) ATP.
17. **Duration of G1 phase in total cell cycle is**
(a) 25-30%, (b) 40-45%, (c) 20-25%.
18. **Duration of S phase in total cell cycle is**
(a) 35-40%, (b) 40-45%, (c) 20-25%.
19. **Duration of M phase in total cell cycle is**
(a) 25-30%, (b) 40-45%, (c) 5%.
20. **Cells in G0 phase**
(a) Terminate the cell cycle,
(b) Suspend the cell cycle,
(c) Enter the cell cycle.
21. **Condensed chromosome with visible centromere is found in**
(a) G2, (b) S. (c) M.
22. **During cell cycle RNA and protein are synthesized in**
(a) G1, (b) S, (c) G2.
23. **No. of DNA molecules perchromosome in G2 is**
(a) 1, (b) 4, (c) 2.
24. **The stage of cell cycle where DNA polymerase is active is**
(a) G1, (b) S, (c) M.
25. **Cyclin is associated to which of the followings?**
(a) CCK, (b) CDK, (c) CK(c)

Answers to MCQs:

1. c, 2. c, 3. a, 4. a, 5. b, 6. a, 7. c, 8. b, 9. a, 10. a, 11. c, 12. b, 13.,c, 14. b, 15. b, 16. c, 17. b, 18. a, 19. c. 20. b, 21. c, 22. a, 23. c, 24. b, 25. a.

Chapter 14

Cancer

In the whole life, during normal growth and differentiation, intricate genetic control system regulates the balance between cell birth and cell death in response to growth promoting, growth inhibiting and death signals and it involves trillions of cell divisions. It is thus inevitable that some of these divisions can be mistimed or misplaced due to genetic damage that deregulates the controlling system of cell division. This uncontrolled cell division is broadly diagnosed as *cancer.*

Cancer is a group of diseases showing an autonomous proliferation of neoplastic cells which have a number of alterations,including mutations and genetic instability. Many forms of cancer are associated with exposure to environmental factors such as tobacco smoking, radiation, alcohol, and certain viruses. Some of these risk factors can be avoided or reduced but there is no known way to avoid the disease entirely.

Cellular functions are controlled by proteins and as these proteins are encoded by DNA organized into genes, molecular studies have shown that cancer is a manifestation of acquired genetic disease. Although there are mechanisms responsible for DNA maintenance and repair, the basic structure of DNA and the order of the nucleotide bases can be mutated. These mutations can be inherited or can occur sporadically, and can be present in all cells or only in the tumor cells. Mutation in a single gene rarely leads to the onset of cancer. Usually a series of mutations in multiple genes give rise to a clone of altered cells that grows into a tumour. Due to uncontrolled division of cells they may either directly invade the adjacent tissues through *invasion* or may be spread to distant sites by blood or lymph which is called *metastasis.* If untreated, cancers eventually cause death. It may affect people at all ages, but risk tends to increase with age. Cancer pathology is ultimately due to the accumulation of DNA mutations that negatively affect expression of tumour suppressor proteins or positively affect the expression of proteins that drive the cell cycle.

Particular substances have been linked to specific types of cancer. Tobacco smoking is

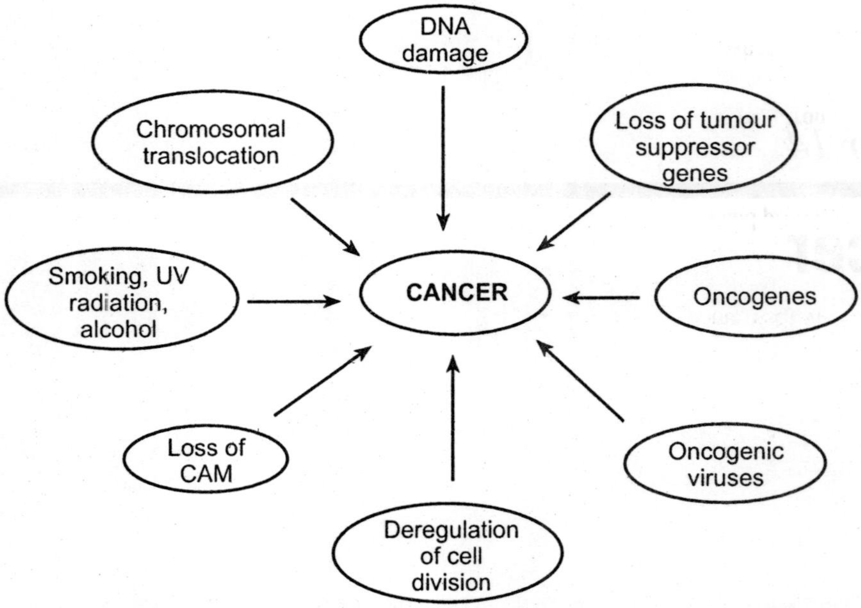

Fig. 14.1 : General causes of cancer.

associated with lung cancer. Prolonged exposure to radiation, particularly UV radiation from the Sun, leads to melanoma and other skin malignancies. Breathing asbestos fibers is associated with mesothelioma. Some point mutations or Chromosomal translocations, forming *Philadelphia chromosome*, are special type of mutation that involves exchanges between different chromosomes.

Damage to DNA, encode defective proteins that perturb cell division. Many mutations caused by radiation, chemicals or physical agents (called *carcinogens)* are required to transform a normal cell into a malignant cell. Certain viruses can insert their DNA into the human genome causing malignancy. Many mutagens are carcinogens, but all carcinogens are not mutagens (e.g. alcohol and estrogen are carcinogens but not mutagens These are thought to promote cancers through their stimulating effect on the rate of cell mitosis. Faster rates of mitosis increasingly leave less opportunity for repair enzymes to repair damaged DNA during DNA replication, increasing the likelihood of a genetic mistake. A mistake made during mitosis can lead to the daughter cells receiving the wrong number of chromosomes, which leads to aneuploidy and may lead to cancer. In order to start uncontrolled divisions, two set of regulatory genes need to be mutated – proto-oncogenes and tumour suppressor genes. *Proto-oncogenes* promote cell growth through a variety of ways. Many can produce hormones, some are responsible for the signal transduction system in the cell controlling a variety of molecular processes. They often produce mitogens, or are involved in transcription of DNA. Mutations in proto-oncogenes can modify them into *oncogenes* showing uncontrolled cell division. The chance of cancer cannot be reduced by removing proto-oncogenes from the genome as they are critical for growth, repair and homeostasis of the body. It is only when they become mutated that the signals for growth become excessive.

Tumor suppressor genes code for anti-proliferation signals and proteins that suppress

Mutation inactivates tumour suppressor genes

↓

Cell proliferates occurs exponentially

↓

Mutation inactivated DNA repair genes

↓

Mutation converts proto-oncogenes to oncogenes

↓

Mutation inactivates several more tumour suppressor genes

↓

Formation of tumour or cancer

Fig. 14.2 : Basic role of mutation in cancer.

mitosis and cell growth. Generally tumor suppressors are transcription factors that are activated by cellular stress or DNA damage. Often DNA damage will cause the presence of free-floating genetic material as well as other signs and will trigger enzymes and pathways which lead to the activation of tumor suppressor genes. The functions of such genes are to arrest the progression of cell cycle in order to carry out DNA repair, preventing mutations from being passed on to daughter cells.

14.1. TYPES OF CANCER

About 200 different types of cancers are known today. These are broadly divided into five categories:

a) ***Carcinoma:*** Cancers of epithelial cells are collectively called carcinoma. Cancers of skin, esophagus, breast and brain belong to this type.

b) ***Sarcoma:*** Cancers of connective tissue belong to this category. Cancers in bone (Osteosarcoma), striated muscle (Rhabedomyosarcoma) and cartilages (Chondrosarcoma) belong to this category.

c) ***Lymphoma:*** Cancers of lymph gland and spleen is called lymphoma. E.g. Hodgkin's disease.

d) ***Leukemia:*** Malignancy of the haemopoetic cells of red bone marrow resulting excessive production of W.B.Cs is called leukemia (Blood cancer).

e) ***Lipoma:*** Cancers on adipose tissue is known as lipoma.

f) ***Melanoma:*** Malignancy of the pigment cells.

g) ***Glioma:*** Malignancy of neuroglial cell of the nervous system.

h) ***Myeloma:*** Malignancy of bone marrow.

14.2. CHARACTERISTIC FEATURES OF A CANCER CELL

Whenever a cell loses control over proliferation, it is usually termed as cancer cell. It has the following peculiarities.

i) ***Cytological features:*** These exhibit large nucleus and nucleolus, haphazardly arranged chromatin networks; increased numbers of ERs and polysomes; reduced number of Golgi bodies and mitochondria appear swollen with less number of cristae.

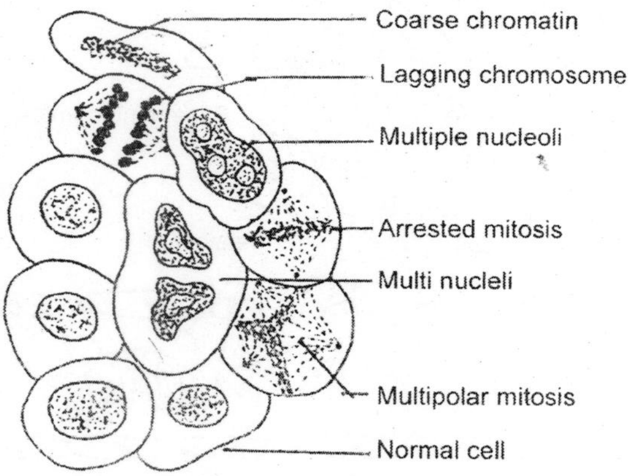

Fig. 14.3 : Distinctive features of normal and cancer cells.

The renal cell carcinoma show abundant mitochondria and microvesicles. Mitochondria often show outpouchings of the outer membranes that has close resemblance with the microvesicles. Some microvesicles contained homogeneous, electron-dense, finely granular matrix, as seen in mitochondria and in some cases even mitochondria were present within some microvesicles. As vesicle formation in normal mitochondriogenesis is noted it is suggested that defective mitochondriogenesis may be the source of renal cell carcinomas.

ii) *Molecular changes in the plasma membrane:*

(a) Phospholipids in the plasma membrane exhibits alterations, in normal cells four types of phospholipids are found GM1a, GM1, GM2 and GM3 but in cancer cells only GM3 type is noted.

(b) Glycoproteins of the molecular weights 46,000; 2,00,000 and 2,40,000 are not found in cancer cells.

(c) Motility of the proteins increases.

(d) In normal cells lectins play major role in the intercellular adhesion but in cancer cells its amount increases and accumulates to form lumps.

(e) The negative charge of the cell membrane of normal cell is -60mV which increases in cancer cells.

(f) Glucose transport through the membrane increases keeping pace with the rapid rate of growth. It enhances the rate of glycolysis and synthesis of lactic acid.

iii) ***Appearance of virus specific antigens:*** Adenovirus or Papovirus infected cancer cells exhibits T-antigen and Epstein-Barr (EB) virus infected cancer cells exhibit EB nuclear antigen on the plasma membrane. These actually cause lysis of the same type of cells but when this property is lost they aggregate with each other.

iv) ***Reduced intercellular adhesion:*** Cells are not united with the neighbouring cells by adhesion molecules as well as extracellular

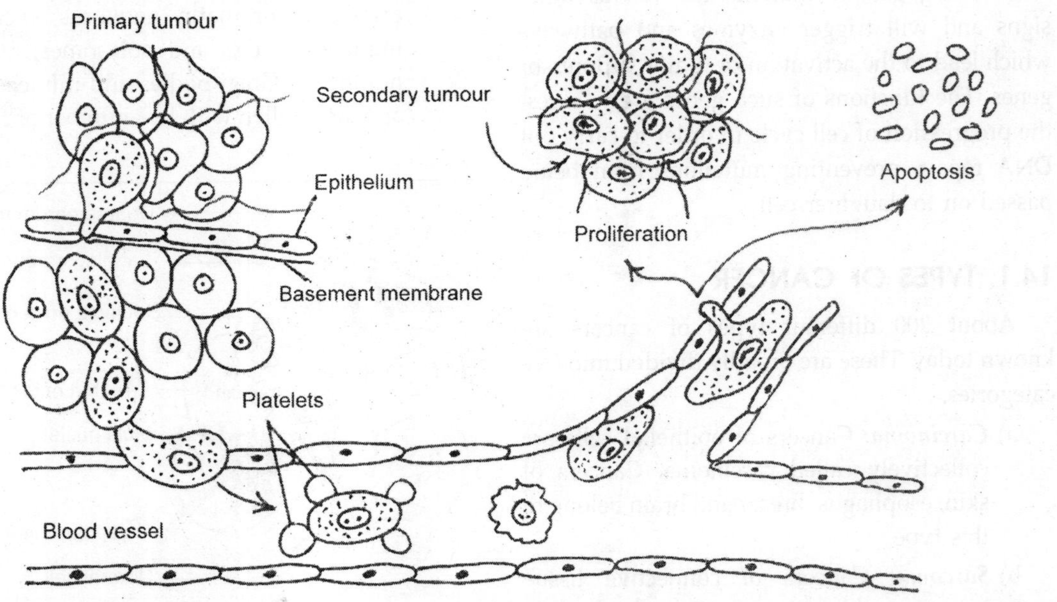

Fig. 14.4 : Metastasis of cancer cells.

matrix. Moreover usually cells exhibit specific intercellular adhesion but tumour cells can adhere with any foreign cell to form a lump.

v) ***Invasiveness:*** Usually a cell from any organ can not enter into another organ but cancer cells secrete a specific type of protease that assists to infect another organ. This quality helps in the invasion of tumour cells. The specific protease, called *cell factor* is required for the synthesis of plasmin which induces cell division in addition to invasion.

vi) ***Contact inhibition:*** Usually if a cell comes in contact with another cell, both of them lose their power of motility and division but cancer cells retain their power of motility as well as division. This property is called contact inhibition.

vii) ***Reduced cAMP requirement:*** Normal cells require about 10% serum for the synthesis of cAMP that is needed for mitosis cell division but cancer cells need very minute amount of cAMP for their division.

viii) ***Disorganization of cytoskeleton:*** Usually the fabric of the cell is maintained by the well organized cytoskeletons but cancer cells show disorganized cytoskeletons particularly actin. It facilitates the formation of blebs or microvilli from the plasma membrane. Alterations in the organization of actin microfilament is associated with cellular morphology, motility and adhesiveness which are the characteristic features of trans formed cancer cells. For this reason actins have become promising targets for chemotherapy.

ix) ***Immortality:*** Usually in a culture medium a normal cell can continue replication up to 50 generations but cancer cells can divide up to indefinite period due to over abundance of telomerase. Some cancer cells from the cervix of Henrieta Lachs (called HeLa cells) is still replicating from 1951 in different laboratory of the world.

x) ***Aneuploidy:*** Inspite of mitotic cell division most of the progeny cells show aneuploidy.

xi) ***Evading apoptosis:*** Molecular biological alterations allow the cancer cells to bypass the apoptotic processes so that they continue to divide with genotoxic damage.

xii) ***Self sufficiency of growth factors:*** Normal cell have to depend upon their micro environment for the supply growth factors for division but cancer cells show self sufficiency of growth factors. They also show insensitivity to antigrowth factors.

xiii) ***Increased rate of cell division:*** Normal cell division has to follow definite rules that obey checkpoints. Cancer cells bypass these checkpoints and employ factors that facilitate uncontrolled cell division. They also exhibit altered ability to differentiate.

xiv) ***Angiogenesis:*** Supply of blood vessels to provide nutrition and disposal of waste products is the primary requisite for cancer cells. Tumour cells synthesize VEGF and FGF that enhance neovascularization.

Table 14.1 : Differences between normal and cancer cells

Normal cell	Cancer cells
A. Cytological features :	
1. Nucleus and nucleoli appear normal in appearance though the structure and number may vary. ERs show normal number, ribosome occurs on the wall of the ERs or in free state. Number of Golgi bodies varies from 5-15. Mitochondia measures 0.5–1 m in breadth with moderate number of cristae.	1. Cancer cells exhibit large nucleus and nucleolus, haphazardly arranged chromatins increased numbers of ERs, ribosomes are attached together in the form of polysomes, Golgi bodies show reduced in number and mitochondria appear swollen with less number of cristae.

Table contd. next page

Normal cell	Cancer cells
2. Phospholipids in the plasma membrane exhibit four isotypes GM1a, GM1, GM2 and GM3.	2. Phospholipids in the plasma membrane show only GM3 type.
3. The notable glycoproteins are of the mol. Wt. 46,000; 2,00,000 and 2,40,000.	3. Glycoproteins of these mol. wts are not found.
4. Lectins play major role in the intercellular adhesion.	4. In cancer cells lectins participate in the formation of lumps.
5. The (–) charge of the cell in the plasma membrane is –60mV.	5. The (–) charge increases in cancer cells.
6. Do not show specific antigens.	6. Virus infected cell show T-antigen or EB nuclear antigen on the plasma membrane.
7. Cytoskeletons show specific arrangements to provide normal structure and functions.	7. Cytoskeletal structures are disorganized and actins are lost.
8. Intercellular adhesion is specific.	8. Cancer cells can adhere with any cell.
9. Need to be anchored firmly with the ECM for growth and differentiation.	9. Cancer cells do not require any firm anchorage and can grow even in a colloid like environment.
B. Physiological features :	
1. Glucose transport through the membrane is moderate.	1. It increases keeping pace with the rapid rate of growth.
2. In culture medium same type of cells can adhere with each other.	2. In culture medium different type of cells can adhere with each other.
3. Usually normal cells can not infect other cells.	3. Cancer cells can enter into other organ by secreting protease.
4. In an artificial medium when normal cells come in contact with other cells they lose the power of locomotion and reproduction.	4. Cancer cells can proliferate in contact with other cells and also in isolated condition.
5. Require much serum for growth and development in artificial medium.	5. Cancer cells require very less amount of serum for their growth.
6. Require much more cAMP for growth.	6. Require very less cAMP for growth.
7. 'S' phase of the cell cycle is prolonged.	7. 'S' phase of the cell cycle is reduced.
8. Secretes no special type of protease.	8. Secretes a special type of protease – plasmin that helps to infect other cells.
9. Shows A, B and C isotypes of aldolase.	9. Show only A and C isotypes.
10. Glycolysis rate is normal.	10. It is very high in such cells.
11. They need to become mature before division.	11. They do not mature before division because most of the genetic factors are lost. So the primitive cells reproduce more quickly and even more haphazardly.
12. Cell signaling regulates all the features.	12. They do not obey cell signaling.
13. In a culture medium they can divide up to 50 generations.	13. They can continue to divide up to indefinite period.
14. The daughter cells are of same type.	14. Daughter cells may show aneuploidy.

14.3. DEVELOPMENT OF AN EPITHELIAL CANCER

(i) An initial mutation in a single cell disrupts signal transduction and cell cycle. The mutated cell divides more frequently than its neighbours. The mutated population of cells generates *dysplasia* which is divided into low and high grade whose cells occupy the bottom layers adjacent to the basement membrane.

(ii) A second mutation occurs in the already affected cell that disrupts the gene expression, modifying metabolism (glucose metabolism becomes anaerobic), growth characteristics, cell shape, and intercellular adhesion. Cell membrane exhibits altered receptor patterns, transport properties and permeability. Altered cytoskeleton causes alterations in cell shape. The doubly mutant cell and its descendants are slightly less specialized than their neighbours.

(iii)A third mutation perturbs cell cycle, impairs DNA repair and broken chromosomes. The descendant cells are less specialized; nuclei appear enlarged with chromatin clumps. The microscopic mass of cell remains within the epithelium bounded by the basement membrane.

(iv)The dividing tumour cells gradually rupture the basement membrane and begin to extend into the surrounding stroma. The tumour cells secrete fibroblast growth factor (FGF) and vascular endothelial growth factor (VEGF) to support neovascularization for the growth of the tumour.

(v) Some cancer cells creep along the associated blood vessels while others squeeze between the lining cells to enter the circulation.

(vi) The cancer cells along with blood and lymph at first invade the lymph nodes and then recolonize to distant sites. In the new sites further mutation and changes in gene expression lead to the new features of the cancerous growth. It is impossible to tell the initial cause for any specific cancer. However, with the help of molecular biological techniques, it is possible to characterize the mutations or chromosomal aberrations within a tumor and rapid progress is being made in the field of predicting prognosis based on the spectrum of mutations in some cases. For example, some tumors have a defective p53 gene. Telomerase mutations remove additional barriers, extending the number of times a cell can divide. Other mutations enable the tumor to grow new blood vessels to provide more nutrients or to metastasize, spreading to other parts of the body. However development of epithelial cancers may involve following steps:

Multi-hit model of cancer induction: From the above description it is evident that several mutations are required for the induction of cancer. Normal genes controlling mitosis and tumor suppressor genes would suppress a mutation limited to one oncogenes. An initial mutation completely isolates the cell from others from the view point of morphological features and genetic alterations affecting the cell cycles but a mutation to only one tumour suppressor gene would not cause cancer either, due to the presence of many "backup" genes that duplicate its functions. The second mutation in one of the progeny cells allows growing uncontrollably and forming a bening tumour. Third mutation allows the tumour to outgrow and overcome the constraints imposed by the tumour micro-environment. It is only when enough proto-oncogenes have mutated into oncogenes, and enough tumor suppressor genes deactivated or damaged, that the signals for cell growth overwhelm the signals to regulate it, that cell growth quickly beomes out of control. An additional mutation leads into metastasis. Often, because these genes regulate the processes that prevent most damage to genes themselves, the rate of mutations increase as one gets older, because DNA damage forms a feed back loop.

This model allows taking two predictions:

1. All the cells in a tumour contain al least some genetic alterations.

2. It takes a long time to accumulate multiple mutations; therefore incidence of cancer increases with age.

Gradual development of cancer can be best studied in colon cancer depicting a clear transformations from polyp to benign adenoma to carcinoma. The first step in colon carcinogenesis involves loss of APC (*A*denomatous *p*olyposis *c*oli) gene leading to the development of polyp. APC is a tumour suppressor gene; it encodes a protein that prevents Wnt signal transduction pathway that inhibits the expression of *c-myc* gene. The second mutation occurs in *ras* gene that leads to even more uncontrolled division.

Further mutation of *p53* gene results in the gradual loss of normal regulation and consequent formation of malignant carcinoma.

Esophagus is lined by squamous epithelium that in response to diverse etiological factors like HCL and duodenal juice (mainly bile) and undergo a series of intermediate cellular changes exhibiting erosion and subsequent replacement of squamous epithelium by columnar epithelium. This condition is called *Barrett's esophagus*. It is a metaplastic condition that slowly acquires through low-grade dysplasia (L.G.D.), high-grade dysplasia (H.G.D.) and ultimately into adenocarcinoma. Recent data in this field led the research workers to conclude that gastric and duodenal reflux collectively lead to severe mucosal damages which result in such esophageal complications.

14.4. GENETIC BASIS OF CANCER

Hereditary cancers constitute about 10% of the total human cancers. Inheritance of the cancer causing genes shows that inheritance of the mutant allele in the germ cell producing cell alone can not form cancer. Moreover somatic mutations and mitotic recombinations also play crucial role in the development of cancer. Specific genetic alterations have been shown to correlate with well-defined histopathological stages of tumor development and progression. Although the significance of mutations to the etiological mechanisms of tumor develop ment has been debated, a causal role for such genetic lesions is now commonly accepted for most human cancers. Thus, genetic lesions represent an integral part of the processes of neoplastic transformation, tumorigenesis, and tumor progression, and as such represent potentially valuable markers for cancer detection and staging.

Normally proto-oncogenes (e.g. *Cdk, ras*) and tumor suppressor genes (e.g. *APC*) encode proteins that regulate cell cycle progression and suppress oncogenesis. Mutations in these two classes of genes play crucial role in the induction of cancer. In fact, conversion of proto-oncogenes to oncogenes and mutated tumor suppressor genes collectively participate in the generation of tumor. Usually, oncogenes are dominant, as they contain gain-of-function mutations, while mutated tumor supperssors are recessive, as they contain loss-of-function mutations. Mutation in one copy of a particular proto-oncogene is enough to convert that gene into a true oncogene, while loss of function mutation needs to happen in both copies of a tumor suppressor gene to render that gene completely non-functional. Loss of function mutation in one copy of a tumor suppressor gene can render the other copy non-functional. This phenomenon is called the *dominant negative effect* and is noted in many p53 mutations.

Currently, about 40 genes are known to be protooncogenes. In 16 of them direct correlation with tumor proliferation was shown, such as *ERBB2 (HER2)*: breast cancer,(by gene amplification); *KRAS*: oesophageal tumors, colon, pancreas cancer (by point mutation); β-Catenin: Pancreatic cancer; Cyclin E: liver tumors; *BRAF*: melanomas; *BCR-ABL*: chronic myeloid leukemia.

Oncogenes were originally identified in viruses causing cancers. These are called *transformation viruses* (retrovirus) as they change the growth pattern of cells from normal to tumor-like. The viral oncogenes are designated with an "v-" prefix, such as *v-src* gene. These are originated as copies of cellular protooncogenes, designated with an "c-" prefix. So *v-myc* is the viral homologue of *C-MYC*.

14.5. CONVERSION OF PROTO-ONCOGENE INTO ONCOGENE THROUGH GAIN-OF-FUNCTION MUTATION

A ***proto-oncogene*** is a normal gene that becomes an oncogene, either after mutation or after increased expression. They encode proteins that help to regulate cell growth and differentiation. They are involved in signal transduction and execution of mitogenic signals. Upon activation, a proto-oncogene becomes a tumor inducing agent, an oncogene by a relatively small modification of its original function. There are four basic activation types:

i) A *point mutation* within a proto-oncogene can cause a change in the protein structure, causing an increase in enzyme activity and a loss of regulation. Point mutation of the 12th codon of *H Ras* gene CGC which normally codes of Glycine amino acid is mutated to CTC that codes for Valine leading to bladder cancer.

ii) A *chromosomal translocation* that fuses two genes together may cause the production of a hybrid or chimeric protein. This type of aberration in a dividing stem cell in the bone marrow leads to adult leukemia. A segment from chromosome 9 carrying *c-abl* gene translocates into 22nd chromosome and fuses with *bcr* (break point cluster region) gene. The resulting hybrid *abl-bcr* produces a fusion protein ABL-BCR which is oncogenic. This protein causes *acute lymphoblastic leukemia* (ALL). Sometimes gene translocation do not produce new genes but places a gene under the regulator that controls another gene. E.g. in patients with *Burkitt lymphoma, c-myc* is moved from chromosome 8 to 14 where it lies next to the regulator that controls the production of IgG production. Thus along with IgG production MYC protein is also produced continuously. A chromosomal translocation that may bring a growth regulatory gene under the control of a different promoter that causes inappropriate expression of the gene.

iii) *Amplification of a DNA segment* (spanning about hundreds of including a proto-oncogene may lead to the formation of numerous (about 100) copies of the existing segment causing over production of the encoded protein. The duplicated DNA may be tandemly present or it may exist as small independent mini chromosome separate from the normal chromosome. This increased gene expression leads to tumour growth. Amplified oncogenes include *c-myc, N-myc, c-abl, c-erbB, c-K-ras, mdm-2* etc.

An ***oncogene*** is a modified gene, or a set of nucleotides that codes for a protein, that increases the malignancy of a tumor cell. Some oncogenes, usually involved in early stages of cancer development increase the chance that a normal cell develops into a tumor cell. Src (pronounced *SARK*) was discovered as the first oncogene by G. Steve Martin (1970) in a chicken retrovirus. Bishop and Varmus (1976) demonstrated that oncogenes were defective protooncogenes.

14.6. ONCOPROTEINS

Growth factors as oncoproteins: Growth factors, or mitogens, are usually secreted by a few specialized cells to induce cell proliferation. If a cell that usually does *not* produce growth factors, suddenly starts to do so (due to the activation of an oncogene), it will thereby induce its own uncontrolled proliferation (*autocrine loop*), as well as proliferation of neighboring cells.

Additionally abnormal growth of endocrine glands often causes ectopic production of growth hormones that have secondary effects on other parts of the body. The *c-Sis* gene (the *v-sis* gene is the oncogene in simian sarcoma virus) encodes the PDGF-B chain. The *v-sis* gene was the first oncogene to be identified as having homology to a known cellular gene. The *int-2* gene (named for the fact that it is a common site of ***int***egration of mouse mammary tumor virus) encodes an FGF-related growth factor. The *KGF* (also called *Hst*) gene also encodes an FGF-related growth factor and was identified in gastric carcinoma and Kaposi's sarcoma cells.

Protein kinases and related proteins as oncoproteins: There are six known classes of protein kinases and related proteins that can become an oncoprotein:

Receptor Tyrosine Kinases: The *c-Fms ("fims")* gene encodes the colony stimulating factor-1 (CSF-1) receptor and was first identified as a retroviral oncogene. The *Flg ("flag")* gene (named because it has homology to the Fms gene, hence *fms-like gene*) encodes a form of the FGF receptor. The *Neu ("new")* gene was identified as an EGF receptor-related gene in an ethylnitrosourea-induced *neu*roblastoma. The conversion of proto-oncogenic to oncogenic Neu requires only a single amino acid change in the transmembrane domain. The *Met* gene encodes the hepatocyte growth factor (HGF)/scatter factor (SF) receptor. The *c-Kit* gene encodes the mast cell growth factor receptor.

The *Trk ("track")* oncogene encode the NGF receptor-like proteins. The first *Trk* gene was found in pancreatic cancer; subsequently two additional Trk-related genes were identified in colon cancer. These three are now identfied as *TrkA, TrkB* and *TrkC*. Normal Trk receptor shows an extra cellular, a transmembrane and a cytosolic domain.

A translocation between Trk gene encoding the extracellular domain with the non muscle

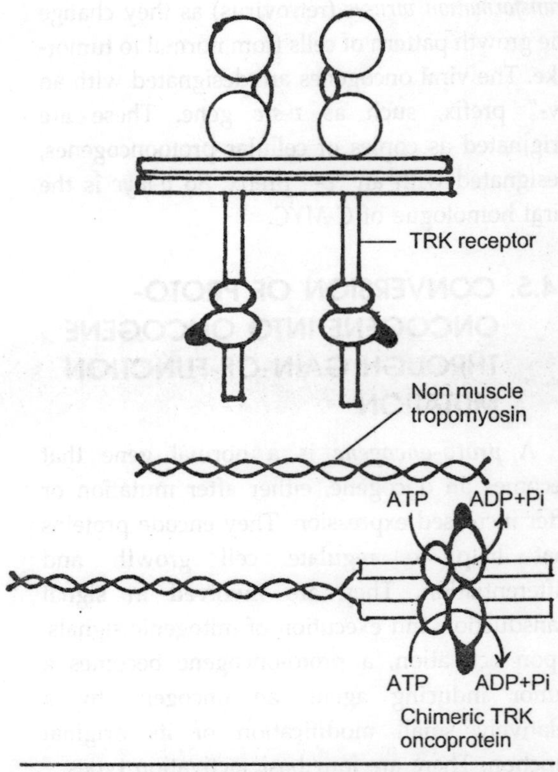

Fig. 14.5 : Translocation between Trk and tropomyosin gene lead to the formation of Trk oncoprotein.

tropomyosin gene encoding N-terminal amino acids results in the formation of a constitutively active cytosolic chimeric Trk oncoprotein Ligand binding to the external domains of the receptor tyrosine kinases (RTK) leads to their dimerization and activation initiating an intracellular signaling pathway that ultimately promotes proliferation. A point mutation changes normal RTK which is constitutively active in the absence of a ligand. E.g. point mutation in Her2 receptor converts it into Neu oncoprotein which is an initiator of some cancers in mouse. Deletion of the extracellular domain of the normal EGF receptor converts it to the constitively active dimeric ErbB oncoprotein.

Membrane Associated Non-Receptor Tyrosine Kinases: The *v-src* gene was the first identified oncogene. The *Lck* gene was isolated from a T cell tumor line (LYSTRA cell kinase) and has been shown to be associated with the CD4 and CD8 antigens of T cells.

Serine/Threonine Kinases: Mammalian Src kinases show SH2 and SH3 domains. Binding of phosphotyrosine 527 to the SH2 domain induces a conformational change to SH3 and kinase domain rendering the kinase catlytically inactive. The kinase inactivity is activated by removing the phosphate on tyrosine 527. At least 8 mammalian proto-oncogenes encode a family of non receptor tyrosine kinase in which tyrosine 527 is missing or altered and are thus constitutively active causing abnormal proliferation of many cells.

G-Protein Coupled Receptors: The *Mas* gene was identified in a mammary carcinoma and has been shown to be the angiotensin receptor - a GPCR.

Membrane Associated G-Proteins: There are three different homologs of the *c-Ras* gene, each of which was identified in a different type of tumor cell. The *Ras* gene is one of the most frequently disrupted genes in colorectal carcinomas.

Nuclear DNA-Binding/Transcription Factors: The *Myc* gene was originally identified in the avian *my*elocytomatosis virus. A disrupted human *c-Myc* gene has been found to be involved in numerous hematopoietic neoplasias. Disruption of *c-Myc* has been shown to be the result of retroviral integration and transduction as well as chromosomal rearrangements. The *Fos* gene was identified in the *f*eline *os*teosarcoma virus. The protein interacts with a second proto-oncogenic protein, Jun to form a transcriptional regulatory complex. The *p53* gene was originally identified as a major nuclear antigen in transformed cells. It is the single most identified mutant protein in human tumors. Mutant forms of the p53 protein interfere with cell growth suppressor effects of wild-type p53 indicating that the p53 gene product is actually a tumor suppressor.

Modification of signal transduction by oncogenes encoded proteins: Many oncogenes encode proteins that aid in transducing signal from an activated receptor to a cellular target.

Ras signaling pathway components: Ras is a key signal transducing component from activated receptors to a cascade of protein kinases. Signal transduction from activated RTK via two adaptor proteins to Ras at first converts it to the active GTP bound form. The active Ras then transmits signal to MAP kinase. It then phosphorylates a number of transcription factors that induces the expression of cell cycle and differentiation specific proteins. A point mutation in *ras*^{D} gene at 12th position substituting any amino acid for glycine can convert normal protein into a constitutively active oncoprotein. This auto activation mutation activate MAP kinase even without ligand binding to induce proliferation leading to tumorigenesis Activated Ras triggers inactive Raf to be converted into active Raf that eventually induces MAP kinase to carry on proliferation. Oncogenes may encode proteins that activate Raf directly so that without involvement of Ras cell proliferation continues. Activation of Ras is also carried out by GRB2 protein. *crk* oncogene produces the protein Crk which con tains one SH2 and SH3 domains as exhibited by GRB2 proteins. These domains play an important role in the signal transduction for proliferation. Over production of Crk protein acts in a similar way that induce proliferation.

Abl protein kinase: Chromosomal translocation fusing a part of *c-abl* gene (which encodes a tyrosine kinase) with a part of *bcr* gene (unknown function) forms the diagonostic *Philadelphia chromosome*. It produces the chimeric oncoprotein Abl-Bcr. The chimeric oncoprotein Abl-Bcr can phosphorelate and activate many unusual intracellular signal transduction pathways (e.g. JAK2 kinase and STAT5 are activated which normally are activated by binding of growth factor to the receptors). If this translocation occurs in the haematopoetic cell in

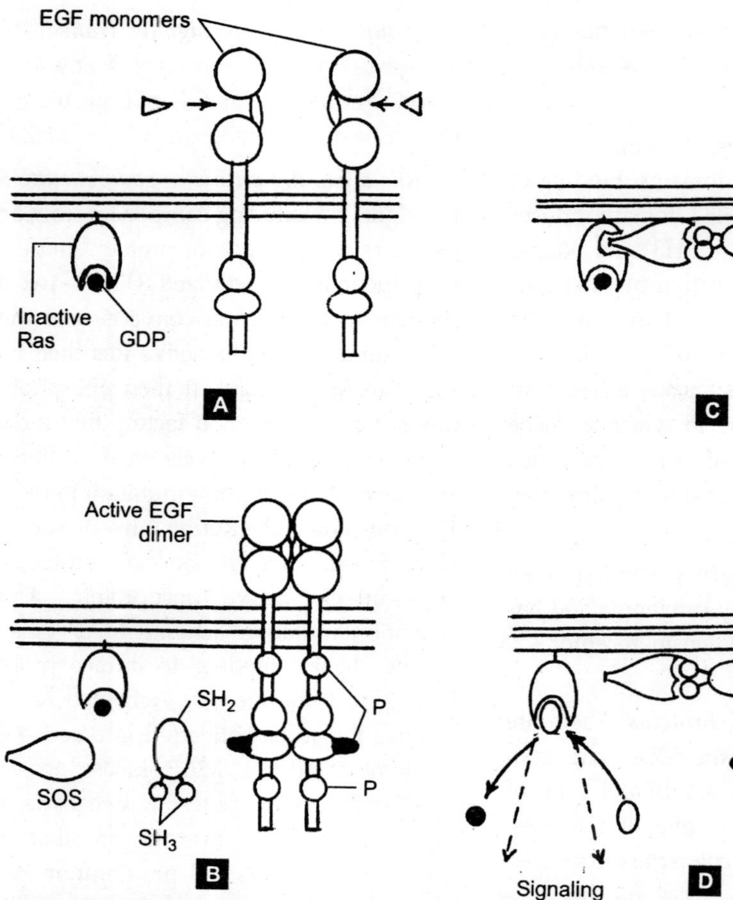

Fig. 14.6 : Activation of Ras following binding of ligand to the receptor and adaptor protein GRB2 to the inactive Ras.

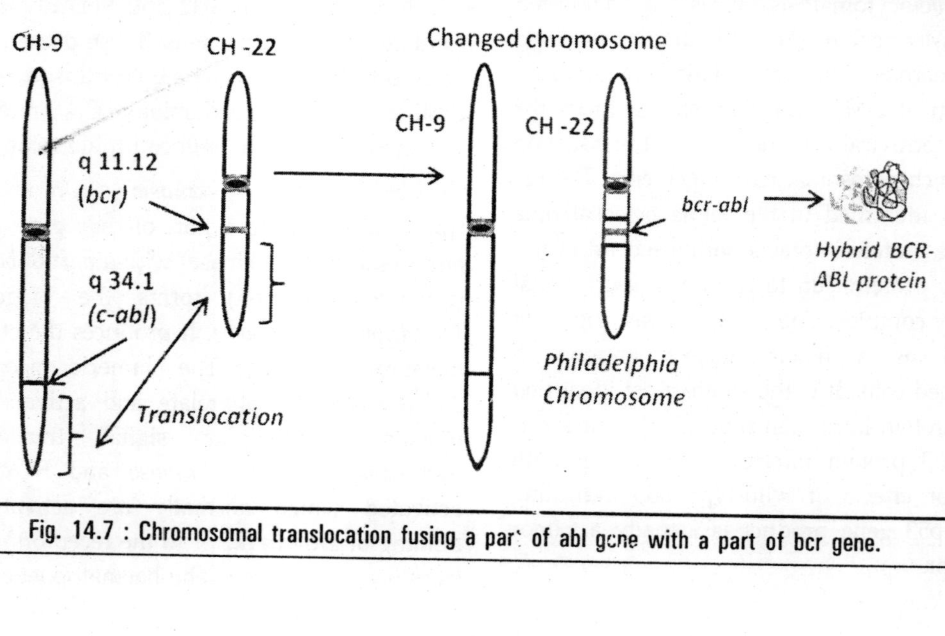

Fig. 14.7 : Chromosomal translocation fusing a part of abl gene with a part of bcr gene.

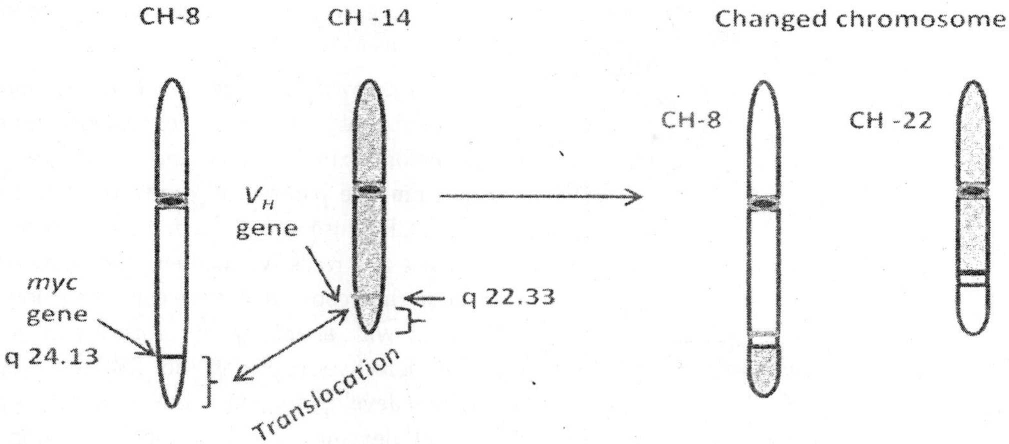

Fig. 14.8 : Translocation of c-Myc and IgH genes causing Burkitt's lymphoma.

the bone marrow it causes human chronic myelogenous leukemia (CML). A second mutation causes acute leukemia. As a possible treatment for CML an inhibitor of Abl kinase-STI-571 (Gleevec) being highly lethal to CML cells has been identified. It is also effective against several other tyrosine kinases.

14.7. ONCOGENES INDUCING MALIGNANCY

All proto-oncogenes and oncogenes cause changes in gene expression by encoding transcription factors. *c-jun* and *c-fos* protooncogenes encode oncoproteins that associate to form a heterodimeric transcription factor – AP1 that may induce transcription of a number of genes though they can show such transactivation individually. In normal cells *c-Fos* and *c-Myc* transcribed mRNAs are unstable but some changes can turn these normal genes into oncogenes. During the maturation of antibody producing lymphocytes, *c-myc* gene can be translocated from chromosome-8 to 14 to a site near the heavy chain antibody gene (V_H) which is normally active in these cells. This translocated *myc* gene become over expressed leading to the high production of Myc protein causing the tumour *Burkitt's lymphoma.*

Overproduction of oncoproteins can also be induced by growth factors. PDGF treatment of 3T3 quicent cells induces 50 fold increase in the production of c-Fos and c-Myc oncoproteins which promote progression through G1 to S phase. In many tumours these transcription factors are over expressed at high and unregulated levels.

14.8. TUMOR SUPPRESSOR GENE

A tumor suppressor gene reduces the probability of transformation of a cell into a tumor cell. Tumor suppressor genes or the proteins they encode, either have a dampenning or repressive effect on the regulation of the cell cycle or promote apoptosis and sometimes do both. The first tumor suppressor protein discovered was the pRb protein in human retinoblastoma; however, recent evidence has also implicated pRb as a tumor survival factor. Another important tumor suppressor is p53 tumor suppressor protein produced by the TP53 gene. A mutation or deletion of such a gene will increase the probability of the formation of a tumor.

Unlike oncogenes, they generally follow the *'two-hit hypothesis,'* which implies that both alleles for a particular gene must be affected before an

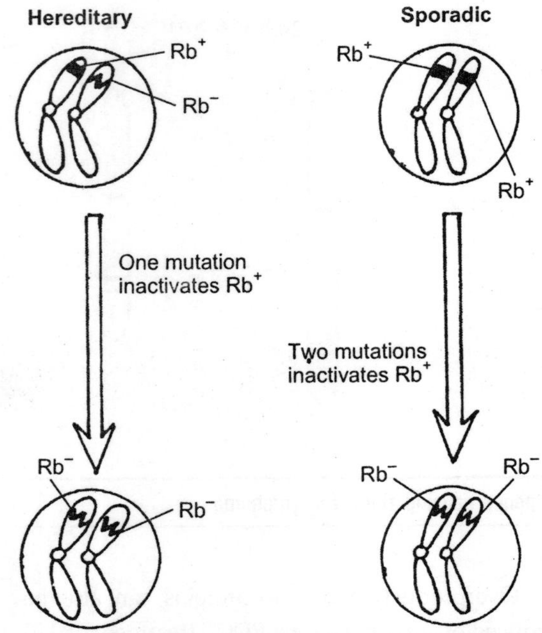

Fig. 14.9 : Inheritance of hereditary and sporadic retinoblastoma.

effect is manifested. This is due to the fact that if only one allele for the gene is damaged, the second can still produce the correct protein. However there are cases where mutations in only one allele (e.g. p53) will cause an effect. The functions of tumor suppressor proteins fall into following categories:

i) Induces receptors or signal transducers (e.g. TGF-β) that inhibit cell proliferation.

ii) Repression of genes those are essential for *continuing* the cell cycle (e.g. *Rb* protein is essential for continuing cell cycle). In response to any type of damage if these genes are not expressed, the cell cycle will be inhibited.

iii) Couples the cell cycle to DNA damage (p53 is such a DNA damage censor). As long as damaged DNA is not repaired the cell stops dividing.

iv) Induces DNA damage repairing enzymes but if damage is irreparable, the cell follows apoptosis.

v) Induces cell adhesion proteins preventing metastasis.

Heterozygocity and increased risk of cancer: The mode of inheritance of mutant tumor suppressors exhibits inheritance of a defective copy from one parent and a normal copy from the other. Because mutations in tumor suppressor genes act in a recessive manner, the suppression of the normal copy creates the cancer phenotype. Children with *hereditary retinoblastoma* inherit a single defective copy of the RB gene. Such children develop retinal tumours in both eyes. For the development of tumour, deletion or mutation of the other allele is essential so that the cell produces no functional Rb protein. As their germ cells contain one normal and one mutant Rb allele, half of their children will get normal and half will get mutant allele. Children who will inherit normal allele will be normal but those who will get mutant allele will have same probability of developing tumours. In other type of retinoblastoma - *sporadic retinoblastoma* children inherit two Rb alleles, both of which exhibit a loss of function mutation. They develop retinal tumors late in life, but in one eye only. Individuals who are heterozygous for *p53* mutations are often victims of Li-Fraumeni syndrome. Similarly, heterozygocity in the *APC* gene is linked to the development of adenopolyposis colon cancer, with thousands of polyps in colon, while mutations in *BRCA1* and *BRCA2* lead to early onset of breast cancer.

Loss of heterozygocity: Inheritance of tumor suppressor gene presents a clear picture of the role of heterozygocity in the development of cancer. In spite of the presence of the normal allele, accidentally it can be lost or inactivated in the somatic cell which is referred to as *loss of heterozygocity (LOH)*. The most common mechanism of LOH involves *missegregation of the chromosome* bearing the gene during cell division which is commonly referred as non-disjunction. Another possibility of LOH is the *mitotic recombination* between a chromatid bearing the

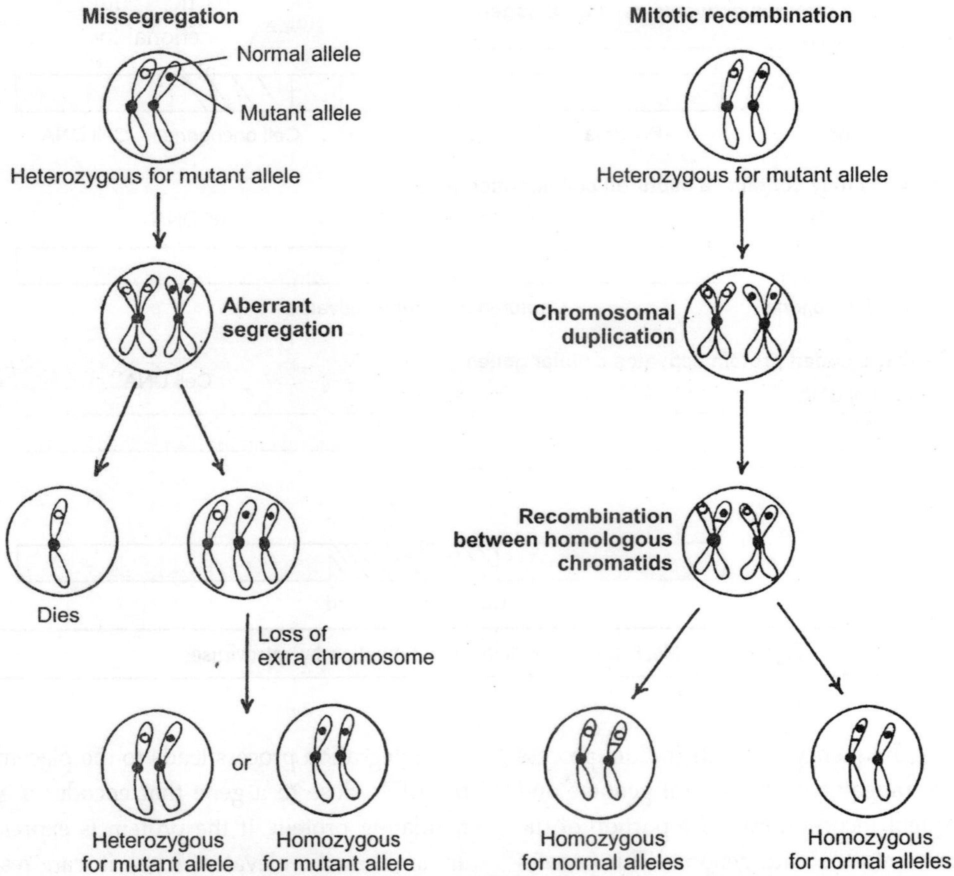

Fig. 14.10 : Processes of the loss of heterozygocity/

normal and mutant allele. The third possibility is the deletion or mutation of the normal allele due to accidental breakage of the chromosome.

14.9. VIRUS AND CANCER

Tumor cells also can arise by non-genetic means through the actions of specific tumor viruses. Tumor viruses are of two distinct types. There are viruses with DNA genomes (e.g. papilloma and adenoviruses) and those with RNA genomes (termed retroviruses). RNA tumor viruses are common in chickens, mice and cats but rare in humans. The only currently known human retroviruses are the human T-cell leukemia viruses (HTLVs) and the related retrovirus, human immunodeficiency virus (HIV).

Cellular transformation by Retroviruses: It can induce the transformed state within the cells by two mechanisms. Both of these mechanisms are related to the life cycle of these viruses. When a retrovirus infects a cell its RNA genome is converted into DNA by the viral encoded RNA-dependent DNA polymerase (reverse transcriptase). The DNA then integrates into the genome of the host cell. Contained within the sequences at the ends of the retroviral genome are powerful transcriptional promoter sequences termed long terminal repeats (LTRs) which promote the transcription of the viral DNA leading to the production of new virus particles.

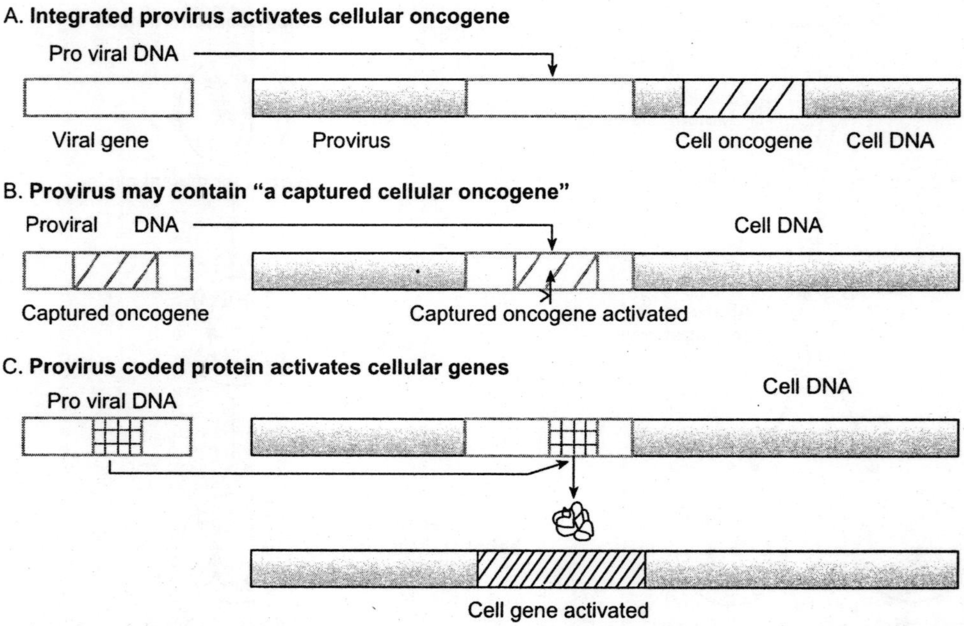

Fig. 14.11 : Mechanism of cellular transformation by retroviruses.

At some frequency the integration process leads to rearrangement of the viral genome and the consequent incorporation of a portion of the host genome into the viral genome. This process is termed *transduction*.

Occasionally this leads to the virus acquiring a gene from the host that is normally involved in cellular growth control. Because of the alteration of the host gene during the transduction process as well as the gene being transcribed at a higher rate due to its association with the retroviral LTRs the transduced gene confers a growth advantage to the infected cell. The end result of this process is unrestricted cellular proliferation leading to tumorigenesis. The transduced genes are termed *viral oncogenes*. Numerous oncogenes have been discovered in the genomes of transforming retroviruses.

The second mechanism by which retroviruses can transform cells relates to the powerful transcription promoting effect of the LTRs. When a retrovirus genome integrates into a host genome it does so randomly. At some frequency this integration process leads to the placement of the LTRs close to a gene that encodes a growth regulating protein. If the protein is expressed at an abnormally elevated level it can result in cellular transformation. This is termed *retroviral integration induced transformation.* It has recently been shown that HIV induces certain forms of cancers in infected individuals by this integration induced transformation process.

Cellular transformation by DNA tumor viruses: A few DNA viruses are oncogenic. They integrate their DNA which contains one or more oncogenes, into the host cell genome. Viral oncogenes permanently transform infected cells. The oncoproteins act in various ways to stimulate growth and proliferations. In most cases, it has been shown to be the result of protein-protein interaction. Proteins encoded by the DNA tumor viruses, termed *tumor antigens or T antigens,* can interact with cellular proteins. This interaction effectively sequesters the cellular proteins away from their normal functional locations within the cell. The predominant types of proteins that are

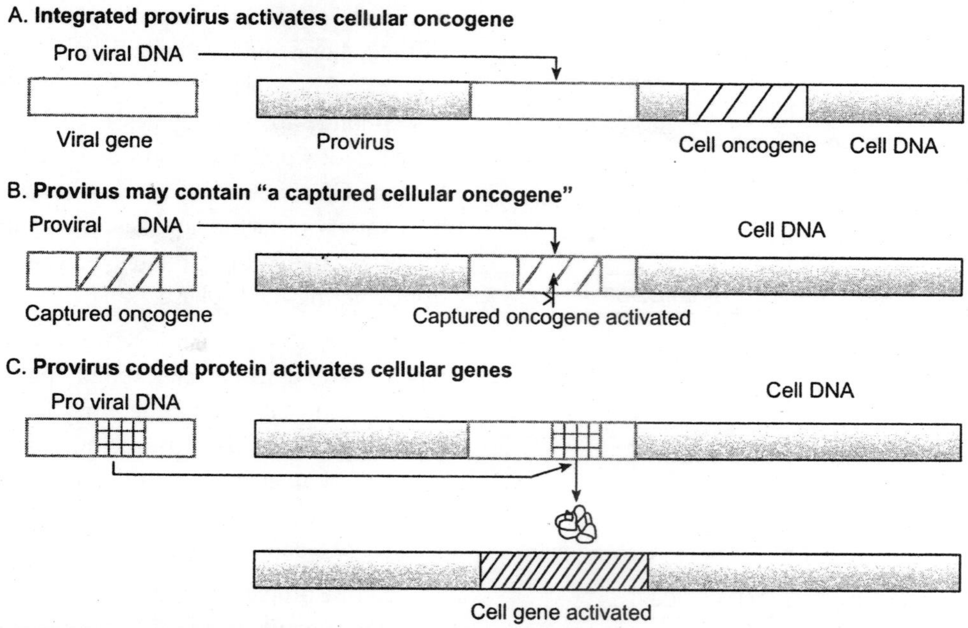

Fig 14.12 : Mechanism of cellular transformation by DNA viruses.

sequestered by viral T antigens have been shown to be of tumor suppressor type. It is the loss of their normal suppressor functions that results in cellular transformation.

Mutations causing loss of cell cycle control: Mutations that disrupt the balance between growth promoting and inhibiting pathways can lead to cancer. TGFβ inhibits proliferation in many cell types. Binding of TGFβ to its receptor induces activation of cytosolic Smad transcription factor. Smad translocates into the nucleus and trans activates the genes p15 and PAI-1 which encode inhibitors of G1-S and ECM degradators respectively. Loss of function mutation in either TGFβ receptor or in Smad cause decreased production of p15 and PAI-1 that allows cell proliferation and increased ECM degradation causing metastasis.

14.10. ABERRANT SIGNALING LEADING TO CANCER

Cell division, cell survival, and cell death are so well balanced that it promotes homeostasis with neither unregulated growth nor inappropriate cell loss. Oncogenesis is the result of the deregulation of this homeostasis due to ever increasing number of interconnecting pathways that result due to activation of oncogenes and inactivation of tumour suppressor genes.

Role of Retinoblastoma (Rb) protein: Rb is a tumor suppressor protein and is one of the mammalian regulators to cross the restriction point. Rb pathway perturbation is a common and significant event in tumorigenesis (Fig. 14.13). Loss of Rb is frequent in retinoblastoma, osteosarcoma, small cell lung carcinoma and bladder carcinoma. Rb remains bound to E2F, a transcription factor and prior to 'S' phase CDKs phosphorylate Rb to free it from its bound state transactivating Di hydrofolate reductase (DHFR), thymidine kinase (TK), Thymidine synthase (TS), DNA polymerase (POL), Cdc-2/Cdk1, Cyclin E, Cyclin A and possibly E2F-1 itself. RB is a target of several pathways. The growth inhibitory factor, TGFβ and Ckis (Cdk-Cyclin kinase

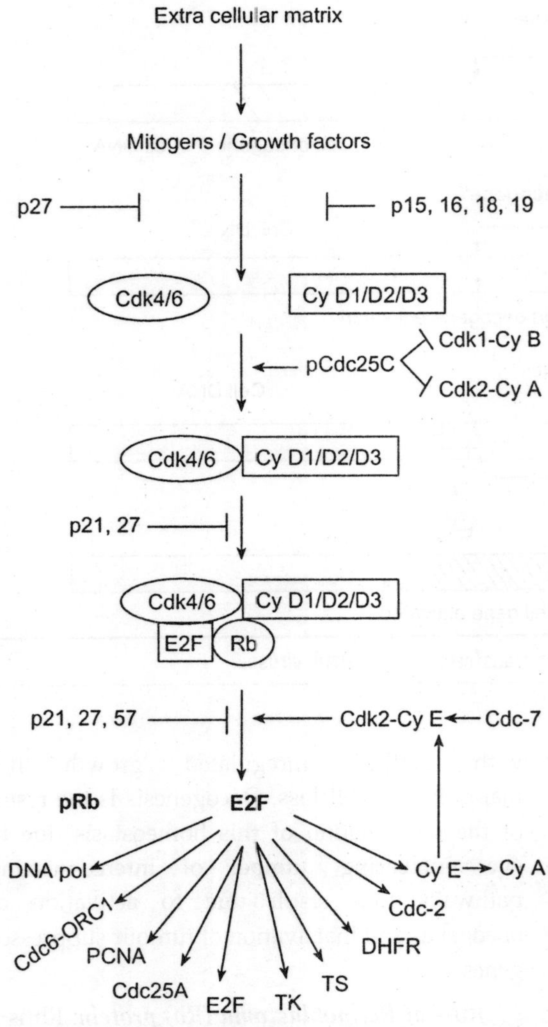

Fig. 14.13 : Misregulation of Rb leading to cancer.

inhibitors) prevent phosphorylation of Rb making it impossible to release E2F. The Ckis fall into two groups - the INK4 and KIP families. INK4 family contains p16, p15, p18 and p19. They are specific for Cdk4/6 only and under stress conditions inhibit the induction of Rb-E2F by Cdk4/6-Cyclin D. KIP families have 3 members - p21, p27 and p57. p21 is a universal Cdk inhibitor, binding to all members of Cdk 2, 4, 6. It inhibits the progression through G_1/S transitional phases. p27 inhibits the kinase activities of a variety of Cyclin/Cdk complexes including Cdk1, 2, 4 or 6 and Cyclin A, E or D. Mutation or loss of p27 is associated with an increased susceptibility to tumorigenesis. p57 inhibits the activities of Cdk2 and 4 and its over expression leads to G1 arrest. Thus Rb activity is regulated by a variety of factors and therefore mutation in RB or its regulating proteins prevents the cell from entering the S phase.

Role of p53 protein: Dysfunction of DNA repair and damage checkpoint pathways is largely responsible for generating instability of cancer cells. Recently mushrooming data expose a nexus of overlapping processes called *DNA damage response pathways (DRPs)* those try to repair DNA damage to prevent cancer.

p53 (the protein is of 53 kDa, hence the name) is a potent regulator of G1 checkpoint. It senses DNA damage, if it is minor, it halts the cell cycle but if it is major then it triggers apoptosis by up regulating Bax or down regulating Bcl-2. Cancer related mutations accumulate in the DNA binding domain of p53, it also binds with E2F inhibiting the expression of the proto-oncogenes like *c-myc*, *c-fos* etc. In response to genotoxic damage, cells become arrested in G1 and always try to repair the damaged DNA before passing to S phase. Cell cycle arrest is mediated by the degradation of Cdc25A whose mRNA is over expressed in human tumors. Human papilloma virus (HPV) synthesizes E5, E6 and E7 proteins. E6 inhibits p53 and E7 inhibits Rb. Thus they act as oncoproteins. Moreover collectively they induce transformation in the cell regulatory proteins. E5 causes sustained activation of PDGF receptor that enhances proliferation of the transformed cells (Fig. 14.14).

Mdm2 is an important negative regulator of p53, it enters into the nucleus in response to a mutagenic stress where it disintegrates p53. In unstressed cell, p53 levels are maintained due to low levels of Mdm2 while in response to DRPs Mdm2 levels go up and collectively inactivate p53. Thus Mdm2 gene is amplified in many

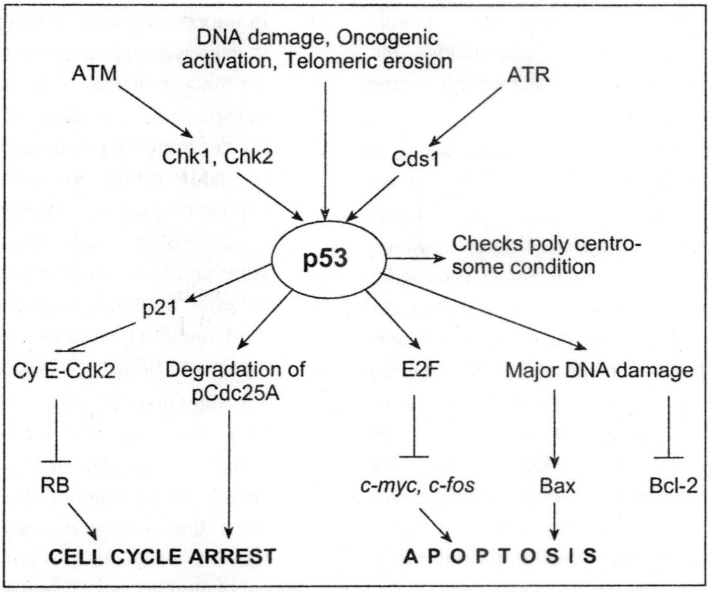

Fig. 14.14 : Role of p53 in the regulation of cell cycle arrest and apoptosis.

sarcomas and other tumours. The relationship of p53 and Mdm2 is not always unidirectional. Phosphorylation of p53 by ATM kinase displaces bound Mdm2 from p53 and Mdm2 is itself transcriptionally activated by p53. Thus Mdm2 functions in an autoregulatory feedback loop with p53.

About 50% human cancers are attributed to the mutation in p53. The mutation product inactivates p53. Duplication of centrosome normally begins at the G1-S transition but in the absence of wild p53, multiple centrosomes appear which eventually lead to aberrant chromosomal segregation during mitosis. In some cells, overexpression of p53 in response to severe DNA damage or chemotherapeutic drugs, apoptosis is induced. Hence cell cycle arrest and apoptosis appear to be alternative p53 induced outcomes. Apoptosis is arguably the most potent natural defense against cancer because it eliminates the premalignant cells that enter S phase inappropriately after genetic sabotage of restriction point controls. If p53 induced G1 arrest does not operate properly damaged DNA replicate and are passed on to daughter cells. p53 may also induce p21 and other two proteins which inhibit cyclinB-Cdk1 complex required to cross G2-M checkpoint. Absecne of p53 thus may arrest the cell in G2.

SUMMARY

1. From birth to death the continuity of life is maintained by cell division that is regulated by an array of factors. Any misregulation in this system leads to deregulated cell proliferation forming tumour or malignant cell. In fact cancer is the outcome of mutations that regulate flawless cell division. Severity of symptoms depends on the site and characters of the malignancy. As proteins are the manifestation of base sequences on the DNA therefore any mutation leading to malignancy can be inherited. Thus sporadic and heritable cancers are known today. However a single mutation rarely leads to cancer and a series of mutations actually causes cancer.
2. The chemicals that cause mutations are called mutagens and mutagens that cause cancer are called carcinogens. Many mutagens are carcinogens but some carcinogens are not

mutagens. Many environmental factors, chromosomal translocation, aneuploidy etc. directly cause cancer. Certain carcinogenic viruses are also known.

3. In order to start uncontrolled divisions, two set of regulatory genes need to be mutated – proto-oncogenes and tumour suppressor genes. Proto-oncogenes promote cell growth through a variety of ways. Mutations in proto-oncogenes can modify them into oncogenes showing uncontrolled cell division. Tumor suppressor genes code for anti-proliferation signals and proteins that suppress mitosis and cell growth. Generally tumor suppressors are transcription factors that are activated by cellular stress or DNA damage. The functions of such genes are to arrest the progression of cell cycle in order to carry out DNA repair, preventing mutations from being passed on to daughter cells.

4. Tumours are localized accumulation of mutated cells. Hyperplasia refers to an excessive rate of cell division but with a normal orderly arrangement of cells within the tissue. Dysplasia is an abnormal type of excessive cell proliferation characterized by loss of normal tissue arrangement and cell structure. It shows a large number of dividing cells, variation in nuclear and nucleolar size and shape, variation in cell size and shape, loss of specialized cell features, loss of normal tissue organization and a poorly defined tumor boundary. The cells composing benign tumours closely resemble and function like normal cells. The cell adhesion molecules that hold tissues together keep the cells localized to the tissue where they originate. b-catenin plays significant role on cancer invasion and metastasis.

5. Whenever a cell loses control over proliferation, it is usually termed as cancer cell. It presents large nucleus and nucleolus, haphazardly arranged chromatin networks; increased numbers of ERs, reduced number of Golgi bodies, swollen mitochondria with less number of cristae. Amount of lectins in the plasma membrane increases to form lumps. Glucose transport through the membrane increases enhancing the rate of glycolysis and synthesis of lactic acid keeping pace with the rapid rate of growth. Most of the cancer cells show aneuploidy and need very minute amount of cAMP for their division.

6. In the development of an epithelial cancer several mutations are required for the induction of cancer. An initial mutation in a single cell disrupts signal transduction and cell cycle. The mutated cell divides more frequently and generates dysplasia. A second mutation disrupts the cytoskeleton, gene expression pattern, modifying metabolism, growth characteristics, cell shape, and intercellular adhesion. Cell membrane exhibits altered receptor patterns, transport properties and permeability. A third mutation perturbs cell cycle, impairs DNA repair and broken chromosomes.

7. The tumour cells secrete fibroblast growth factor (FGF) and vascular endothelial growth factor (VEGF) to support neovas- cularization for the growth of the tumour. Some cancer cells creep along the associated blood vessels while others squeeze between the lining cells to enter the circulation to get dispersed to distant locations. The first phase of this dispersal is called invasion and second stage is called metastasis.

8. The prime requisite of malignant growth is constant supply of nutrients which is aided by neovascularization / angiogenesis. Formation of new blood capillaries from the old ones involves production of angiogenic growth factor that diffuses into the surrounding tissues and bind to specific receptors located on the endothelial cells that produce VEGF leading to their enhanced proliferation of the endothelial cell that eventually form vessels. VEGF gene induces neovascularization in and around tumor that augment metastasis

9. Hereditary cancers constitute about 10% of the total human cancers. Inheritance of the cancer causing genes shows that proto-oncogenes (e.g. ras) and tumor suppressor genes (e.g. APC) encode proteins that regulate cell cycle progression and suppress oncogenesis. Mutations in these two classes of genes play crucial role in the induction of cancer. In fact, conversion of proto-oncogenes to oncogenes and mutated tumor suppressor genes collectively participate in the generation of tumor. Loss of function mutation in one copy of a tumor suppressor gene can render the other copy non-functional.

10. A protooncogene is a normal gene that becomes an oncogene, either after mutation or increased

expression. They encode for proteins that help to regulate cell growth and differentiation. An oncogene is a modified gene, or a set of nucleotides that codes for a protein, that increases the malignancy of a tumor cell. Some oncogenes, usually involved in early stages of cancer develop ment; increase the chance that a normal cell develops into a tumor cell, possibly resulting in cancer. Overproduction of oncoproteins that induce cell proliferation can also be induced by growth factors. There are six known classes of protein kinases and related proteins that can become an oncoprotein.

11. Many oncogenes encode proteins that aid in transducing signal from an activated receptor to a cellular target. It then phosphorylates a number of transcription factors that induces the expression of cell cycle and differentiation specific proteins. Oncogenes may encode proteins that activate Raf directly so that without involvement of Ras, cell proliferation continues. Normal c-Abl protein causes branching of filamentous actin and extension of cell processes thus controlling cytoskeleton and cell shape. Chromosomal translocation fusing a part of c-abl gene with a part of bcr gene forms the Philadelphia chromosome producing the chimeric oncoprotein Abl-Bcr that activates many unusual intracellular signal transduction pathways and causes human chronic myelogenous leukemia.

12. c-jun and c-fos proto-oncogenes encode oncoproteins that associate to form the transcription factor - AP1 that may induce transcription of a number of genes. Some changes can turn these normal genes into oncogenes. Translocation of c-myc gene from chromosome-8 to 14 to a site near the heavy chain antibody gene in the lymphocytes leads to high production of Myc protein causing the tumour Burkitt's lymphoma.

13. A tumor suppressor gene reduces the probability of transformation of a cell into a tumor cell. They induces receptors or signal transducers that inhibit cell proliferation, repress genes which are essential for continuing the cell cycle even in response to any genotoxic damage, act as DNA damage censor, induce DNA damage repairing or apoptotic enzymes, induce cell adhesion proteins preventing metastasis

14. Heterozygocity often increases the risk of cancer. Because mutations in tumor suppressor genes act in a recessive manner, the suppression of the normal copy creates the cancer phenotype. In hereditary retinoblastoma children inherit a single defective copy and a normal copy of the RB gene. For the development of tumour, deletion or mutation of the other allele is essential. In sporadic retinoblastoma both Rb alleles exhibit a loss of function mutation. They also develop retinal tumours but late in life in one eye only. Heterozygocity in the APC gene is linked to the development of adenopolyposis colon cancer.

15. Tumour cells also can arise by non-genetic means through the actions of specific tumor viruses. A few DNA viruses are oncogenic. They integrate their DNA that contains one or more oncogenes, into the host cell genome. Viral oncogenes permanently transform infected cells. Proteins encoded by the DNA tumor viruses, termed tumor antigens or T antigens, can interact with cellular proteins. This interaction effectively sequesters the cellular proteins away from their normal functional locations within the cell.

16. Estrogens are regarded as the most crucial carcinogenic hormone due to their well known growth promoting effect by accelerating the rate of cell proliferation.. It exerts its effect through a multistep process. Constant use of oral contraceptives, early menarchy and late menopause, late age of first child delivery, all these increase the exposure of cells to circulating estrogen and thereby enhance the risk of breast cancer. Phytoestrogens could have an anti estrogen effect suppressing the estrogen responsive genes - c-fos, c-jun and the cytokines IL-1a and TNFα.

17. Tumor suppressors, growth signaling pathways and oncoproteins are all involved in DNA damage response pathway. Dysfunction of DNA repair and damage checkpoint pathways is largely responsible for generating instability of cancer cells. Any mutation in p53 can increase the risk of cancer. This compound then recruits the protein like serine/threonine kinase- Akt which through Mdm2 promotes cell survival. β-catenin is a multifunctional protein involved in cell-cell adhesion and b-catenin signaling has been proposed to act as inducer of cell proliferation in

different tumors. Over 90% of the colorectal cancers have muta tions in APC or β-catenin genes. are needed for the activity of other tumour suppressor genes.

18. Rb, a tumor suppressor protein is one of the regulators to cross the restriction point. Rb pathway perturbation is a common and significant event in tumorigenesis. Rb remains bound to E2F, a transcription factor and prior to 'S' phase Rb becomes free from E2F transactivating many genes needed for DNA replication. Under stress conditions Ckis and KIPs inhibit the separation of Rb and E2F arresting cell cycle

19. p53 senses DNA damage and is a potent regulator of G1 checkpoint halting the cell or triggering apoptosis by up regulating Bax or down regulating Bcl-2. It also binds with E2F inhibiting the expression of the proto-oncogenes - c-myc, c-fos etc. Cell cycle arrest is mediated by the degradation of Cdc25A. Mdm2 is a negative regulator of p53; it disintegrates p53 in the nucleus but phosphorylation of p53 by ATM displaces bound Mdm2 from p53 and Mdm2 itself is activated by p53. PTEN inhibits the entrance of Mdm2 into the nucleus, protecting p53. In PTEN mutation, p53 is not protected so that damaged DNA leads to cancer. If p53 induced G1 arrest does not operate properly replicated damaged DNA is passed on to daughter cells. It then induces p21 that inhibit cyclinB-Cdk1 complex required to cross G2-M checkpoint. Loss of p53 thus may arrest G2-M phase.

20. β-catenin is involved in cell-cell adhesion and as an inducer of cell proliferation. The levels of cytoplasmic b-catenin is kept low due to its association with several proteins - APC, GSK -3β and axin/conductin, which direct excess β-catenin to proteasome mediated degradation. Activation of the Wnt pathway inhibits GSK-3β which translocates β-catenin to the nucleus where it associates with the TCF-1/Lef-1 transcription factors and activates a series of genes, some of which are involved in cell proliferation

21. Swi/Snf complex a chromatin-remodeling complex is involved in the transcription of many genes. Its several subunits either possess intrinsic tumor suppressor activity or are needed for the activity of other tumour suppressor genes. It represses the E2F gene, thereby inhibiting progression through the cell cycle. Any mutation can lead to altered regulation of many down stream genetic pathways exerting unregulated growth promoting effects

22. The targeting of a tumour suppressor or oncoprotein is regulated at multiple points involving nuclear transport proteins. It accelerates cell proliferation as well as inhibits apoptosis. The NF-κB, in normal cell forms a complex with the inhibitor IκB that prevents translocation of NF-κB to the nucleus but when IκB is phosphorylated, nuclear import of NF-κB accelerates cell proliferation as well as inhibits apoptosis. In normal cells, growth factors signal through PI3K and AKT but PI3K activation can be inhibited by PTEN. This results nuclearization of the transcription factor FOXO and p27. FOXO transactivates genes that promote cell cycle arrest and apoptosis while p27 inhibits E2F mediated transcription leading to cell cycle arrest. Cancer cells lack PTEN export of FOXO from the nucleus and cytoplasmic localization of p27 can no longer activate cell cycle arrest and proapoptotic genes.

QUESTIONS

Long questions:

1. Describe the diagnostic features of a cancer cell.
2. Describe the steps of the generation of epithelial cancer.
3. What do you mean by tumour angionesis? Describe the process with the help of diagram.
4. How retroviruses affect a host cell to initiate tumorigenesis?

Short questions:

1. What do you mean by cancer?
2. Classify cancers according to its occurrence.
3. Differentiate invasion and metastasis.
4. Name some environmental factors that cause cancers.
5. What is the importance of Philadelphia chromosome in cancer?
6. Differentiate mutagens and carcinogens.
7. How aneuploidy causes cancer?
8. Differentiate proto-oncogenes, oncogenes and tumour suppressor genes.
9. Differentiate benign and malignant tumours.

10. Describe different views explaining metastasis.
11. Differentiate hyperplasia and dysplasia.
12. Name the enzyme that helps in invasion of cancer cells.
13. What do you mean by Barrett's esophagus?
14. What do you mean by 'dominant negative effect'? Explain with example.
15. How a proto-oncogene is activated to form oncogene?
16. Name some proto-oncogenes, oncogenes and tumour suppressor genes.
17. How Ras signaling is related with cancer?
18. How translocation leads to cancerous growth? Give examples.
19. Name some cancers that occur due to heterozygocity.
20. How hereditary retinoblastoma differs from sporadic retinoblastoma?
21. How phytoestrogens prevent carcinogenesis?
22. Name the genes that are transactivated by E2F.
23. Name the factors that prevent dissociation of Rb and E2F.
24. Name the members of INK and KIP family members.
25. Name the proteins that participate in DNA damage response pathway.
26. Name the proteins that participate in the proteasomal degradation of cytosolic β-catenin.
27. How Wnt-β-catenin signaling promotes oncogenesis?
28. What do you mean by retroviral integration induced transformation?

Fill up the blanks:

1. Wart is a common ——.
2. Cancer cells exhibit —— nucleus and nucleolus.
3. Amount of —— in cancer cells increases and helps to form lumps.
4. Tumour cells secrete —— and —— to support neovascularization.
5. —— is an inhibitor of Abl kinase.
6. *p53* mutations are often victims of —— syndrome.
7. —— is a carcinogenic hormone.
8. Before G1-S checkpoint, Rb remains associated with ——.
9. Rb is a tumour —— protein.
10. The most common DNA damage sensor is ——.
11. ATM and ATR propagate signal via —— and —— respectively.
12. β-catenin-TCF/Lef1 association activates —— oncogenes.
13. Swi/Snf is a —— complex.
14. Brg1 encodes —— complex.
15. NF-kb is prevented from entering into the nucleus by ——.
16. The nature of T-antigen resemble tumour —— type.

MCQs:

1. **The cell adhesion protein**
   (a) integrin,
   (b) α-catenin,
   (c) β-catenin is an oncoprotein.
2. **The number of ERs in cancer cell is**
   (a) reduced, (b) increased, (c) distorted.
3. **Mitochondria in a cancer cell appear**
   (a) constricted, (b) fragmented, (c) swollen.
4. **The rate of**
   (a) ATP synthesis,
   (b) glycolysis,
   (c) electron transport increases.
5. **Cancer cell shows disorganized**
   (a) actin, (b) microtubule, (c) intermediate filament.
6. **The most important negative regulator of p53 is**
   (a) p27, (b) ATM, (c) Mdm2.
7. **Entrance of Mdm2 into nucleus is prevented by**
   (a) ATR, (b) PTEN, (c) ATM.
8. **Esophageal cancer can also be referred as**
   (a) lipoma, (b) carcinoma, (c) lymphoma.
9. **Cancer of the fat cells belongs to the group**
   (a) lipoma, (b) sarcoma, (c) lymphoma.
10. **Breast cancer belongs to the group**
    (a) sarcoma, (b) lymphoma, (c) carcinoma.

Answers to MCQs:

1. c, 2. b, 3. c, 4. b, 5. a, 6. c, 7. b. 8. b, 9. a, 10. a.

Chapter 15

Cell Signaling

One of the characteristic features of the living organisms is the power of exitability in response to the sensa tion received from the environment or neighbouring cells. In multicellular organisms cells must depend upon the neighbouring cell for performing the normal vital activities. Unicellular organisms also respond to the stimuli that emanate from the environment for food capturing, escaping from enemies and for sexual purpo ses. Motile cells respond to chemical attractants by migrating towards the source of the chemical. In lower organisms (yeast, slime moulds, protozoans etc.), the secretion of pheromones attract the sexual partners from a distant place. The hormone, adrenaline stimulates cellular energy metabolism and growth factors stimulates cells to duplicate their genome. Cells tune their activity to adapt to changing environmental condi tions. Intercellular signaling is most remarkably noted during the developmental periods since each step of the ontogeny is dependent upon a finely tuned interaction of factors that originate from neighbouring cell.

The complexity of signaling system is still to be clearly deciphered. Each cell receives hundreds of different signals that may be soluble, bound to extra cellular matrix, or to the plasma membrane of the neighbouring cell but how a cell always responds to the specific combination of signals for survival, growth and differentiation is still a matter of enigma to the cell biologists.

15.1. GENERAL FEATURES OF CELL SIGNALING

Cell gets signal from the environment or it may originate from the cytosol. In adults the effects of signal transduction is transient and it fades away as the source of signal ceases to operate but during embryonic development the effect of signal transduction is longlasting and it initiates a cascade of developmental events and continues upto the end of developmental programme. However signaling may be extracellular, intracellular or intercellular.

15.2. TYPES OF SIGNALING

Extracellular signaling: This type of signal transduction involves the binding of "extracellular" signaling molecules to receptors

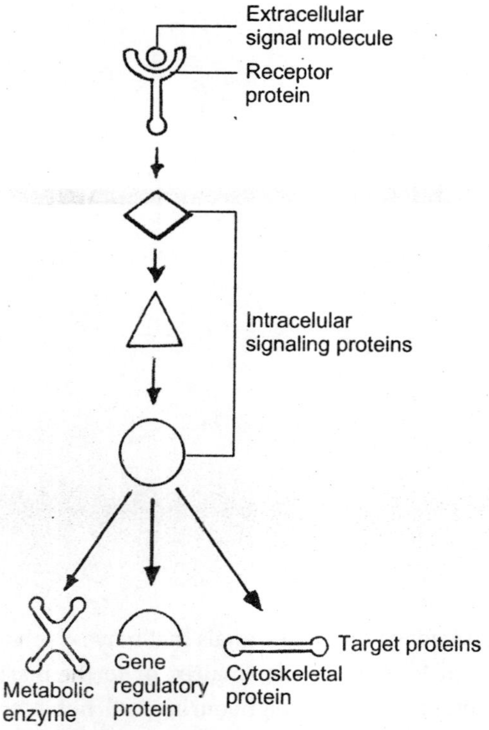

Fig. 15.1 : Principle of cell signaling. a- metabolic enzymes, b- gene regulatory proteins, c- cytoskeletal proteins that alter metabolism, gene expression and cell shape/movement respectively.

that face outwards from the membrane and trigger events inside. This type of intercellular or environmental communication depends upon the reception of extracellular signaling or stimuli by numerous *receptors* present on the cell surface. Cell surface receptors bind in one hand with the signaling molecule and in other hand with many intracellular signaling proteins like kinases, phospholipases, GTP binding proteins and many other proteins with which they interact. Such extracellular chemicals are called *ligands* that transduce the stimulus [signaling molecules (amino acids, peptides, proteins, lipid molecules, acetyl choline etc.)] into a chemical signal by orchestrating an array of intracellular biochemical rections that control metabolic as well as behavioral activities. Stimulus-receptor binding takes place via a change in the shape or conformation of the receptor. They typically respond only to the specific molecule or "ligand" for which they have affinity. Some molecules that are even only slightly different tend to have no effect and act as inhibitors. At the end of each intracellular signaling pathway there is a target protein (gene regulatory protein, ion channels, metabolic regulator, parts of cytoskeletons etc.) that sets out definite biochemical reaction. This *transduction* step converts one type of signal (stimulus) into another form (messenger) and carries out biochemical alterations through effectors that ultimately regulate the behavior of the cell (Fig. 15.1).

Most extracellular chemical signals are hydrophilic in nature, therefore they are unable to penetrate the oily barrier that surrounds cells so part of the transduction process involves transfer of information across the cell membrane which is accomplished by the integral proteins of the cell membrane. However a few stimuli like light, steroid hormones and gases penetrate the plasma membrane and react directly with the effectors or messengers which ultimately set out definite biochemical alterations.

Intracellular signaling: The responses of a cell to a signal varies in different cells on the basis of the intracellular machinery e.g. acetylcholine stimulates the contraction of skeletal muscle but decreases the contraction of cardiac muscle. Intracellular signaling in strict sence refers to the series of reactions that follow reception of extracelular stimulus. Intracellular signaling molecules include heterotrimeric G protein, small GTPases, cyclic nucleotides such as cyclic AMP (cAMP) and cyclic GMP (cGMP), Ca^{++}, phosphoinositide derivatives, such as Phosphatidyl inositol triphosphate (PIP_3), Diacylglycerol (DAG) and Inositol triphosphate IP3), and various protein kinases and phosphatase. Some of these are also called *second messengers* [the extracelluar signaling molecules are called *first messengers*]. In such cases the chain of steps is referred to as a *"signaling cascade"* or *"second messenger pathway"* and the result is that a small stimulus elicits a large response. These chemicals may be water soluble (Ca^{++} and cAMP) or lipid soluble (e.g. Diacylglycerol). Many proteins can also acts as second messenger by binding with Ca^{++}, by exchange or hydrolysis of a bound nucleotide or phosphorylation or dephosphorylation. The intracellular signal proteins are variously defined as *relay proteins* (pass the message to the next signaling component in the cascade); *messenger proteins* (carries the message from cytosol to nucleus); *adaptor proteins* (links different signaling proteins); *amplifier proteins* (these are actually enzymes that amplify the signal either by producing large amount of intracellular mediators or by activating many down stream signaling proteins); *transducer proteins* (converts the signal into different forms); *bifurcation protein* (spreads the signal from one pathway to other); *integrator protein* (receives signals from more than one pathways and integrate them before relaying the signal onwards); *latent gene regulatory proteins* (activated beneath the plasma membrane and then migrate to the nucleus to cause gene activation); *modulator proteins* (they activate or inhibit or modify the activity of intracellular signaling proteins and there by regulate the strength of signaling proteins); *anchoring proteins* (specify the location of signaling proteins by tethering them either to any organelle or to cytoskeleton); *scaffold proteins* (integrate many signaling proteins in a functional complex and hold them in a specific location) (Fig. 15.2).

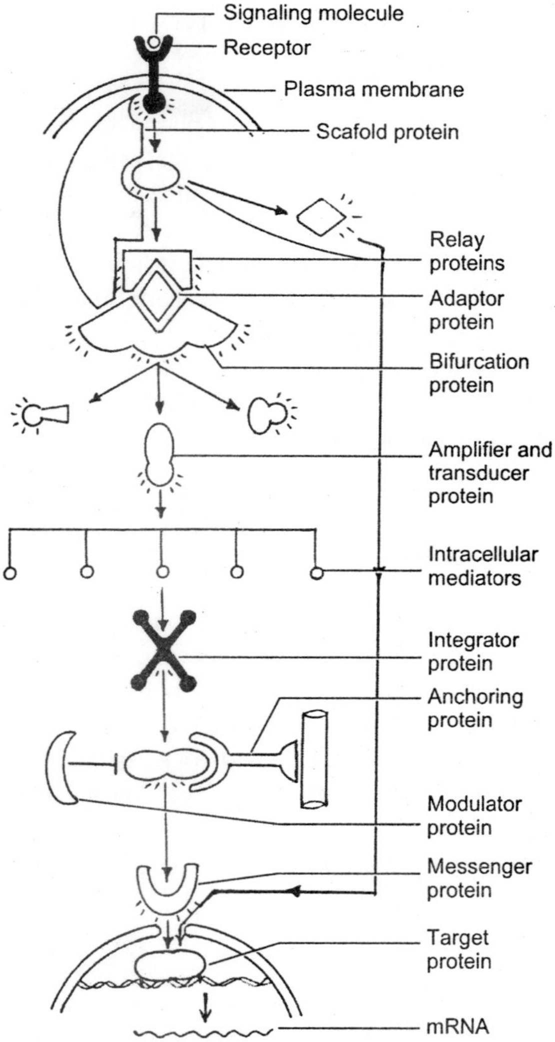

Fig. 15.2 : Different kinds of intracellular signaling proteins and their interactions.

Some intracellular signaling proteins, on receipt of the signal switch from an inactive to active form. Most of such activation is carried out by the phosphorylation. Once phosphorylated, it again phosphorylates the next protein of the cascade. During this course it may amplify or spread the signal to other pathway. Additionally GTP binding proteins (G proteins) also play crucial role in relaying the signal. They switch over from GDP bound form (inactive) to GTP bound (active) form. They also help in vesicular traffic.

A single type of extracellular signal, acting through a single type of receptor may activate multiple parallel signaling pathway. In fact G-protein linked or enzyme linked receptors usually activate the same signaling pathway.

Most intracellular scaffold proteins organize the signaling proteins into signaling complexes that relay the signal with maximum efficiency and precision and speed. In some cases the cytoplasmic tail of these receptors are phosphorylated and these phosphorylated amino acids act as docking sites for the assembly of other signaling proteins. Activation of receptors also modify the phospholipid molecules of the adjacent domains of the plasma membrane that recruit specific intracellular signaling molecules to that region. However all such signaling complexes are formed transiently and rapidly disassemble after the dissociation of the extracellular ligand.

Many intracellular signaling proteins are provided with binding domains that bind the signaling compounds in multiple combinations that determine the route followed by the signaling pathway. Two such domains are Src homology 2 (SH2) and phosphotyrosine binding domain (PTB). These bind to the phosphorylated tyrosine on activated receptors or intracellular signaling protein. SH3 bind to a proline rich sequence and Plekstrin homology domain (PH) bind to phosphorylated inositol phospholipid in the plasma membrane in response to extra cellular signal (Fig. 15.3).

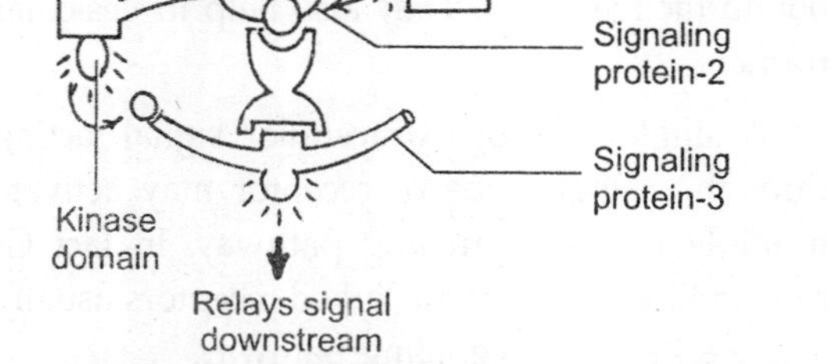

Fig. 15.3 : Schematic representation of signaling pathway using modular binding domains.

Some intracellular cell signaling proteins cluster in specific microdomains of the plasma membrane –the *lipid raft.* It mediates signal transduction with maximum speed and efficiency. Response to extracellular signaling also depends upon the concentration of the signaling molecule. In some cases response does not begin until a threshold concentration of the molecule reaches. In case of hormonal response, as the concentration rises, more than one intracellular effector molecule bind to the target molecule to induce a response. Sometimes the intracellular signaling molecule activates one enzyme and inhibits another enzyme e.g. adrenalin binds to GPCRs that cause increased concentration of cAMP which in one hand activates an enzyme that promotes glycogenolysis while simultaneously inhibits glycogenesis.

The effects of extracellular signaling may persist even after the disappearance of the signal. Protein kinase is activated by Ca^{++} to phosphorylate itself. This autophosphorylation keeps the kinase active long after Ca^{++} returns to normal level.

Intercellular signaling: Intercellular communication is accomplished by extracellular signaling that takes place in complex multicellular organisms. It is of following types (Fig. 15.4).

a) *Endocrine signaling* are produced by endocrine cells and travel through the blood to reach all parts of the body.

b) *Paracrine signaling* targets only cells in the vicinity of the emitting cell. The conductance of neurotransmitters to another nerve cell or from nerve cell to muscle cell, regulation of development by growth factors occurs via this type of signaling. Sometimes, in lieu of directly affecting, the molecule bind to the ECM but subsequently is released in an active form inducing other cells. Sometimes signaling molecules form a concentration gradient and diffuse to the target cell.

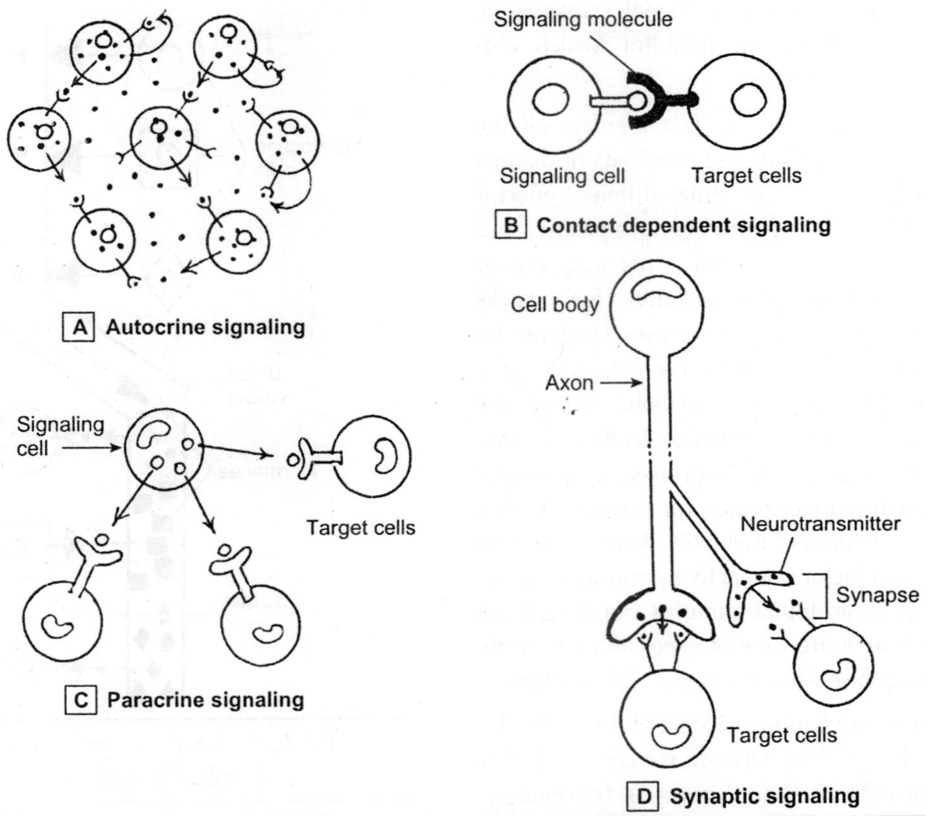

Fig. 15.4 : Diagramatic representation of Autocrine (A), Contact dependent (B),Paracrine (C), and synaptic signaling (D).

c) *Autocrine signaling* involves cellular response by the molecules released by themselves. They affect only cells that are of the same cell type as the emitting cell. This type of signals is found in immune cells.

d) *Juxtacrine signaling* are transmitted along cell membranes via protein or lipid components integral to the membrane and are capable of affecting either the emitting cell or adjacent cells.

e) *Contact dependent signaling* is highly important during developmental stages. In such cases the receptors project into the extra cellular space while still remaining attached to the membrane. Such signaling sends multiple signaling pathways intracellularly to direct the differentiation in a particular direction by affecting specific protein synthesis.

f) *Synaptic signaling* is highly effective in multicellular organisms involving transmission of electrical impulses through the axon through exocytosis of neurotrans mitters.

g) *Signaling through gap junctions* involves coordination of cellular activities through inter cellular exchange of Ca^{++} and cAMP.

Target cells can use a variety of intracellular mechanisms by which they can reversibly adjust their sensitivity to a signal. A prolonged exposure to a stimulus decreases cellular response (called *desensitization*) in a variety of ways. i) Ligand binding may induce their endocytosis followed by their destruction in lysosome. It is called *receptor down regulation.* ii) Signal transduction can be blocked by an inhibitor which also inactivates the receptor.

Signaling pathways in fact regulate all the cellular processes. Cells use hundreds of distinct signaling pathways that utilize different effector systems. Most of the pathways have positive and negative *feed back loops* that either augment or inhibit responses since negative feed backs inactivate the receptors. Responses triggered by signal transduction include the activation of a gene, the production of metabolic energy, cell locomotion etc. Gene activation leads to further effects, since genes are expressed as proteins, many of which are enzymes, transcription factors or other regulators of metabolic activity. Because transcription factors can activate still more genes in turn. Thus an initial stimulus can trigger the expression of entire suite of genes and a panoply of physiolgical events via signal transduction.

Signaling by hormones: Most of the molecules that enable signaling between the cells or tissues within an individual are known as "hormones." These modify the activity of the organism and are called *agonists.* Hormone-initiated signal transduction takes the following steps:

a) Biosynthesis of a hormone.
b) Storage and secretion of the hormone.
c) Transport of the hormone to the target cell.
d) Recognition of the hormone by the membrane receptor protein, leading to a conformational change.
e) Relay and amplification of the signal that leads to defined biochemical reactions within the target cell. The reactions of the target cells, in turn, send a signal to the hormone-producing cell leading to the down-regulation of hormone production.
f) Removal of the hormone (Fig. 15.5).

Hormones and other signaling molecules may exit the sending cell by exocytosis or other means of membrane transport. Its recipients may be of one type or several (insulin triggers diverse and

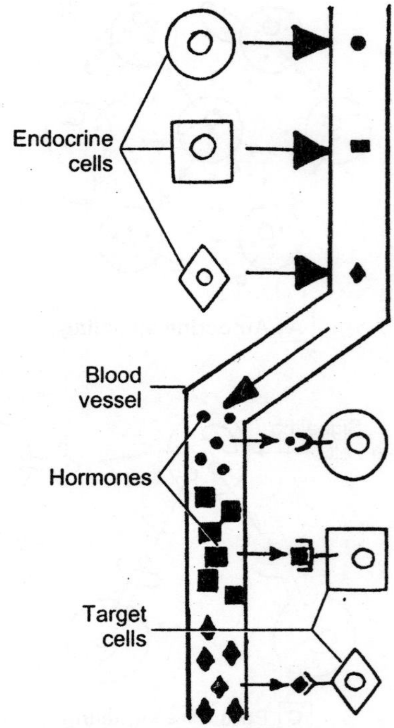

Fig. 15.5 : Schematic representation of endocrine signaling.

systemic effects). A cell can have several different receptors that recognize the same hormone, but activate different signal transduction pathways or different hormones and their receptors can invoke the same biochemical pathway. Different tissue types can answer differently to the same hormone stimulus. There are two classes of hormone receptors, *"membrane-associated receptors"* and *"intracellular or cytoplasmic"* receptors.

15.3. SIGNALING MOLECULES

Signal transduction involves a variety of molecules that transmit information between cells. These molecules act as ligands that bind to the receptors expressed by the target cells. The signaling molecules vary from gases to proteins; some carries signal to long distances while others may act locally to convey information between neighbouring cells. Some molecules able to cross the plasma membrane and bind to intracellular cytosolic or nuclear receptors while some only bind to the cell surface receptors. Following sections provide a brief idea about the varieties of signaling molecules:

Steroid hormones as signaling molecule: Testosterone, oestrogen, progesterone, corticoste roids and ecdysone belong to this group of molecules that are basically synthesized from cholesterol. Thyroid hormones, vitamin D_3 and retinoic acid are structurally and functionally distinct from the steroid hormones but they share a common mechanism of action on the target cells. Due to their hydrophobic character, they diffuse across the plasma membrane and bind to any intracellular receptor belonging to steroid receptor superfamily. These are basically transcription factors showing domains for ligand binding, DNA binding and transcriptional activation. Ligand binding may cause a conformational change in the receptor allowing it to bind to the regulatory DNA sequence and transactivate specific gene. Some even can not bind to DNA in the absence of hormone. The effect of binding to DNA may vary with the presence or absence of hormones. For example, in the absence of thyroxin, the hormone receptor acts as repressor but hormone binding may convert it to activator that stimulates thyroxin inducible genes.

Nitric Oxide and Carbon Monooxide as signaling molecule: Nitric oxide (NO) is a paracrine signaling molecule in the nervous, immune and circulatory system. NO is a free radical that diffuses through the plasma membrane and affects nearby cells. NO is made from arginine and oxygen by the enzyme NO synthase. The first step of its activity is the release of acetylcholine from the nerve cells in the blood vessel wall, it then acts on the endothelial cells to stimulate NO synthesis which then diffuses to neighbouring smooth muscle cells where it reacts with Fe bound to the enzyme guanylyl cyclase. It produces the second messenger cyclic guanosine monophosphate (cGMP) which induces muscle cell relaxation and blood vessel dilation. NO serves three main functions–relaxation of blood vessels; regulation of exocytosis of neurotransmitters and cellular immune response. It is responsible for dilation of the blood vessels during penile erection. Additionally, medical use of nitroglycerine in the treatment of heart disease is based on its conversion to NO that dilates the coronary blood vessels and increases blood flow to the heart.

Blood vessel wall → Acetyl choline → Endothelial cells → NO synthesis → Smooth muscle cells of the neighbouring cell → Fe bound to guanylyl cyclase → cGMP → Muscle cell relaxation→ Blood vessel dilation.

Carbon monooxide (CO) acts as a neurotransmitter and mediator of blood vessel dilation.CO can stimulate guanylyl cyclase that causes blood vessel dilation in the same way as in the case of NO.

Neurotransmitters as signaling molecules: These are diverse groups of hydrophilic molecules (acetylcholine, dopamine, adrenaline, serotonin, histamine, glutamate, glycine and γ-amino butaryc acid). The released neurotrans-mitters diffuse across the synaptic cleft and bind to the target cell surface receptors. Many such receptors (acetylcholine receptors) are ligand gated ion channels and neurotrans mitters binding to these receptors induces a conformational change that opens ion channels resulting in ion flux in the target cell. Other neurotransmitters are coupled to G proteins that are involved in many intracellular responses.

Neurotransmitters → Diffuse the synaptic cleft → Bind to target cell surface receptors → Causes change in ion balance or transduces signals via GPCRs.

Peptide hormones and Growth factors: This group shows peptide hormones (insulin, glucagon, STH, FSH, LH, prolactin etc.)

neuropeptides (enkephalins and endorphins) and many polypeptide growth factors (NGF, EGF, PDGF etc.) that play critical role in the cell proliferation during embryonic development and in the adult organism. Members of another large group of polypeptide growth factor (*Cytokines*) regulate the development and differentiation of blood cells and control the activities of lymphocytes during immune response. Some membrane anchored growth factors remain associated with the plasma membrane and play crucial role during direct cell-cell interactions. These signaling molecules can not cross the plasma membrane and act by binding to the cell surface receptors only. Abnormalities in growth factor signaling are the basis for variety of diseases including cancers.

Ca^{++} as a signaling molecule: Ca^{++} acts as a signaling molecule. Within the cytosol, free Ca^{++} is the active form that paticipates in various activities. Therefore, the level of cytosolic free Ca^{++} is usually very low; it is stored within organelles, usually the ER (sarcoplasmic reticulum in muscle cells), where it is bound to calreticulin.

Activation of Ca^{++}: To become active, Ca^{++} has to be released from the ER into the cytosol. There are two combined receptors / ion channel proteins that perform the task of the transport of Ca^{++}: A variety of receptors activate phospholipase C (PLC) to produce IP_3 from PIP_2. IP3 diffuses into the ER to open IP3 receptors releasing Ca^{++} to the cytosol. The *ryanodine receptor* is similar to the IP_3 receptor and are stimulated to transport Ca^{++} into the cytosol by recognizing Ca^{++} on its cytosolic side, thus establishing a feedback mechanism; a small amount of Ca^{++} in the cytosol near the receptor will cause it to release even more Ca^{++}. It is especially important in neurons and muscle cells. In heart and pancreas cells, another second messenger (cyclic ADP ribose) takes part in the receptor activation. Excess cytosolic Ca^{++} is released outside via store operated channels and calcium pumps. Low Ca^{++} levels in the cytoplasm induces voltage sensitive L-type channels to allow Ca^{++} to enter into the cytosol to replenish the loss (Fig. 15.6).

The localized and time-limited activity of Ca^{++} in the cytosol is also called *Ca^{++} wave.* The

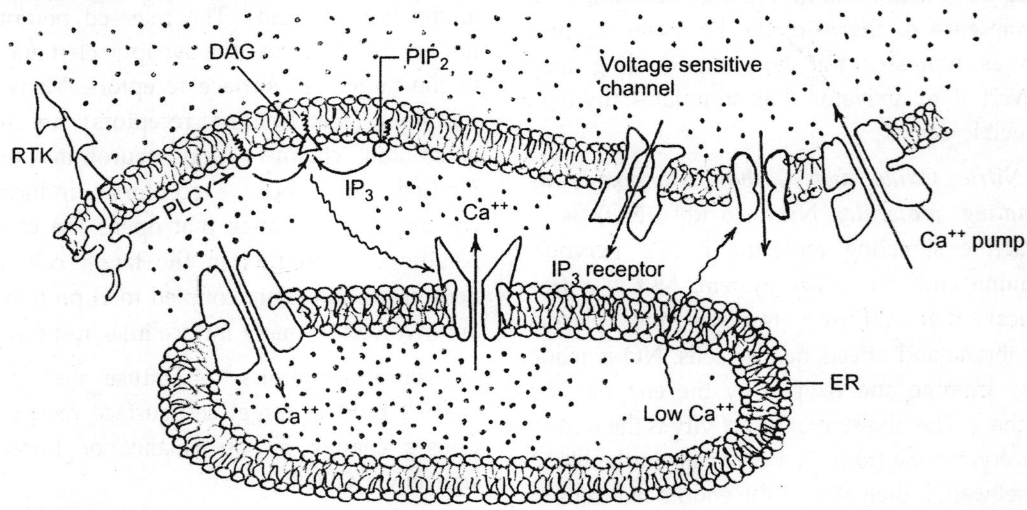

Fig. 15.6 : Schematic pathways of Ca^{++} release and uptake.

production of the wave is done by the feedback mechanism of the ryanodine receptor and the activation of phospholipase-C by Ca^{++}, which leads to the production of inositol triphosphate (IP_3), which in turn activates the IP_3 receptor.

Functions of Ca^{++}: Ca^{++} is used in a multitude of processes, among them the most important are muscle contraction, release of neurotransmitter from nerve endings, vision in retina cells, proliferation, secretion, cytoskeleton management, cell motion, gene expression, apoptosis and metabolism. The three main pathways that lead to Ca^{++} activation are G protein regulated pathways; receptor-tyrosine kinases pathways regulated; ligand- or current-regulated ion channels.

There are two ways in which Ca^{++} can regulate the activity of the enzymes - direct recognition of Ca^{++} by the protein and binding of Ca^{++} in the active center of an enzyme. One of the best studied interactions of Ca^{++} with a protein is the regulation of calmodulin by Ca^{++}. Calmodulin itself can regulate other proteins or be part of a larger protein like phosphorylase kinase. The Ca^{++}/calmodulin complex plays an important role in proliferation, mitosis and neural signal transduction.

Lipophilic signaling molecules: One group of lipophilic second messenger molecules consists of inositol triphosphate and diacylglycerol. Others are ceramide and lysophosphatic acid.

15.4. CLASSIFICATION OF RECEPTORS

Signal-transducing receptors are of three general classes:

Membrane receptors: Plasma membrane is the limiting boundary of the cell that is the main border of the cell that receives all the signals from outside as well as delivers all the products of the cell to the outside. Reception of extracellular signals is restricted to numerous proteins, called receptors. 20 different families of proteins serve as receptors. As most of the stimuli are hydrophilic in nature (except steroid hormones and nitric oxide or carbon-mono-oxide gases) they cannot directly cross the lipid bilayer, rather they are received by transmembrane receptors to elicit the biochemical response. Transmembrane receptors span the thickness of the plasma membrane with one external end (*extracellular domain*) and one internal end (*intra cellular domain*) (Fig. 15.7).

Receptors with enzymatic activity: Receptors that penetrate the plasma membrane and have intrinsic enzymatic activity include ***tyrosine kinases*** (e.g. PDGF, insulin, EGF and FGF

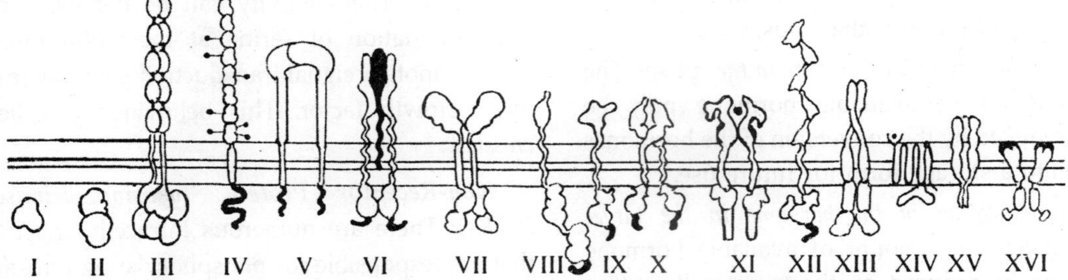

Fig. 15.7 : Sixteen classes of signal transduction receptors. I-Cytoplasmic steroid receptor, II-Cyto -plasmic guanylyl cyclase, III- Cadherin, IV- Selectin. V- Integrin, VI- Sphingomyelinase linked receptor, VII- Receptor guanylyl cyclase, VIII- Receptor tyrosine phosphatase, IX- Receptor serine threonine kinase, X- Tyrosine kinase linked receptor, XI- Cytokine, XII- Receptor tyrosine kinase. XIII- Two component receptors, XIV- GPCR, XV- Ligand gated ion channel, XVI- Voltage gated ion channel.

receptors), ***tyrosine phosphatases*** (e.g. CD 45 - *cluster determinant-45*) protein of T cells and macrophages), ***guanylate cyclases*** (e.g. natriuretic peptide receptors) and ***serine/threonine kinases*** several families of receptors lack intrinsic enzyme activity, yet are coupled to intracellular tyrosine kinases by direct protein-protein interactions.

Receptors that migrate to the nucleus: There are some receptors that upon ligand binding migrate to the nucleus where the ligand-receptor complex directly affects gene transcription.

Nuclear receptors: Nuclear receptors are soluble proteins localized within the nucleoplasm. Nuclear receptors are ligand-activated transcription activators. The typical ligands for nuclear receptors are lipophilic hormones (e.g. testosterone, progesterone and cortisol) and derivatives of vitamin A and D. On binding with the receptors, the ligand will pass through the nuclear membrane and activates the transcription of certain gene and thus synthesis of a protein. The key value for the signal strength is the hormone concentration, which is regulated by:

(i) *Biosynthesis and secretion of hormones in the endocrine tissue.* The hypothalamus receives stinulus to produce *releasing factors* for hormone production which, in turn, activate endocrine organs to produce hormones for the target tissues. This cascade *amplify* the original signal that reached the hypothalamus.

(ii) *Availability of the hormone in the cytosol.* The amount of available hormone may be reduced by the conversion of the hormones into a storage form for future use.

(iii) *Modification of the hormone in the target tissue.* The amount of available hormone can be reduced by the target cell so that they fail to trigger the receptor.

(iv) *Hormone responsive elements (HREs)* The nuclear receptors that were activated by the hormones bind to a receptor-specific DNA sequences located in the promoter region of the genes that are activated by the hormone-receptor complex. As this enables the transcription of the target gene, these hormones are also called *inductors of gene expression.* The transactivation of gene is much slower than signals that directly affect the existing proteins. Therefore the effects of hormones that use nuclear receptors are usually long-term. All nuclear receptors have a similar, modular structure:

N-AAAABBBBCCCCDDDDEEEEFFFF-C

where CCCC is the DNA-binding (showing Zinc fingers) and EEEE the ligand-binding domains. The Zn fingers stabilize DNA binding by holding contact to the phosphate backbone of the DNA. It also contains structural elements responsible for transactivation.

Steroid receptors: *Steroid receptors* are a subclass of nuclear receptors, located primarily within the cytosol. In the absence of steroid hormone, the receptors cling together in a complex called *aporeceptor complex,* which also contains *chaperone proteins / heatshock proteins / Hsps*). The *Hsps* are necessary to activate the receptor by folding in such a way that the signal sequence becomes accessible for its passage into the nucleus.

Sometimes the transactivation domain of steroid receptors becomes hidden so it cannot activate transcription thus repressing the gene expression. Their activity can be enhanced by phosphorylation of serine at their N-terminal end, by another signal transduction pathway (e.g. by a growth factor. This behaviour is called *crosstalk.*

Non-Receptor Protein Tyrosine Kinases (PTKs): There are numerous intracellular PTKs that are responsible for phosphorylating tyrosine residues of a variety of intracellular proteins on following activation of cellular growth and proliferation signals. There is now recognized two distinct families of non-receptor PTKs. The archetypal PTK family is related to the Src protein. The Src protein is a tyrosine kinase first identified as the transforming protein in Rous sarcoma virus. Subsequently, a cellular homolog was identifed as c-Src. Numerous proto-oncogenes were identified as the transforming proteins carried by retroviruses. The second family is related to the Janus kinase (Jak).

Most of the proteins of both families of non-receptor PTKs couple to cellular receptors that lack enzymatic activity themselves. This class of receptors includes all of the cytokine receptors (e.g. the interleukin-2 (IL-2) receptor as well as the CD4 and CD8 cell surface glycoproteins of T cells and the T cell antigen receptor (TCR). This mode of coupling receptors to intracellular PTKs suggests a split form of RTK.

Another example of receptor-signaling through protein interaction involves the insulin receptor (IR). This receptor has intrinsic tyrosine kinase activity but does not directly interact, following autophosphorylation, with enzymatically active proteins containing SH2 domains (e.g. PI-3K or PLC-γ). The principal IR substrate is a protein termed IRS-1 which contains several motifs that resemble SH2 binding consensus sites for the catalytically active subunit of PI-3K. These domains allow complexes to form between IRS-1 and PI-3K which suggests that IRS-1 acts as a docking or adapter protein to couple the IR to SH2 containing signaling proteins. Growth factor receptor-binding protein 2, Grb2. have been identified as an additional adapter proteins.

Non-Receptor Serine/Threonine Kinases: There are several serine/threonine kinases that function in signal transduction pathways. The two more commonly known are cAMP-dependent protein kinase (PKA) and protein kinase C (PKC). Additional serine/threonine kinases important for signal transduction are the mitogen activated protein kinases (MAP kinases).

15.5. G-PROTEIN COUPLED RECEPTORS (GPCRS)

The largest family of membrane receptors showing 7 transmembrane a helices use trimeric GTP-binding proteins to relay extracelluar signals inside the cell as they have domains facing the extra cellular side of the cell membrane that recognize and bind particular signal molecules (ligands). These classes of receptor are termed as *seven helix receptors* or *G-protein coupled receptors (GPCRs)* (Fig. 15.8). All G protein linked receptors consists of a single polypeptide chain that crosses the plasma membrane seven times and are therefore called *serpentine receptors.* Rhodopsin, olfactory receptors belong to this group. GPCRs may dimerize or form oligomeric complexes within the membrane. Ligand binding may promote oligomerization, which may in turn affect activity of the receptor.

In *Caenorhabditis elegans* 790 genes code for GPCRs. In mammals only olfactory cell encodes 500-1000 different GPCRs. Other cells code for about 500 such receptors (Pollard and Earnshaw, 2004). Their general structure is more or less same in all cases having N terminus showing 5-600 and C terminus with 12-300 amino acids. Association of receptor and ligand varies in different cases. Soluble chemical ligands are receptor specific and bind in a small pocket lined by variable number of amino acids. The light absorbing pigment – 11-cis retinal is bound covalently with rhodopsin. Neurotransmitters bind between helices about 1/3rd of the membrane width. Peptide hormones in addition to binding with the exposed amino acids bind deep in the helical pocket. The clotting enzyme thrombin activates its receptor on platelets by ptoteolysis not by direct binding. The large ligands (LH, FSH, TSH) and small ligands (glutamate, γ amino butaric acid, Ca^{++} etc.) bind with high affinity to the extracellular domains of the receptors. GPCRs exist in resting and activated state. Activation due to ligand binding

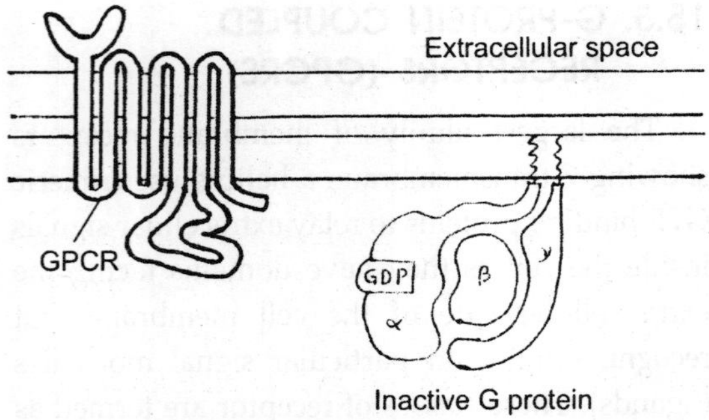

Fig. 15.8 : Structure of a GPCR and G-protein.

involves structural rearrangement of the helices particularly in the cytoplasmic side to create a site for binding target G proteins.

Phosphorylation of the C terminal ends may inactivate GPCRs. A class of protein kinases – *G protein coupled receptor kinases* (GRKs) phosphorylate serines or threonines on the C terminal cytoplasmic tails of receptors which facilitates binding of a protein – *arrestin* that inactivates the receptor by blocking the interaction of the receptor with G protein. Arrestin even causes removal of some GPCRs by endocytosis. Moreover two strategies sometimes acting on the same receptor may provide negative feed back.

During hormone recognition, when the extracellular domain recognizes the hormone, the whole receptor undergoes a structural shift that affects the intracellular domain, leading to further action. In this case the hormone itself does not pass through the plasma membrane into the cell. The important value for the strength of the signal relayed by the receptor is the concentration of the hormone-receptor complex, which is defined by the affinity of the hormone for the receptor, the concentration of the hormone and of course, the concentration of the receptor. The concentration of the circulating hormone is the key value for the strength of the signal, since the other two values are constant. For fast reaction, the hormone-producing cells can store prehormones, and quickly modify and release them if necessary. Also, the recipient cell can modify the sensitivity of the receptor, for example by phosphorylation also, the variation of the number of receptors can vary the total signal strength in the recipient cell.

GPCRs are classified into three categories:

i) GPCRs that modulate adenylate cyclase activity leading to the production of cAMP as the second messenger. Receptors of this class include the glucagon and odorant molecule receptors. Increase in the production of cAMP leads to an increase in the activity of PKA in the case of glucagon receptors. In the case of odorant molecule receptors the increase in cAMP leads to the activation of ion channels. In contrast to increased adenylate cyclase activity, the adrenergic receptors are coupled to inhibitory G-proteins (Gi) that repress adenylate cyclase activity upon receptor activation.

ii) GPCRs that activate phospholipase C-b leading to hydrolysis of polyphosphoinositides (e.g. PIP_2) generate the second messengers - diacylglycerol (DAG) and inositoltrisphosphate (IP_3). This class of receptors includes the angiotensin, bradykinin and vasopressin receptors.

iii) A novel class of GPCRs is the photoreceptors that are coupled to a G-protein termed *transducin* that activates a phosphodiesterase which leads to a decrease in the level of cGMP. The drop in cGMP results in the closing of a Na^+/Ca^{++} channel leading to hyperpolarization of the cell.

15.6. G-PROTEINS

The G proteins are heterotrimeric with three subunits - α, β, and γ, attached to the cytoplasmic face of the plasma membrane. The β and γ subunits have covalently attached lipid anchors, which is inserted into the plasma membrane, binding a G-protein to the cytosolic surface of the plasma membrane. a subunit - Gα shows a GTP binding domain and can hydrolyze it to GDP + P_i. In unstimulated state α, β and γ subunits remain united but in stimulated state, a subunit binds GTP. This causes conformational change which results in the dissociation of β and γ subunits. It results in the formation of α and βγ subunits. α subunit itself shows conformational changes that allow it to interact with the target protein. The βγ subunit exhibit no conformational change but its surface that masks with the a subunit becomes available for interaction with a second set of protein – either enzyme or ion channel. G-βγ inhibits Gα. Once the α subunit hydrolyses the GTP in to GDP and Pi it reassociates with βγ attaining inactive form. The GTP site in Gα consists of loops that extend out from the edge of β-sheet. Three switch domains have been identified, that change position when GTP substitutes for GDP. A G-protein that activates cAMP formation is called a stimulatory G-protein or Gsα which is activated by receptors for the hormones epinephrine and glucagon. The Gs, activates adenylate cyclase when it binds GTP while an inhibitory Giα, inhibits adenylate cyclase when it binds GTP. Adenylate Cyclase is a transmembrane protein, with cytosolic domains forming the catalytic site of GTPase.

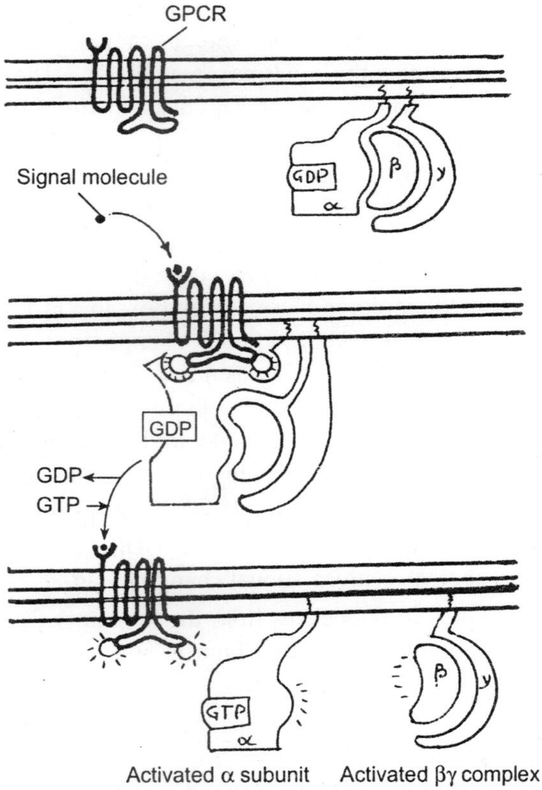

Fig. 15.9 : Structural changes of activated G-protein.

GPCR interacting proteins: Various GPCR-interacting proteins (GIPs) modulate receptor function. Effects of GIPs may be related with altered ligand affinity, promoting receptor dimerization or oligomerization, control of receptor localization, including transfer to or removal from the plasma membrane, promoting close association with other signal proteins

G-Protein Regulators: The activity of G-proteins with respect to GTP hydrolysis is regulated by a family of proteins termed *GTPase activating proteins (GAPs)*. The proto-oncogenic protein, Ras, is a G-protein, involved in the genesis of numerous forms of cancer. Of particular clinical significance is the fact that oncogenic activation of Ras occurs with higher frequency than any other gene in the development of colorectal cancers. Regulation of Ras GTP ase activity is controlled by *rasGAP*. There are several other GAPs like *neurofibromin* encoded by the neurofibromatosis type-1 (NF1) [tumor suppressor] gene.

Signaling through G protein coupled receptors (GPCRs): GPCRs are activated by hormones, neurotransmitters and many local mediators. The signaling molecules range from proteins, peptides, amino acids and fatty acids.

Overview of the signal transduction through G protein:

i) Initially the G-protein a subunit has bound GDP and the α, β, ...γ subunits are complexed together.

ii) When the extracellular domain of GPCR recognizes the hormone, it cannot pass through the plasma membrane. The strength of the signal relayed by the receptor is the concentration of the hormone-receptor complex, which is defined by the affinity of the hormone for the receptor, the concentration of the hormone and of course, the concentration of the receptor. Hormone binding to GPCR causes a conformational change in the receptor that is transmitted to the G-protein. The nucleotide-binding site on Gα becomes more accessible to the cytosol, where GTP is usually higher than GDP. Gα releases GDP and binds GTP (Fig. 15.10).

iii) Substitution of GTP for GDP causes another conformational change in Gα.

iv) Gα-GTP dissociates from the inhibitory βγ subunit complex and bind to and activate adenylate cyclase.

v) Adenylate cyclase catalyzes synthesis of cAMP.

vi) cAMP-dependent Protein Kinase-A catalyzes phosphorylation of various cellular proteins, altering their activity. Replacement of GDP with GTP activates the G proteins which stimulates the target effectors and amplify the signal as a result of multiple interactions. If extracellular stimulation sustains, then the signaling slows down its response .It is called *adaptation / attenuation / desensitization / tachyphylaxis / tolerance.*

GPCR signaling has revealed some novel features of their signal transduction pathways.

(i) A class of GPCR- bradykinin type 2 receptor (B2R) may couple to different classes of G proteins.

(ii) Different signaling pathways emanating from one or several GPCRs exhibit extensive cross-talk, resulting in positive or negative signal modulation.

(iii) GPCRs can generate mitogenic signals.

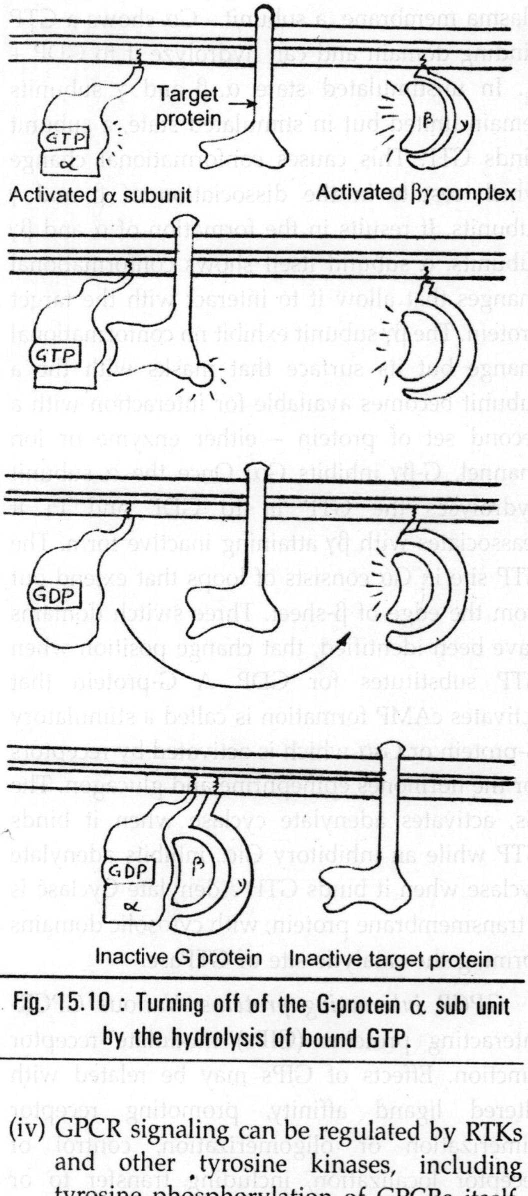

Fig. 15.10 : Turning off of the G-protein α sub unit by the hydrolysis of bound GTP.

(iv) GPCR signaling can be regulated by RTKs and other tyrosine kinases, including tyrosine phosphorylation of GPCRs itself, of G proteins, and of downstream molecules (members of the protein kinase C family).

(v) GPCRs also elicit signals when resident at other subcellular sites including the endoplasmic reticulum, Golgi apparatus, and the nucleus (Boivin *et al*, 2008).

G proteins and the subsequent regulation of a diverse variety of effector systems and agonist activation of GPCRs also initiates processes that are involved in the feedback desensitization of GPCR responsiveness, the internalization of GPCRs and the coupling of GPCRs to heterotrimeric G protein-independent signal transduction pathways. GPCR desensitization occurs as a consequence of G protein uncoupling in response to phosphorylation by both second messenger-dependent protein kinases and G protein-coupled receptor kinases (GRKs). GRK-mediated receptor phosphorylation promotes the binding of β-arrestins, which not only uncouple receptors from heterotrimeric G proteins but also target many GPCRs for internalization in clathrin-coated vesicles. β-arrestin-dependent endocytosis of GPCRs involves the direct interaction of the C-terminal tail domain of β-arrestins with both β-adaptin and clathrin.

GPCR activation also initiates a series of molecular events that leads to GPCR-kinase-mediated receptor phosphorylation and the binding of β-arrestin proteins to the intracellular face of the receptor that may target many GPCRs for endocytosis in clathrin-coated pits. Several GPCRs are internalized with β-arrestin and are either dephosphorylated in early endosomes and recycled back to the cell surface as fully functional receptors or retained in early endosomes and targeted for degradation in lysosomes. Neuron-enriched endosomal protein of 21 kDa (NEEP21) plays an important role in the recycling mechanism at the level of early endosomes to promote sorting of receptors toward a recycling pathway. However the movement of GPCRs through these intracellular membrane compartments is regulated by a variety of Rab GTPases. GPCRs may directly control their own targeting in different intracellular compartments. In addition to signaling via heterotrimeric G-proteins, GPCRs can also signal by interacting with various small G-protein superfamilies like Ras, Rho/Rac/Cdc42, Rab, Sar1/Arf and Ran families. They regulate down stream effector pathways to control many eukaryotic cell functions by activating Ras, Rho GTPases RhoA, Rabs, ARFs and ARF GEFs (guanine nucleotide-exchange factors). GPCR-regulatory proteins may influence GPCR signaling and intracellular trafficking.

Turn off of the signal: It can be done by many ways.

i) Gα hydrolyzes GTP to GDP + P_i. The presence of GDP on Gα causes it to rebind to the inhibitory bg complex
ii) Phosphodiesterase catalyzes hydrolysis of cAMP to AMP.
iii) Hormone concentration may cause desensitization of receptors. Some receptors are phosphorylated via specific receptor kinases. The phosphorylated receptor may then bind to a protein β-arrestin, that promotes removal of the receptor from the membrane by clathrin-mediated endocytosis.
iv) Protein phosphatase catalyzes removal of phosphates by hydrolysis that were attached to proteins via Protein Kinase A.

G protein signaling by cAMP: cAMP is an intracellular mediator; its concentration is 10^{-7}M

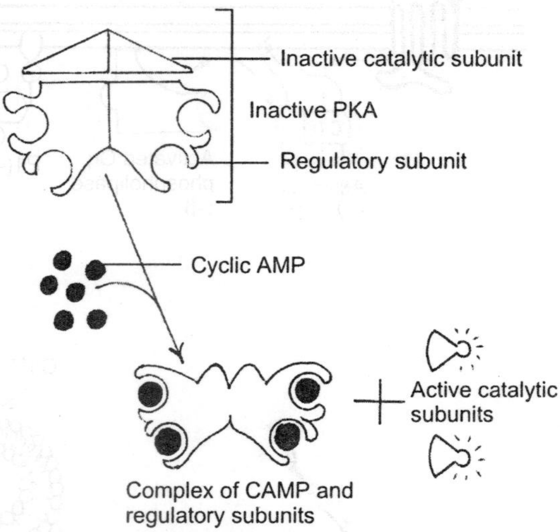

Fig. 15.11 : Activation of CAMP dependent protein kinae A.

but an extra cellular signal can cause its 20 fold increase within seconds. Such a rapid synthesis is however balanced by its rapid break down. cAMP is synthesized from ATP by a cell membrane bound enzyme – *adenylyl cyclase* and is rapidly destroyed by cAMP phosphodiesterases. All receptors that act via cAMP are coupled to stimulatory G protein (Gs) which activates adenylyl cyclase and thereby increases cAMP concentration. Another G protein- inhibitory G protein (Gi) inhibits adenylyl cyclase by directly regulating the ion channel. In addition to activating the ion channel, cAMP exerts its effect mainly by activating cAMP dependent protein kinase (PKA). It catalyzes the transfer of terminal phosphate group from ATP to serine/threonine of selected target protein regulating their activity. Inactive PKA has two catalytic and one regulatory site. cAMP binds in the regulatory sites causing their dissociation from the complex and there by activation of PKA. cAMP mediated response may be rapid (e.g. in skeletal muscle activated PKA phosphorylates an enzyme that triggers glycogenolysis and simultaneously inhibits gluconeogenesis) or slow (described in latter section).

Inositol phosphatases signaling pathway activated by G protein or IP_3 / DAG pathway: Many G protein linked receptors exert their effects by another G protein that activates plasma membrane bound enzyme – phospholipase C-β. It acts on an inositol phospholipids – *phosphatidyl inositol 4,5-biphosphate [$PI(4,5)P_2$]* that occurs in the inner half of the plasma membrane. The receptors that operate through this pathway mainly activate a G protein – Gq that in turn activates phospholipase to form inositol 1,4,5-triphosphate (IP_3) and 1, 2- DAG. IP_3 rapidly diffuses in the cytosol while DAG remains associated with the membrane. IP_3 and DAG acts as second messenger. Therefore the downstream events involving these second messengers constitute ***IP_3 / DAG pathway*** (Fig. 15.12).

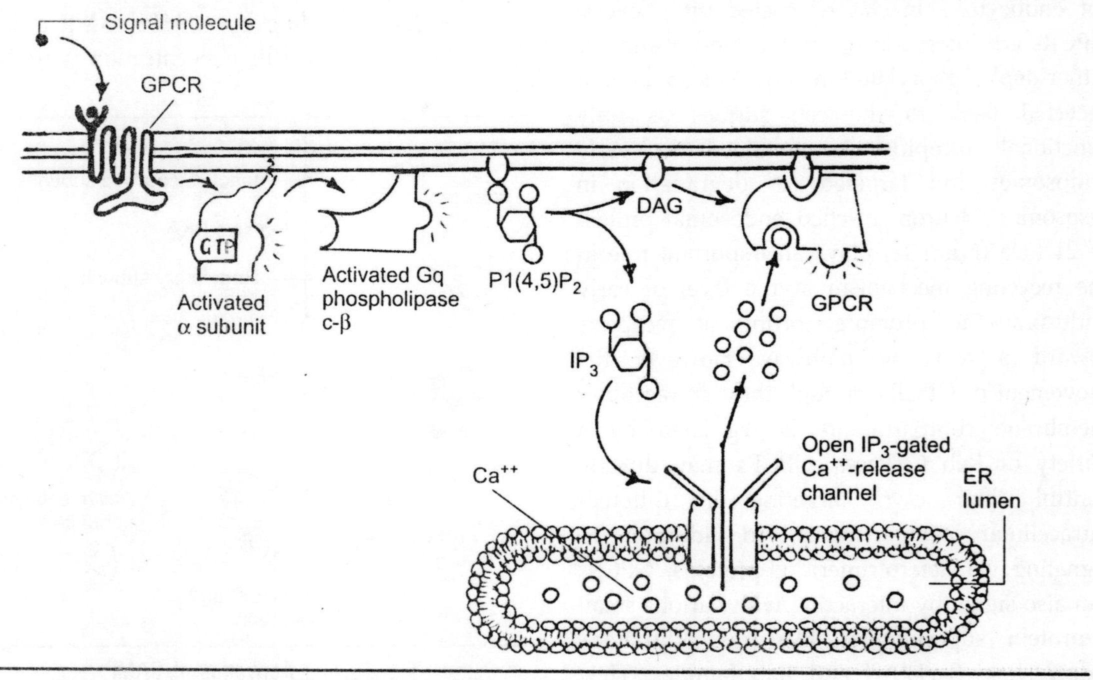

Fig. 15.12 : Major steps of inositol phospholipids pathway.

Mostly intracellular Ca^{++} is stored inside the mitochondria or ER. Cells employ many mechanisms to keep the levels of Ca^{++} up to a certain limit - 0.2µM. But binding of a hormone to the GPCR lead to the activation of phospholipase C. Cleavage of PIP_2 by phospholipase C produces IP_3 and DAG. IP_3 opens IP_3 gated Ca^{++} release channels in the ER membrane. Ca^{++} stored in the ER is released to raise cytosolic Ca^{++} level. DAG follows two pathways - it can be cleaved to release arachidonic acid that synthesizes other small lipid messenger called *eicosanoids*. They augment pain sensation. For this reason most of the painkillers inhibit their synthesis. DAG mostly activates serine/threonine protein kinase C (PKC). Rise in cytosolic Ca^{++} causes PKC to bind to the inner leaflet of the plasma membrane where it is activated by DAG. It results in cellular growth, glycogen metabolism and transactivation of various genes.

Regulation of Ion channels by G proteins: G proteins may directly activate and inactivate ion channels in the plasma membrane altering ion permeability that alters the excitability of the membrane. Acetylcholine receptors activate Gi proteins whose a subunit inhibits adenylyl cyclase while bg subunits bind to the K^+ channels to open them. It makes the cell harder to depolarize that reduces the rate and contraction of the cardiac muscle. Other G proteins stimulate channel phosphorylation by PKA/PKC showing production and destruction of cyclic nucleotides that directly activate or inactivate ion channels. The cyclic nucleotide gated ion channel has a crucial role in smell or vision.

Olfactory signal transduction mediated by G protein: The chemosensory processes of smell and taste have evolved into highly specialized biochemical and electrophysiological pathways. Olfaction allows humans to detect more than 10,000 air borne odorants.

Olfactory neurons have the ability to be replaced every 60 days throughout the adult life. When the progenitors mature they move into more apical region of the epithelium extending apical processes comparable to the dendrites of other neurons. About 12 sensory cilia sprout from these dendritic processes and axon project from the base of the neuron to a secondary neuron in the olfactory bulb.

Olfactory neurons have three specialized zones – *apical sensory cilia*, specialized for responding to extra cellular odorants [odorant receptor, G_{olf}, provided with adenylyl cyclase and cyclic nucleotide gated ion channels allowing extra cellular odorants to control membrane potential], *cell body* contains the protein synthesizing machinery and *plasma membrane* showing pumps and channels.

Odorant receptors comprise a family of related GPCRs. Odorant binding causes conformational changes in the receptor. Activated receptors catalyze the rapid exchange of GDP to GTP on a specialized G protein – $G_{olf.}$ This then dissociates from Gbg subunit. It is followed by signal amplification and allows diverse range of ligands and their receptors to converge on a common intracellular pathway. GTP - G_{olf}a activates adenylyl cyclase. Each enzyme generates 100 cAMP molecules. cAMP binds to cyclic nucleotide gated ion channels that admit the entry of enough Na+ and Ca++ indirectly Cl^-. All the ions directly depolarize the membranes that trigger an axon potential which travel along the axon to the secondary neuron of the brain.

Odorant → Odorant receptor → G*olf* → G *olf* a → Adenylyl cyclase → cAMP → Cyclic nucleotide gated ion channels open → Entrance of Na^+, Ca^{++} and Cl^- → Depolarization of the axolemma → Action potential → Nerve impulse to the brain.

Multiple negative feed back loops terminate stimulation. GPCRs are desensitized by protein kinases that phosphorylate the receptor with which bind the protein, arrestins. Simultaneously

Gbg activates the odorant receptor kinase (ORK). Ca^{++} binds calmodulin which activates cAMP phosphodiesterase (PDE) that rapidly converts cAMP to 5′-AMP. It ultimately inhibits cAMP gated channels and causes breakdown of cAMP.

Photosignaling mediated by G protein: Visual transduction responses are the fastest G-protein linked signaling pathway in vertebrates. Human retina contains two types of photoreceptors – rods and cones that are involved in non color dim vision and colored bright vision respectively. The rod cell shows outer and inner segments that are connected by a vestigial cilium, a cell body and a synaptic region. The vesicles in the synaptic region contain the neurotransmitter - glutamate that activates the next neuron to pass a chemical signal to a retinal nerve cell that relays the signal along the visual pathway. The outer segment contains about 1000 closed discs whose membrane contains the pigment rhodopsin while the plasma membrane surrounding the discs contains cGMP gated Na^{+} channels. Rhodopsin, is associated with GPCR-Gt that associate with the cytoplasmic face of the disc membrane and a covalently attached light absorbing chromophore – 11-*cis* retinal. Human rod cells contain about 4 x 10^{7} rhodopsin molecules. In the resting condition production of cGMP keeps the cGMP gated Na^{+} channels open in the plasma membrane allowing an influx of Na^{+} and Ca^{++} inside the cell. On the other hand synaptic terminal secretes the neurotransmitter, glutamate. Ca^{++} leaves the outer segment via a Na^{+}-Ca^{++} exchange carrier while Na^{+} leaves the inner segment via a Na^{+} pump. Absorption of photon activates one rhodopsin allowing it to catalyze the exchange of GDP bound on many molecules of transducin for GTP. It results in the liberation of free Gtα that activates cGMP phosphodiesterase (PDE). The activated PDE then converts cGMP to GMP. The decreased level of cGMP leads to dissociation of cGMP from the nucleotide gated channels in the plasma membrane of the cell that closes the channels. The membrane then becomes hyperpolarized and transmits the impulse to the brain via optic nerve.

Absorption of photon → Activation of rhodopsin → Gt-GDT → Gt-GTP → Free Gt a → Activation of cGMP-PDE → cGMP → GMP → Dissociation of cGMP from the nucleotide gated channel → Closure of the channels → Hyperpolarization of the membrane → Impulse transmission to the brain.

GPCR mediated activation of gene transcription: GPCR mediated intracellular signal transduction pathway may have short term effect by the modulation of enzymes or proteins but it may lead to gene transcription that is related with cell proliferation or differentiation. The *tubby* gene, expressed in certain areas of brain seems to control gluttonous feeding behavior and is responsible for obesity (Fig. 15.15). Tubby protein (showing a transcriptional

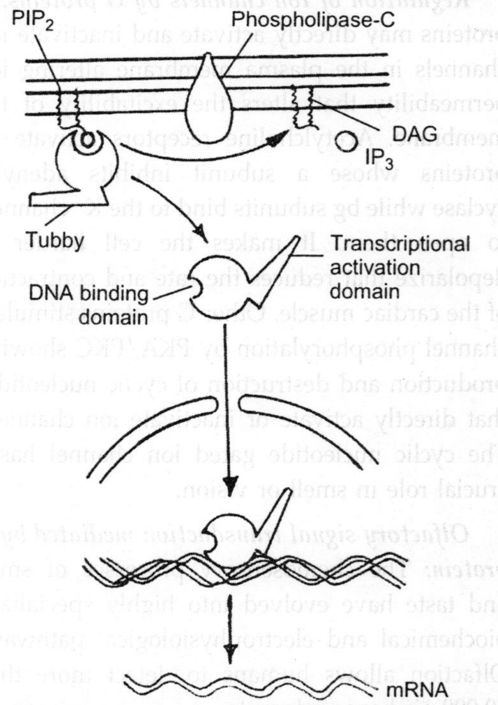

Fig. 15.13 : Activation of tubby transcription factor following ligand binding to GPCR.

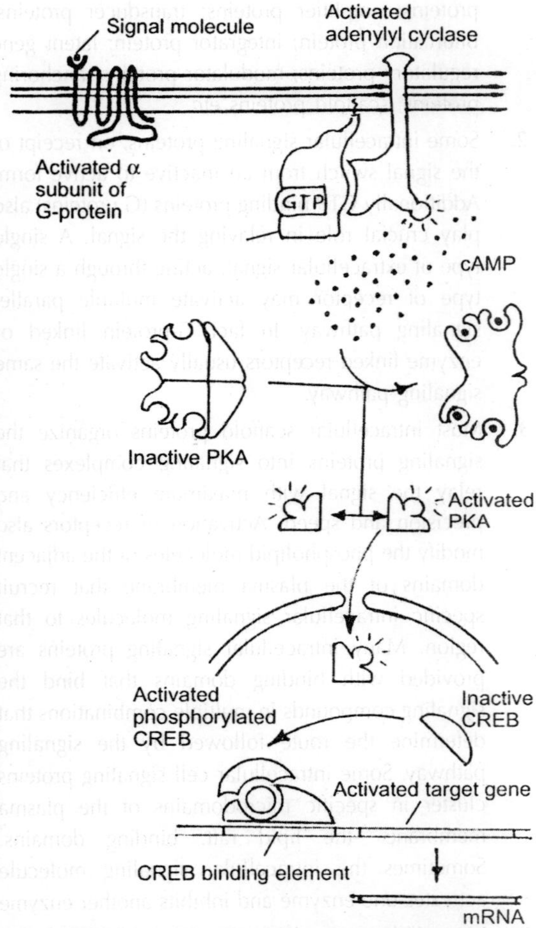

Fig. 15.14 : G protein mediated cAMP regulated transactivation of gene.

activation domain and a DNA binding domain) is found near the plasma membrane; it tightly binds to PIP2. Hormone binding to GPCR activates phospholipase C leading to the hydrolysis of PIP_2 and release of tubby protein which then enters into the nucleus and transactivates the gene.

Transactivation of genes by cAMP mediated by G-protein: cAMP mediated action may be slow due to activation of certain genes. The regulatory site of the genes regulated by cAMP contains cAMP responsive element – CRE to which binds a phosphorylated transcription factor – CREB. The activated GPCR in turn activates adenylyl cyclase leading to an increase in cAMP and subsequent release of the catalytic subunit of PKA that translocate into the nucleus. There it phosphorylates serine-133 on CREB protein which interacts with a coactivator CBP that links CREB to the basal transcription machinery leading to gene transcription (Fig. 15.14).

Gene expression by several kinases activated by Arrestin and mediated by GPCRs: β-arrestin binds to phosphorylated serine and tyrosine residues of the GPCRs. It transduces signals from activated receptors by binding to and activating

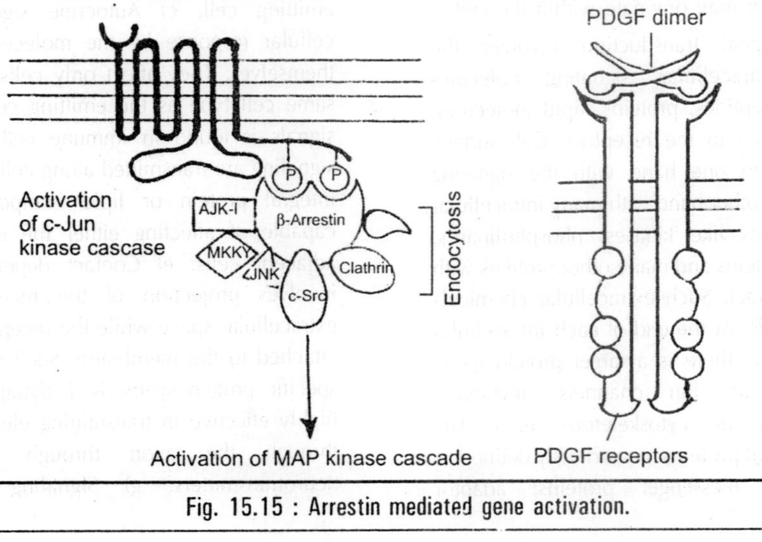

Fig. 15.15 : Arrestin mediated gene activation.

several cytosolic protein kinases like cSrc that activates the MAP kinase pathway leading to phosphorylation of key transcription factors for the synthesis of proteins needed for cell division. Interaction of β-arrestin with three proteins including JNK-1 (Jun N-terminal kinase) results in phosphorylation and activation of the c-Jun transcriptional factor needed for the synthesis of growth promoting enzymes and other proteins that help cells to respond against some stress conditions (Fig. 15.15).

Occurrence of GPCRs in the nuclear membrane: The action of several peptides and drugs is thought to be dependent on their interactions with specific cell surface GPCRs and ionic transporters such as channels and exchangers. Nuclear membrane also possesses Ca^{++} channels, Na^+-H^+ exchanger and endothelin-1 and angiotensin II receptors. Stimulation of these receptors induced an increase of nuclear Ca^{++}. There is compelling evidence that nuclear membrane located GPCRs exhibit intracrine mode of genomic regulation and elicit signals (Boivin *et al,* 2008).

SUMMARY

1. In multicellular organisms cells must depend upon the neighbouring cell for performing the normal vital activities. Cell gets signal from the environment or it may originate within the cell.

   Extracellular signal transduction involves the binding of "extracellular" signaling molecules (amino acids, peptides, proteins, lipid molecules, acetylcholine etc.) to the receptors. Cell surface receptors bind in one hand with the signaling molecule and in other hand with many intracellular signaling proteins like kinases, phospholipases, GTP binding proteins and many other proteins with which they interact. Such extracellular chemicals are called ligands. At the end of each intracellular signaling pathway there is a target protein (gene regulatory protein, ion channels, metabolic regulator, parts of cytoskeletons etc.) The intracellular signal proteins are variously defined as relay proteins; messenger proteins; adaptor proteins; amplifier proteins; transducer proteins; bifurcation protein; integrator protein; latent gene regulatory proteins; modulator proteins; anchoring proteins; scaffold proteins etc.

2. Some intracellular signaling proteins, on receipt of the signal switch from an inactive to active form. Additionally GTP binding proteins (G proteins) also play crucial role in relaying the signal. A single type of extracellular signal, acting through a single type of receptor may activate multiple parallel signaling pathway. In fact G-protein linked or enzyme linked receptors usually activate the same signaling pathway.

3. Most intracellular scaffold proteins organize the signaling proteins into signaling complexes that relay the signal with maximum efficiency and precision and speed. Activation of receptors also modify the phospholipid molecules of the adjacent domains of the plasma membrane that recruit specific intracellular signaling molecules to that region. Many intracellular signaling proteins are provided with binding domains that bind the signaling compounds in multiple combinations that determine the route followed by the signaling pathway. Some intracellular cell signaling proteins cluster in specific microdomains of the plasma membrane –the lipid raft. binding domains. Sometimes the intracellular signaling molecule activates one enzyme and inhibits another enzyme

4. Intercellular signaling involves a) Endocrine signaling, produced by endocrine cells b) Paracrine signaling, targets only cells in the vicinity of the emitting cell. c) Autocrine signaling involves cellular response by the molecules released by themselves. They affect only cells that are of the same cell type as the emitting cell. This type of signals is found in immune cells. d) Juxtacrine signaling are transmitted along cell membranes via integral protein or lipid components and are capable of affecting either the emitting cell or adjacent cells. e) Contact dependent signaling involves projection of the receptors into the extracellular space while the receptors still remain attached to the membrane. Such signaling affects specific protein synthesis. f) Synaptic signaling is highly effective in transmitting electrical impulses through the axon through exocytosis of neurotransmitters. g) Signaling through gap

junctions involves coordination of cellular activities through intercellular exchange of Ca^{++} and cAMP.

5. The molecules that enable signaling between the cells or tissues within an individual animal or plant are known as "hormones." Hormone-initiated signal transduction involves i) Biosynthesis of a hormone. ii) Storage and secretion of the hormone. iii) Transport of the hormone to the target cell that in turn may down-regulate hormone production. iv) Recognition of the hormone by the membrane receptor protein, leading to a conformational change. v) Relay and amplification of the signal vi) Removal of the hormone. There are two classes of hormone receptors, "membrane-associated receptors" and intracellular or "cytoplasmic" receptors. Signaling pathways in fact regulate all the cellular processes. Most of the pathways have positive and negative feed back loops that either augment or inhibit responses. Responses include the activation of a gene, the production of metabolic energy, cell locomotion etc.

6. Signaling molecules may only bind to the cell surface receptors, or may cross the plasma membrane and bind to cytosolic or nuclear receptors. Steroid hormones diffuse across the plasma membrane and bind to any intracellular receptor. Ligand binding may cause a conformational change in the receptor allowing it to bind to the regulatory DNA sequence and transactivate specific gene. Nitric oxide (NO) is a paracrine signaling molecule in the nervous, immune and circulatory system. Acetylcholine acts on the endothelial cells to stimulate NO synthesis that diffuses to neighbouring smooth muscle cells where it produces cyclic guanosine monophosphate (cGMP) that induces muscle cell relaxation and blood vessel dilation. Carbon monooxide also acts as a mediator of blood vessel dilation. The released neurotransmitters diffuse across the synaptic cleft and bind to the target cell surface receptors inducing a conformational change that opens ion channels resulting in ion flux in the target cell. Peptide hormones, neuropeptides, many polypeptide growth factors and cytokines regulate the development and differentiation of blood cells; control the activities of lymphocytes during immune response. They cannot cross the plasma membrane and act by binding to the cell surface receptors only. Ca^{++} remains bound with calreticulin and stored in ER and prior to participating in different activities; it is released into the cytosol. Ca^{++} is used in muscle contraction, release of neurotransmitter, vision in retina cells, proliferation, secretion, cytoskeleton management, cell motion, gene expression, apoptosis and metabolism. Ca^{++} can directly recognize protein or bind with the active center of an enzyme One of the best studied interactions of Ca^{++} with a protein is the regulation of calmodulin by Ca^{++}. The Ca^{++} / calmodulin complex plays an important role in proliferation, mitosis and neural signal transduction.

7. Receptions of extracellular signals is restricted to numerous proteins, called receptors. Some membrane anchored growth factors remain associated with the plasma membrane and play crucial role during direct cell-cell interactions. 20 different families of proteins serve as receptors. Transmembrane receptors span the thickness of the plasma membrane of the cell with one external end (extracellular domain) and one internal end (intracellular domain). Receptors are of three general classes: i) Receptors that are coupled, inside the cell, to GTP-binding and hydrolyzing proteins (termed G-proteins). ii) Receptors that penetrate the plasma membrane and have intrinsic enzymatic activity include tyrosine kinases, tyrosine phosphatases, guanylate cyclases and serine/threonine kinases. iii) Receptors that upon ligand binding migrate to the nucleus to affect gene transcription.

8. G-Protein Coupled Receptors (GPCRs) is the largest family of membrane receptors showing 7 transmembrane a helices to relay extracelluar signals inside the cell. These classes of receptor are also called seven helix receptors. All G protein linked receptors consists of a single polypeptide chain that crosses the plasma membrane seven times and are therefore called serpentine receptors. G-protein shows three sub units a, b, and g. bind with high affinity to the extracellular domains of the receptors. The β and γ subunits are inserted into the plasma membrane, α subunit shows a GTP binding domain. In unstimulated state a, b and γ subunits remain united but in stimulated state, α subunit binds GTP and dissociates from β and γ subunits. α subunit interacts with the target protein. Various GPCR-interacting and activating proteins modulate receptor function.

9. Initially the G-protein a subunit has bound GDP and the α, β, ... γsubunits are complexed together. Hormone binding causes Gα to releases GDP and bind GTP. Gα-GTP dissociates from the bg subunit complex and bind to adenylate cyclase that synthesizes cAMP. cAMP-dependent Protein Kinase-A phosphorylates various cellular proteins, altering their activity. Different signaling pathways emanating from one or several GPCRs exhibit extensive cross talk, resulting in positive or negative signal modulation. GPCRs can generate mitogenic signals and can be regulated by RTKs and other tyrosine kinases and downstream molecules. GPCR activation also initiates a series of molecular events that leads to GPCR-kinase-mediated receptor phosphorylation and the binding of β-arrestin proteins to the intracellular face of the receptor that may target many GPCRs for endocytosis in clathrin-coated pits. GPCRs can also signal by interacting with various small G-proteins super family like the Ras, Rho/Rac/Cdc42, Rab, Sar1/Arf and Ran families. They regulate downstream effector pathways to control many eukaryotic cell functions. An extra cellular signal can cause 20-fold increase of cAMP within seconds. All receptors that act via cAMP are coupled to stimulatory G protein that activates adenylyl cyclase and thereby increases cAMP concentration. Many GPCRs exert their effects by another G protein - Gq that activates membrane bound phospholipase C-β to produce IP_3 and DAG that act as second messenger. IP_3 opens IP_3 gated Ca^{++} channels in the ER membrane to raise cytosolic Ca^{++} level. DAG can be cleaved to release arachidonic acid or activates serine/threonine protein kinase C enhancing cellular growth, glycogen metabolism and transactivation of various genes.
10. Olfactory neurons show apical sensory cilia, G_{olf}, with cyclic nucleotide gated ion channels and cell body showing pumps and channels. Odorant receptor a modified GPCR catalyzes rapid exchange of GDP to GTP on G_{olf} that dissociates from Gbg subunit and activates adenylyl cyclase that generates cAMP molecules. It binds to cyclic nucleotide gated ion channels that admit the entry of Na^+, Ca^{++} and Cl^-. These ions depolarize the membrane that trigger an axon potential which travel to the brain.
11. Visual transduction responses are the fastest G-protein linked signaling pathway invertebrates. Human retina contains rods and cones that show outer and inner segments, a cell body and a synaptic region. The outer segment contains the pigment rhodopsin, and its plasma membrane contains cGMP gated Na^+ channels. Rhodopsin is associated with GPCR - transducin (Gt) and a chromophore – 11-cis retinal. In the resting condition cGMP keeps the cGMP gated Na^+ channels open allowing an influx of Na^+ and Ca^{++}, the synaptic terminal secretes the neurotransmitter glutamate, Ca^{++} leaves the outer segment while Na^+ leaves the inner segment via a Na^+ pump. Photon activates one rhodopsin exchanging GDP for GTP. It liberates free Gtα activating cGMP phosphodiesterase (PDE). The activated PDE converts cGMP to GMP leading to dissociation of cGMP from the nucleotide gated channels and hyperpolarization of the membrane.
12. GPCR mediated signal transduction may lead to the expression of tubby gene that controls gluttonous feeding behavior. Activation of certain genes may modulate cAMP mediated action. β-arrestin may bind to phosphorylated serine and tyrosine residues of the GPCRs. It transduces signals from activated receptors to activate the MAP kinase pathway leading to enhance cell division.
13. All RTKs show four major domains - a) an extra cellular ligand binding domain; b) a transmembrane domain; c) an intra cellular tyrosine kinase domain; d) an intracellular regulatory domain. Such receptors bind EGF, PDGF etc. and in turn phospho rylates and activates effector proteins that control cellular proliferation. The cytoplasmic protein kinase domain phosphorylates additional tyrosines in the receptor and effector proteins. Many receptors that have intrinsic tyrosine kinase activity contain tyrosine residues that upon phosphorylation interact with other proteins of the signaling cascade. The domains of other proteins are termed SH2 domains. Another protein-protein interaction domain identified in many signal transduction proteins is related to a third domain in c-Src and identified as the SH3 domain. The interactions of SH2 domain containing proteins with RTKs or receptor associated tyrosine kinases leads to tyrosine phosphorylation of the SH2

containing proteins. Many cell surface bound signal proteins like ephrins also act as ligands of these receptors. Binding of a ligand to the ligand-binding domain of the extra cellular part of the receptor activates the intracellular tyrosine kinase domain. Activated kinase transfers a phosphate group from ATP to selected tyrosine side chain both on the receptor protein and on the intracellular signaling protein. The docking proteins may be enzymes like PLC-γ that increases cytosolic Ca^{++} levels; cytosolic tyrosine kinase Src that phosphorylates other signaling proteins on tyrosine or PI 3-kinase which generates specific lipid molecules in the plasma membrane to attract other signaling proteins. The intracellular signaling proteins that bind to phosphotyrosine have either SH2 or SH3 domain or PTB (phosphotyrosine binding domain) that help them to bind either to phosphorylated tyrosine or to many intracellular signaling proteins that have been phosphorylated on tyrosines. Some signaling proteins consists of entirely of SH2 and SH3 domains and functions as adaptors to bind phosphotyrosine and other effector proteins. Such adaptor proteins couple activated receptor to the downstream signaling protein Ras that acts as a transducer and diversifies the signal on multiple pathways. Some activated RTKs recruit another adaptor protein – Shc that binds both to activated receptor and to Grb-2.

14. Ras belongs to the super family of monomeric GTPases. It transduces signal for proliferation or differentiation. Ras triggers a series of down stream serine/threonine phosphorylation that relay the signal to the nucleus to transactivate specific gene. The major kinases in this cascade are MAP kinase. Ultimate targets of the MAP kinases are several transcriptional regulators and the proto-oncogenes Fos, Myc and Jun. Once activated, MAP kinase phosphorylates other kinases and even gene regulatory proteins within the nucleus. Thus it activates the transcription of a set of immediate early genes some of which again activates other gene regulatory proteins. Thus Ras-MAPK signaling pathway transduces signal from the cell surface to the nucleus and affects the pattern of gene expression. MAPK are inactivated by dephosphorylation from either tyrosine or threonine or both providing a form of negative feed back.

15. Jak-STAT signaling pathway involves a diverse class of enzyme-linked cytoplasmic receptors that rely on cytoplasmic kinases called Jaks that mainly receive information from local mediators (called cytokines) and hormones for signal transduction. These in turn activate the latent gene regulatory membrane associated proteins called Stats. Activated Stats migrate to the nucleus and trasactivate specific gene.

16. Protein tyrosine phosphatases (PTP) occur as both cytoplasmic and transmembrane protein. They remove phosphates from phosphotyrosines. Some of the PTPs act as receptors and are called receptor like tyrosine phosphatases (RTP). They play crucial role in the activation of T and B-lymphocytes. Some participate in cell adhesion. Cytoplasmic tyrosine phosphatases help in blood cell development

17. TGF-β super family is categorized as enzyme linked receptors. During development they regulate cell proliferation, differentiation, ECM production and apoptosis. Activated TGF-β and activin phosphorylates Smad that dissociates from the receptor migrates to the nucleus and recruits other gene regulatory protein and transactivate specific target gene.

18. Signal transduction during animal development depends on regulated proteolysis. Signaling through Notch receptor protein involves inhibition of neighbouring nerve cell by each precursor nerve cell. It is called lateral inhibition, regulated by a single pass protein called delta (or delta like ligand – serrate) that is received by neighbouring notch receptor. On binding with the delta protein the cytoplasmic portion of the notch receptor is cleaved and migrates to the nucleus and transactivate notch responsive gene required for neural differentiation.

β-catenin serves in intercellular adhesion and in the regulation of latent gene. Wnt protein controls many aspects of development. Wnt binds to both a frizzled receptor and a co-receptor protein called low-density lipoprotein receptor related protein (LRP). Phosphorylated β-catenin is rapidly degraded with the help of a GSK,3β, APC and axin. Unphosphorylated β-catenin accumulates in the cytoplasm and translocates to the nucleus where β-catenin transactivates the Wnt target genes like c-myc that stimulates cell growth and proliferation leading to cancer.

Hedgehog proteins are a family of signal molecules acting as local mediators during development. Any abnormality in the Hedgehog pathway during development can be lethal and lead to cancer.

QUESTIONS

Long questions:

1. What do you mean by intracellular signaling molecules? Classify them.
2. How Ca^{++} remains stored in the ER and how it is released in to the cytosol. Mention its roles.
3. Describe the structure of G protein and signal transduction through G proteins.
4. How olfactory and visual signal transduction are mediated by G-proteins.
5. Describe the steps and malregulation of Wnt signaling.

Short questions:

1. What do you mean by ligands?
2. Which chemicals act as signaling molecules?
3. Name the chemicals that can penetrate the plasma membrane directly.
4. Name some intracellular signaling molecules?
5. Distinguish between first and second messenger?
6. What are the roles of relay proteins, messenger proteins, adaptor proteins, amplifier proteins, transducer proteins and integrator proteins, bifurcation proteins, latent gene regulatory proteins, modulator proteins, anchoring proteins and scaffold proteins?
7. Name a protein that switch over from inactive to active form during signal transduction.
8. What are the roles of SH and PTB domains?
9. Name a signaling molecule that has antagonistic effects.
10. What do you mean by desensitization?
11. What do you mean by feed back inhibition?
12. How NO, mediate signal transduction?
13. What do you mean by calcium wave?
14. What are the characteristic features of nuclear receptors? How they act?
15. What do you mean by aporeceptor complex?
16. What do you mean by cross talks?
17. What do you mean by desensitization?
18. Name some nuclear envelope GPCRs.
19. What do you mean by kinase insert?
20. What do you mean by bi-directional signaling?
21. Why MAPK are called ERKs?
22. What do you mean by lateral inhibition?
23. Name the receptors that are associated with Hedgehog signaling.
24. What are the roles played by ROS?

Fill up the blanks:

1. The part of the plasma membrane that participates in signal transduction is called ———.
2. ——— signaling targets the neighboring cells of the emitting cell.
3. Hormones are also called ———.
4. ——— diffuses across the plasma membrane and bind to any intracellular receptor.
5. NO is a ———.
6. NO is synthesized from the amino acid ———.
7. Many neurotransmitters mediated cellular response are coupled to ——— proteins.
8. Abnormalities in the cytokine signaling may lead to ———.
9. GPCRs are also called ——— receptors.
10. Glucagon, angiotensin, vasopressin and bradykinin receptors are ———.
11. PDGF, insulin, EGF and FGF receptors are ——— receptors.
12. Activin and TGF-β receptors belong to serine / threonine kinase family.
13. Mitogen activated protein kinases (MAP kinases) belong to ——— family.
14. GPCRs modulating adenylate cyclase activity produces ——— as the second messenger.
15. GPCRs that activate PLC-β hydrolysis generate DAG and IP_3 as ———.
16. Ras is a ——————— that is related with the development of many cancer.
17. Neurofibromin is a ———.
18. cAMP is synthesized by a cell membrane bound enzyme ———.
19. Eicosanoids are associated with -——— sensation.

20. ——— facilitates intracellular signal transduction along multiple routes.
21. The MAPKs that phosphorylate tyrosine and threonine are called ———.
22. MAPK kinase in the mammalian Ras signaling pathway is called ———.
23. MEK is activated by MAPKK kinase that is called ———.
24. ——— are a type of cytoplasmic kinases that receive information from cytokines and hormones.
25. Jaks activate the latent gene regulatory membrane associated proteins ———.
26. The molecule that regulates salt and water balance and dilates blood vessels is ———.
27. The basal part of notch receptor is cleaved by ——— and is translocated to the membrane associated ——— protein.
28. Wnt protein transduce signal through a cytoplasmic protein ———.
29. Mutation in GSK-3β and APC lead to ——— cancer.
30. Cytoskeletal reorganizations are involved in ——— signaling in plants.
31. During osmotic stress, plant cells may increase the production of ———.
32. During osmotic stress, plant cells trigger ——— release due to cleavage of PIP_2.
33. Osmotic stress is related with ——— signaling in plants.

MCQs:

1. **Nerve impulse propagation involves**
(a) autocrine, (b) paracrine, (c) juxtacrine signaling.
2. **Signaling in immune cells involves**
(a) autocrine, (b) paracrine, (c) juxtacrine signaling.
3. **Signal transduction via membrane proteins and lipids is called**
(a) autocrine, (b) paracrine, (c) juxtacrine signaling.
4. **The second messenger in NO mediated signaling is**
(a) Ca^{++}, (b) cAMP, (c) cGMP.
5. **Ca^{++} in the ER/SR remains bound to**
(a) calmodulin, (b) calreticulin, (c) cGMP.
6. **Glucagon receptors are**
(a) GPCRs,
(b) tyrosine kinase receptor,
(c) nuclear receptor.
7. **In stimulated G protein, Gα binds with**
(a) ATP, (b) GTP, (c) GDP.
8. **Proto-oncogenic serine/threonine kinase in Ras signaling pathway is**
(a) MEK, (b) Raf, (c) My(c)
9. **One of the targets of MAPK is**
(a) Myc, (b) CREB, (c) Tubby.
10. **Notch signaling is best known during the development of**
(a) heart, (b) nerves, (c) kidney.
11. **Notch receptor is generated in**
(a) ER, (b) Mitochondria, (c) Golgi bodies.
12. **Free β-catenin is degraded by a serine/threonine kinase**
(a) GSK-3β, (b) axin, c AP(C)
13. **In the nucleus**
(a) APC,
(b) groucho,
(c) axin inactivates Wnt responsive genes.
14. **A *Wnt* target gene is**
(a) Fos, (b) Jun, (c) Myc
15. **Cold, drought or salinity stress in plants, cause**
(a) Ca^{++}, (b) Na^{+}, (c) K^{+} influx into the cytosol.
16. **Stress response in plants may be mnediated by**
(a) NADP, (b) FADH, (c) ROS without Ca^{++}.

Answers to MCQs:

1. b, 2. a, 3. c, 4. c, 5. b, 6. a, 7. b, 8. b, 9. a, 10. b, 11. c, 12. a, 13. b, 14. c, 15. a, 16.c.

Glossary

A

α-actinin: It is important actin filament cross-linking protein forming cortical network that help to bind actin filaments to the plasma membrane. Looser association of actin and α-actinin allows myosin to be incorporated in the stress fibres. a-actinin holds the ends of the stress fibre in the *focal contacts* at the plasma membrane.

A-, P- and E sites: These are three specific sites in the ribosome that are associated with amino acid incorporation. Amino acyl tRNA carrying a specific amino acid at first binds in acceptor or A site and then is translocated to peptidyl of P site. After peptide linkage formation between these amino acids the first amino acid is translocated to the exit site or E site from which the tRNA leaves the ribosome.

ABC Transporters: ABC ("ATP Binding Cassette") transporters are transmembrane proteins that expose a ligand-binding domain at one surface and an ATP-binding domain at the other surface. ATP provides energy to pump the ligand across the membrane.

Abzyme: An antibody selected for its ability to catalyze a chemical reaction by binding to and stabilizing the transition state intermediate.

Acentric chromosome: Applied to a chromatid or a chromosome when it lacks a centromere. This condition may arise in an inversion heterozygote as a result of crossing over between a normal and an inverted segment that does not include the centromere.

Activation energy: Energy needed to initiate a chemical reaction.

Actin filament / Microfilament: Actin is a globular protein that polymerizes helically to form *actin filaments.* These are being the thinnest of the cytoskeletal filaments, are also called *microfilaments.*

Actin related proteins: More than 60 families of proteins regulate the assembly and dynamics of actin based structures in the cytoplasm. These are called actin *related proteins*

Acid-base catalysis: Enzymes contain several kinds of functional groups capable of acting as general acids or bases. Acid-base catalysis provides a means for catalyzing at neutral pH where the concentration of H^+ and OH^- are very low.

Acrocentric chromosome: Chromosome where centromere occurs near the telomere. Such chromosome show a long arm and a short arm.

Active site : A small portion of the enzyme that participates in the catalysis by binding with the substrate and contributes in the chemical transformation of the substrate.

Active Transport: Pumping of molecules or ions through the membrane against their concentration gradient with the help of energy.

Achiasmatic meiosis: Usually chiasma formation is the normal rule in meiosis but in many cases like male *Drosophila, Panorpa* etc. chiasma formation does not take place. This is called *achiasmatic meiosis.*

Acetylcholine: A chemical neurotransmitter that is released from the ends of certain types of nerve fibres when they are stimulated. It transmits a signal to an adjacent nerve or muscle cell by binding to receptors on the target cell surface.

Acetylcholine receptor channel: A type of receptor for the neurotransmitter acetylcholine, found on skeletal muscle cells and some nerve cells. Also known as the nicotinic acetylcholine receptor, it is made up of a number of protein subunits that form an ion channel through the cell membrane. The channel is opened by the binding of acetylcholine to extra cellular sites on the protein.

Action potential: The sequential, electrochemical polarization and depolarization that travels across the membrane of a nerve cell (neuron) in response to stimulation. It involves rapid influx of Na^+ so that the initial membrane potential becomes +30 mV.

Adapter: A device used to fit smaller tubes or centrifugal devices in the rotor cavities.

Adhesion molecules: In multicellular organisms, most of the cells are bound together by these molecules.

Adult stem cells: The stem cells that are found in various adult tissues that give rise to differentiating cells and also act as a repair system for the body, replenishing adult tissues.

Agarose: It is purified from agar. Different purities of agarose with different melting properties are available. High purity low melt agarose is used for the extraction of DNA.

Agarose gel electrophoresis*:* It is a biochemical and molecular biological method used to separate DNA or RNA molecules facilitated their movement through an agarose matrix with an electric field. Shorter molecules move faster and migrate further than longer ones.

Aleuroneplasts: The plastids that mainly stores proteins.

Allosome: Other than autosomes, all chromosomes of an organism are called allosome.

Allosteric enzymes: The tetrameric enzyme, one whose catalytic activity is altered by binding of specific ligands at sites other than the substrate binding site.

Allozyme: Form of an enzyme that differs in amino acid sequence, as shown by electrophoretic mobility or some other property, from other forms of the same enzyme and is encoded by one allele at a single locus.

Amyloplasts: The plastids that store starch.

Anaphase-A: It is characterized by the poleward separation of the sister chromatids. Up to metaphase the attraction between the sister chromatids is maintained by cohesin. At M-A transition the centromeric proteins MAD and BUB activates APC to release the enzyme - *separase* that cleaves the Sccl subunits of cohesin leading to its destruction and allowing the spindles to pull the chromatids towards opposite poles.

Anaphase B: It involves slight movement of the spindle poles towards the cortex. As the chromatids reach the poles, the motor molecules of the cortical cytoplasm - dynein pulls the spindle poles towards the cortex. Keeping pace with this pulling tension the kinesins - BimC at the mid zone of the inter polar fibers slowly glides, as a result both poles move towards the cortex slightly.

Analyte: It is the substance which is to be separated during chromatography.

Analytical chromatography: It is used to determine the existence and possibly also the concentration of analyte(s) in a sample.

Angiogenesis: It refers to formation of new blood vessels to provide nutrition and disposal of waste products that is the primary requisite for cancer cells. They synthesize VEGF and FGF that enhance neovascularization.

Ankyryn: It is important actin filament cross-linking protein forming cortical network that help to bind actin filaments to the plasma membrane.

Annexins: They bind in a Ca^{++}-dependent manner to membrane phospholipids and are supposed to regulate dynamic changes in membrane architecture. They also regulate the binding between secretory vesicles and the membrane that proves that they are the ideal mediators in cell membrane and cytoskeleton interactions.

Antiport Pumps: These carriers drive ion in one direction providing the energy for the active transport of other molecule or ion in the opposite direction.

Antivitamin: Some chemical compounds inhibit the absorption or actions of vitamins. E.g. Avidin, a protein in egg whites inhibits the absorption of biotin

Apicoplasts: The Apicomplexa, a phylum of obligate parasitic protozoa also harbors a complex plastid called *apicoplast.* It is no longer capable of photosynthesis, but is an essential organelle and a promising target for antiparasitic drug development.

Apoenzyme: The protein part of an enzyme.

Arp2/3 complex: They are important actin capping proteins. They jointly cap the pointed end of the actin filament. They also participate in the formation of branched filament.

Astral fibres: The small fibers that occur around each pole, are called *astral fibers,* these remain connected with the cortical fibers of the cell.

Asymmetric karyotype: remarkable differences between smallest and largest chromosomes and fewer metacentric and mostly acrocentric chromosomes are called asymmetric karyotype.

Autolysis: Disintegration of a cell due to the rupture of the membrane of the lysosome is called autolysis.

Autophagosome: During autophagy, a membrane cisterna known as the *isolation membrane* (IM or *phagophore*) appears and curves around the membrane-bound organelles along with part of the cytoplasm that forms a unique double membrane vesicle, the *autophagosome* that fuse with the lysosome. Within the autophagosome the dying organelles are gradually degraded.

Autophagy: Cellular organelle can be senile or damaged rendering putrefactive damage to the cell. Lysosome mediated degradation of the non functional cellular organelles is called autophagy

Autosome: Those chromosomes that regulate the vital activities of an organism are called autosomes.

B

Bacterial artificial chromosome: A vector used to clone DNA fragments of 100 to 300 kb insert size (average of 150 kb) in *E. coli* cells. Based on the naturally occurring F-factor plasmid found in the bacterium *E. coli.*

Balbiani rings: Some of the puffs of the polytene chromosomes appear are fuzzy and large called Balbiani rings.

Barrett's esophagus: Esophagus is lined by squamous epithelium that in response to diverse etiological factors like HCL and duodenal juice and undergo a series of intermediate cellular changes exhibiting erosion and subsequent replacement of squamous epithelium by columnar epithelium. This condition is called *Barrett's esophagus.*

Benign tumour: Tumours that are localized to a definite area are called benign tumours. (e.g. warts, benign skin tumours). A fibrous capsule completely covers the tumour. The cell adhesion molecules that hold tissues together keep the cells localized to the tissue where they originate.

Binding energy: The energy obtained from non-covalent interaction between energy and substrate is used to lower the activation energy. It is called binding energy.

Binding site: Adjacent to the active site, it here is a site which assists in the binding of the substrate called binding site.

Biphasic action of the inhibitor of allosteric enzyme: Allosteric enzymes may be subject to normal compe titive inhibition. A low concentration of competitive inhibitor would increase the reaction velocity while at higher concentrations of the inhibitor it slows the reaction rate. This is the biphasic effect of the inhibitor of the allosteric enzymes.

Biofilm*:** Bacteria may produce micro colonies within the slime layer called biofilm which protects them from the environment. ***Bivalents: In zygotene, the homologous chromosomes are arranged side by side to ensure genomic and physical contacts between them. This process of lateral arrangement is called *pairing* and the paired homologous chromosomes are called *bivalent.*

***Branching enzyme*:** An enzyme that catalyzes the creation of branch points in glycogen (in plants, amylopectin). Its deficiency causes glycogen storage disease type IV. 1,4-±-glucan is a branching enzyme.

Breakage and reunion: The general mode by which recombination occurs. DNA duplexes are broken and reunited in a crosswise fashion according to the Holliday model.

Bright field microscopy: Simplest light microscopy technique in which sample is illuminated via transmitted light that is viewed from the above.

Bouquet or polarized arrangement: In the leptotene stage, the Telomeres of all chromosomes aggregate near the centrosome associated nuclear membrane. This meiosis-specific telomere clustering is defined as the *'Bouquet arrangement' or 'Polarized arrangement'* due to its resemblance to a bouquet of cut flowers.

Bundle sheath cells: Special cells found in leaves of C4 plants in which the 4 carbon acids formed initially in the mesophyll cells are decarboxylated and CO_2 is fixed by Rubisco. These cells are characterized by thick walls which increase CO_2 con-centrations near the Rubisco sites, that prevent CO_2 from leaking out of the cells and enhance carbon fixation over photorespiration.

Burkitt lymphoma: A cancer of the lymphatic system manifested by tumours in the jaw, often associated with a translocation bringing a specific oncogenes next to a novel regulatory element.

Buttressing group: The part of the active site where substrate is held before conversion.

C

C-banding: Specific chromosomal bands obtained with Giemsa staining showing constitutive heterochromatin regions, so it stains centromeres and other such regions.

cDNA library: It refers to a complete or nearly complete, set of all the mRNAs contained within a cell. As mRNA is unstable and is easily degraded by RNases (which can be found even in the sweat secreted on the skin), reverse transcriptase is used to form a DNA copy of each mRNA strand. cDNA or reverse transcribed mRNAs are collectively known as the library.

C-3 plants: After fixation of CO_2 by RUBISCO, it adds CO_2 to the five carbon sugar ribulose 1,5-biphosphate to form two molecules of three carbon compound - *Phosphoglycerate.* Due to the formation of this 3-carbon compound as the first formed stable compound in the CO_2 fixation these plants are called *C-3 plants.*

C4 plants: In C4 plants, the first stable products of carbon assimilation are 4-carbon carboxylic acids produced. Cellular differentiation occurs which spatially regulates carbon assimilation. The 4 carbon acids are formed in the mesophyll cells and are then transferred to special cells (bundle-sheath cells) where they are decarboxylated and the CO_2 released is then fixed by Rubisco in the PCR cycle.

Cajal bodies: In plant and human nuclei, in the interchromosomal domains, in the vicinity of nucleoli, roughly round *Cajal bodies* (formerly known as *Coiled bodies*) are noted. These are involved in mRNA proce ssing, snRNP biogenesis and in the trafficking of snoRNPs and snRNPs, telomerase RNP biogenesis

Caldesmon: These proteins together with tropomyosin regulate actin-myosin interaction mediated by Ca^{++}. It also reorganizes actin filaments during mitosis.

Calnexin: It is a type of chaperone that promotes folding of the entire polypeptide into proper conformation with the help of Ca^{++}. It monitors the folding of the glycoproteins by repeated addition and release of glucose by glucosyltransferrase and glucosidases respectively and protects the nascent glycoprotein from premature misfolding and degradation.

Cap Z: They are important actin capping proteins and bind the '+' end of the actin filament in muscles independently of Ca^{++} but mediated by cofilin and profilin.

Capsules: Polysaccharide or polypeptide layer outside of the cell wall of a bacteria.

CARMIL: It is an actin filament capping protein may play an important role actin assembly.

Carcinoma: Cancers of epithelial cells are collectively called carcinoma. Cancers of skin, esophagus, breast and brain belong to this type.

Carcinogens: The physical agents that cause cancer are called *carcinogens.*

Cardiolipin: The inner membrane of mitochondria contains this specific lipid that renders it virtually impermeable to any metabolite.

Catastrophe: The microtubules may grow steadily and shrink rapidly by disassembly at the plus end. The rapid disassembly is referred to as catastrophe.

Caveolae: Invaginated parts of membranes rich in free cholesterol, glycoshingolipid, arachidonic acid, cave G-proteins, integrins and flotillin, regulatig growth, cell division, adhesion, signal transduction, cholesterol trafficking, capillary permeability and hormonal response.

Cdks: These are the principal protein that drives the molecular events of cell cycle but unless they are tightly bound to cyclin they have no kinase activity. They integrate intra and extra cellular signals to ensure smooth coordination of the cell cycle events.

Cell: A protoplasmic mass, surrounded by a selectively permeable membrane that depends upon its neigh- bouring cells for its existance and activities.

Cell cycle: The cycle of cell growth, replication of the genetic material and nuclear and cytoplasmic division. The cell cycle is divided into a G1 (G for gap) phase in which the cell grows; an S (S for synthesis) phase where the DNA is replicated, a second gap phase - G2 and an M phase (for Mitosis) in which the replicated DNA (organized into chromosomes) is separated and the cell divides into 2

Cell theory: Advocated by Schleiden and Schwann (1839) states that (i) Cell is the unit of all organisms; (ii) all cells are derived from preexisting cells; (iii) all vital functions of an organism occur within cells and (iv) cells contain the hereditary information that is transmitted to the next generation of cells.

Cell plate: At the end of telophase, numerous electron dense particles accumulate in the interface of two cells forming a plate like structure called cell plate.

Cellulose: It is a polymer of glucose – typically consisting of 1,000 to 10,000 1-4-b-D-glucose residues. It is the major component of primary and secondary wall layers

Cell sap: The central vacuole of a plant cell remains filled up by the watery structure that differs from that of the surrounding cytoplasm called cell sap.

Cell wall: In plant cells, outside the plasma membrane there is a three layered structure called cell wall having middle lamella, primary cell wall and secondary cell wall.

Centromere: A kinetochore; the constricted region of a nuclear chromosome, to which the spindle fibers attach during division.

Coenocytes: Multinucleated plant cells.

Centromere: The central part of the primary constriction where the DNA fibers remain united is called centromere. It is a chromosomal locus that ensures binding of the chromosomes with the kinetochore microtubules leading to the delivery of chromatids to each daughter cell during cell division.

Centromere marker: Locus located close to a centromere.

Centromeric fission: Creation of two-chromosomes from one by splitting the centromere.

Centrosome directed pathway: In this pathway of spindle fibre formation MT assembly is nucleated by MTOC with the involvement of γ-tubulin in association with an additional protein XMAP215.

Centrosphere: The pericentriolar material surrounding the paired centrioles is called *centrosphere.*

Centrifuge machine: An apparatus, used to separate different cellular components on the basis of shape, size and density in response to the centrifugal force.

Checkpoints: Flawless progression of cell cycle means completion of a phase that triggers the initiation of the next phase. If any flaw is noted it automatically inhibits the entry into the next phase. These halts are technically known as *checkpoints* that monitor extra cellular signals, DNA damage DNA replication etc.

Chemiosmosis: The energy released as electrons passing down the gradient from NADH to oxygen is harnessed by three enzyme complexes of the respiratory chain (I, III, and IV) to pump protons (H^+) against their concentration gradient from the matrix into the intermembrane space of the mitochondrion. As their concentration increases there, a strong diffusion gradient is set up and the protons flow down their gradient through the ATP synthase to be used for the synthesis of ATP. The process is called chemiosmosis.

Chiasma: The synaptinemal complexes begin to disintegrade in the diplotene, so that the homologous chromosomes begin to separate but at the point of contact of the crossing over they remain united so that due to constant repulsion it looks like an 'X' shaped structure called *chiasma* (plural, *chiasmata)* observed between non-sister chromatids. They represent physical crossovers between DNA molecules.

Chromatosome: Nucleosome with H1 and associated DNA showing a total length of 166 bp DNA constitutes the unit chromatin subunit called chromatosome

Chromocentre: The heterochromatic regions of all

the chromosomes in a cell with polytene chromosome are fused in the center to form an undifferentiated structure, called chromocentre

Chondriome: In higher plants all mitochondria in a cell collectively is called *chondriome.*

Chloroplast: The plastids with green chlorophyll pigments and responsible for photosynthesis.

Chromatids: Towards the end of prophase, the chromonemata gradually condensed to form chromatids. Thus chromonema and chromatids are same structure in different phases of cell cycle.

Chromomeres: Polytene chromosomes and leptotene chromosomes, due to non-uniform condensation, show numerous beads like structures throughout the longitudinal axis of the chromosomes. These are called chromomeres.

Chromoplasts: The plastids containing various coloured pigments (orange coloured carotene, lycopene and luteol pigments or yellow coloured xanthophylls) meant for the colouration of leaves, fruits etc. Synthesis of sugars from CO_2 is the most extensively studied biosynthetic pathway in plant cells. The first phase of this pathway is the fixation of CO_2.

Chromonemata: In interphase and early periods of prophase, in the nucleoplasm the paired, tiny, thread like paired identical structures, joined at one point are called chromonemata

Chromosome directed pathway: In this pathway of spindle fibre formation, chromosomes play an active role. During interphase, in the nucleoplasm aster promoting activator (APA), importin α and β unite to form a multiprotein complex. In prometaphase, APA and importins become separated. As a result, the APA becomes active in promoting microtubule assembly.

Chromosomal passenger proteins: These proteins play important roles in chromatin modifications, kinetochore-microtubule interactions, chromosome bi-orientation, stability of the bipolar spindle, mitotic checkpoint function, assembly of the central spindle and cytokinesis. These are INCENP (inner centromere protein), Aurora B kinase, survivin, telophase disc 60 (TD-60) and borealin. These are all mutually interdependent and are required for a number of key functions during mitosis.

Chromosome scaffold: In all chromosomes the central fabric with whom the loops of DNA fibers remain attached is called *chromosome scaffold.* It is made up of non-histone proteins and *DNA topoisomerase II.*

Chromosome territory-interchromatin compartment model: As per this model, nucleoplasm shows different domains specified for specific structures. The transcription complexes occupy interchromatin compart ments. Active genes are positioned on the surface of the chromatin domains that line the interchromatin compartment. Each chromosome occupies its own territory. Chromosomes with low gene density and inactive X chromosomes in females are located at the nuclear periphery while chromosomes with high gene density are located in the interior of the nucleus.

Chromatogram: It is the visual output of the chromatograph. in the case of an optimal separation, different peaks or patterns on the chromatogram correspond to different components of the separated mixture.

Chromatograph: It is equipment that enables a sophisticated separation e.g. gas chromatographic or liquid chromatographic separation.

Chromatography: A physical method of separation in which the components to be separated are distributed between two phases, one of which is stationary (stationary phase) while the other (the mobile phase) moves in a definite direction.

Uses of different chromatography:

i) ***Column chromatography:*** Used *for separation of the mixture of tissue lipids.*

ii) ***Thin layer chromatography:*** Used for the sepoaration of amino acids, nucleotides and low mol. wt. components.

iii) ***Ion exchange chromatography:*** Used for the purification of insulin and plasma fractionation.

iv) ***Gel filtration chromatography:*** Used for determining molecular weight of proteins.

v) ***Affinity chromatography:*** Used for the separation of immunoglobulin, enzymes and mRNA.

Cis Golgi network (CGN): In the cis part of the Golgi complex, the intermediate vesicles, tubules released from the ER form a network like structure called Cis Golgi network.

Cloning: Artificial multiplication of a desired gene by plasmid mediated multiplication in a bacterium or in a PCR machine.

Coenzyme: An organic molecule, such as a vitamin, that binds to an enzyme and is required for its catalytic activity

Cohesin: It consists of five subunits - Scc1, Scc3, Smc1, Smc 3 and Pds1p proteins. Smc1 and Smc3 associate to form a 'V' shaped structure. Scc1 forms a triangular ring that holds the sister DNA molecules following DNA replication, Cohesins bind along the chromosomal arms and around the centromere.

Coenocyte: Multinucleate plant cell. E.g. *Voucheria*

Cofactor: The non-protein part of a conjugated enzyme. They can be either inorganic (*e.g.*, metal ions and iron-sulfur clusters) or organic compounds (e.g., flavin and heme).

Cold sites: Specific regions of the histone core around H3-H3, H2A-H4 and H4-H2A interhistone interfaces, termed "cold sites", which retain a significant fraction of contacts with adjoining residues indicating their functional role in nucleosome organization.

Competitive inhibition: Such inhibitors resemble the substrates so closely that in lieu of substrates the inhibitor binds the active site forming enzyme-inhibitor (EI) complex.

Complex-I: The largest of the electron transport chain proteins that accepts electrons from NADH and passes them to the next electron carrier, coenzyme Q.

Complex-II: One of the electron transport chain proteins that transfers electrons from $FADH_2$ onto coenzyme Q.

Complex-III: One of the electron transport chain proteins that accepts electrons from ubiquinone and passes them on to cyt-c.

Complex-IV: One of the electron transport chain proteins that accepts electrons from cytochrome c and passes them to oxygen to form water (H_2O).

Condensin: It seems to be the key molecular component underlying chromatin condensation and segregation. It occurs at every 5-10 kb of DNA and binds directly throughout the central axis of the chromosome. They result supercoil formation in the presence of topoisomerase I and knots in the presence of topoisomerase II.

Congregation: During metaphase the kinetochore fibers elongate and shorten and align the chromosomes in one plane that is exactly half way between the two poles. This type of chromosomal movement for proper alignment is called congregation.

Cofilin: It binds to ADP-actin units and promotes their dissociation from the '-' end and prevents their reassembly during nucleation of actin filament. It an important actin filament severing protein.

Conjugated enzymes: These enzymes show one protein and one non-protein part. The protein part is called *apoenzyme* while the non-protein part is called *cofactor.*

Conserved Oligomeric Golgi (COG) complex: It is a peripheral Golgi protein associated with the that participate in protein trafficking between ER and Golgi, retrograde vesicular transport and to maintain normal Golgi structure and function.

Constitutive enzyme: Those enzymes which are produced constantly, irrespective of environmental conditions or demand.

Constitutive secretory pathway: As per this pathway some proteins are produced and secreted in a constitutive or constant manner. Irrespective of getting signal for release these proteins are budded off from the TGN and exported directly to the plasma membrane for exocytosis. Such proteins are used just outside the cell like extra cellular matrix.

Contact sites: The contact points between outer membrane and the inner boundary membranes of the mitochondria.

Copy-choice hypothesis: An incorrect hypothesis that stated that recombination resulted from the switching of the DNA-replicating enzyme from one DNA homologue to the other.

Copy-choice model: A model of the mechanism for crossing over, suggesting that crossing over occurs during chromosome division and can occur only between two supposedly new nonsister chromatids; the experimental evidence does not support this model.

Corona: The outermost layer of the kinetochores showing motor proteins that move chromosomes during mitosis is called corona.

Corral model: As per this model some specific parts in the cytoskeleton beneath the membrane called '*corral*' which binds the proteins and regulated intercellular protein translocation.

Crinophagy: It involves direct fusion of lysosomes with the secretory vesicles to degrade non-functional secretory proteins.

Cristae: Tubular, lamellar or triangular projections from the cristae membrane of the mitochondria.

Cristae junctions: The tubular junctions of the inner boundary membrane and cristae.

Cristae membrane: The membrane that forms in close apposition to the inner membrane of the mitochondria.

Crossing over: Throughout the length of the non-sister chromatids of the tetrad, at the recombination nodules the chromatids break and reciprocally unite. This phenomenon is called *crossing over.*

Cryopreservation: It involves freezing a specimen by liquid nitrogen or liquid helium at a temperature, in which water forms vitreous (non-crystalline) ice.

Crystalloids: Reserve proteins in the aleurone grains of castor contain polyhedrral crystals that are called crystalloids which that remain in close association with a round granular part (cantaining calcium and magnesium phosphate) called *globoids.*

Cyanophycin: It is an interesting protein found in the cytosol of blue green algae that sequesters nitrogen for future use.

Cyanophycean starch: The cytosol of blue green algae is also loaded with small grains of polymerized carbohydrate (product of photosynthesis) called *cyanophycean starch.*

Cyclin: The protein that actively regulate the cell cycle; typically synthesized and degraded during the cell cycle to regulate the activity of a cyclin-dependent kinase. Cyclin B is the large protein subunit of mitosis promoting factor (MPF); cyclin B is synthesized and degraded during the cell cycle to regulate MPF activity. Cyclin B plus cyclin dependent kinase 1 (cdk1, aka cdc2, aka p34 kinase) together form an active MPF protein. There are many other cyclins that regulate the cell cycle.

Cystoliths: These are spindle shaped depositions of $CaCO_3$ crystals in the lithocysts.

Cytoskeleton: Cytoplasm of almost all cells are traversed by three types of protein polymers called cytoske letons. These include actin filaments, microtubules and intermediate filaments.

Cytokinesis: The division of the cytoplasm of a cell into two daughter cells and it occurs during telophase.

D

D-loop: Configuration found during DNA replication of chloroplast and mitochondrial chromosomes wherein the origin of replication is different on the two strands. The first structure formed is a displacement loop or D-loop.

Dark field microscopy: It involves illumination of the sample with light that will not be collected form the image so that a dark background with bright objects on it will be produced.

Dark reaction: During photosynthesis conversion of CO_2 into carbohydrates occurs in the absence of light, it is called dark reaction.

Default pathway: Some proteins do not need to be packaged into vesicles and are delivered immediately and directly to the cell surface. This is called *non-selective constitutive pathway or default pathway*

DNA damage response pathway: A Dysfunction of DNA repair and damage checkpoint pathway is largely responsible for generating instability of cancer cells. The pathways those try to repair DNA damage to prevent cancer are called *DNA damage response pathways (DRPs)*

Deplasmolysis: Plasmolysed plant cells if placed in hypotonic solution then water enters into the cytoplasm and the cell again attains the normal form, this condition is called deplasmolysis.

Depolarization: Due to polarization of the axolemma positive charges causes closure of nearby voltage gated Na^+ channels. Just after their closure, the K^+ channels open and K^+ exits the axon to normalize the resting potential. It is called depolarization.

Desmin: It is an important type III IF protein found in muscle cells, connecting Z disks and is responsible for stabilizing sarcomeres. It is also found connecting desmosomes in cardiac muscle.

Deuteroplasm: The nonliving substances produced by the protoplasm constitutes deuteroplasm.

Dialyzing membrane – Such membranes allow the passage of different molecules in response to high hydrostatic pressure (e.g. Basal membrane of the endothelial cells).

Dicentric chromosome: Chromosome with two centromeres.

Diffusion limit: The theoretical maximum for the specificity constant is called the diffusion limit; it is about 10^8 to 10^9.

Diplosomes: Each centrosome consists of a pair of orthogonally connected *centrioles;* collectively called *diplosomes.*

Docking protein: Responsible for attaching (docking) a ribosome to a membrane by interacting with a signal particle attached to a ribosome destined to be membrane bound.

Double crossover: Two crossovers occurring in a chromosomal region under study.

Dynein: It is a microtubule associated motor protein involved in a variety of intracellular transport processes. They move any vesicle towards the minus end of the MT. Sometime MTs are reported to be destabilized by some proteins. Such severed MTs are reported to be transported down the axon by dynein.

Dystrophin: It is the most abundant important actin filament cross-linking protein present in the muscle forming cortical network that help to bind actin filaments to the plasma membrane.

***Dysplasia*:** It is an intermediate step in the development of cancer. It refers to excessive cell proliferation characterized by loss of normal tissue arrangement and cell structure. The most severe cases of dysplasia are referred to as "*carcinoma in situ*" which refers to an uncontrolled growth of cells that remains in the original location and shows no propensity to invade other tissues.

E

ECM: In multicellular organisms each cell is surrounded by an array of fibres and macromolecules that form a surrounding medium with which the cell remains anchored and carry out different transport activities.

Elaioplasts: The plastids that store fats.

Electron transport chain: Synthesis of most of the ATP is coupled to the reoxidation of NADH and $FADH_2$ by O_2 in a process involving the electron transport chain (ETC) that consists of many carriers which translocates protons (H^+) from the matrix to the perimitochondrial space.

Elementary particles / *Oxysomes / Farnandez Moran sub units / F1 particles / ATP synthese or F_1F_0 complex:* These are stalked granule like structures on the wall of the cristae; consist of a base (80A° diam.), a 50A° long stalk and a head (100A° diam). These are the sites of ATP production in mitochondria.

Embryonic stem cells: In mammals, stem cells that are isolated from the inner mass of blastocyst and give rise to any differentiating cell are called *embryonic stem cells.*

Emiocytosis: It is forceful expulsion of substances from the cell.

Endocytosis: It involves engulfment of solid or liquid substances through the plasma membrane. In response to some specific signals, plasma membrane forms vesicles called endosome around the substances that is pinched off into the cytosol later on.

Endosmosis: Osmosis showing entrance of water from the surrounding medium or extracellular medium into the cell.

Endocytosis: It involves engulfment of solid or liquid substances through the plasma membrane. In response to some specific inductions plasma membrane forms vesicles - endosome around the substances that is pinched off into the cytosol later on.

Endomysium: The membrane that invests the fasciculae are called endomysium.

Endomitosis: A process by which chromosomes replicate without nuclear or cellular division that results in cells with many copies of the same chromosome. In such a modified cell cycle, the chromatin duplicates its DNA content during the G1 and S stages, but instead of passing to the G2 stage, the nucleus initiates a new G1 phase, thus starting a new cycle of chromatin duplication. This type of cell division is termed as endomitosis. Polytene chromosomes are produced by such mitotic cycle during the interphase stage.

Endoreduplication: It is the duplication of the genome without mitosis. It is a common process in eukaryotes and results in an increase in the nuclear DNA content, permitting amplification of the genome of specialized cells. In this cycle the chromatid number is duplicated, but do not segregate, and after various cycles, larger and thicker chromosomes are produced.

End replication problem: Synthesis of the lagging strand of DNA double helix requires a short primer that will be removed in due course. At the extreme end of a chromosome, there is no way to synthesize this region when the last primer is removed. Therefore, the lagging strand of telomeric DNA is always shorter than its template by at least the length of the primer. This is the so-called *"end-replication problem"*.

Endospore: Dormant, highly resistant, thick walled structure that protect the bacterium during environmental stress.

Endosome: The endocytosed macromolecules are initially taken up by small irregularly shaped organelle known as *early endosome* from where some of the materials are recycled back to the plasma membrane via *sorting endosomes.* The rest of the materials are delivered to *late endosome* and ultimately fuse with the lysosome.

Epimysium: Whole muscle is is invested by a connective tissue membrane called epimysium.

Equational or homotypic division: Meiosis II is basically homologous with mitosis cell division producing two identical haploid cells i.e. the replica of the meiosis I daughter cells. It is therefore called *equational or homotypic cell division.*

ER overload response: This signaling pathway leads to apoptosis of cells in response to the accumulation of the misfolded protein.

Ergastic substances: These are non living materials found in cells.

ER-Golgi intermediate compartments (ERGIC): The intermediate site in between ER and Golgi bodies. From this site the vesicles transit from ER to the Golgi along MTs using the dynein/dynactin motor proteins and fuse with the cis cisterna of the Golgi apparatus or with the wall of other respective organelle.

Etioplasts: The chloroplasts that have not been exposed to light.

Euchromatin: The parts of chromatin that are loosely coiled. These portions stain lighter than other portions and are regularly transcribed indicating the presence of active genes.

Exocytosis: It involves expulsion of secretory substances to the outside produced in the cell.

Exosmosis: Osmosis showing expulsion of water from the cell into the extracellular or surrounding medium.

Extensin: A structural cell wall protein, forming covalent bonds with other extensin proteins through the amino acid tyrosine. They can wrap around other cell wall constituents "knitting" the wall together. Cells that have thick, hard walls and are exposed to mechanical wounding, infection etc. are rich in extensin.

F

F-actin: Polymerized G-actin that forms a double helix that is produced in the presence of a metal cation (as of calcium) and ATP. F-actin forms part of the cytoskeleton.

Facilitated diffusion: Diffusion being facilitated by transmembrane proteins.

Fasciculae: The blocks of muscle fibres that are invested by epimysium are called fasciculae.

Fascin: This protein might mediate polymerization of actin filaments in to bundles.

Fast – twitch fibre / Type II Fibers: Contains few mitochondria, low in myoglobin. rich in glycogen and depend on glycolysis for ATP production. They fatigue easily because of the formation of lactic acid. They dominate in muscles used for rapid movement.

Fatigue: When muscleular activity is continued for a long period, ATP supply is stopped. In that condition myosin heads contact with the actin molecules but can not swivel back. It results ***muscle fatigue.***

Feedback Inhibition: In multistep reaction if the products of the last step get accumulated it inhibits the action of the first step; ***Filamin:*** The bundles formed by actin filaments can form networks regulated by filamin. It cross links actin filaments in the cortex of many cells and anchors them to the integrins.

Filaggrin: It is a common intermediate filament associated protein that **bundles keratin filaments in differentiating cells of the epidermis to provide toughness to the outermost layer of the skin.** it is called *feed back inhibition.*

Filensin: It is an important type IV IF is found in lens fibers of the eye.

Fimbriae: These are short hair like involved in the adherence of bacteria to different substrates.

Fimbrin: This protein might mediate polymerization of actin filaments in to bundles noted in the filopodia. ***Flex movement*** : Alternate contration and relaxation of the phospholipid tails.

Flip flop movement: Lateral movement of the lipids is called flip-flop movement.

Fluorescence microscope: This microscope works on the principle that certain dyes and minerals when bombered with UV light (which is used as light source) shows fluorescence. The object to be viewed is coated with dyes and minerals and illuminated with UV radiations.

Flurochrome: These are certain dyes used in fluorescence microscopy. Commonly used fluorochromes are fluorescein (green) or rhodamine (red).

Formin: It helps to nucleate actin filaments with variable efficiencies, as high as one actin filament / formin complex. It coordinates actin filaments and microtubules.

Fluidity of plasma membrane: Plasma membrane due to its unique lipoid nature and presence of random distribution of proteins appears fluidish which is basically responsible of the variety of functions performed by the membrane.

Fluid mosaic model: Plasma membrane shows fluidish phospholipids into which proteins are randomly distributed.

Fragmin / Severin: These are reported from molds and show Ca^{++} dependent severing activity.

G

G-1: It is the most vital phase of cell cycle in which the cell responds to extra cellular signals either by advancing towards another division or withdrawing from the cycle into a resting phase (G0). Thus G1 progression normally relies on stimulation by mitogens and can be blocked by antiproliferative cytokines.

G-actin: A globular monomeric form of actin produced in solutions of low ionic concentration.

Gametic meiosis: In higher plants and animals, the germ cells divide to form spermatogonia (in male) and oogonia (in female) that undergo meiosis to form sperms and ova respectively.

Gap periods: The main divisional phase i.e. mitosis or meiosis (*M phase*) remain separated from the DNA replication phase (*S or Synthesis phase*) by two preparatory or gap periods, one preceding the S phase - *Gap-1 or G-1* and other succeeding the S phase - *Gap-2 or G-2* phase.

G-banding: Specific chromosomal bands obtained with Giemsa staining showing a series of lightly and darkly stained bands - the dark regions tend to be heterochromatic and AT rich while the light regions are euchromatic and GC rich.

Genome: Haploid sets of chromosomes represent the genome of the organism.

Gated channels: Transmembrane channels may be regulated by different components . If they are regulated by ligands, these are called *ligand gated channels,* if by mechanical agitation, these are called *mechani-cally gated channels,* if by voltage, these are called voltage gated channels.

Gelsolin: It is an actin capping protein that binds either to pointed or barbed end and checks the addition or dissociation of subunits. In Ca^{++} level they act as severing protein.

Gems: In close association with Cajal bodies, there are highly condensed structures called *Gems* that play an essential role in cytoplasmic snRNP biogenesis.

Generative disc: During the division of centrosome, the daughter centriole moves slightly away from the mother and initiates the formation of new daughters at the onset of 'S' phase. At first a disc like structure is formed at right angle to the proximal end of each mother centriole called *generative disc* that acts as the precursor of the daughter centriole.

GERL: The trans faces of Golgi complex remain intimately associated with ER and lysosomes forming a large complex called *GERL.*

Golgi blobs: In mammalian cells, at the onset of cell division, Golgi complex in the pericentriolar region, are fragmented into smaller elements called *'Golgi blobs'* which undergo further fragmentation in latter stages of division. Such fragmentation is essential for flawless division because in the absence of fragmentation, cell gets arrested in G_2.

Golgin: The cisternae of Golgi complex remain connected with each other by electron dense structures by a protein called *golgin* that maintains a dynamic exchange between the membrane and the cytoplasm.

Glycosphingolipid enriched membranes (GEM): These are rich in cholesterol, glycoshingolipid, low in other phospholipids. GEM is merely a caveolae without caveolin-1. Their main function is signaling.

GERL: An association of Golgi, ER and lysosome.

Glycocalyx: A network of glycoprotein and polysaccharides present outside the cell wall of bacteria. It protects them from phagocytosis. It allows them to form *biofilms*- a coating on the teeth or rocks or to the membrane of any eukaryotic organs or other bacteria.

Glycolysis: Inside the cell, glucose is broken down to make ATP in two pathways. The first pathway occurs in the cytoplasm requires no oxygen and is called glycolysis in which glucose is broken down into pyruvate.

Glypiated proteins: Some proteins are inserted in the ER lumen showing long N terminus in the ER lumen and stop transfer sequence in the membrane with C terminal end in the cytosol. A glycosyl phosphatidyl inositol- GPI molecule already remains attached with the ER membrane. The inserted polypeptide is cleaved near the stop transfer sequence and the cut C terminal end is attached with the N terminal end of the GPI anchor. Such anchored proteins are called glypiated proteins (e.g. alkaline phosphatase).

Lysosomal storage disease: As a result of direct consequence of the failure of the enzymes to be transported from ER for lysosomal functions so that residual bodies in the lysosomes are not properly liberated outside the cell.

Grana: Within the stroma of the chloroplast there are stacks of *thylakoids,* called a *grana.*

Gross photosynthesis (Pg): The total fixation of energy by photosynthesis.

Growth factor: A serum protein that stimulates cell division when it binds to its cell-surface receptor

H

H^+/K^+ ATPase: Gastric parietal cells transport H^+ from a concentration of about 4 x 10^{-8} M within the cell to a concentration of about 0.15 M in the gastric juice to secrete gastric juice.

Hatch and Slack pathway or C-4 pathway: Many angiosperms (*e.g.* corn, maize, sugarcane, sorghum etc.) utilize a two-step pathway to fix CO_2. It involves two types of cells, *mesophyll cells* that are adjacent to the air spaces and *chloroplast containing bundle sheath cells* surrounding the vascular bundle tissues (called Krantz anatomy). In the mesophyll cells *phospho enol pyruvate (PEP) carboxylase* reacts with CO_2 to from *oxaloacetic acid* (OAA) that is converted to *malic acid or aspartic acid* (4 carbon compounds). These are then transferred to the cells surrounding the vascular bundle having enzymes of Calvin cycle. Here the malic acid is converved to pyruvic acid and CO_2. Pyruvic acid is transferred back to the mesophyll cells where it is converted to PEP and repeat the cycle.

Harlequin chromosome: Sister chromatids that stain differently, so that one appears dark and the other light (harlequin-like).

Hay flick limit: Each time a cell divides, the telomeres get shorter and until finally they become so short that damage begins to happen to the "interior" DNA and chromosomes start to stick to one another on the ends. The number of times a cell divides before going into this decline is called the *"Hay flick limit."*

Hemicellulose: Hemicellulose is abundant in primary walls but is also found in secondary walls. These molecules are hydrophilic and branched.

Hensen's zone: At the centre of the I band there is less dense portion called H band (Hensen's Zone).

Heterochromatin: The parts of chromatin that are tightly coiled are called *heterochromatins.* The heterochromatins stain deeper than the euchromatins. It is of two types - *constitutive heterochromatins* (which remain permanently condensed) and *facultative heterochromatins* (which remain condensed in some stages but in decondensed state in other stages).

Heteroduplex: A DNA double helix formed by annealing single strands from different sources; if there is a sequence difference between the strands, the heteroduplex may show single strand loops or bubbles (unpaired regions).

Heterophagosome: Aggregation of primary lysosomes around any bacteria or foreign organisms.

Hilum: First formed grain in the carbohydrate type of reserve food is represented by a minute particle called hilum around which concentric layers of starch are deposited.

Heterophagy: Lysosome mediated degradation of the extra cellular invaders (bacterium) or macromolecules are called *heterophagy.*

Histone protein: Histone proteins are one of the important chromatin forming protein around which DNA helices wound to form nucleosome. Four histones are encountered in the octamer H2A, H2B, H3 and H4 having 129, 125, 135 and 102 amino acids respectively.

The incoming and outgoing ends of DNA are clamped by H1 that contains about 224 amino acids (rich in lysine).

Holliday junction: Paired bivalents in the pachytene sub stage consist of four chromatids of which two non-sister chromatids participate in recombination. At a particular point an enzyme *recombinase* creates a nick in one strand of a DNA double helix. The enzyme pulls apart the nicked strand from its complementary strand and anneal to one strand of the double helix on the opposite chromatid. It follows a second nick in the one strand of the other helix and the nicked strand is pulled apart and anneal to the remaining strand of the first helix. The crossed strands thus form a cross-*strand exchange* called Holliday junction. It can be 'pulled' by other recombinases along the four-stranded structure. This phenomenon is called ***"branch migration"***. Initially the Holliday structure is sustained by hydrogen bonding but subsequently the nicks are sealed by ligase and the 'X' shaped structure is slipped towards one of the telomeres and the connected pair of duplexes is called a ***'joint molecule'***. At the site of recombination one helix juxtaposes with another helix of the other DNA molecule. This portion is called ***heteroduplex DNA.***

Holocentric chromosome: Chromosome with many diffused centromeres.

Holoenzyme: The apoenzyme together with its cofactor(s) is called holoenzyme.

Homogeniger: A glass made apparatus to grind the tissue by manual compression and pressure.

Homologous chromosome: In a sexually reproducing organism each set of chromosome exhibit two identical copies; father and mother contribute one of the copies. Such genetically identical paired chromosomes are called homologous chromosomes.

Hot spot hypothesis: As per this hypothesis ER makes intimate contacts with mitochondria via calcium-dependent -caveolae/raft domain of the membranes. Ca^{++} is transported from ER to mitochondria with the involvement of Bax. Mitochondria accumulate Ca^{++} at microdomains and affect both Ca^{++} release from the ER and Ca^{++} entry across the plasma membrane, thereby regulating intracellular Ca^{++} signal.

Hydrolases: These enzymes catalyze the hydrolysis of various bonds.

Hydrostatic ('Water-stopping') pressure: If the pressure in the compartment into which water is flowing, during osmosis s raised to the equivalent of the osmotic pressure, movement of water will stop. This counter pressure is called hydrostatic pressue.

Hypervitaminosis: Accumulation of fat soluble vitamins leads to excess vitamin storage within the body.

Hyperplasia: It is an intermediate stage in the generation of cancer. It refers to an excessive rate of cell division but with a normal orderly arrangement of cells within the tissue.

Hypotonic solution: The solution out side the cell shows lower concentration of solutes.

I

Idiogram: The chronological arrangement of the photographs of the chromosomes exhibiting a definite diagram is called idiogram.

Immunofluorescent antibody labelling: Fluorescence microscopy is specially used to detect specific proteins or other molecules in the cell. The antibodies are coupled chemically to a fluorochrome and then used to trace the desired proteins in the cells. The antibody mixed dyes can be gathered on the surface of the microorganisms and can be traced easily. This method is called immuno-fluorescent antibody labelling.

Inducible enzyme: Those enzymes whose production can be stimulated by another compound, often a substrate or a structurally related molecule.

Induced pluripotent stem cells: Certain adult cell types can be "reprogrammed" into other cell types using a well-controlled process of genetic modification. E.g. pancreating insulin-producing α cells, that are lost or damaged in diabetes, could be genetically reprogrammed to secrete insulin.

Inner boundary membrane: The outermost layer of the inner membrane of the mitochondria.

Interchromatin granules: In the interchromosomal spaces bright speckles like *interchromatin granules* are noted that contain aggregates of small nuclear ribonucleoproteins (snRNPs) and other protein factors involved in RNA processing.

Interkinesis: The abbreviated interphase that occurs between meiosis I and II. No DNA replication occurs here.

Intermediate filaments (IF): These are one of the cytoskeletal structures are found in all vertebrates. In epidermal cells and the axons of neurons, IFs occur abundantly. Their principal role is to reinforce cells and to organize them into tissues. IFs provide mechanical support for the plasma membrane where it comes in contact with other cells or with ECM. They are prominent in cells that withstand mechanical stress.

Intermediate filament associated proteins (IFAPs): They cross-link IF with one another, forming a bundle or a network and with other cell structures, including the plasma membrane.

Interphase: In the cell cycle, S, G1 and G2 phases are collectively called *interphase.*

Interpolar fibres: The long fibers that connect two spindle poles are called *inter polar fibers,* in the mid line, these fibers overlap with each other.

Intracellular signaling Each cell responds to different chemical signals that either is transported into the cell or is generated within the cell to perform definite chemical reactions.

Irreversible inhibition: Such type of inhibitors binds permanently covalently to the active site and inactivates them so that the tertiary structures cannot be restored.

***Isomerases*:** These enzymes catalyze isomerization changes within a single molecule.

Isotonic: The solutions, inside and outside the cell showing equal concentration of solutes.

Intracellular signaling: Each cell responds to different chemical signals that either is transported into the cell or is generated within the cell to perform definite chemical reactions of gene activation.

Isozyme: Enzymes that differ in amino acid sequence yet catalyze the same reaction. Isozymes differ in biochemical properties (e.g. electropheric mobility) and kinetic properties (Km, rate constant etc.). They are encoded by different genetic loci.

K

K_{cat}: It is the number of substrate molecules handled by one active site per second.

***Ki*:** It is the dissociation constant of EI complex. Low Ki is essential for enzyme activity while high K_i decreases it.

Karyology: the study of whole sets of chromosomes is known as karyology.

Katanin: It is a MT severing protein. They break thirteen bonds, one for each protofilament; they separate the protofilaments and also release the MTs from the MTOC. At the end of mitosis katanin performs the role of depolymerization of MTs.

Kleptoplast: Some dinoflagellates take up algae as food and keep the plastid of the digested alga to profit from the photosynthesis; after a while the plastids are also digested. These captured plastids are known as *kleptoplastids.*

Karyopherins: Nuclear pores act as the important gates through which different moles are either exported, imported. The importable and exportable proteins bear specific sequences called nuclear *localization signal or sequence - NLS* and *nuclear export sequence – NES.* Specific carrier proteins called karyopherins recognize these signals.

Karyotype: Every species has a definite number, size and shape of chromosomes. This species-specific characteristic feature of the chromosomes is called karyotype.

Kinetochore: A proteinaceous structure on each side of the centromere that act as special site for spindle microtubule attachment.

Keratins: The most diverse classes of IF proteins. Some of its isoforms are specific for "hard" epithelial tissues like nails, hair, and wool while some are found in the epithelial lining of internal body cavities.

Kinesin: It is a double-headed microtubule associated motor protein involved in a variety of intracellular transport processes through the modulations of α and β tubulin structure. They move any vesicle towards the plus end of the MT.

Kinetochore fibres: The spindle fibers that remain connected with the kinetochores are called *kinetochore fibers.* These fibres actually pull the chromosomes/ chromatids towards the pole.

***Km value or Michaelis-Menten constant*:** It is the substrate concentration required by an enzyme to reach one-half of its maximum velocity.

***Krantz anatomy*:** Characteristic of C4 plants. Plant veins are encased by thick-walled photosynthetic bundle-sheath cells that are surrounded by thin-walled mesophyll cells.

Krebs cycle/Citric acid cycle/TCA cycle: In aerobic organisms, it is a part of the metabolic pathway involved in the chemical conversion of carbohydrates, fats and proteins into CO_2, water and to supply the raw material for ATP production. It starts with acetyl-CóA adding acetyl to oxalic acid, creating citric acid.

L

Lampbrush chromosomes: Chromosomes of amphibian oocytes having loops suggestive of a lampbrush. Large chromosomes found in amphibian eggs, with lateral DNA loops, extending from chromomeres, producing a brush like appearance under the microscope.

Leaflet of plasma membrane: Plasma membrane consists of two portions each showing distince proteins and lipid carrying out distince functions. These two portions of plasma membrane are called leaflets.

Leptotene stage: The first stage of prophase 1 of meiosis in which chromosomes become distinct.

Leucoplasts: The plastids that contain specific stored constituents.

Leukemia: Excessive production of W.B.Cs is called leukemia (Blood cancer).

Ligases: These enzymes join two molecules with covalent bonds.

Light-compensation point :The amount of light at which photosynthesis and respiration balance (i.e. net photosynthetic rate of a plant = 0). For a plant like wheat this is about 5 Wm-2.

Light reaction: During photosynthesis, (i) absorption of light; (ii) electron transport leading to the genera- tion of oxygen from water and simultaneously reduction of $NADP^+$ to NADPH and generation of proton motive force; (iii) synthesis of ATP take place in presence of light. These are called light reactions.

Linker DNA: 56 bp long DNA in between two nucleosomes is called "linker DNA".

Linker histone: (= H1 histone protein)

Lipoma: Cancers on adipose tissue is known as lipoma.

Liposome: Due to amphipathic nature of the membrane components, they spontaneously form circular bi -layers with hydrophilic portions facing external and hydrophobic ends facing core side surrounding aque ous compartment.It is called *liposome.*

Lock and key model: Emil Fischer (1894) opined that both the enzyme and the substrate show specific complementary geometric shapes that fit exactly into one another. It is referred to as *lock and key* model.

Loop domain model: Various cytological studies suggest that from the central fabric of the chromatin, loops of DNA arise and provide the structural basis for *loop domain model of chromatin.* These are best seen in lampbrush chromosome. In normal somatic chromosomes also, loops of chromatin radiate from the central chromatid axis. DNA loops are attached with the sca -ffold at specific sites called *S/MARs* (scaffold/**m**atrix **a**ttachment **r**egions), *locus control regions (LCRs) and boundary elements (BE).*

Lyases: These enzymes cleave various bonds by means other than hydrolysis and oxidation.

Lymphoma: Cancers of lymph and spleen is called lymphoma.

Lysosomal secretion: Some secretory cells store their secretory proteins in a dual-function organelle, called a *secretory lysosome.* These fuse with the plasma membrane in response to extra cellular stimuli releasing their contents by exocytosis. This is called *lysosomal secretion or defecation.*

Lysosomal storage diseases: Lysosomal storage diseases are caused by the accumulation of improper macromolecules (proteins, polysaccharides, lipids) in the lysosomes. This is caused by the malfunction of the lysosomes to manufacture proper enzyme needed for their breakdown.

MCS (Membrane contact sites): These are special tethering complexes between two organelles in eukaryotes.

M-line: At the centre of the H zone, there is a line where myosin filaments unite. It is called M –line.

MPF - maturation/mitosis promoting factor: Heterodimeric protein that initiates prophase of mitosis and alters the activity of diverse proteins that function in mitosis by phosphorylation; active MPF contains both cyclin B (large) and cyclin-dependent kinase 1 (small) subunits. Activity is regulated partly by the synthesis and degradation of cyclin B subunit.

mRNA gate: The most striking feature of the smaller subunit of mitoribosome is the presence of a triangular gate like structure that covers the mRNA entry site called *mRNA gate.*

mt DNA): Mitochondria have their own genome and it consist of a single circular molecule of DNA with more guanosine and cytosine than other nucleosides but there is no histone associated with the DNA. The genome of human mitochondria contains 16,569 base pairs of DNA organized in a closed circle.

Macroautophagy: It is the major lysosomal catabolic pathway for macromolecules and organelles in eukaryotic cells. It involves engulfment of large volume of cytoplasm with various organelles into a vacuole with which lysosomes merge delivering acid hydrolases into the lumen of *autophagosome.*

Malignant tumour: The tumour that grow and divide rapidly and invade neighboring tissues through blood or lymph and establish secondary areas of proliferation by *metastasis.*

Marker: It a commercially available mixture of proteins having defined molecular weights, typically stained so as to form visible, coloured bands in Western blot technique.

Membrane skeleton fence model: It describes that membrane skeleton provides structural integrity by restricting the move ment of proteins.

Mesosomes: The plasma membrane of prokaryotes forms vesicular invaginations into the cytoplasm. These are called *mesosomes or peripheral body or chondroid.* These are also involved in DNA replication and segregation, cell wall synthesis, or increased enzymatic activity.

Metacentric chromosome: Chromosome with median centromere.

Metalloenzyme: Enzymes requiring different metal ions as the prosthetic group.

Metaplasia: The transformation of one differentiated cell type into another

Metastasis: At the last stage of cancer, the cancerous cells may either invade the adjacent tissues through *invasion* or may be spread to distant sites by blood or lymph that is called metastasis.

Microautophagy: It involves encirclement of small volumes of cytoplasm called multivesicular bodies followed by subsequent degradation by lysosome. When yeasts were starved for glucose, several enzymes and membrane transporters appear. Upon addition of glucose, these enzymes are enclosed in multivesicular bodies and are degraded by the fusion of lysosomal vesicles.

Microbodies: Peroxisomes, glyoxysomes and glycosomes are related organelles, and are collectively known as *microbodies* found in different organisms.

Microsome: In response to the centrifugal force the produced tiny fragmented part of SER or RER (100 nm diam), which are mainly used to study the structure and functions of ER are called microsomes.

Microtubule: It is a hollow cylindrical cytoskeletal structure formed by the assembly of 13 protofilaments.

Microtubule-associated proteins – MAPs: These are diverse classes of cell and tissue specific proteins that bind along the sides of microtubules e.g. kinesin, dynein which move along the MT in opposite direction.

Middle lamella: It is a colloidal plate like structure; highly elastic consists of mainly protein and calcium and magnesium pectates It is formed during early telophase in the equatorial plane of the mitotic apparatus by the fusion of phragmoplasts.

Minimal cell model: *Mycoplasma* contains the absolute minimum structures (plasma membrane, ribosomes and a highly coiled circular chromosome) necessary for survival, furnishing model organisms for the essential functions of all living cells.

Mini nuclei/Karyomeres: In telophase, after segregation of the cellular organelles, nuclear envelope begins to assemble. The dephosphorylated nuclear membrane proteins bind to the decondensing chromatins and fuse with each other to form double membrane around each chromatid. The assembly of NPC traversing the inner and outer nuclear membranes follows assembly of nuclear membrane. Reassembly of other structures form *'mini nuclei' or 'karyomeres'.* Fusion of these mini nuclei generates daughter nucleus.

Mitogen: Any substance that causes cells to begin division (mitosis).

Mitoplast: The inner mitochondrial membrane and the matrix are called *mitoplasts* that help to learn about the functional compartmentation of mitochondria.

Mitotic crossing over: It is a rare type of genetic recombination that may occur in some types of somatic cells during mitosis. It may occur in organisms that do not have a cycle of sexual reproduction. It results in the production of homozygous allele combinations in all heterozygous genes those are located on that chromosome arm distal to the crossover. Thus when a mitotic crossover occurs, genes that were previously recessive are expressed creating a new phenotype. Mitotic crossover is known to occur in some asexually reproducing fungi and in normal human cells, where the event may allow normally recessive cancer-causing genes to be expressed and thus predispose the individual to the development of cancer.

Monopolar attachment: Attachment of both kinetochores of a chromosome with the kinetochore microtubules emanating from the same pole.

Micro propagation : It is a collection of techniques meant for asexual propagation where a very small piece of tissue (shoot apex, leaf section, or even an individual cell) is allowed to propagate in a sterile (aseptic) culture medium.

Mitoribosomes: Ribosomes that occur inside the mitochondria are called *mitoribosomes.*

Mitotic spindle checkpoint: In prometaphase, some centromeric proteins MAD and BUB are known to regulate mitotic spin- dle checkpoint to form mitotic checkpoint complex that monitors proper attachment of the kinetochores with the kMTs. MAD mainly inactivates anaphase-promoting complex (APC) up to M-A transition. If attachment of kinetochore with the spindle fibers is failed the cell cycle becomes arrested in M-A transition. Even a single unattached kinetochore can inhibit the APC. CENP-E is proposed to be an important mechanosensor of kinetochore tension and its depletion leads to a failure of checkpoint activation.

Mixed inhibition: This type of inhibition resembles the non-competitive, except that the EIS-complex has residual enzymatic activity.

MPF: At the onset of M phase, Cdc-2 unites with Cyclin B1 to form maturation-promoting factor (MPF) that induces cell to cross G2/M and enter into M phase.

Monocentric: Chromosome with one centromere.

Motor end plate: A minute pad at the neuromuscular junction at the end of the axon of a motor neuron on a muscle fibre terminates into called motor end plates.

Movements of phospholipids: Phospholipid molecules show different movements like *Flex movement* (con traction and relaxation of the fatty acid chains), *lateral diffusion movement* (lateral side wise movement), *rotation* around the axis etc. to render the membrane fluid.

MTOC: The mass of protein encircling the centrioles is called Microtubule Organizing Centre (MTOC). It contains centrin, pericentrin, ninein, cenexin and other proteins.

Multi-hit model of cancer induction: Several mutations are required for the induction of cancer. Normal genes controlling mitosis and tumor suppressor genes would suppress a mutation limited to one oncogenes. An initial mutation completely isolates the cell from others but a mutation to only one tumour suppressor gene would not cause cancer, due to the presence of many "backup" genes. The second mutation in one of the progeny cells allows growing uncontrollably and forming a bening tumour. Third mutation allows the tumour to outgrow and overcome the constraints imposed by the tumour microenvironment. It is only when enough proto-oncogenes have mutated into oncogenes and enough tumor suppressor genes are deactivated or damaged and the signals for cell growth overwhelm the signals to regulate it, and cell growth quickly becomes out of control. An additional mutation leads into metastasis.

Multipotency – It involves differentiation of a cell into a number of closely related family of cells. E.g. haematopoetic cells can differentiate into skeletal, muscles, liver cells and all blood cells.

Multiscale model of nucleosome: As per this model, histone tails bind to nucleosomal DNA via strong salt-bridge interactions suggesting a mechanism of chromatin structural organization whereby histone tails regulate inter- and intranucleosomal assemblies via binding with nucleosomal DNA. Histone tails is proposed to have a direct functional role in stabilizing higher-order chromatin structure, mediated by salt-bridge interactions with adjacent DNA.

Multivesicular bodies: The materials collected in early endosome are delivered to *multivesicular bodies* (MVB) and ultimately to *late endosomes* where the materials first come in contact with the acid hydrolases. Thus MVB are an intermediate compart-ment between endosomes and lysosomes.

N

NLS, NES and NRS: Nuclear pores act as the important gates through which different moles are either exported, imported. The importable proteins bear a *nuclear localization signal or sequence - NLS* that is recognized by specific carrier proteins – *adapters.* The exportable proteins also bear a *nuclear export sequence – NES.* Another signal *nuclear retention signal (NRS)* is present on a number of proteins that need not transport beyond the nucleus e.g. immature RNAs.

Na⁺-K⁺ Pump: Most abundant and important of all membrane pumps. It is a complex pump, simultaneously pumping 3 Na^+ out of the cell and 2 K^+ into the cell for each molecule of ATP split generating a potential difference across the cell membrane. This is called the *membrane potential.*

Nebulin: The accessory protein that regulates the length of each actin filament is called nebulin.

Nematolysosomes: In prolonged starvation, lysosomes appear long, thread like called *nematolysosomes.*

Net photosynthesis (Pn) - The actual rate of production of new biomass . This is essentially gross photosynthesis minus the rate of respiration.

Neuromuscular junction: The junction of motor neuron and muscle.

Neuroglandular junction: Synapse between an axon and a glandular cell.

Neurotransmitter: In response to the stimulus from the presynaptic knob a chemical (e.g. acetyl choline) is released which sets out ionic changes in the post synaptic membrane.

Neuromodulator: These are additional neurotransmitter that are released and modify the trans mission of an impulse at a synapse either prolonging or inhibiting it. These are responsible for mood and emotion. The additive drug cocaine often act as neuromodulator.

Non-competitive inhibition: Non-competitive inhibitors bind to the enzyme at any other site than active site causing a conformational change to the enzyme and thereby inhibiting enzymatic catalysis

Nuclear lamina: Inner nuclear membrane is followed by a filamentous scaffold structure called *nuclear cortex or nuclear lamina.* It maintains the shape of the nucleus, spacing of nuclear protein complex, organization of heterochromatins.

Nuclear pore: Nuclear envelope contains numerous pores for nucleocytoplasmic translocation. Each pore shows three annuli - a *luminal ring* embedded in the nuclear envelope, it is flanked by an outer *cytoplasmic* and an inner *nuclear ring* which is anchored in the nuclear lamina. Three rings are joined by 8 *radial spoke* like structures which anchor the rings with the nuclear envelopes. Eight filaments project out both from cytoplasmic and nuclear rings. The nuclear filaments are joined at their outer ends to form a *terminal ring.* The nuclear ring, filaments and the terminal ring are collectively known as *nuclear basket.*

Nuclear skeleton: Throughout the nucleoplasm, anastomosing filaments, made up of lamin vimetin and cytokeratins remain scattered discontinuously showing large voids that constitute

nuclear skeleton. It mainly maintains the structural basis of nuclear organization as well as the distribution of chromosomes in the nucleoplasm.

Nuclease hypersensitive sites: Except 15 nucleosomes surrounding the centromeric DNA, majority of nucleosomes are not precisely positioned. Many specific sites of DNA throughout its length lack nucleosomes that are called nuclease hypersensitive sites.

Nucleation: The process of *actin polymerization* that starts with the weak binding of two actin monomers and addition of a third stabilizes the complex forming a trimer. This is called nucleation.

Nucleosome: The basic unit of eukaryotic chromosome structure; a ball of eight histone molecules wrapped about by two coils of about 147 base pairs of DNA. Arrangement of DNA and histones forming regular spherical structures in eukaryotic chromatin.

Nucleoid: Large circular DNA molecule, localized in the center of the prokaryotic cell or mitochondria or chloroplast.

Nucleolar organizer region: However in those cases where more than two secondary constrictions per diploid set are present at least two are involved in the synthesis of rRNA that ultimately give rise to nucleolus. This part is thus called nucleolar organizer region (NOR).

Numerical aperture: The resolution power of microscope depends upon the wavelength of illuminating agent and light gathering capacity of objective lens, called numerical aperture (NA). The greater the numerical aperture, the greater is the resolution power.

O

Octamer: These are bead like structures situated at 56 bp apart and basically shows 8 histone protein molecules - H2A, H2B, H3 and H4 (each represented by 2 molecules) forming *nucleosome core particle or octamer or tetrasome.*

Oligopotency – It involves differentiation of a cell into only a few cells (lymphoid or myeloid stem cells). e.g. Myeloid stem cell can form five types of cells (Monocytes, macrophages,eosinophils, neutrophils, and RBC).

Oncogenesis: The progression of cytological, genetic, and cellular changes that culminate in a malignant tumor.

Optical microscopy: It involves use of visible wave lengths of light passing through or reflected from the subject through a single or a series of lenses. The resolution power of human eye is 0.1nm and that of light microscope is 200 nm.

Osmotic pressure: Diffusion of water across a membrane during osmosis generates a pressure called *osmotic pressure.*

Osmosis: It involves transport of water across the membrane driven by a difference in solute concentrations on the two sides of the membrane till the attainment of equilibrium.

Oxidoreductases: These enzymes catalyze oxidation/reduction reactions.

Oxysomes: Stalked particles present on the wall of the cristae that are involved in ATP production.

P

p53: A 53KDa protein the product of a tumour suppressor gene. The loss of this protein due to mutation is a primary event in the formation of many types of cancer (breast, colon, lung, leukaemia, liver). P53 is involved in regulating the activity of some other genes (e.g. P21, a cell division kinase inhibitor). It also prevents cells entering cell division. P53 levels are increased after DNA is damaged by UV and ionizing radiation and cells are arrested in cell division until either the damage is repaired or they die by apoptosis (programmed cell death).

Paranemic coiling: Parallel arrangement of chromonemata is called paranemic coiling.

Pectic acid: It is the major component of middle lamella but also found in primary walls. They form salts and salt bridges with Ca++ and Mg++. They seem to be involved in cell elongation.

Pectin: It is the major component of middle lamella but also found in primary walls. It binds to cellulose microfibrils and may be of considerable significance in the modeling of primary cell walls.

Pellet: Hard-packed concentration of particles in a tube or rotor after centrifugation.

Pelleting time: time taken to pellet a given particle. T = K/S where T= pellet time in hours. K = K-factor of the rotor, and S = sedimentation coefficient.

Pericentromere: The region around or near the centromere (also known as kinetochore) of a nuclear chromosome.

Perichromatin fibrils: RNA transcription and processing occur at about 10,000 discrete sites on the surface of condensed chromatins called *perichromatin fibrils* spread throughout the nucleus. These sites show hnRNAs, RNA polymerase II, other associated factors and splicing factors.

Perimitochondrial space: A double-layered membrane invests the mitochondria. Each layer is separated by a 60-80A° wide space called perimitochondrial space / outer chamber.

Perimysium: The membrane that invests each fasciculus is called *perimysium.*

Peripheral reticulum: The thylakoid membranes are connected with the inner chloroplast membrane through many small vesicles or tubules called *peripheral reticulum.*

PEP carboxylase (PEPCASE) - The enzyme responsible for the carboxylation of CO_2 in the mesophyll cells of C4 plants into oxaloacetate.

Pexophagy: Degradation of peroxisome by autophagy is called *"pexophagy".*

Phagocytosis: If the endocytosed substance is solid with size range 1-2 m diam. then the process is called phagocytosis.

Phagolysosome: During heterophagy, fusion of heterophagosome with lysosome produces a hybrid vacuole called a *phagolysosome.*

Phagophore: During autophagy, a membrane cisterna called *isolation membrane* or *phagophore* appears and curves around the membrane-bound organelles along with part of the cytoplasm that forms a double membrane vesicle, the *autophagosome* that fuse with the lysosome.

Phragmoplast: In late anaphase, two antiparallel arrays of MTs, actin filaments and Golgi derived vesicles called *phragmoplast* accumulate over the central spindle. The vesicles contain cell wall materials that gradually accumulate along the equator where they fuse and form the new cell wall called *cell plate*.

Phakinin: It is an important type IV IF is found in lens fibers of the eye.

Phase contrast optical microscopy: It involves a shift in the light passing through a transparent specimen showing differences in refractive index. It enhances the contrast between cells and environment and between organelles and their surroundings. It is commonly used to observe living or unstained cells and various organelles in the cell.

Philadelphia chromosome: Chromosomal translocation containing *c-abl* gene with *bcr* gene forms the diagonostic *Philadelphia chromosome*. It produces the chimeric oncoprotein Abl-Bcr. Normal c-Abl protein causes branching of filamentous actin and extension of cell processes thus controlling cytoskeleton and cell shape. The Abl-Bcr oncoprotein can phosphorelate and activate many unusual intracellular signal transduction pathways. If this translocation occurs in the haematopoetic cell in the bone marrow it causes human chronic myelogenous leukemia (CML). A second mutation causes acute leukemia.

Photorespiration: Uptake of oxygen in the light by Rubisco (which also acts as an oxygenase) and the consumption of carbohydrate. This process competes with photosynthetic CO_2 uptake and carbohydrate synthesis. Photorespiration increases in dry hot climates when conditions limit CO_2 availability and favor high oxygen concentrations.

Photosystems (PS): In the process of photosynthesis, light is absorbed by a multiprotein complex called photosystem. It is differentiated by the wave length of light to which it is most reactive. PS-I and PS-II receive light of 700nm and 680nm wavelength respectively. PS consists of a reaction center showing chlorophyll *a* and many accessory pigments and proteins that constitute antenna system. The pigments absorb light of various wavelengths and pass along their gained energy to the reaction center.

Photosynthesis: It involves production of glucose with the help of sunlight, water and chrlorophyll. It is divided into four stages – (i) absorption of light; (ii) electron transport leading to the generation of oxygen from water and simultaneously reduction of $NADP^+$ to NADPH and generation of proton motive force; (iii) synthesis of ATP and (iv) conversion of CO_2 into carbohydrates.

Photon: A photon is a particle of radiation with a discrete amount of energy.

Pilus (plural pili): A conjugation tube; a hollow hair-like appendage of a donor *Escherichia coli cell* that acts as a bridge for transmission of donor DNA to the recipient cell during conjugation.

Pinocytosis: If the endocytosed substance is liquid having 0.1–0.2μ diam. then the process is called pinocytosis.

Plasmid: Autonomously replicating extra chromosomal rings of DNA present in monerans and some other organisms. Some plasmids can temporarily associate with the nucleoid and are known as *episomes.* Plasmids carry non-vital genes which may or may not be useful to bacteria. Plasmids which do not confer any useful trait are called cryptic plasmids. Important plasmids are- F-plasmid, R-plasmid, Col-plasmid etc.

Plasmid suicide vector: A plasmid that cannot replicate in a particular host.

Plasmodesmata: Plant cells are interconnected by plasmodesmata allowing free exchange of ions and signaling molecules.

Plasmolysis. In plant cells due to excessive exosmosis the cytoplasm shrinks so that the cell squeezes. This condition is called plasmolysis.

Plastid nucleoid: Plastid DNA exists as large protein-DNA complexes associated with the inner envelope called *'plastid nucleoids'*. Each nucleoid particle contains more than 10 copies of DNA.

Plectin: It is a common intermediate filament associated protein that interacts with other cytoskeletal proteins, including spectrin, MAPs, and lamin B. It also cross-links MTs and vimentin.

Plectonemic coiling: Spiral arrangement of chromonemata is called plectonemic coiling.

Pluripotency- It involves differentiation of totipotent cells into any cell of the three germ layers.

PMCA: It transports Ca^{++} away from the cytosol into the ER lumen or from the cell to keep cytosolic Ca^{++} low suitable for cell signaling.

Polarization: In response to a stimulus, Na^{+} channels open in the axolemma and Na^{+} rushes into the axoplasm which becomes positive; it is called polarization.

Polarization microscope: This microscope is used to observe highly ordered objects like crystals or bundles of parallel filaments. Thus cytoskeletons can be observed by this microscope.

Pole ward flux model: As per this model, kinetochores anchor kMTs and chromatids are pulled toward the pole by the depolymerization of kMTs at their '-' ends. Thus due to continuous disassembly of MTs either at the '+' or '-' ends the MTs shorten in length but to restore the binding between kinetochore and MTs the kinetochore glide poleward by its motor units.

Polymeric network model: As per this model plasma membrane acts as polymeric network showing diffu sion of different components in horizontal and vertical directions.

Polysomes: During protein synthesis many ribosomes assemble with one mRNA strand and are collectively known as polyribosomes or polysomes.

Porin: A transmembrane pore, present in the lipid layer of the plasma membrane in Gram(-) bacteria. It allows passage of the useful nutrients through the barrier of the outer membrane.

PrPc and PrPsc: PrPC is a normal prion protein found on the membranes of cells. The infectious isoform of PrPC, known as PrPSc. It is able to convert normal PrP^{C} proteins into the infectious isoform by changing their conformation.

Pre assembled pore complex: The nuclear pore complex is not exclusively found in the nuclear envelope but is also found to be associated with ER cisternae called *annulate lamellae.* These are called *pre assembled pore complex* to be needed by rapidly proliferating cells.

Preprophase band: In plant cell, early in mitosis, a band of microtubules and actin filaments forms a band around the equator above the nucleus called *preprophase band.*

Primary cell wall: On each side of the middle lamella, protoplasmic secretions accumulate to form cellulose microfibrils meshed in to a matrix of pectin, lignin and hemicelluloses.

Primary constriction: Throughout the longitudinal axis of the chromosome one or more constricted, faintly stained parts are noted, called primary constriction.

Primary and secondary transporters: Pumps transduce electromagnetic or chemical energy directly into transmembrane concentration gradients; these are usually referred as *primary transporters.* Some carriers use ion gradient created by pumps to drive uphill movement of other ions or solutes. These are called *secondary transporters.*

Proprotein: Proteins, synthesized as inactive precursors that are processed by limited proteolysis are termed, proproteins.

Proton leak: Protons normally are pumped from the matrix to inter mitochondrial space and translocates back to the matrix via ATPases but under certain conditions, protons re-enter the matrix without contributing to ATP synthesis. This process, known as *proton leak* or *mitochondrial uncoupling,* that results in the release as heat. This mechanism generates heat specially in the "brown fat" of newborn or hibernating mammals.

Proto-oncogene: The non-activated form of a cellular oncogene in an untransformed cell. A gene that, when mutated or otherwise affected, becomes an oncogene.

Premeiotic pairing: Sometimes pairing of the homologous chromosomes occurs in leptotene substage that is called *premeiotic pairing.*

Preparative chromatography: It is used to purify sufficient quantities of a substance for further use, rather than analysis.

Preprotein: After entrance of nascent polyleptide into the ER lumen, signal peptide is cleaved and immediately proteolyzed. Therefore it is never found in mature protein. Proteins that contain a signal peptide are called preproteins.

Preproprotein: Some preproteins that are further proteolyzed after secretion are called preproproteins.

Pre ribosome: In the nucleoli, during the biogenesis of ribosomes, the proteins and rRNA unite to form *pre ribosomes* (S value - 90S) that mainly gives rise to 80S ribosomes.

Primary constriction: Throughout the longitudinal axis of the chromosome one or more constricted, faintly stained parts are noted, called primary constriction.

Primary lysosomes: The nascent lysosomes that have not participated in any digestion process are called primary lysosomes. They fuse with autophagosome or heterophagosome to carry out digestion.

Procentrioles: Centrosome duplication is semi-conservative in that one centriole is new and the other is inherited from the parental centrosome. In the early S phase, at the proximal end of both old and new centrioles, a disc like structure is formed that begins to nucleate a ring of nine MTs called *'procentrioles'*.

Profilin: This protein stimulates the exchange of bound ADP for ATP during nucleation of actin filament..

Protofilaments: Along the microtubule axis, tubulin heterodimers are joined end-to-end to form with alternating α and β subunits to yield a helical arrangement to tubulin heterodimers called protofilaments.

Provitamin: Some chemicals are converted within the body to a vitamin. These are called provitamins. E.g. Panthenol is converted in the body to Pantothenic Acid (Vit. B5). B-carotene is converted into Vit. A.

Pseudocentrosome: In the chromosome directed spindle fibre formation pathway, in prometaphase, various MT activating proteins (MAPs), transported by '-'end directed kinesin concentrate in each pole, that is known as *'pseudocentrosomes'*. It generates MT assembly as in the centrosome directed pathway.

Pseudosynapsis: Sometimes synapsis occurs in between non-homologous chromosomes, it is called *pseudosynapsis.*

Puff: A swelling at a site along the length of a polytene chromosome; the site of active transcription. A diffuse uncoiled region in a polytene chromosome where active transcription takes place.

Pyroptosis: It is a form of programmed cell death associated with antimicrobial responses during inflammation. In this process, immune cells that recognize certain danger signals within themselves produce cytokines, swell, burst and die. This releases the cytokines, attracts other immune cells to fight the infection and contributes to inflammation.

Q

Q_{10} : On increasing the temperature the rate of reaction will increase initially and then declines. For the increase of 10°C the rate of reaction increases 2-3 times. This is called temperature coefficient (Q_{10}).

Q banding: A staining technique in which metaphase chromosomes are stained with quinacrine mustard to produce temporary fluorescent Q bands on the chromosomes indicating the A-T rich zones.

Quadripotency- It involves differentiation of a cell to form four types of cells. e.g. Mesenchymal progenitor cells can form cartilage cells, fat cells, stromal cells, bone-forming cells.

Quality control system: It is monitoring system of the ER involving the delivery of only properly folded proteins to the Golgi apparatus for further enzymatic transformation. Incompletely assembled oligomers are retained and misfolded proteins are degraded.

R

R-banding: Specific chromosomal bands which are the reverse of G-banding (the R stands for "reverse"). The dark regions are euchromatic (G-C rich regions) and the bright regions are heterochromatic (A-T rich regions).

Rafts: Flat tightly packed structure rich in cholesterol and sphingolipid and a specific protein-flotillin. Rafts are very unstable, gelsolin an actin-binding protein links actin cytoskeleton and rafts. These contain several signaling molecules and play an important role in receptor-medicated signal transduction, endocytosis, cholesterol transport, and pathogen entry

Raphides : These are elongated bundles of the crystals of calcium oxalate within a plant cell.

RER: Endoplasmic reticulum with attached ribosomes. Mainly involved in protein synthesis.

Reaction specificity: The property that an enzyme always mediates a specific reaction is called reaction specificity e. g. an amino acid can undergo transamination, deamination, decarboxylation etc but every trans formation must be mediated by specific enzyme.

Reactive oxygen species (ROS): Molecular oxygen is ideal as a terminal electron acceptor and normally it is reduced to produce water but in 1-2% cases, due to the reduction of coenzyme Q in complex III, a highly reactive free radical is formed as an intermediate (Q˙). Oxygen is reduced to give potentially harmful intermediates - the superoxide anion and peroxide. These compounds are collectively referred to as *reactive oxygen species* or ROS and are very harmful to cellular components.

Recombination: In molecular biology *recombination* generally refers to the process by which genetic variation is found at two different places in a continuous piece of DNA that becomes disassociated at the time of exchange between two chromosomes. The process leading to such molecular recombination is called chromosomal crossing over.

Recombination nodules: Large protein complexes on the synaptonemal complex between two synapsed homologous chromosomes. There is evidence that genetic recombination events in meiosis are catalyzed by these recombination nodules. It is divided into early and late nodules. Early nodules appear in the early zygotene and late nodules are converted into *chiasma* during diplotene.

Red cell ghost: For studying the membrane structure and physiology, usually mammalian RBC is treated with hypotonic solution resulting endosmosis leading to swelling and rupture of RBC. The ruptured membrane is called red cell ghost. In case of mild haemolysis, permeability of the membrane can be restored by certain treatment, such ghosts are called ***resealed ghost*** but in case of drastic haemolysis the resulting ghost is called ***white ghost***.

Reductional or heterotypic division: Meiosis-I shows complex remodeling of chromosomes leading to the formation of two haploid cells. This phase is thus called *reductional or heterotypic division.*

Refractory period: During depolarization, there is a large outward movement of K^+ causing a reversal of the membrane potential towards negative inside. During this brief period, neuron is unable to conduct an impulse. This period is called refractory period.

Regulated secretory pathway: As per this pathway the secretory vesicles containing a high concentration of specific protein fuse with the plasma membrane and are released into the blood only when they receive the correct "signal". Endocrine cells, neurons and digestive enzyme forming cells follow this pathway.

Resin: It is a viscous secretion of many coniferous trees, valued for its uses as varnishes, adhesives, for the production of incense and perfume etc.

Resting membrane potential: Difference in ionic concentration that exists between inside and outside the axon determines the resting potential. The concentration of K+ in the axoplasm and outside the axon are 140 4 and mEq /L while that of Na+ are 14 and 142 m Eq/ L.mEq/L respectively.

Restriction point: Progression through G1 is controlled by many growth factors. To regulate the phase flawlessly, in mid G1, there is a point after which cells become independent of growth factors. This special point is called *restriction point.* Disruption of this point is a common biological feature in human cancer.

Residual bodies: Autophagosomes retain some post-degradative undigested materials within the vacuole, called *residual bodies.*

Resolution or resolving power (RP) of a microscope: Resolution is the ability of a microscope to distinguish two closely related points. It indicates the size of the smallest object that can be distinguished by a particular lens system. For determining the resolution, NA (the size of the cone of light entering the lens system) is multiplied by 2 and the result is converted in mm. It can be increased by shortening the wave length of the visible source and increasing the the size of the cone of light i.e. NA. RP is expressed in terms of resolution limit (RL) which is the shortest distance between the two points that can be distinguished.

Retention time: It is the characteristic time it takes for a particular analyte to pass through the system (from the column inlet to the detector) under set conditions.

Retrograde transport process: Many proteins are transported flawfully to the Golgi apparatus but the associated KDEL receptor bring back the protein to the ER from pre-Golgi intermediates or the Golgi following retrograde transport process

Reverse transcriptase: It is a DNA polymerase enzyme that transcribes single-stranded RNA into double-stranded DNA. It is also called RNA-dependent DNA polymerase.

R_f value: R_f values are calculated by dividing the distance the pigment travels up the paper by the distance the solvent travels. R_f values are standard for a given compound.

Ribosomal DNA (rDNA): rDNA is a set of tandemly repeated genes coding for preribosomal RNA. They are the templates for nucleolus formation; these segments of the chromosomes are called nucleolar organizer region or NORs.

Ribozyme: A ribozyme is a catalytic or autocatalytic RNA molecule that catalyzes either their own cleavage or cleavage of other RNAs, , for instance, self-splicing RNA molecules in *Tetrahymena.* but they have also been found to catalyze the aminotransferase activity of the ribosome.

Rigor mortis: In the absence of Ca++ and ATP myosin head binds to actin firmly and never dissociates. This is the *"rigor mortis"* state giving rigidity to the muscle.

Robertsonian fusion: Fusion of two acrocentric chromosomes at the centromere.

Rotation: During diplotene, as the repulsion continues the bivalents rotate 90° or 180° which is called *rotation.*

Robertsonian fusion: Fusion of two acrocentric chromosomes at the centromere.

RPM: Revolutions Per Minute (Speed).

RUBISCO: The inner membrane of the chloroplast encloses a jelly like substance called *stroma.* It contains different enzymes of which the most notable in C_3 plants is RUBISCO – Ribulose 1, 5-biphosphate carboxylase oxygenase. In C_4 plant, RUBISCO is absent in the stroma It is a bifunctional enzyme that can fix CO_2 or molecular O_2, which leads to photosynthesis or photo respiration, respectively. Rubisco is the most abundant enzyme on earth.

S

S-phase: The phase of the cell cycle in which DNA replicates is called S or synthesis phase.

Saltatory conduction: It is a type of galloping nerve impulse propagation. In myelinated axons, the myelin (impermeable to ions) is interrupted at the nodes of Ranvier. Action potential through the axon appears to jump move faster and is only regenerated at the nodes of Ranvier

Sarcoma: Cancers of connective tissue belong to this category. Cancers in bone, muscle and cartilages belong to this category.

Sarcomere: The part of the myofibril between two Z lines is called a *sarcomere.*

Satellite body: The tiny chromosomal part, which lies beyond the secondary constrictions are called satellite body. Such chromosomes are called SAT - chromosomes.

Satellite DNAs: Non-coding DNAs shows large amounts of repetitive DNA sequences. Some repetitive DNAs are clustered in discrete areas (e.g. around centromere) and are called satellite DNA.

Scaffold: The eukaryotic chromosome structure remaining when DNA and histones have been removed; made from non-histone proteins. The central framework of a chromosome to which the DNA solenoid is attached as loops; composed largely of topoisomerase.

Scaffold protein: A protein whose main function is to bring other proteins together for them to interact. These proteins usually have many protein binding domains (like WD40 repeats). Below are some experiments that can be done to see if a protein acts as a scaffold in vivo:

Secondary constriction: Near the telomeric end, there may be one colored constricted part, called secondary constriction. Usually in one genome their number is one or more than one.

Semi conservative model of nucleosome assembly: As per this model, the nucleosides are disrupted at the replication fork; the parental H3-H4 tetramers are segregated equally in to daughter DNA strands as H3-H4 dimer with which the newly synthesized H3-H4 dimers are assembled by the action of H1.

Scaffold protein: A protein whose main function is to bring other proteins together for them to interact. These proteins usually have many protein binding domains (like WD40 repeats). Below are some experiments that can be done to see if a protein acts as a scaffold in vivo:

SDS-PAGE: (sodium dodecyl sulfate polyacrylamide gel electrophoresis) is a technique used in biochemistry, genetics and molecular biology to separate proteins according to their electrophoretic mobility. Separtion of protein is carried out in Poly Acrylamide Gel in the presence of negatively charged detergent Sodium dodecyl sulphate (SDS)

Search and capture model: In centrosome directed microtubule assembly pathway the microtubules grow and shrink continuously and capture kinetochores. At first two kinetochores of a chromosome become attached with the sides of same kMT and then slides to the '+' end of the kMT. Gradually this configuration changes and two kinetochores of chromatids get attached with the kMTs emerged from opposite poles so that they can segregate properly.

Secondary cell wall: On the inner side of the primary cell wall cellulose, hemicellulose, lignin, suberin, cutin, tanin, wax, and various salts of calcium and silica are deposited to form secondary cell wall. This layer consists of external (S1) internal (S3) and middle (S2) sublayers.

Secondary constriction: Near the telomeric end, there may be one colored constricted part, called secondary constriction. Usually in one genome their number is one or more than one.

Secondary lysosomes: Primary lysosomes may fuse with each other or with auto/heterophagosomes to take part in digestion. These are called secondary lysosomes.

Secretory lysosome: Lysosomes are specialized for their role in intracellular protein degradation. Some secretory cells store their secretory proteins in a dual-function organelle, called a *secretory lysosome*. These store additional secretory products, and fuse with the plasma membrane in response to extra cellular stimuli releasing their contents by exocytosis. This is called *lysosomal secretion or defecation.*

Selectively permeable membrane: These membrane allow the passage of selective ions or molecules (e.g. plasma membrane of most of the animal cells).

Semipermeable membrane: Such membranes allow the passage of only water or solutes (e.g. plasma membrane of the fertilized eggs or swim bladder of fishes).

SERCA: It pumps Ca^{++} from the cytosol to the ER lumen.

Sequential model: As per this model, a unit of the enzyme may occur as R, T or RT forms. The substrates binding to R or T form may influence the activity of substrates bound to other state. RT or hybrid state influences the activity of enzymes' action in completely different way.

Selfish DNA: These are blocks of repetitive non-functional DNAs that are tolerated as long as they do not move into genes that are essential for life. Such DNAs are called selfish DNA.

Sex chromosome: The chromosomes that regulate sex determination are called sex chromosomes.

Signal hypothesis: The major mechanism whereby proteins that insert into or cross a membrane are synthesized by a membrane-bound ribosome. The first thirteen to thirty-six amino acids synthesized, termed a signal peptide, are recognized by a signal recognition particle that draws the ribosome to the membrane surface by interaction with a docking protein. The signal peptide may later be removed from the protein.

Signal patch: Acid hydrolase derived from ER and modified by Golgi complex is targeted to lysosome. Golgi complex receives lysosomal enzymes from the RER in the cis region. There, it has a phosphate radical attached to the mannose-6-phosphate (M6P). M6P is added to the appropriate glycoproteins of the hydrolases that leave the RER. The specific signal for adding M6P to the oligosaccharide involves the presence of a cluster of amino acids on the protein surface called *signal patch*.

Signal sequence: The N-terminal sequence of a secreted protein, with 13-36 specific amino acids that regulate the transloca-ion of the nascent polypeptide chain into the ER.

Simple enzymes: These enzymes consist only of protein. E.g. Trypsin, Pepsin etc.

SINES and LINES: Some repetitive DNAs are dispersed through out the genome that are composed of short interspersed repeated sequences (SINES) and long interspersed repeated sequences (LINES). Most of the SINES and LINES elements appear to be non functional.

Sister chromatid: During mitosis or meiosis, the chromosome consists of a pair of two exactly identical "sister" chromatids joined at the central point of each chromatid, called the centromere.

Sister-chromatid exchange: An event similar to crossing over that can occur between sister chromatids at mitosis or at meiosis.

Slow-Twitch fibre / Type I Fibers: The muscle fibres that are loaded with mitochondria, rich in myoglobin and maintain posture of the body are called slow-twitch fibres. They are resistant to fatigue.

Svedberg unit: A unit of sedimentation rate during centrifugation. Abbreviation is S as in 50S.

SER: Endoplasmic reticulum withour attached ribosomes, mainly involved in lipid systhesis.

Slime layer: Some bacteria produce slime layer to adhere among themselves forming a colony.

Solenoid structure: The supercoiled arrangement of DNA in eukaryotic nuclear chromosomes produced by coiling the continuous string of nucleosomes (about 6 nucleosomes per turn) having 30 nm diam.

SPF: At the initiation of S phase, E2F is released from Rb and joins with CyclinA and E-Cdk-2 complex (S-Cdk) that acts as S phase promoting factor (SPF).

Specificity constant or Kcat / Km: The efficiency of an enzyme can be expressed in terms of *catalytic capacity of the enzyme*. It also reflects the affinity of an enzyme with the substrate.

Spectrin: It is a long flexible protein that forms a two dimensional web held by short actin filaments. It links this web to the peripheral membrane proteins of the plasma membrane.

Spindle fibres: As the nuclear membrane breaks down, the microtubules get attached with the condensed chromosomes, now these are called *spindle fibers* and the daughter centrioles in each pole is called *spindle pole*. Four different types of spindle fibers are noted -the small fibers that occur around each pole, are called *astral fibers*; long fibers that connect two poles are called *inter polar fibers*; those fibers which remain connected with the kinetochores are called *kinetochore fibers* and the fibers which connect the chromosomal arms are called *chromosomal fibers*.

Sporic meiosis: In some thallophytes, meiosis takes place at some indeterminate stage producing haploid mega and micro-spores that undergo repeated mitosis forming male and female gametophytes respectively.

Stathmin: It is a MT destabilizing protein noted in some cancer cells.

Stem cells: After the development is completed some cells in different organs even in the adult retain the power of differentiation. These are called stem cells.

Structural maintenance of chromosome protein (SMC proteins): It is an important scaffold protein essential for the regulation of the structure of chromosome

Symport Pumps: These carriers drive ions and the molecule in the same direction.

Stereospecificity or Optical specificity: The property of enzymes (except *isomerases*) acting only on one isomer is called stereospecificity. e.g. L- and D-amino acid oxidases act on L and D amino acids respectively.

Stromules: Sometimes long thin protuberances called *stromules* extend from the main plastid body into the cytosol that interconnects several plastids.

Sub metacentric: Chromosome where centromere occurs near the center.

Substrate analogue: The inhibitor that resembles the substrate, is called substrate analogue.

Substrate specificity: Some enzymes act only on one substrate; this is called *absolute substrate specificity*; e.g. glucokinase acts only on glucose to give glucose-6-phosphate. Some enzymes act on structurally related substrates having specific groups or bonds; e.g. trypsin hydrolyses peptide bonds involving arginine or lysine. This is called *relative substrate specificity*. Some enzymes act on closely related substrates; e.g. hexokinase acts on hexose sugars. This is called *broad specificity*.

Super solenoid: During coiling, the solenoid forms *super solenoid* (70 nm diam).

Supernatant: The clarified liquid above the pellet.

S-value: the sedimentation coefficient is a number that gives information about the molecular weight and shape of the particle. S-value is expressed in **Svedberg units**. The larger the S-value, the faster the particle separates.

Symmetry or concerted model: The enzyme has two different conformational states, **R** and **T**; T has low and R has high affinity for the substrate. The substrate binds preferentially tc the R state, than T state.

Synapsis: During zygotene, the two chromosomes of a bivalent become so intimately paired that every minor part of the chromosomes becomes tightly wound up having an interchromosomal gap - 1000A°. This is called *synapsis* or *syndesis* or *synizesis*. It may proceed from the telomere towards the centromere or may proceed from the centromere towards the telomere that are called proterminal and *procentric synapsis* respectively.

Syncytium: Multinucleate animal cell. E.g. *Opalina*, Striated muscle cells.

Synaptonemal complex: A proteinaceous complex that apparently mediates synapsis during zygotene stage and then disintegrates. A complex structure that unites homologous chromosomes during the prophase of meiosis.

Symmetric karyotype: The karyotype showing less difference between smallest and largest chromosomes and presence of mostly metacentric chromosomes is called symmetric karyotype.

Symport pumps: These carriers drive ions and the molecule in the same direction.

Synapse: It is a junction between the terminal branches of the axon of one neuron and the dendrites of another neuron. It may be *axodendritic* – in between the axon of the pre synaptic neuron with the dendrites of the post synaptic neuron; *axo-axonal* – association of the axon of the presynaptic neuron and axon of the post synaptic neuron; *axo-somatic* - axon of the presynaptic neuron with the muscle and *dendrodendritic* – association of the dendrite of a neuron with the dendrites of the receiving neuron.

Syncytium: Multinucleate animal cell. E.g. *Opalina*, Striated muscle cells.

T

T Cell Lymphoma: A cancer of the lymphatic system where the cells that have become cancerous are a type of white blood cell called T Lymphocytes.

TGN: Tubular structure present at the anterior portion of the Golgi apparatus.

T-banding: It visualizes the telomeres.

Telocentric: Chromosome where centromere occurs at the telomere

Telomeres: The special nucleoprotein complexes found at the end of all linear chromosome.

Telomerase: It is a naturally occurring enzyme which re-lengthens telomeres. It is present in high concentrations in the fast-growing embryo's stem cells, and declines with age.

Terminal web: Actin filaments that extend longitudinally in bundles in the microvilli bind with the horizontal filaments to form a web like structure called *"terminal web"* that are connected by spectrin molecules.

Terminalization: Due to continued repulsion in the diplotene and diakinesis, the chiasmata gradually move towards the end of the bivalents. It is called *terminalization.*

Tetrad: During the early periods of pachytene, each chromosome of a bivalent shows two chromatids; the four chromatids of a bivalent are called *tetrad.* The chromatids produced from one chromosomes are called *sister chromatids* while those are produced from different chromosomes are called *non-sister chromatids.*

Tetrasome: *(=Octamer):* 8 histone protein molecules - H2A, H2B H3 and H4 (each represented by 2 molecules) forming *nucleosome core particle* is called *octamer or tetrasome*

Thylakoids: Within the stroma of the chloroplast there are stacks of flattened disks called *thylakoids.* A stack of thylakoids is called a *granum.*

Thymosins: These are the most abundant actin binding protein in some animal cells. They have higher affinity for ATP-actin monomers than ADP-monomer. They, in fact inhibit actin polymerization.

Titin: The protein molecules that fills up the gap between myosin filaments and the Z disc.

Transitional vesicles: Proteins released from ER that are destined to be delivered to the Golgi bodies are enclosed in minute vesicles, coated with a protein called COPII called transitional vesicles.

Trans Golgi network (TGN): In the trans region of Golgi complex there is a network of tubular structures that form an anastomosing structure, called Trans Golgi network (TGN). They remain connected with other similar structures of neighbouring Golgi.

Threshold for stimulus: When a stimulus is applied to the nerve it causes a rapid change in the membrane potential. Voltage gated Na^+ channels open and the membrane potential raises about 20 mV above the resting potential; this potential is called the "threshold for stimulus".

Treadmili effect: The rate of action monomer assembly at both ends differ; the barbed end is ten times preferred over the pointed end where disassembly is more favoured. This is called *treadmill effect.*

Tropomodulin: They are important actin capping proteins. They cap the '-' end of the stable actin filaments.

Tropomyosin: They can specify the functional capacity of the actin microfilaments and determine their stability by increasing the tensile strength of actin filament. In skeletal muscle it is an important component of the Ca^{++} sensitive regulatory machinery controlling the interaction of actin with myosin.

Troponin / Tropomyosin: Each actin filament is invested by along and thin chain, called ***tropomyosin*** molecules. At its tip the binding sites of globular actin is blocked a protein called ***troponin.*** In response to Ca^{++} the blockage site is removed so that it becomes accessible to myosin.

TOM and TIM complex: There are distinct translocators in the outer (*Translocators **O**uter **M**embrane -**TOM** complex*) and the inner membrane (*Translocators **I**nner **M**embrane - **TIM** complex)* of the mitochondria that mediate translocation of metabolites in between inner and outer chambers of mitochondria.

Tonofilament: Intermediate filaments are cross-linked with each other to form this bundle like structure.

Tonoplast: The membrane encircling the central vacuole in plant cell is called tonoplast.

Totipotent cell: Capacity of a cell to develop the whole body.

Totipotency: It involves differentiation of a cell into embryonic / extra embryonic cell that can form a complete, organism. Cells produced by the first few divisions of the zygote are totipotent.

Transdifferentiation: Some adult stem cell types can differentiate into cell types which are different from predicted lineage (i.e., brain stem cells can differentiate into blood cells or blood-forming cells that differentiate into cardiac muscle cells, and so forth). This phenomenon is called *transdifferentiation.*

Translocon: It is a protein conductance channel formed in the ER membrane through which the newly synthesized protein is introduced into the lumen of the ER. It is formed by Sec61α, β and γ subunits forming a ring.

Transferases: These enzymes mediate transfer a functional group.

Tripotency – It involves differentiation of a cell to form three types of cells. e.g. Glial-restricted precursor cell forms two types of astrocytes and oligodendrocytes.

Tubulin: The protein which when polymerized forms microtubules, necessary for (among other processes) karyokinesis.

Tumor suppressor genes: These genes code for anti-proliferation signals that suppress cell growth. Generally these are transcription factors that are activated by cellular stress or DNA damage. Often DNA damage will trigger enzymes and pathways leading to the activation of tumor suppressor genes. e.g. p53 and Rb are typical tumour suppressor genes.

T-tubule: The invagination of sarcolemma at every 2μ gap at the point of Z disks showing voltage-gated Ca^{++} channelsis. ***Turnover number:*** The number of substrate molecules converted to product / enzyme molecule / second..

Two hit hypothesis: Unlike oncogenes, the tumour suppressor genes generally follow the *'two-hit hypothesis,'* which implies that both alleles for a particular gene must be affected before an effect is manifested. This is due to the fact that if only one allele for the gene is damaged, the second can still produce the correct protein.

Two-strand double crossover: A double crossover that involves only two of the four chromatids of a tetrad.

Two-start model: According to this model of chromatin structure compacted nucleosome arrays are stabilized by the introduction of disulfide cross-links and chromatin fibers show two stacks of nucleosomes. During coiling the flat surfaces of the octamers remains are arranged parallel to the long axis of the solenoid filament.

U

Undershoot: During depolarization of axolemma K^+ exits from the cell gradually but not immediately so that K^+ continues to exit even after the membrane has fully repolarized decreasing the membrane potential below the normal resting membrane potential of the cell for a brief moment; this dip of hyperpolarization is called the undershoot.

Unequal crossing over: Usually crossing over is the process that occurs when two homologous chromosomes break and then reconnect reciprocally exchanging some portion of their DNA. If result is an exchange of genes if they break at the same place or locus but if they break at slightly different loci, the result can be a duplication of genes on one chromosome and a deletion on the other. This is known as an unequal crossover.

Uncompetitive inhibition: The inhibitor cannot bind to the ES-complex forming enzyme-inhibitor-substrate (EIS) complex that is enzymatically inactive.

Uniport pumps: These carriers mediate transport of a single solute.

Unit membrane concept: This model describing the chemical structure of plasma membrane showing fixed protein-lipid-protein is not only applicable to the plasma membrane but also to all intracellular membranes. For this reason it is called *Unit membrane.*

Unusual organelle: In each animal cell a cloud of amorphous multiprotein complex surrounds centrosome, that constitute microtubule-organizing center (MTOC). It is referred as *'unusual organelle'* that acts as a template for nucleation of tubulin subunits to generate spindle microtubules.

V

VASP: It is an actin adapter protein that interacts with membrane protein and facilitates actin filament assembly by inhibiting barbed end capping.

V_{max} : The maximum enzyme reaction velocity attained when all the active sites remain filled up with the substrates.

Villin: Polymerization of actin filaments in to bundles might be mediated by this protein in the microvilli of intestinal cells.

Vimentin: It is an important type III IF protein associated with MTs. It is expressed in leukocytes, blood vessel endothelial cells, some epithelial cells, and mesenchymal cells such as fibroblasts. Vimentin filaments support cellular membranes, help to keep nucleus and other organelles in defined places within the cell.

Virion: A virus particle, (DNA or RNA) which is essential for infection.

Virusoid: Viroid like molecules that get into the cell as passengers inside a conventional plant virus. The molecules consist of single-stranded RNA whose ends are covalent bonded to form a circle.

W

Wilm's tumour: Childhood kidney cancer due to inactivation of a tumour suppressor gene.

Wiskott-Aldrich syndrome causing protein – WASp: It is an actin adapter protein that activates Arp2/3 complex to nucleate new filament on the side of the existing filament.

Y

Yeast artificial chromosome (YAC): Originating from a bacterial plasmid; a YAC contains additionally a yeast centromeric region (CEN); a yeast origin of DNA replication (ARS); a cluster of unique restriction sites; and a selectable marker and a telomere region at the end of each arm. YACs are capable of cloning very large pieces of DNA, more than 1 megabase long. Much of the work on the Human Genome Project depended on YACs. YACs have also been very useful in making transgenic mice, since very large pieces of DNA can be inserted.

Z

Z-scheme: Flow of electrons from PSII to PSI during photosynthesis via carriers move down an electron transport chain successively at lower energy levels so that at each step some energy is released. This flow of electrons exhibit "Z" like appearance that is called Z-scheme

Zigzag model: As per this model of chromatin structure nucleosomes are arranged in a zigzag fashion forming a ribbon like structure which coil up to form a helical 30-nm filament with 12 nucleosomes / turn.

Zinc finger: In the chromatin two types of non-histone proteins are noted; *sequence non-specific non-histone proteins* (bind non-specifically to the DNA), and *sequence specific non-histone proteins* (bind to specific short DNA sequences). Binding of sequences specific non-histone proteins with DNA involves recognition of particular base sequence with specific amino acid residues. These proteins have a variety of functions like folding of DNA in to distinct domain, replication, transcription etc. The first discovered sequence specific non-histone protein is *transcription factor III A* (TF IIIA). It forms 9 folds and each fold shows a central Zn atom. Such folds with central Zn atom is called *Zinc finger.*

Zone of exclusion: Clear zone around the Golgi apparatus.

Z-line: The centre of the I bands where both filaments do not overlap is called *Z line* or *Z disc.*

Z-phase: Disperson of a small proportion of nuclei of the chromocentre. At the end of the S phase the chromatins appear dispersed with minimal density.

Zymogens: **Inactive** precursors of the proteolytic enzymes. These prevent unwanted protein degradation, and enable spatial and temporal regulation **of** proteolytic activity.

Zygotene: The second stage in prophase 1 of meiosis, characterized by the synapsis or pairing, of homologous chromosomes to form a bivalent.

Zygotic meiosis: In lower plants meiosis cell division follows fertilization producing haploid cells that form the adult gametophytic generation.

Index

A

B

C

D

E

F

G

H

I

J

K

L

M

N

O

P

Q

R

S

T

U

V

W

Z